EVALUATION OF HSDPA AND LTE

EVALUATION OF HSDPA AND LTE

FROM TESTBED MEASUREMENTS TO SYSTEM LEVEL PERFORMANCE

Sebastian Caban, Christian Mehlführer, Markus Rupp and Martin Wrulich

Vienna University of Technology, Austria

A John Wiley & Sons, Ltd., Publication

This edition first published 2012

Registered office

John Wiley & Sons Ltd, The Atrium, Southern Gate, Chichester, West Sussex, PO19 8SQ, United Kingdom

For details of our global editorial offices, for customer services and for information about how to apply for permission to reuse the copyright material in this book please see our website at www.wiley.com.

Library of Congress Cataloging-in-Publication Data

Evaluation of HSDPA and LTE : from testbed measurements to system level performance / Sebastian Caban ... [et al.].
p. cm.
Includes bibliographical references and index.
ISBN 978-0-470-71192-7 (cloth)
1. Long-Term Evolution (Telecommunications) 2. Packet transmission (Data transmission) I. Caban, Sebastian.
TK5103.48325.E93 2012
621.382 – dc23

2011023059

A catalogue record for this book is available from the British Library.

ISBN: 9780470711927 (H/B)
ISBN: 9781119954699 (ePDF)
ISBN: 9781119954705 (oBook)
ISBN: 9781119960881 (ePub)
ISBN: 9781119960898 (mobi)

Typeset in 10/12pt Times by Laserwords Private Limited, Chennai, India

Printed and bound in Singapore by Markono Print Media Pte Ltd

Contents

About the Authors

Sebastian Caban was born in Vienna, Austria, on March 13, 1980. After attending a technical high school and a mandatory service in the Austrian Armed Forces, he started studying electrical and communication Engineering at the Vienna University of Technology. He received his masters degree in 2005 and his Ph.D. degree in 2009 as "sub-auspiciis presidentis rei publicae." In between, he studied a year at the University of Illinois at Urbana Champaign, USA. In 2005, Sebastian Caban also started to study economics at the Vienna University to relax from engineering. He completed a 5-year masters degree in business administration in 2010. Nonetheless, Sebastian Caban's work and research interest at the Vienna University of Technology focuses on developing measurement methodologies and building testbeds to quantify the actual performance of wireless cellular communication systems (UMTS, LTE) in realistic scenarios. Next to 12 national scholarships and prizes, Sebastian Caban was awarded the "Förderpreis 2006" of the German Vodafone Stiftung für Forschung for his research. E-mail: scaban@nt.tuwien.ac.at.

Christian Mehlführer was born in Vienna, Austria, in 1979. In 2004 he received his Dipl.-Ing. degree in electrical engineering from the Vienna University of Technology. Besides his diploma studies he worked part time at Siemens AG, where he performed integration tests of GSM carrier units. After finishing his diploma thesis on implementation and real-time testing of space–time block codes at the Institute of Communications and Radio-Frequency Engineering, Vienna University of Technology, for which he received the Vodafone Förderpreis 2006 (together with Sebastian Caban), he started his doctoral thesis at the same institute. In 2009, he finished his Ph.D. about measurement-based performance evaluation of WiMAX and HSDPA with summa cum laude. His research interests include experimental investigation of MIMO systems, UMTS HSDPA, WLAN (802.11), WiMAX (802.16), and the upcoming 3GPP LTE system. Currently, he is working as a consultant for Accenture. E-mail: chmehl@nt.tuwien.ac.at.

Markus Rupp was born in 1963 in Völklingen, Germany. He received his Dipl.-Ing. degree in 1988 at the University of Saarbrücken, Germany and his Dr.-Ing. degree in 1993 at the Technische Universität Darmstadt, Germany, where he worked with Eberhardt Hänsler on designing new algorithms for acoustical and electrical echo compensation. From November 1993 until July 1995 he had a postdoctoral position at the University of Santa Barbara, California, with Sanjit Mitra, where he worked with Ali H. Sayed on a robustness description of adaptive filters with impact on neural networks and active noise control. From October 1995 until August 2001 he was a member of Technical Staff in the Wireless Technology Research Department of Bell-Labs at Crawford Hill, NJ, where he worked on various topics related to adaptive equalization and rapid implementation for IS-136, 802.11, and UMTS. Since October 2001 he has been a full professor for Digital Signal

Processing in Mobile Communications at the Vienna University of Technology, where he served as Dean from 2005 to 2007. He was associate editor of *IEEE Transactions on Signal Processing* from 2002 to 2005, is currently associate editor of *JASP EURASIP Journal of Advances in Signal Processing*, and *JES EURASIP Journal on Embedded Systems*. He is a Senior Member of IEEE and elected AdCom member of EURASIP since 2004 and serving as president of EURASIP from 2009 to 2010. He has authored and co-authored more than 350 scientific papers, including 15 patents on adaptive filtering and wireless communications. E-mail: mrupp@nt.tuwien.ac.at

Martin Wrulich was born in 1980 in Klagenfurt, Carinthia. He received his Dipl.-Ing. degree from Vienna University of Technology in March 2006, which was awarded by the 'Diplomarbeitspreis der Stadt Wien'. Furthermore, he received three times a scholarship from the faculty of electrical engineering and information technology for excellent academic performance. He started his doctoral studies in the area of system-level modeling and network optimization at the institute of communications and radio-frequency engineering in March 2006. In 2009, he received a Marshallplan scholarship to join the group of Prof. Paulraj at Stanford University, after which he finished his PhD 'sub-auspiciis presidentis rei publicae'. He also acted as a TPC member for the IEEE broadband wireless access workshop in 2009. Currently he is working as a consultant for McKinsey&Company, Inc. E-mail: mwrulich@gmail.com

About the Contributors

Josep Colom Ikuno was born 1984 in Barcelona, Spain. He received his bachelor in Telecommunications Engineering from the Universitat Pompeu Fabra in Barcelona in 2005 and his Master's degree also in Telecom Engineering from the Universitat Politècnica de Catalunya in 2008, both with honors. Since beginning of 2008 he is with the Vienna University of Technology in the Mobile Communications Group, where he works towards his Ph.D thesis in the field of LTE link and system level simulation, modeling and optimization. His past and present research interests include video coding, OFDM systems, and accurate, low-complexity modeling of the physical layer for system level simulation of wireless networks, with emphasis on the integration of real operator data in the simulations and models. E-mail:jcolom@nt.tuwien.ac.at

José Antonio García-Naya was born in A Coruña, Spain, in June 1981. He obtained his master degree and his Ph.D. degree in Computer Engineering from the University of A Coruña in 2005 and 2010, respectively. Since 2005 he is with the Group of Electronic Technology and Communications (GTEC) at the Department of Electronics and Systems, University of A Coruña, Spain. From December 2007 to October 2008 he was a recipient of a Maria Barbeito post-degree grant from the Galician Government. Between 2008 and 2009 he was at the Institute of Communications and Radio-Frequency Engineering, Vienna University of Technology, Austria. He has been an Assistant Professor at the University of A Coruña since 2008. His research interests are in the field of digital communications, with special emphasis on prototyping of wireless communication equipment, hardware and software design of multiple antenna testbeds. E-mail: jagarcia@udc.es

Qi Wang was born in Beijing, China, in November 1982. In 2001, she received the Bachelor degree in telecommunication engineering from Beijing University of Posts and Telecommunications. From August 2005 to October 2007, she studied in Linköping University in Linköping, Sweden, where she obtained the Master degree in computer science and engineering. Since November 2007, she works as university assistant at the Institute of Communications and Radio-Frequency Engineering of the Vienna University of Technology. Her research interests are in the field of wireless communications, in particular, synchronization aspects of OFDM systems. E-mail: qwang@nt.tuwien.ac.at.

Stefan Schwarz was born in 1984 in Neunkirchen, Austria. His interests in electronics and telecommunications were inspired while attending a technical high school. During this time, he was awarded the 'GIT-Preis' endowed by the 'Austrian Electrotechnical Association' for his thesis. He received a BSc degree in Electrical Engineering in 2007 and a Dipl.-Ing. degree in Telecommunications in 2009, both from Vienna University of Technology with highest distinction. Three times he received a scholarship from the faculty of electrical engineering and information technology for excellent academic performance, as well as the 'Würdigungspreis' from the Austrian federal department of science and research for the same reason. Since October 2009, he works as project assistant at the institute of telecommunications of the Vienna University of Technology, towards his Ph.D degree. His research interests are in wireless communications, currently focusing on link and network optimization and channel state information acquisition. E-mail: sschwarz@nt.tuwien.ac.at

Preface

When you cannot measure it,
when you cannot express it in numbers,
your knowledge is of a meager
and unsatisfactory kind;
it may be the beginning of knowledge,
but you have scarcely, in your thoughts,
advanced to the stage of science.

William Thomson, Lord Kelvin (1894)

At the beginning of the 1990s, several research groups around the world were working on adaptive filters for hands-free telephones. At that time the first experimental DSP boards were available and well suited for audio experiments. While many research groups focused only on simulations, a few also implemented real-time experiments with such DSP boards. While such boards needed to be programmed in assembler, it was very tedious to make a program run and at the same time very time consuming, and many posed the question, "Why would experimental work be worth at all?" Why would it not be good enough to perform MATLAB® simulations? In simulations one can be so much faster, have everything under control, and measurement precision is not an issue at all. However, experiments showed that simulations were not capturing the true nature of the problem. Some designs sounded artificial owing to their phase response degradation, some showed smaller Echo Return Loss Enhancement (ERLE) but simply sounded better. The "Quality of Experience" as we call it today was more important than the performance metric based on a Euclidian distance of ERLE. Also, rapid changes in the environment as they occur when the local speaker moves while speaking could never be simulated realistically. A simple linear combination from a starting impulse response to a final one was not capturing the motion. In conclusion, for this problem experimental work was crucial for its success, a success that we all experience today: hands-free telephones simply work!

At the beginning of the 1990s, similar problems occurred when designing the first cellular systems: the wireless channel and the multitude of elements in the transmission chain were largely mismodeled. At best we captured the qualitative behavior, never the true quantities. Even ray-tracing simulations, which are very demanding on the complexity, showed errors of 15 dB. Nevertheless, the efforts in experimental work at that time became marginal compared with simulation efforts. New transmission standards such as Universal Mobile Telecommunications System (UMTS) were entirely built on simulations. A 200 Mio investment for a new line of base stations was entirely built on simulations and not backed up by some physical demonstrations. By the time prototypes or demonstrators were built, the assembly line for the new-generation base station was already operational.

What is the reason that such a change in behavior has taken place? The first reason is certainly money, or, to be more precise, expected financial success. Whoever is first on the market makes the biggest profit. In wireless cellular systems the golden rule is "being six months early doubles profit, being six months late results in break-even." Achieving results by simulations is a faster method than by experiments and, therefore, being favored. But there are other aspects that need to be mentioned. While in experiments you can never control all parameters, in simulations you can. Changing parameters as desired is possibly the greatest asset of simulations. Building a new setup for experiments can take a couple of weeks, if not months; changing two parameters in a simulation takes a minute. Next to flexibility there is also the aspect of time, as simulations are usually relatively fast. We use the word "relatively" for good reasons. Once the freedom exists to change parameters at free will, the requests for more and more simulations never stop. Also note that cellular systems become more and more complex and their complexity grows much faster than the available number crunching complexity on future computers (Shannon beats Moore). A third aspect is accuracy. Models become more and more detailed to achieve higher accuracy, but at the same time they become more complex with even more parameters. We can thus expect in the future that simulations will not give us as much benefit as they have given us in the past.

A strong argument for simulations has always been their flexibility. This has gone as far as some of our fellow researchers stating that only simulations can offer the potential to investigate a large variation of scenarios and thus allow for general statements, while experimental research is always limited to two or three setups that will never allow for generalization. We would like to argue strongly against such statement. First of all, models reflect the reality, but they are by far not reality. A good model is a simple model with very few parameters. As soon as the model requires several tens or even hundreds of parameters, it cannot be handled properly any more. A simulation including a variation of all parameters would not be feasible, and thus a conclusion about generality is not possible. This is exactly the situation we are in currently when dealing with modern cellular systems including multiple antennas at the transmitter and receiver end. Simple tapped delay line models (Pedestrian A, B) are not accurately reflecting reality; complex models (SCA, Winner) may include up to 100 parameters. Let us summarize our discussion with the following words:

> Measurements reflect reality
> while
> simulations reflect the simulation environment.

We thus consider experimental work and validation to be of the utmost importance. Simply putting all faith in simulation results can only be good for the inhabitants of "SimCity," but not for people living in the real world.

On the other hand, the drawbacks of experimental results are striking as well. Their little flexibility even over the design parameters of transmitter and receiver is often painful. In particular, when going for a prototype, the time for development might be as long as for the entire product and still not be as flexible as a simulation. A fair compromise seems to be our concept of testbeds. In testbeds the true physical channel is involved, as it is the part that is known the least. RF-front ends are typically replaced by relatively precise (but expensive) measurement equipment to allow for later evaluation of low-cost designs. The transmit signals are generated from a simulation environment such as MATLAB in order to ensure identical data for simulation as well as experiments. Typically, receive signals are past processed in a non-real-time fashion. Such a testbed system thus provides almost the same flexibility as the simulation; only the wireless channel must be selected by the experimental setup. A further point is the automation of the measurement process utilizing $XY\Phi$-positioning tables. Taking the manual aspects out and at the same time measuring several thousand locations (typically quarter wavelength spaced) makes the approach very valuable. Most time goes into the preparation of the experiment and the final postprocessing, allowing a larger measurement campaign, including postprocessing in 2–3 months, which is not so much longer than a modern simulation task of the same proportion. As computers become faster the simulation times will decrease, and so will the postprocessing task. As the received signals are fed back into the MATLAB® chain, the results of the testbed measurement can be directly compared with simulation results, allowing the study of many effects in the more flexible simulation environment.

Our studies follow the idea that the transmission chain can be described by different kinds of losses. The first loss is named the "channel state information loss" or "feedback loss." This is the difference between the channel capacity as described by Shannon and the mutual information of the channel. It can be interpreted as the loss a transmission must face when not knowing the channel at the transmitter end. As we show for Single-Input Single-Output (SISO) and cross-polarized antennas, such channel loss can be neglected. Only for four and more antenna systems does the channel loss take on considerable values. The second loss we encounter is called the "design loss." If we take into account the part of the transmit power that we put in pilots and synchronization, or equivalently its corresponding part in bandwidth, then we can compute the so-called "achievable mutual information." Owing to the design of the cellular system, even with perfect modulation and coding schemes, we will never exceed this bound. The difference between achievable mutual information and mutual information we thus call the design loss. Some cellular systems are well designed, such as High-Speed Downlink Packed Access (HSDPA), with small amounts in design loss, whereas Orthogonal Frequency-Division Multiplexing (OFDM) systems, such as Worldwide Inter-operability for Microwave Access (WiMAX) and Long-Term Evolution (LTE), exhibit large design losses. The final loss we encounter is the "implementation loss." This is the difference between the measured throughput and the achievable mutual information. Such a loss is due to imprecise channel knowledge, to nonperfect modulation and coding schemes, to self- interference, as this is very strong in Wideband Code-Division Multiple Access (WCDMA) systems such as HSDPA, and all other known or unknown implementation aspects (oscillator phase noise, IQ imbalances, and many more). Our testbed approach allows the measurment of the first two losses, the channel loss and the design loss, perfectly. For the implementation loss we obtain a measure based on our realization including high-precision hardware. A low-cost product can expect to experience a higher implementation loss; thus, our measures provide a good idea of what potentially could be achieved. It is very educative to study the differences in design and implementation losses, as systems with higher design loss often compensate this with lower implementation loss. At the end, what counts is the sum of both. Let us end this discussion by another *bon mot* (Jan L. A. van de Snepscheut (1953–1994)).

> In theory, there is no difference between theory and practice.
> But, in practice, there is.

How to read this book

Part One is a quick brush up on the cellular standards: High Speed Downlink Packet Access (HSDPA) and UMTS Long Term Evolution (LTE). Given the limited space in a book, the two chapters provide a compressed description of the physical layer with smaller add-ons of the upper layers. Basically everything required to understand the following four parts is provided for the reader. Experts can simply skip the chapters, whereas the engineer in the field working in a specific area may find a quick introduction into other standards. Students will find a comprehensible approach without becoming overloaded with too many details.

In Part Two we explain the concept of testbeds, as opposed to prototypes or demonstrators. We provide the challenges in building such testbeds and explain what to expect from them. Essentially, it turns out that simply combining subsystems does not make a testbed. The testbed explained here offers a very high flexibility, as essentially signals directly from MATLAB® can be transmitted and similarly received signals can directly be fed back into the MATLAB® simulator. This, in turn, offers a rigorous comparison of simulation and measurement results. Specifically, we focus on the measurement techniques, including evaluation of the measurement quality by statistical inference techniques as well as measurement automation, as only this step assures high-quality results in a moderate time frame. Finally, some first measurement examples are provided for indoor and outdoor measurements validating the measurement method.

Part Three reports on link-layer evaluation of cellular systems. We mostly focus on results obtained during several measurement campaigns in alpine and urban environments and show experimental results for HSDPA, and to some extent for WiMAX. In particular, we focus on questions of antenna distance when employing several antennas at the base station and antenna selection techniques, as our measurements show an entirely different picture than summation results of most publications, which are based on too simplistic channel assumptions. An entire chapter is related to the question of frequency synchronization, as this is a crucial issue typically not handled at all in the literature. As measurement results of LTE were not yet available in sufficient amounts at the time of writing this book, we relied on simulation results here. This part thus allows a direct comparison of the three technologies, not only in terms of throughput, but also from an information theoretical approach, by comparing the measured performance with mutual information and channel capacity. Moreover, we thoroughly investigated the reasons for the differences found; that is, we provide a detailed analysis of design and implementation losses. An important aspect also addressed here that is missing in most older literature is transmission with cross-polarized antenna systems. They not only provide higher capacity at a substantially smaller footprint, but they also offer substantial advantages for the signal processing part as the channel losses – that is, the differences between capacity and mutual information – become very small, offering to design and operate systems with very little feedback information. In theory, only the knowledge of the receiver SNR is required at the transmitter to achieve most of the capacity. The challenge is to design systems that achieve such quality.

In Parts Four and Five we take a different approach than in the previous parts; that is, the evaluation of cellular systems by simulation. In Part Four we explain how simulators are working at the link and system levels. Here, a particular focus is on system-level simulations, as such experiments would hardly be possible employing testbeds, since they require several base stations and a multitude of users. Particular focus is on the evaluation of such simulator tools. We demonstrate how this can be achieved by either comparing with cumulative density functions of measurements or by co-simulation of link and system-level parts. A simulator pair for link and system levels is introduced that was designed for co-simulation so that the link-level simulator can provide crucial information for the system level. Here, as in the previous chapters on measurements, we used modern statistical inference techniques, such as bootstrapping algorithms, to provide not only the average value (measured or simulated), but also always to complement our results with confidence intervals to make the statements much more meaningful.

Reproducibility has become an increasingly important issue in recent years. As systems become more and more complex, and thus complicated, it becomes more and more difficult to repeat results from others, and even reproducing our own results is often difficult after some time has passed. To facilitate reproducibility, therefore, we have launched open access[1] "Vienna LTE Simulators" that can be downloaded, including many examples from our publications, and made to run on your own PC. After several ten thousands of downloads we also installed a forum to support the growing community. This gave us a lot of feedback and continuously smaller errors are found and corrected.

In Part Five we present evaluation results obtained from system-level simulators. On the example of HSDPA, we show how difficult and, at the same time, how complex such simulations are. With the available system-level simulator, several optimizations were performed: network performance prediction, Radio Link Control (RLC)-based stream number decision, content aware scheduling, Common Pilot CHannel (CPICH) power optimization. Finally, the last chapter deals with topics of LTE advanced standards, as they are being discussed at time of writing this book.

Sebastian Caban, Christian Mehlführer, Markus Rupp, Martin Wrulich

[1] The open access, free of charge is granted for purely academic users. If a company or a project with a monetary flow is involved, a fee is required. The fee is used to further support these efforts.

Acknowledgments

This book is the outcome of many years of research and teaching in the field of signal processing and wireless communications.

We would like to thank many people from A1 (former Mobilkom and Telekom), as well as Kathrein for their steady support over a longer period of time: W. Wiedermann, W. Müllner, W. Karner, A. Ciaffone, T. Ergoth, W. Weiler, T. Baumgartner, J. Peterka, A. Kathrein, G. Schell, R. Gabriel, and J. Rumold.

A very special thank you goes to Christoph F. Mecklenbräuker, who supported many of our efforts via his Christian Doppler Laboratory "Wireless Technologies for Sustainable Mobility." The HSDPA simulator was jointly developed with the ftw (Forschungszentrum Telekommunikation Wien). Many thanks for the support.

The work reported in this book required a lot of people that are usually not mentioned in publications, but without their administrative help we could not have done it: J. Auerböck, N. Hummer, W. Schüttengruber, B. Wistawel. Many thanks for your help.

Such a book would never have been possible without the constant support by the many helpful people from Wiley: T. Ruonamaa, A. Smart, S. Barclay, M. Cheok and including, G. Vasanth of Laserwords and P. Lewis. We cannot thank you enough.

List of Abbreviations

2G	2nd Generation
3GPP	3rd Generation Partnership Project
ACK	ACKnowledged
ALMMSE	Approximate Linear Minimum Mean Square Error
AMC	Adaptive Modulation and Coding
ARQ	Automatic Repeat reQuest
AWGN	Additive White Gaussian Noise
BCa	Bias-Corrected and accelerated
BCCH	Broadcast Control Channel
BCH	Broadcast Channel
BICM	Bit-Interleaved Coded Modulation
BER	Bit Error Ratio
BLEP	Block Error Probability
BLER	Block Error Ratio
CA	Content Aware
CAPEX	Capital Expenditure
CB	Code Block
CC	Chase Combining
CCCH	Common Control Channel
CDD	Cyclic Delay Diversity
cdf	cumulative density function
CDI	Channel Direction Indicator
CDMA	Code-Division Multiple Access
CFO	Carrier Frequency Offset
CL	Closed Loop
CLMI	Closed-Loop Mutual Information
CLSM	Closed-Loop Spatial Multiplexing
CM	Coded Modulation
CoMP	Coordinated Multi-Point
COST	European COoperation in the field of Scientific and Technical research
CP	Cyclic Prefix
C-Plane	Control Plane
CPICH	Common Pilot CHannel
CPU	Central Processing Unit
CQI	Channel Quality Indicator

CRC	Cyclic Redundancy Check
CSI	Channel State Information
CSI-RS	CSI Reference Signal
CTC	Convolutional Turbo Code
CVQ	Channel Vector Quantization
CW	Codeword
DCCH	Dedicated Control Channel
DCH	Dedicated Channel
DFT	Discrete Fourier Transform
DHCP	Dynamic Host Configuration Protocol
DiffServ	Differentiate Service
DL	Downlink
DL-SCH	Downlink Shared Channel
DMRS	Demodulation Reference Signals
DoFs	Degrees of Freedom
DPC	Dirty Paper Coding
DPCCH	Dedicated Physical Control CHannel
DPDCH	Dedicated Physical Data CHannel
DS	Double-Stream
DSA	Dynamic Sub-carrier Allocation
DSTTD-SGRC	Double Space–Time Transmit Diversity with Sub-Group Rate Control
DTCH	Dedicated Traffic Channel
D-TxAA	Double Transmit Antenna Array
ECR	Effective Code Rate
EDGE	Enhanced Data Rates for Global system for mobile communications Evolution
EESM	Exponential Effective Signal to Interference and Noise Ratio Mapping
eNodeB	Evolved base station
EPC	Evolved Packet Core
ESM	Effective Signal to Interference and Noise Ratio Mapping
EUTRA	Evolved Universal Terrestrial Radio Access
E-UTRAN	Evolved Universal Terrestrial Radio Access Network
EVehA	Extended Vehicular A
FBI	FeedBack Information
FDD	Frequency Division Duplex
FEC	Forward Error Correction
FFO	Fractional Frequency Offset
FFT	Fast Fourier Transform
FIFO	First In, First Out
FPGA	Field-Programmable Gate Array
FTP	File Transfer Protocol
GGSN	Gateway-General packet radio service Support Node
GOF	Goodness Of Fit
GOP	Group Of Pictures

GPRS	General Packet Radio Service
GPS	Global Positioning System
GSCM	Geometry-based Stochastic Channel Model
GSM	Global System for Mobile communications
HARQ	Hybrid Automatic Repeat reQuest
HSDPA	High-Speed Downlink Packet Access
HS-DPCCH	High-Speed Dedicated Physical Control CHannel
HS-DSCH	High-Speed Downlink Shared CHannel
HSPA	High-Speed Packet Access
HS-PDSCH	High-Speed Physical Downlink Shared CHannel
HSS	Home Subscriber Server
HS-SCCH	High-Speed Shared Control CHannel
HSUPA	High-Speed Uplink Packet Access
IC	Interference Cancelation
ICI	Inter-Carrier Interference
IR	Incremental Redundancy
IDE	Integrated Development Environment
IEEE	Institute of Electrical and Electronics Engineers
IFFT	Inverse Fast Fourier Transform
IFO	Integer Frequency Offset
IMS	IP Multimedia Subsystem
IMT	International Mobile Telecommunications
IP	Internet Protocol
IR	Incremental Redundancy
ISI	Inter-Symbol Interference
ISM	Industrial, Scientific and Medical
ISO	International Standard Organization
ITU	International Telecommunication Union
LAN	Local Area Network
LDPC	Low Density Parity Check
LEP	Link Error Prediction
LLR	Log-Likelihood Ratio
LMMSE	Linear Minimum Mean Square Error
LMMSE-MAP	Linear Minimum Mean Square Error Maximum A Posteriori
LOS	Line-Of-Sight
LS	Least Squares
LTE	Long-Term Evolution
LTE-A	Long-Term Evolution-Advanced
MAC	Medium Access Control
MAC-d	Medium Access Control dedicated
MAC-hs	Medium Access Control for High-Speed Downlink Packet Access
MAP	Maximum A-Posteriori
maxCI	Maximum Carrier-to-Interference Ratio
MB	MacroBlock
MBMS	Multicast Broadcast Multimedia Services

MCCH	Multicast Control Channel
MCH	Multicast Channel
MCS	Modulation and Coding Scheme
MI	Mutual Information
MIESM	Mutual Information Effective Signal to Interference and Noise Ratio Mapping
MIMO	Multiple-Input Multiple-Output
MISO	Multiple-Input Single-Output
ML	Maximum Likelihood
MME	Mobility Management Entity
MMSE	Minimum Mean Square Error
MoRSE	Mobile Radio Simulation Environment
MPEG	Moving Picture Expert Group
MRC	Maximum Ratio Combining
MSE	Mean Square Error
MTCH	Multicast Control Channel
MTU	Maximum Transfer Unit
MU	Multi-User
MU-MIMO	Multi-User MIMO
MVU	Minimum Variance Unbiased
NACK	Non-ACKnowledged
NAL	Network Abstract Layer
NAS	Non-Access Stratum
NBAP	NodeB Application Part
NDI	New Data Indicator
N-SAW	N Stop And Wait
NSN	Nokia Siemens Networks
NodeB	Base station
OFDM	Orthogonal Frequency-Division Multiplexing
OFDMA	Orthogonal Frequency-Division Multiple Access
OLMI	Open-Loop Mutual Information
OLSM	Open-Loop Spatial Multiplexing
OOP	Object-Oriented Programming
OPEX	Operational Expenditure
OSI	Open Systems Interconnection
PAPR	Peak-to-Average Power Ratio
PARC	Per-Antenna Rate Control
PBCH	Physical Broadcast Channel
PC	Personal Computer
PCCC	Parallel Concatenated Convolutional Code
PCCH	Paging Control Channel
PCCPCH	Primary Common Control Physical Channel
PCH	Paging Channel
PCI	Precoding Control Indicator
PCIsock	Peripheral Component Interconnect
PCRF	Policy and Charging Rules Function

PDCCH	Physical Downlink Control Channel
PDCP	Packet Data Convergence Protocol
PDSCH	Physical Downlink Shared Channel
pdf	probability density function
PDP	Packet Data Protocol
PDU	Packet Data Unit
PDN	Packet Data Network
PedA	Pedestrian A
PedB	Pedestrian B
PER	Packet Error Ratio
P-GW	PDN Gateway
PHY	Physical
PF	Proportional Fair
PMCH	Physical Multicast Channel
PMI	Precoding Matrix Indicator
PN	Pseudo Noise
PPS	Pulse Per Second
PSCH	Primary Synchronization Signal
PSD	Power Spectral Density
PSK	Phase-Shift Keying
PSNR	Peak Signal to Noise Ratio
PRACH	Physical Random Access Channel
PUCCH	Physical Uplink Control Channel
PUSCH	Physical Uplink Shared Channel
QAM	Quadrature Amplitude Modulation
QBICM	Quantized BICM
QCIF	Quarter Common Intermediate Format
QoE	Quality of Experience
QoS	Quality of Service
QPP	Quadratic Permutation Polynomial
QPSK	Quadrature Phase Shift Keying
QSBICM	Quantized and Shifted BICM
RACH	Random Access Channel
RAID	Redundant Array of Independent Disks
RAN	Radio Access Network
RAS	Receive Antenna Selection
RB	Resource Block
RE	Resource Element
rvidx	redundancy version index
RE	Resource Element
RF	Radio Frequency
RFO	Residual Frequency Offset
RI	Rank Indicator
RLC	Radio Link Control
RMS	Root Mean Square
RN	Relay Node

RNC	Radio Network Controller
ROI	Region Of Interest
RR	Round Robin
RRC	Radio Resource Control
RRCfilt	Root Raised Cosine
RRE	Remote Radio Equipment
RRM	Radio Resource Management
RS-CC	Reed–Solomon Convolutional Code
RS232	Recommended Standard 232
RTP	Real-time Transport Protocol
RV	Redundancy Version
RX	Receiver
SAE	System Architecture Evolution
SBICM	Shifted BICM
SC-FDMA	Single-carrier FDMA
SCH	Synchronization Channel
SCM	Spatial Channel Model
SCME	Spatial Channel Model (SCM) Extension
SDMA	Spatial Division Multiple Access
SDU	Service Data Unit
SER	Symbol Error Ratio
SGSN	Serving-General packet radio service Support Node
S-GW	Serving Gateway
SIC	Successive Interference Cancelation
SID	Size Index Identifier
SIMO	Single-Input Multiple-Output
SINR	Signal to Interference and Noise Ratio
SISO	Single-Input Single-Output
SM	Spatial Multiplexing
SNR	Signal to Noise Ratio
SQP	Sequential Quadratic Programming
SS	Single-Stream
SSCH	Secondary Synchronization Signal
SSD	Soft Sphere Decoder
STBC	Space–Time Block Code
STMMSE	Space–Time Minimum Mean Squared Error
STTD	Space–Time Transmit Diversity
SU	Single-User
SU-MIMO	Single-User MIMO
SVD	Singular Value Decomposition
TAS	Transmit Antenna Selection
TB	Transport Block
TBS	Transport Block Size
TDD	Time Division Duplex
TFC	Transport Format Combination
TFCI	Transport Format Combination Indicator

TFT	Traffic Flow Template
TOS	Type Of Service
TPC	Transmit Power-Control
TrCH	Transport CHannel
TSN	Transmission Sequence Number
TTI	Transmission Time Interval
TX	Transmitter
TxAA	Transmit Antenna Array
TxD	Transmit Diversity
UDP	User Datagram Protocol
UE	User Equipment
UL	Uplink
ULA	Uniform Linear Array
UL-SCH	Uplink Shared Channel
UMTS	Universal Mobile Telecommunications System
U-Plane	User Plane
USB	Universal Serial Bus
UTRA	Universal mobile telecommunications system Terrestrial Radio Access
UTRAN	Universal mobile telecommunications system Terrestrial Radio Access Network
V-BLAST	Vertical Bell Laboratories Layered Space–Time
VCEG	Video Coding Expert Group
VCL	Video Coding Layer
VehA	Vehicular A
VHSIC	Very High Speed Integrated Circuits
VoIP	Voice over IP
WB-CLMI	Wideband Closed-Loop Mutual Information
WB-CLMI-LR	Wideband Closed-Loop Mutual Information with Linear Receiver
WCDMA	Wideband Code-Division Multiple Access
WiMAX	Worldwide Inter-operability for Microwave Access
WLAN	Wireless Local Area Network
ZF	Zero Forcing

Part One

Cellular Wireless Standards

Introduction

Since the introduction of the Global System for Mobile communications (GSM) in 1991, the ecosystem of equipment vendors, application providers, and service enablers has grown significantly. Moreover, the development and availability of wireless broadband techniques now allows for an even tighter integration of web-based services on mobile devices. In the context of wireless networks, the key parameters defining the application performance include the data rate and the network latency. Some applications require only low bit rates of a few tens of kilobits per second but demand very low delay, as in Voice over IP (VoIP) and online games [4]. On the other hand, the download time of large files is only defined by the maximum data rate, and latency does not play a big role in this application.

In terms of penetration, mobile wireless telephony surpassed fixed-line volumes in 2004, whereas broadband coverage is still lagging behind [1]. Current surveys forecast a maximum penetration of approximately 25 % for European markets [5]. In Indonesia actually, High-Speed Downlink Packet Access (HSDPA) broadband access surpassed fixed broadband access in 2008. A recent study by Ericsson [2] claims that there are already 5 billion subscribers worldwide (July 2010), with a daily growth rate of 2 million; for 2020, some 50 billion subscribers are predicted. HSDPA played a big, if not the most important, role in the success of mobile broadband services. The dramatic push of the typical data rates for most services, as well as the achievable peak data rate compared with the Universal Mobile Telecommunications System (UMTS), together with the lowered latency, drew a lot of attention from customers. In addition, the rapid decline in prices [3] and the low-cost hardware to connect conveniently to HSDPA networks are very attractive for end users. Nowadays, no or only little effort is required to adapt internet applications to the mobile environment.

Essentially, HSDPA for the first time is a broadband wireless access with seamless mobility and extensive coverage. Together with High-Speed Uplink Packet Access (HSUPA), HSDPA forms the so-called High-Speed Packet Access (HSPA), which can be deployed on top of the existing Wideband Code-Division Multiple Access (WCDMA) UMTS networks, either on the same carrier (technically, this requires the split of the spreading code resources in the cell) or – for high capacity and high bit rate – on another carrier, thus consequently denoting a pure HSDPA operation. The ever-increasing demand for HSDPA broadband wireless access, however, drives mobile network operators to allocate their spectrum resources in a progressive way towards HSPA usage.

In this first part of the book, a short introduction of two wireless cellular standards is presented: HSDPA in Chapter 1 and the downlink mode of UMTS Long-Term Evolution (LTE) in Chapter 2. Given the limited space, the two chapters provide a compressed description of the physical layer with smaller add-ons of the upper layers, providing the reader with sufficient background information to understand the rest of the book. Experts can simply skip the chapters, whereas the engineer in the field working in a related area may find a quick introduction into such standards. Students of wireless telecommunications will find an easy-to-follow approach without becoming overloaded by too many details.

In Chapter 1 we particularly focus on the means for radio resource management, scheduling, as well as Multiple-Input Multiple-Output (MIMO) techniques. Already in Single-Input Single-Output (SISO) HSDPA, a so-called Channel Quality Indicator (CQI) has been introduced to allow the selection of optimal rate and modulation schemes depending on the observed channel situation. A small amount of feedback is thus returned to the transmitter, adjusting the data rate on the channel link. The aspect of multiple antennas comes with the notion of a precoding matrix which extends the concept of feedback information with a so-called Precoding Control Indicator (PCI). The topic on scheduling includes the functions of the Medium Access Control (MAC), in particular its advanced version Medium Access Control for High-Speed Downlink Packet Access (MAC-hs), as well as Radio Resource Management (RRM), and also covers the aspect of Hybrid Automatic Repeat reQuest (HARQ) in the form of Incremental Redundancy (IR) as applied in the HSDPA standard.

Such principles are also employed in LTE in Chapter 2, focusing more on the differences from the first chapter. In contrast to HSDPA, sets of subcarriers are combined in the form of Resource Blocks (RBs) in LTE. Adaptive Modulation and Coding (AMC) is being employed by means of the CQI as before and also the now-called Precoding Matrix Indicator (PMI) which takes on the role of the PCI parameter. But additionally, LTE now offers a Rank Indicator (RI) to select different modes of operation. The most common modes of operation, namely transmit diversity, Open-Loop Spatial Multiplexing (OLSM), and its counterpart Closed-Loop Spatial Multiplexing (CLSM), are explained.

References

[1] Commission of the European Communities (2008) 'Progress report on the single European electronic communications market (14th report),' Technical Report SEC (2009) 376, Commission of the European Communities. Available from http://eur-lex.europa.eu/LexUriServ/LexUriServ.do?uri = COM:2009:0140:FIN:EN:PDF.

[2] Ericsson (2010) 'Mobile subscriptions hit 5 billion mark,' Technical report, Ericsson. Available from http://www.ericsson.com/thecompany/press/releases/2010/07/1430616.

[3] Grinschgl, A. and Serentschy, G. (2007) 'Kommunikationsbericht 2007,' Technical report, Rundfunk und Telekom Regulierungs-GmbH. Available from http://www.rtr.at/de/komp/KBericht2007/K-Bericht_2007.pdf.

[4] Svoboda, P., Karner, W., and Rupp, M. (2007) 'Traffic analysis and modeling for World of Warcraft,' in *Proceedings of IEEE International Conference on Communications (ICC)*, pp. 1612–1617. Available from http://publik.tuwien.ac.at/files/pub-et_12119.pdf.

[5] Wittig, H. (2008) 'The power of mobile broadband – implications for European telcos and equipment vendors,' Technical report, J.P. Morgan Securities Ltd.

1

UMTS High-Speed Downlink Packet Access

1.1 Standardization and Current Deployment of HSDPA

The 3rd Generation Partnership Project (3GPP) is the forum where standardization has been handled from the first Wideband Code-Division Multiple Access (WCDMA) Universal Mobile Telecommunications System (UMTS) specification on. It unites many telecommunication standard bodies to coordinate their regional activities in order to establish global standards aligned down to bit-level details. The 3GPP was created in December 1998, with the original scope of producing technical specifications and technical reports for 3G mobile systems, both for Frequency Division Duplex (FDD) and Time Division Duplex (TDD) modes. The scope was subsequently amended to include the maintenance and development of Global System for Mobile communications (GSM), General Packet Radio Service (GPRS), and Enhanced Data Rates for Global system for mobile communications Evolution (EDGE) specifications. The specifications themselves are published regularly, with major milestones being denoted as "Releases" [20].

Figure 1.1 shows the chronological development of the 3GPP standardization releases. During the work on Rel'4, it became obvious that some improvements for packet access would be needed [30]. Rel'5, which was finished in June 2002, thus introduced a high-speed enhancement for the Downlink (DL) packet data services: High-Speed Downlink Packet Access (HSDPA).[1] The innovation that happened for HSDPA was quite tremendous, including changes in the physical layer, the Medium Access Control (MAC) layer, and slight changes in the core network. In March 2005, 3GPP finished its work on Rel'6, specifying the Uplink (UL) pendant of HSDPA, called High-Speed Uplink Packet Access (HSUPA).

Multiple-Input Multiple-Output (MIMO) was already of interest during the work on Rel'5 and Rel'6; however, the feasibility studies up to that point concluded that the benefits of it were limited to the extent that the additional complexity could not be justified. Finally, after a long and detailed study discussing many proposals [3], MIMO

[1] The terms Rel'5 HSDPA and Single-Input Single-Output (SISO) HSDPA will be used interchangeably.

Evaluation of HSDPA and LTE: From Testbed Measurements to System Level Performance, First Edition.
Sebastian Caban, Christian Mehlführer, Markus Rupp and Martin Wrulich.

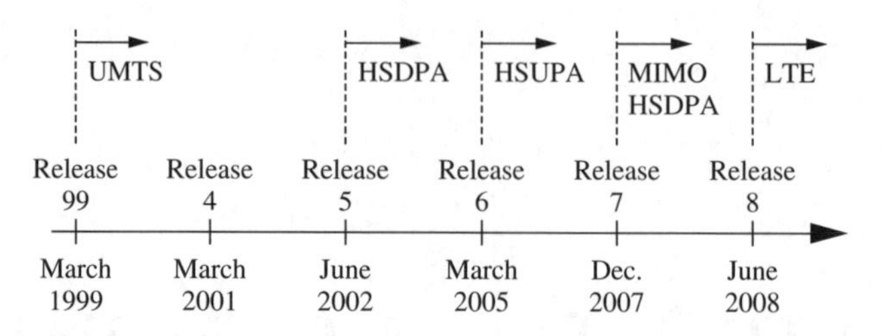

Figure 1.1 Chronological development of the 3GPP standardization releases including important evolutionary steps.

was included in Rel'7 in December 2007. Besides this revolutionary step, 3GPP also added many other improvements, among which some of the most important are:

- the specifications for new frequency bands;
- the utilization of linear Minimum Mean Square Error (MMSE) receivers to meet the performance requirements on the wireless link;
- an optimization of the delay in the network, for example by introducing "continuous connectivity" to avoid setup delays after idle time;
- the definition of 64-Quadrature Amplitude Modulation (QAM) as a higher order modulation scheme for SISO HSDPA;
- the specification of a flexible Radio Link Control (RLC) Packet Data Unit (PDU) assignment; and
- an investigation of the benefits of a direct tunnel for the User Plane (U-Plane) data in High-Speed Packet Access (HSPA) networks.

In June 2008, Rel'8 was published. It contains the next logical step in the evolution of wireless networks, called Long-Term Evolution (LTE). The most notable innovations in the Radio Access Network (RAN) in this release are:

- the redevelopment of the system architecture, called System Architecture Evolution (SAE);
- a definition of network self-organization in the context of LTE;
- the introduction of "home" base stations;
- a study of interference cancellation techniques;
- the first steps towards circuit-switched services for HSDPA; and
- 64-QAM modulation for MIMO HSDPA.

As of 2010, approximately 300 HSDPA networks are operating in 130 countries [27].

1.2 HSDPA Principles

The introduction of HSDPA in the 3GPP Rel'5 implied a number of changes in the whole system architecture. As a matter of fact, not only was the terminal affected, but so were the base station as well as the Radio Network Controller (RNC) [1]. In the 3GPP terminology, a base station is also often called a NodeB. A Base station (NodeB) contains multiple "cells" (that means sectors) in a sectorized network.

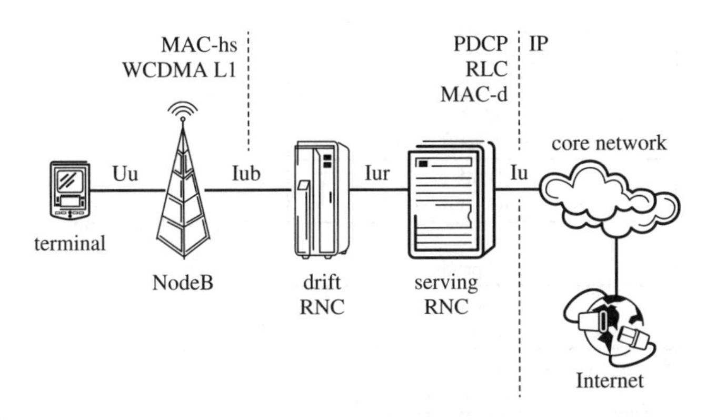

Figure 1.2 Network architecture for HSDPA including the protocol gateways.

1.2.1 Network Architecture

The principal network architecture for the operation of HSDPA is still equal to the architecture for UMTS networks [29]; see Figure 1.2. Basically, the network can be split into three parts: (i) the User Equipment (UE) or terminal connected via the Uu interface; (ii) the Universal mobile telecommunications system Terrestrial Radio Access Network (UTRAN), including everything from the NodeB to the serving RNC; and (iii) the core network, connected via the Iu interface, that establishes the link to the internet. Within the UTRAN, several NodeBs – each of which can control multiple sectors – are connected to the drift RNC via the Iub interface. This intermediate RNC is needed for the soft handover functionality in Rel'99, and is also formally supported by HSDPA. However, for HSDPA, soft handover has been discarded, thus eliminating the need to run user data over multiple Iub and Iur interfaces to connect the NodeB with the drift RNCs and the serving RNC. This also implies that the HSDPA UE is only attached to one NodeB; and in the case of mobility, a hard handover between different NodeBs is necessary. In conclusion, the typical HSDPA scenario could be represented by just one single RNC [30]. We will thus not distinguish further between those two.

For the HSDPA operation, the NodeB has to handle DL scheduling, dynamic resource allocation, Quality of Service (QoS) provisioning, load and overload control, fast Hybrid Automatic Repeat reQuest (HARQ) retransmission handling, and the physical layer processing itself. The drift RNC, on the other hand, performs the layer-two processing and keeps control over the radio resource reservation for HSDPA, DL code allocation and code tree handling, overall load and overload control, and admission control. Finally, the serving RNC is responsible for the QoS parameters mapping and handover control.

The buffer in the NodeB in cooperation with the scheduler enables having a higher peak data rate for the radio interface Uu than the average rate on the Iub. For 7.2 Mbit/s terminal devices, the Iub connection speed can somewhere be chosen around 1 Mbit/s [30]. With the transmission buffer in the NodeB, the flow control has to take care to avoid potential buffer overflows.

In the core network, the Internet Protocol (IP) plays the dominant role for packet-switched interworking. On the UTRAN side, the headers of the IP packets are compressed by the Packet Data Convergence Protocol (PDCP) protocol to improve the efficiency of

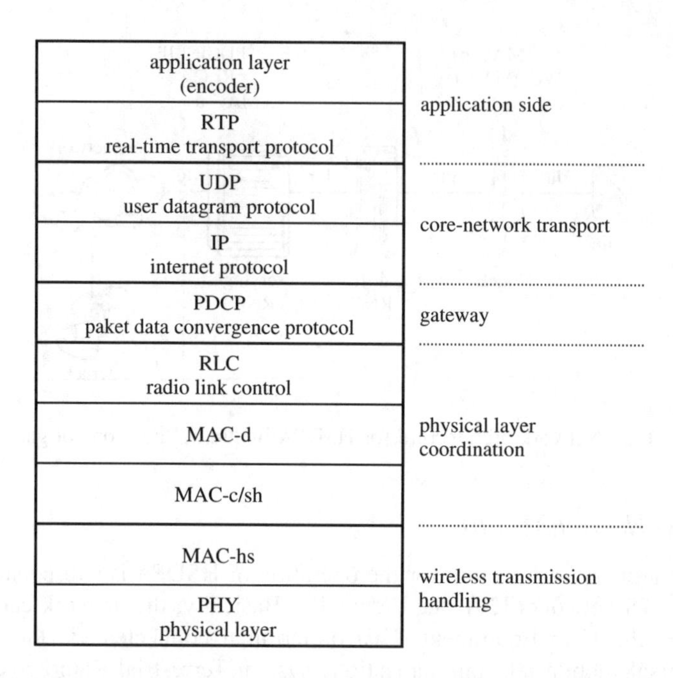

Figure 1.3 Protocol design in UMTS from the application to the physical layer.

small packet transmissions, for example Voice over IP (VoIP). The RLC protocol handles the segmentation and retransmission for both user and control data. For HSDPA the RLC may be operated either (i) in unacknowledged mode, when no RLC -layer retransmission will take place, or (ii) in acknowledged mode, so that data delivery is ensured. The MAC layer functionalities of UMTS have been broadened for HSDPA; in particular, a new MAC protocol entity, the Medium Access Control for High-Speed Downlink Packet Access (MAC-hs) , has been introduced in the NodeB. Now, the MAC layer functionalities of HSDPA can operate independently of UMTS, but take the overall resource limitations in the air into account. The RNC retains the Medium Access Control dedicated (MAC-d) protocol, but the only remaining functionality is the transport channel switching. All other functionalities, such as scheduling and priority handling, are moved to the MAC-hs. Figure 1.3 illustrates the stack of protocols in UMTS-based networks including their purpose.

For the operation of HSDPA, new transport channels and new physical channels have been defined. Figure 1.4 shows the physical channels as employed between the NodeB and the UE, as well as their connection to the transport channels and logical channels utilized [6]. The physical channels correspond in this context to "layer one" of the Open Systems Interconnection (OSI) model, with each of them being defined by a specific carrier frequency, scrambling code, spreading code (also called "channelization code"), and start and stop timing. For communication to the MAC layer, the physical channels are combined into the so-called transport channels. In HSDPA-operated networks, only two transport channels are needed: (i) the High-Speed Downlink Shared CHannel (HS-DSCH),

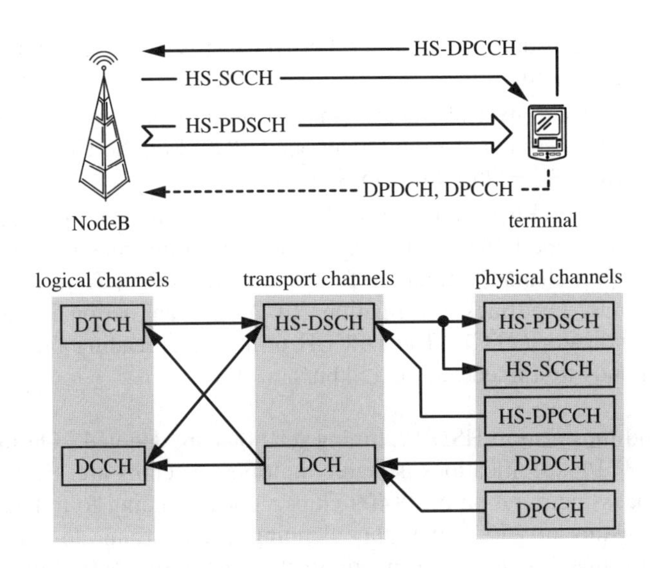

Figure 1.4 Communication channel design of HSDPA. DL information is illustrated by arrows pointing towards the physical channels, and vice versa.

responsible for carrying the DL information (both data and control), and (ii) the Dedicated Channel (DCH), responsible for carrying the UL information (again, both UL and control). Finally, the logical channels are formed in the MAC to convey the information to the higher layers, headed by the RLC layer. In particular, for the HSDPA operation, there are two logical channels: (i) the Dedicated Traffic Channel (DTCH), carrying data, and (ii) the Dedicated Control Channel (DCCH), carrying control information.

1.2.2 Physical Layer

HSDPA utilizes the same WCDMA-based transmission scheme as UMTS, with the basic idea of utilizing spreading codes to introduce individual physical channels [30]. Furthermore, scrambling codes are used to distinguish different NodeBs. However, for HSDPA, many important innovations have been included in the WCDMA context. Let us start by first elaborating on the physical channels utilized, as depicted in Figure 1.4 [6].

HS-PDSCH: The High-Speed Physical Downlink Shared CHannel is used to carry the data from the HS-DSCH. Each HS-PDSCH corresponds to exactly one spreading code with fixed spreading factor SF = 16 from the set of spreading codes reserved for HS-DSCH transmission. The HS-PDSCH does not carry any layer-one control information.

HS-SCCH: The High-Speed Shared Control CHannel carries DL signaling related to the HS-DSCH transmission. It also has a fixed spreading factor SF = 128 and carries a constant rate of 60 kbit/s.

HS-DPCCH: The High-Speed Dedicated Physical Control CHannel carries UL feedback signaling related to the HS-DSCH transmission and to the HS-SCCH control information. The spreading factor of the HS-DPCCH is SF = 256, which corresponds to a fixed rate of 15 kbit/s.

DPDCH: The Dedicated Physical Data CHannel is used to carry the UL data of the DCH transport channel. There may be more than one DPDCH on the UL, with the spreading factor ranging from SF = 256 down to SF = 4.

DPCCH: The Dedicated Physical Control CHannel carries the layer-one control information associated with the DPDCH. The layer-one control information consists of known pilot bits to support channel estimation for coherent detection, Transmit Power-Control (TPC) commands, FeedBack Information (FBI), and an optional Transport-Format Combination Indicator (TFCI). The DPCCH utilizes a spreading factor of SF = 256, which corresponds to a fixed rate of 15 kbit/s.

In WCDMA, and thus also in HSDPA, timing relations are denoted in terms of multiples of chips. In the 3GPP specifications, suitable multiples of chips are: (i) a "radio frame," which is the processing duration of 38 400 chips,[2] corresponding to a duration of 10 ms; (ii) a "slot" consisting of 2560 chips, thus defining a radio frame to comprise 15 slots; and (iii) a "sub-frame," which corresponds to three slots or 7560 chips, which is also called Transmission Time Interval (TTI), because it defines the basic timing interval for the HS-DSCH transmission of one "transport block."

The general HSDPA operation principle brings a new paradigm in the dynamical adaptation to the channel as utilized in UMTS. Formerly relying on fast power control [29], the NodeB now obtains information about the channel quality of each active HSDPA user on the basis of physical-layer feedback, denoted the Channel Quality Indicator (CQI) [4]. Link adaptation is then performed by means of Adaptive Modulation and Coding (AMC), which allows for an increased dynamic range compared with the possibilities of fast power control. The transport format used by the AMC (that is, modulation alphabet size and effective channel coding rate) is chosen to achieve a chosen target Block Error Ratio (BLER).[3] This rises the question why a system should be operated at an error ratio that is bound above zero. Two main arguments for this basic ideology of link adaptation are that: (i) In practical systems the UE has to inform the network about the correct or incorrect reception of a packet. Since this information is needed to guarantee a reliable data transmission, it makes no sense to push the error probability of the data channel lower than the error ratio of the signaling channel. Furthermore, by fixing the BLER to a given value, (ii) the link adaptation has the possibility to remove outage events of the channel.

On top of the link adaptation in terms of the modulation alphabet size and the effective coding rate, HSDPA can take excessive use of the multi-code operation, up to the full spreading code resource allocation in a cell. The other new key technology is the physical-layer retransmission. Whereas retransmissions in UMTS are handled by the RNC, in HSDPA the NodeB buffers the incoming packets and, in case of decoding failure, retransmission automatically takes place from the base station, without RNC involvement. Should the physical-layer operation fail, then RNC -based retransmission may still be applied on top. Finally, as already mentioned, the NodeB is now in charge of the

[2] In 3GPP WCDMA systems, the chip rate is fixed at 3.84 Mchips/s, utilizing a bandwidth of 5 MHz.
[3] One block denotes one TTI in HSDPA.

Table 1.1 Fundamental properties of UMTS and HSDPA [30]

Feature	UMTS	HSDPA
Variable spreading factor	No	No
Fast power control	Yes	No
Adaptive modulation and coding	No	Yes
Multi-code operation	Yes	Yes, extended
Physical-layer retransmissions	No	Yes
NodeB-based scheduling and resource allocation	No	Yes
Soft handover	Yes	No

scheduling and resource allocation. Table 1.1 summarizes the key differences between UMTS and HSDPA [30].

1.2.2.1 HSDPA User Data Transmission

Together with the various new features listed in Table 1.1, Rel'5 HSDPA also came along with the support of higher order modulation, in particular 16-QAM. This made it necessary not only to estimate the phase correctly – as in the UMTS case – but also to evaluate the amplitude. HSDPA utilizes the Common Pilot CHannel (CPICH) for this purpose, which offers the phase information directly and can be used to estimate the power offset to the HS-DSCH power level. This suggests that the NodeB should avoid power changes during the TTI period [30]. The user allocation due to the NodeB-based scheduling happens every TTI, adapting the modulation and coding to the channel quality information as received from the UE.

The coding for the HS-DSCH is turbo-coding of rate 1/3,[4] and there is only one transport channel active at a time; thus, fewer steps in multiplexing/de-multiplexing are needed [5]. Each TTI, a Cyclic Redundancy Check (CRC) field is appended to the user data to allow for the detection of decoding failure. To gain as much as possible from eventually necessary retransmissions, the UE stores the received data in a buffer. When the NodeB retransmits exactly the same chips, the UE can combine the individual received bits in a maximum ratio combining sense, called "soft combining." However, in HSDPA, the HARQ stage is capable of changing the rate matching between retransmissions, thus tuning the redundancy information. The relative number of parity bits to systematic bits varies accordingly, which can be exploited by the receiver in the UE to achieve an additional "incremental redundancy" gain when decoding the packet [22, 34, 46].

1.2.2.2 HSDPA Control Information Transmission

Control information for the DL of HSDPA is exchanged on the HS-SCCH and the HS-DPCCH. The HS-SCCH carries time-critical signaling information which allows the terminal to demodulate the assigned spreading codes. Its information can be split into two parts, with the first part comprising the information needed to identify the relevant

[4] This is due to the fact that the turbo-encoder used is based on the 3GPP Rel'99 turbo-encoder.

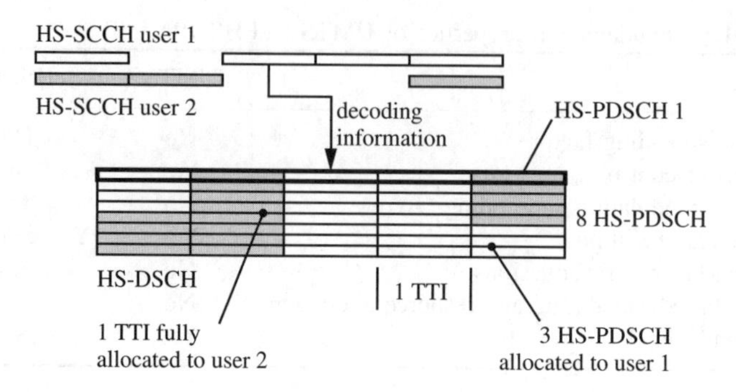

Figure 1.5 Multi-user transmission on the HS-DSCH. Every user needs their own HS-SCCH to obtain the necessary decoding information.

spreading codes in the HS-DSCH. The second part contains less urgent information, such as which HARQ process is being transmitted if the transmission contains new data; or if not, which redundancy version has been used for channel encoding as well as the transport block size. For every simultaneously active user in the cell there is one distinct HS-SCCH necessary, such that the individual terminals know their relevant decoding information, as depicted in Figure 1.5. The individual HS-SCCHs are scrambled with user-specific sequences, which depends on the user identification number managed by the NodeB. To avoid errors due to wrong control information, the second part of the HS-SCCH is protected by a CRC.

For the link adaptation and retransmission handling, HSDPA needs UL physical-layer feedback information. This UL information is carried on the HS-DPCCH. It contains the HARQ information regarding the decoding of the last received packet and the CQI information of the current channel state. The CQI reporting frequency is controlled by a system parameter signaled by higher layers. The evaluation of the CQI report is defined in an abstract way in [4], which leaves the practical implementation open to the vendors. A practical implementation is reported in [36].

1.2.2.3 HSDPA Timing Relations

HSDPA is synchronous in terms of the terminal response for a packet transmitted in the DL. The network side, on the other hand, is asynchronous in terms of when a packet or a retransmission for an earlier transmission is sent. This provides the NodeB scheduler the necessary freedom to act according to its decision metric, buffer levels, and other relevant information. The terminal operation times between the different events are specified accurately from the HS-SCCH reception in [4, 6]; in particular, the transmission of the HS-DPCCH is accurate within a 256 chip window, corresponding to any variations due to the need for symbol alignment. Figure 1.6 depicts the timing relations between the main DL HSDPA channels.

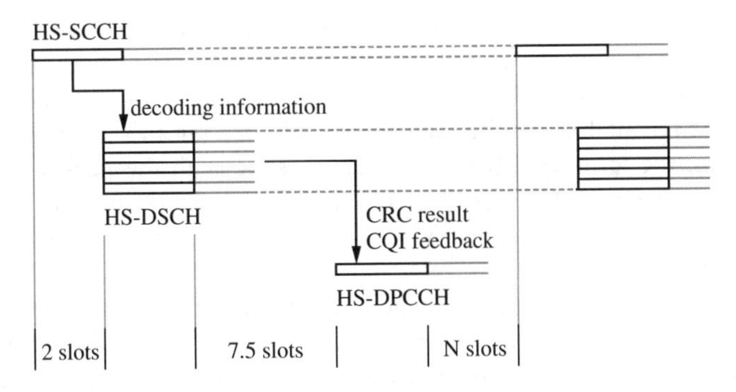

Figure 1.6 HSDPA timing relations between the main physical channels.

1.2.2.4 Terminal Capabilities

The support of the HSDPA functionality is optional for the terminals. Furthermore, the requirements of the high-speed data transmission, in particular the higher order modulation, the code-multiplex of different spreading codes, the minimum interval time between two consecutive utilized TTIs for data transmission, and the dimensioning of the buffer storing the received samples for retransmission combining, put tough constraints on the equipment. Accordingly, the 3GPP defined 12 UE "capability classes," effectively specifying the maximum throughput that can be handled by the device [4, 8]. The highest capability is provided by category 10, which allows the theoretical maximum data rate of 14.4 Mbit/s. This rate is achievable with one-third rate turbo coding and significant puncturing (performed in the rate-matching stage), resulting in a code rate close to one. For a list of terminal capability classes, see [30] for example.

1.2.3 MAC Layer

The HSDPA MAC layer comprises three key functionalities for resource allocation: (i) the scheduling, (ii) the HARQ retransmission handling, and (iii) the link adaptation [2]. Note that the 3GPP specifications do not contain any parameters for the scheduler operation, which is left open for individual implementation to the vendors. Similar arguments hold for the other two functionalities, with the exception that (ii) and (iii) have to obey more stringent restrictions owing to the interworking with the UE side.

For the operation of the HARQ retransmissions, the MAC-hs layer has to consider some important points. In order to avoid wasting time between transmission of the data block and reception of the ACKnowledged (ACK)/Non-ACKnowledged (NACK) response, which would result in lowered throughput, multiple independent HARQ "processes" can be run in parallel within one HARQ entity. The algorithm in use for this behavior is an N Stop And Wait (N-SAW) algorithm [14]. However, the retransmission handling has to be aware of the UE minimum inter-TTI interval for receiving data, which depends on the UE capability class.

The terminal is signaled some MAC-layer parameters, with the MAC-hs PDU consisting of the MAC-hs header and the MAC-hs payload. It is built by one or more MAC-hs Service Data Units (SDUs) and potential padding if they do not fit the size of the transport block available. As the packets are not arriving in sequence after the MAC HARQ operation, the terminal MAC layer has to cope with the packet reordering. The MAC header, therefore, includes a Transmission Sequence Number (TSN) and a Size Index Identifier (SID) that reveals the MAC-d PDU size and the number of MAC PDUs of the size indicated by the SID.

The MAC-d is capable of distinguishing between different services by means of MAC-d "flows" [40], each of which can have different QoS settings assigned. In the MAC-hs, every MAC-d flow obtains its own queue, to allow different reordering queues at the UE end. Note however that only one transport channel may exist in a single TTI, and thus in a single MAC-hs PDU. This means that flows from different services can only be scheduled in consecutive TTIs [14, 30].

1.2.4 Radio Resource Management

The Radio Resource Management (RRM) algorithms are responsible for transferring the physical-layer enhancements of HSDPA to capacity gain while providing attractive end-user performance and system stability. At the RNC, the HSDPA-related algorithms include physical resource allocation, admission control, and mobility management. On the NodeB side, RRM includes HS-DSCH link adaptation, the already explained packet scheduler, and HS-SCCH power control.

In order for the NodeB to transmit data on the HS-DSCH, the controlling RNC needs to allocate channelization codes and power for the transmission. As a minimum, one HS-SCCH code and one HS-PDSCH code have to be assigned to the NodeB. The communication between the RNC and the NodeB follows the NodeB Application Part (NBAP) protocol [7]. In the case that both HSDPA and UMTS traffic are operated on the same carrier, the RNC can assign and release spreading codes to HSDPA dynamically to prevent blocking of UMTS connections.

Another scarce DL resource is the transmission power. The power budget for a cell consists of the power needed for common UMTS channels, as the CPICH power for UMTS transmissions, and power for the HSDPA transmission. In principle, there are two different possibilities for allocating the power of HSDPA in the base station DL power budget [39]:

1. The RNC can dynamically allocate HSDPA power by sending NBAP messages to the NodeB, which effectively keeps the HSDPA power at a fixed level, whereas the UMTS power varies according to the fast closed-loop power control.
2. The other option is that no NBAP messages are sent and the base station is allowed to allocate all unused power for HSDPA.

The behavior is illustrated in Figure 1.7. Note that some safety margin has to be reserved in order to account for unpredictable variations of the non-HSDPA power; for example, due to the fast power control of UMTS.

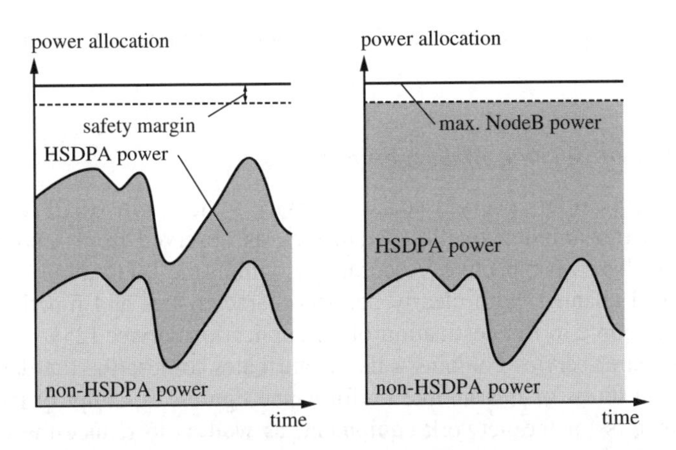

Figure 1.7 HSDPA power allocation principles. *Left*: the explicit power allocation by means of NBAP messages. *Right*: the fast NodeB -based power allocation.

The HS-DSCH link adaptation at the NodeB is adjusted every TTI. In order for the UE to determine the current transmit power utilized by the HS-DSCH, the RNC sends a Radio Resource Control (RRC) message, which contains the power offset Γ in relation to the CPICH. A simple link adaptation algorithm would follow the CQI values reported by the UE directly. However, when a change in Γ occurs, the NodeB has to take that into account and remap the CQI value accordingly. The HS-SCCH is the exception to the AMC link adaptation of HSDPA. This channel is power controlled, where the 3GPP specifications do not explicitly specify any power control mechanism. Some ideas about the implementation of a power control for the HS-SCCH can be found in [30].

The last RRM algorithm is the packet scheduler, which determines how to share the available resources among the pool of users eligible to receive data. There are numerous scheduler ideas established, the most famous and well-known are:

- The round-robin scheduler, which assigns the physical resources equally amongst all active users in the cell.
- The maximum carrier-to-interference ratio scheduler, which assigns the physical resources to the user with the best carrier-to-interference ratio in order to maximize the cell throughput.[5]
- The proportional fair scheduler trying to balance between throughput maximization and fairness. More details on "fairness" can be found in [10, 19, 44].

Besides these allocation strategies, the scheduler can schedule multiple users within one TTI. Up to 15 HS-PDSCHs may be used by the NodeB, which gives the scheduler the freedom to maximize spectral efficiency when scheduling more than one user at once. On the other hand, the scheduler may require a multi-user policy, if there are many

[5] In fact, the maximum carrier-to-interference ratio scheduler tries to maximize the achievable sum-throughput for every TTI, which for a single-user scheduler results in scheduling the user with the maximum carrier-to-interference ratio.

HSDPA users that require a low source data rate but strict upper bounds on the delay; for example, VoIP.

1.2.5 Quality of Service Management

The UMTS network offers every associated service – thus, also HSDPA – a possibility to define certain transmission quality parameters, as observed from a core-network- or application-layer-based perspective. Accordingly, within the UTRAN, a "bearer service" defining a logical channel, with clearly defined characteristics and functionality, can be set up from the source to the destination of an application service [25].

The UMTS bearer service contains a list of attributes to describe the definition of the service quality in terms of measurable performance figures. To simplify the handling of the QoS parameters for the network equipment, as well as to reduce the complexity of the processing of these parameters, only four different QoS classes have been defined by 3GPP [9]. The four classes are:

Conversational class: This class works in a real-time fashion, thus preserving the time relation between information entities of the stream, which corresponds to a conversational pattern with stringent and low delay; for example, voice traffic.

Streaming class: Although this class also works in a real-time fashion it imposes less stringent delay constraints; for example, video streaming traffic.

Interactive class: This class works on a best-effort basis and, thus, represents a request – response pattern with the goal to preserve the payload content; for example, web-browsing traffic.

Background class: This class also works on a best-effort basis, where the destination is not expecting the data within a certain time but the payload still has to be preserved; for example, telemetry or email traffic.

These classes then define the attributes associated with the UMTS bearer service [9]; in particular:

- the traffic class and the source statistics descriptor;
- the maximum bit rate and the guaranteed bit rate;
- the maximum SDU size, the SDU format information, the SDU error ratio, and the delivery of erroneous SDUs;
- the residual bit error ratio and the transfer delay; and
- the delivery order, the traffic handling priority, and the allocation/retention priority.

By defining these attributes, the traffic classes can be accurately specified by means of their logical channel requirements for successful operation of the application service. Especially in HSDPA, the MAC-hs entities allow for linking the QoS information established in the UTRAN with the physical layer, which falls into the category of cross-layer optimization techniques.

1.3 MIMO Enhancements of HSDPA

MIMO techniques are regarded as the crucial enhancement of today's wireless access technologies to allow for a significant increase in spectral efficiency. They have drawn attention since the discovery of potential capacity gains [21, 43], which has motivated the scientific community to consider the potential of MIMO to increase user data rates, system throughput, coverage, and QoS with respect to delay and outage. Basically, MIMO as a technique offers two gain mechanisms: (i) "diversity," including "array gain" [38], and (ii) "spatial multiplexing gain".[6]

The spatial multiplexing gain can be observed by investigating the ergodic channel capacity of an MIMO link. Consider an i.i.d. Rayleigh block fading channel $\mathbf{H}$ between N_T transmit and N_R receive antennas with a receive antenna Signal to Noise Ratio (SNR) ϱ.[7] The capacity of the MIMO channel when no Channel State Information (CSI) is available at the transmitter,

$$C_E = \mathbb{E}\{C(\mathbf{H})\} = \mathbb{E}\left\{\log_2 \det\left(\mathbf{I} + \frac{\varrho}{N_\mathrm{T}}\mathbf{H}\mathbf{H}^H\right)\right\} = \mathbb{E}\left\{\sum_{i=1}^{r} \log\left(1 + \frac{\varrho}{N_\mathrm{T}}\lambda_i\right)\right\}, \quad (1.1)$$

can be interpreted to unravel multiple scalar spatial pipes, also called "modes," between the transmitter and the receiver. The number of nonzero eigenvalues λ_i of $\mathbf{H}\mathbf{H}^\mathrm{H}$, defined by the rank r of the matrix $\mathbf{H}$, corresponds to the number of parallel independent data "streams" that can be transmitted simultaneously. This specifies the spatial multiplexing gain of the channel. In general, the capacity increase due to the utilization of MIMO can be approximated by [26]

$$C_E \approx \min\{N_\mathrm{T}, N_\mathrm{R}\} \log_2\left(\frac{\varrho}{N_\mathrm{T}}\right), \quad (1.2)$$

where $\min\{N_\mathrm{T}, N_\mathrm{R}\}$ defines the spatial multiplexing gain. Thus, every 3 dB increase in receive SNR results in $\min\{N_\mathrm{T}, N_\mathrm{R}\}$ additional bits of capacity at high SNR. This behavior is depicted in the left part of Figure 1.8. Clearly, the spatial multiplexing gain – and thus the number of simultaneously transmitted data streams – in the $N_\mathrm{T} \times N_\mathrm{R} = 4 \times 2$ case is the same as in the 2×2 case. It can be observed that an additional array gain – noticeable by the shift to the left – can be observed. The array gain, however, diminishes quickly with the number of transmit antennas, thus effectively limiting the benefits of only adding spatial resources at the transmitter side.

Diversity, on the other hand, describes the error probability decrease as a function of the SNR. MIMO channels can offer significant improvement in terms of diversity. If

[6] More recently, the spatial multiplexing gain of MIMO systems has been studied in the more general framework of the "Degrees of Freedom (DoFs)" [32], which can also be extended to whole networks.

[7] Assuming that the channel matrix $\mathbf{H}$ is normalized; that is, $\mathbb{E}\left\{|h_{i,j}|^2\right\} = 1$ with i, j denoting the channel coefficient index and an average path gain of one.

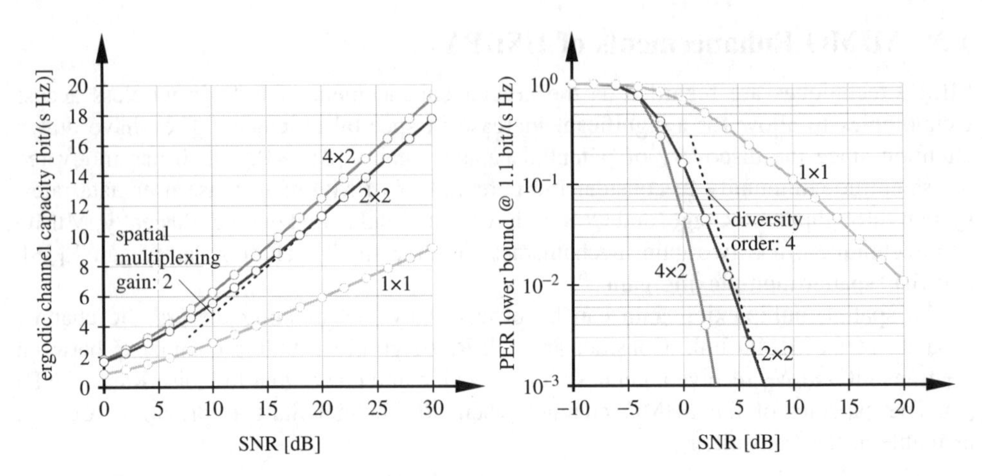

Figure 1.8 Ergodic channel capacity and PER lower bound for i.i.d. Rayleigh block fading MIMO channels without CSI at the transmitter. The spatial multiplexing gain and the diversity orders can be observed by determining the slopes of the curves.

perfect channel codes are considered and the transmitter has no CSI again, the Packet Error Ratio (PER) will equal the outage probability for that signaling rate. This also defines the "outage capacity" of MIMO channels. Thus, for a system with unity bandwidth transmitting packets with a bit rate R, the PER can be lower bounded by

$$\Pr\{\text{packet error}\} \geq \Pr\{C(\mathbf{H}) < R\} \tag{1.3}$$

The magnitude of the slope of the PER curve can then be used to define the diversity order of the MIMO system. It has been shown that the achievable slope can be up to $N_\text{T}N_\text{R}$ for fixed-rate transmissions [42], which directly implies that every antenna deployed can increase the diversity order. The right part of Figure 1.8 illustrates the diversity order, which can be read off the slope of the PER lower bound; that is, the orders of magnitude gained by an increase of 10 dB in the receive SNR [45]. The rate at which the lower bound is evaluated to be 1.1 bit/(s Hz) was chosen because this is the expected spectral efficiency of MIMO enhanced HSDPA. A rigorous definition of the terms spatial multiplexing gain and diversity can be found in [48].

The spatial multiplexing and the diversity gain are strongly influenced by antenna correlation at the transmitter and the receiver side, as well as path coupling, also called the "bottleneck scenario" [38]. Accordingly, receiver antenna deployment is not the main focus, since the average mobile terminal will only have limited space and battery capacity, impeding complex signal-processing algorithms. Furthermore, additional Radio Frequency (RF) frontends needed for multi-antenna operation would increase the terminal costs significantly. Thus, in mobile communications, emphasis has been put on transmitter antenna processing techniques; in particular:

- spatial multiplexing schemes, such as for example Vertical Bell Laboratories Layered Space–Time (V-BLAST) techniques [21], trying to extract the gain in spectral efficiency of the MIMO channel; or

- diversity schemes, such as for example Alamouti space–time coding [11], trying to extract the diversity gain offered by the MIMO channel.

Both classes of schemes can in principle be operated in "closed-loop" mode – utilizing feedback information provided by the UE – or "open-loop" mode, which does not rely on feedback.

Current insights [33] suggest that, for variable-rate link-adapted systems with HARQ, retransmission spatial multiplexing is the best choice, even for high UE speed scenarios. However, Space–Time Transmit Diversity (STTD) techniques are considered to be very useful for signaling channels which operate at a fixed rate and without retransmissions, called "unacknowledged" mode. Other Space–Time Block Codes (STBCs) are not considered for WCDMA MIMO extensions at the moment, but are discussed for future generation wireless networks. One example is the Golden–STBC, which achieves full rate and full diversity [12].

1.3.1 Physical Layer Changes for MIMO

Focusing on the FDD mode of HSDPA, 3GPP has considered numerous proposals for the MIMO enhancement [3]. The most promising ones were (i) Per-Antenna Rate Control (PARC), (ii) Double Space–Time Transmit Diversity with Sub-Group Rate Control (DSTTD-SGRC), and (iii) Double Transmit Antenna Array (D-TxAA). In late 2006, the 3GPP finally decided in favor of D-TxAA to be the next evolutionary step of the classical SISO HSDPA, which (among many other changes) established the 3GPP Rel'7. D-TxAA offers the flexibility to exploit the MIMO gains, spatial multiplexing, and diversity and tries to benefit from channel quality adaptability by means of closed-loop feedback.

This scheme focuses on two transmit and two receive antennas. An overview of the D-TxAA scheme is depicted in Figure 1.9. Channel coding, interleaving, as well as spreading and scrambling – the Transport CHannel (TrCH) processing – are implemented as in the non-MIMO mode, with a "primary transport block" always being present. However, now the physical layer supports the transport of a "secondary transport block" to a UE within one TTI. This simultaneous transmission of two codewords aims at the spatial multiplexing gain of the MIMO channel. The precoding, on the other hand, is designed to extract (at least some) of the diversity offered. The precoding weights for the Transmit Antenna Array (TxAA) operation of the transmitter are determined from a quantized set by the UE [4] and serve as "beamforming" to maximize the receive Signal to Interference and Noise Ratio (SINR). In the case of a single-stream transmission, the primary precoding vector, determined by $[w_1, w_2]^{\mathrm{T}}$, is utilized for the transmission of the primary transport block. In a double-stream transmission, the secondary precoding vector $[w_3, w_4]^{\mathrm{T}}$ is chosen orthogonal to the primary one. This has the benefit of increasing the separation of the two streams in the signal space [15], and also keeps the amount of feedback bits constant for single- and double-stream transmission modes. The two CPICHs are utilized for the channel estimation, receive power evaluation, and serving NodeB evaluation.

The evaluated feedback in the form of the Precoding Control Indicator (PCI) is fed back to the NodeB within a composite CQI/PCI report on the HS-DPCCH, which implies a change in the meaning of the feedback bits compared with Rel'5 HSDPA[6]. In order

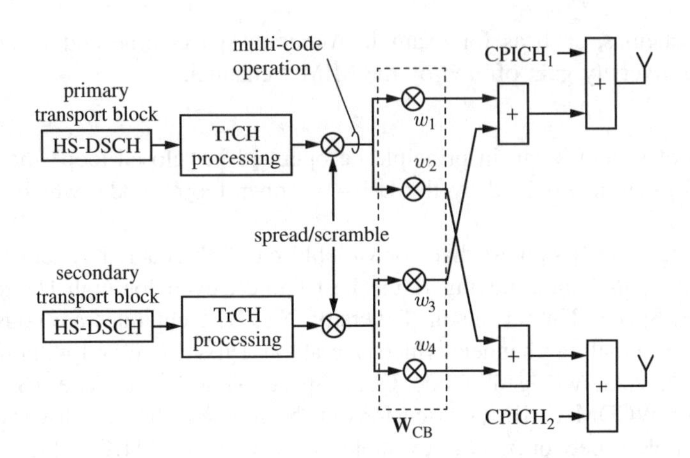

Figure 1.9 Overview of the D-TxAA transmitter scheme for MIMO-enhanced HSDPA.

for a particular UE receiver to adapt to the precoding utilized by the NodeB for this user at the time of the transmission, the currently employed precoding weights are signaled on the HS-SCCH.

Theoretically, D-TxAA allows for a doubled data rate compared with the actual possible SISO HSDPA data rates [37], which can be utilized in high SNR regions. In principle, the scheme can also be extended to systems with more than two transmit antennas on the transmit and/or the receiver side.

With the introduction of a second simultaneously transmitted stream, 3GPP also had to update the CQI reporting to support individual reports for the individual streams. This ensures an independent link adaptation. Accordingly, the UE has to report two CQI values, depending on the channel quality as seen by the corresponding stream. In practice, the signaling of the two CQI values is performed in a combined way; thus, only one equivalent CQI value is fed back to the NodeB [4]. To evaluate the Transport Format Combination (TFC) for the consecutive transmission to this particular user, the network utilizes a mapping from the CQI value to the TFCI, which specifies the transmission parameters. These mappings, however, depend on the UE capabilities. The 3GPP thus introduced two new UE capability classes in Rel'7, namely 15 and 16. For both UE classes, the mapping tables for the single-stream and the double-stream transmission are defined in [4]. An excerpt of mapping Table I of the double-stream transmission mode for UEs of capability class 16 is given in Table 1.2, which holds for the CQI of both streams. The Transport Block Size (TBS) specifies the number of bits transported within one TTI. It has to be noted that, in contrast to the Rel'5 SISO HSDPA mapping tables, where obviously only one stream is transmitted, a CQI value of zero does not denote an "out of range" report. As a matter of fact, in unfavorable channel conditions the network would be wise to switch to a single-stream transmission long before the UE becomes out of range. Furthermore, the TFCs for CQI values zero and one appear to be equal. However, in the case of a CQI zero report, the NodeB would increase its transmission power by 3 dB [4], which is not the case for the CQI one report. Finally, note that all TFCs request a utilization of 15 multiplexed spreading codes, which leaves no room for any other data channels operated

Table 1.2 Excerpt of the CQI mapping Table I [4], utilized in double-stream transmissions for MIMO-capable UEs

CQI	TBS [bit]	Nr. Codes	Modulation
0	4 581	15	4-QAM
1	4 581	15	4-QAM
2	5 101	15	4-QAM
⋮	⋮	⋮	⋮
6	11 835	15	4-QAM
7	14 936	15	16-QAM
8	17 548	15	16-QAM
⋮	⋮	⋮	⋮
14	27 952	15	16-QAM

in parallel. Although the UE is assuming the transmission of data on all 15 spreading codes for the evaluation of the CQI report, the network does not have to stick to this restriction. Based on the RNC information, the NodeB may as well assign only a subset of the 15 available spreading codes for the D-TxAA operation, which then requires a remapping of the TFC for the UE CQI reports. In addition, multi-user scheduling could still be performed by allocating each stream to each individual user. This transmission mode is sometimes called Multi-User (MU) MIMO HSDPA, but is currently not included in the 3GPP standard.

As elaborated in [31], the most interesting property of MIMO HSDPA for network operators is the fact that most of the 3GPP Rel'7 enhancements are expected to be software upgrades to the network, excluding, of course, the need for multiple antennas at the NodeB.

1.3.2 Precoding

The 3GPP technical recommendation [3] specifies a precoding codebook for D-TxAA MIMO HSDPA. The precoding coefficients for double-stream operation are defined as

$$w_1 = w_3 \triangleq \frac{1}{\sqrt{2}}, \tag{1.4}$$

$$w_2 \in \left\{ \frac{1+\mathrm{i}}{2}, \frac{1-\mathrm{i}}{2}, \frac{-1+\mathrm{i}}{2}, \frac{-1-\mathrm{i}}{2} \right\}, \tag{1.5}$$

$$w_4 \triangleq -w_2, \tag{1.6}$$

where w_1 and w_2 are utilized by stream one and w_3 and w_4 are utilized by stream two. Accordingly, the precoding matrix $\mathbf{W}_{\mathrm{CB}}$ is given by

$$\mathbf{W}_{\mathrm{CB}} = \begin{bmatrix} w_1 & w_3 \\ w_2 & w_4 \end{bmatrix} = \begin{bmatrix} \mathbf{w}_1 & \mathbf{w}_2 \end{bmatrix} = \mathbf{W}_{\mathrm{CB}}(w_2), \tag{1.7}$$

which is fully determined by the choice of the precoding weight w_2 in this case. It also has to be noted that, given the precoding weights of (1.4)–(1.6), the precoding matrix is "unitary" in the case of a double-stream transmission (see also Theorem 12.1 in Chapter 12). For the single-stream transmission, only the precoding vector $\mathbf{w}_1$ is utilized. With this definition of the precoding codebook, it remains to elaborate how the "best precoding" choice can be determined by the UE. First, it is necessary to define a metric in order to be able to introduce a measure for the term "best." For this purpose, many different figures of merit could be used; for example, the capacity, achievable maximum rate, or the BLER performance.

The question of the metric to be optimized for the precoding in MIMO systems as well as the algorithm solving for the metric has been investigated by researchers in great detail. Many approaches have been proposed; for example, based on interference alignment ideas [16], utilizing game-theoretic concepts [24], introducing the energy efficiency into the problem [13], taking into account that the channel is only imperfectly known [28], applying Dirty Paper Coding (DPC)[8] techniques [18], aiming for robust solutions [41], or trying to exploit the benefits of joint precoding and scheduling [23]. A lot of research has also tried to include coordination among the transmitters; that is, the NodeBs. In HSDPA, such approaches are currently not supported by the network; however, in LTE the X2-interface provide means to allow for such algorithms (see Section 15.2.4 for more details). Despite all these research efforts, the 3GPP recommends the utilization of the SINR as the underlying metric to decide upon the precoding vector choice [3, 4]. Only a coordination with the scheduler regarding the precoding utilization would be possible. Defining the SINR as the cost function to be maximized, it is still unclear "which" SINR should be considered. Given the structure of MIMO HSDPA, see also Figure 1.9, the SINR metric can be evaluated

- before the equalization, thus directly at the receive antennas; or
- after the equalization, which is much more complex to be evaluated.[9]

For the UE to utilize the post-equalization SINR as a decision metric for the precoding, the receive filter would have to be evaluated for all – or at least a set of – precoding possibilities. Given the limited computational power and battery constraints in wireless mobile devices, such calculations are intractable. For the sake of completeness, it should be pointed out that there may be possibilities to assess the post-equalization SINR as a metric by a suitable low-complex representation. If done so, this also allows for the possibility of joint precoding and link-adaptation feedback reporting [35]. Owing to its anticipated lower computational complexity, the case of the pre-equalization SINR metric is considered here. Consider the samples at receive antenna n_r given by

$$y_i^{(n_\mathrm{r})} = \mathbf{h}_\mathrm{W}^{(n_\mathrm{r})} \mathbf{x}_i + \mathrm{n}_i, \tag{1.8}$$

where $\mathbf{h}_\mathrm{W}^{(n_\mathrm{r})} \triangleq (\mathbf{H}_\mathrm{W})_{(n_\mathrm{r},:)}$ is defined to be the n_rth row of the equivalent MIMO channel matrix $\mathbf{H}_\mathrm{W}$.[10] Considering that there are multiple streams transmitted to the receiver, the

[8] DPC is a technique for pre-destorting the transmit signal to cancel the interference at the receiver end [17].
[9] Note that the post-equalization SINR builds the basis for the system-level model in Chapter 12 with the despreading gain included.
[10] The equivalent channel matrix is defined in Equation (12.2). More details of the model are provided in Equation (12.5).

above equation can be rewritten as

$$y_i^{(n_\mathrm{r})} = \sum_{n=1}^{N_S} \mathbf{h}_\mathrm{w}^{(n_\mathrm{r},n)} \mathbf{x}_i^{(n)} + \mathrm{n}_i, \tag{1.9}$$

where $\mathbf{h}_\mathrm{w}^{(n_\mathrm{r},n)}$ is composed of the columns with indices[11] $n + N_\mathrm{S}[1, L_\mathrm{h}]$ from the vector $\mathbf{h}_\mathrm{w}^{(n_\mathrm{r})}$. The channel vector entries of $\mathbf{h}_\mathrm{w}^{(n_\mathrm{r},n)}$ correspond to the transmit chips of stream n and, correspondingly, $\mathbf{x}_i^{(n)}$ is defined to contain only the transmit chips of that particular stream. With these definitions, the pre-equalization SINR of stream n on receive antenna n_r is given by

$$\rho^{(n,n_\mathrm{r})} = \frac{\mathbb{E}\left\{\left|\mathbf{h}_\mathrm{w}^{(n_\mathrm{r},n)} \mathbf{x}_i^{(n)}\right|^2\right\}}{\mathbb{E}\left\{\sum_{\substack{m=1\\m\neq n}}^{N_S} \left|\mathbf{h}_\mathrm{w}^{(n_\mathrm{r},m)} \mathbf{x}_i^{(m)}\right|^2\right\} + \mathbb{E}\{|\mathrm{n}_i|^2\}} \tag{1.10}$$

For simplifying this equation, consider

$$\mathbb{E}\left\{\left|\mathbf{h}_\mathrm{w}^{(n_\mathrm{r},n)} \mathbf{x}_i^{(n)}\right|^2\right\} = \mathbf{h}_\mathrm{w}^{(n_\mathrm{r},n)} \mathbb{E}\left\{\mathbf{x}_i^{(n)} \left(\mathbf{x}_i^{(n)}\right)^\mathrm{H}\right\} \left(\mathbf{h}_\mathrm{w}^{(n_\mathrm{r},n)}\right)^\mathrm{H} = \mathbf{h}_\mathrm{w}^{(n_\mathrm{r},n)} \left(\mathbf{h}_\mathrm{w}^{(n_\mathrm{r},n)}\right)^\mathrm{H}, \tag{1.11}$$

where the data chips are assumed to be i.i.d. uncorrelated with unit variance, $E_\mathrm{c} = 1$. Note also that the stream-specific equivalent MIMO channel row $\mathbf{h}_\mathrm{w}^{(n_\mathrm{r},n)}$ is equal to

$$\mathbf{h}_\mathrm{w}^{(n_\mathrm{r},n)} = \mathbf{H}_{(n_\mathrm{r},:)} (\mathbf{I} \otimes \mathbf{w}_n) = \mathbf{w}_n^\mathrm{H} \left(\mathbf{H}^{(n_\mathrm{r})}\right)^\mathrm{H} \mathbf{H}^{(n_\mathrm{r})} \mathbf{w}_n, \tag{1.12}$$

with $\mathbf{H}^{(n_\mathrm{r})}$ defining the $\mathbb{C}^{L_\mathrm{h} \times N_\mathrm{T}}$ frequency-selective MIMO channel matrix that contains the channel coefficients from all transmit antennas to receive antenna n_r. Accordingly, Equation (1.10) becomes

$$\rho^{(n,n_\mathrm{r})} = \frac{\mathbf{w}_n^\mathrm{H} \left(\mathbf{H}^{(n_\mathrm{r})}\right)^\mathrm{H} \mathbf{H}^{(n_\mathrm{r})} \mathbf{w}_n}{\sum_{\substack{m=1\\m\neq n}}^{N_S} \mathbf{w}_m^\mathrm{H} \left(\mathbf{H}^{(n_\mathrm{r})}\right)^\mathrm{H} \mathbf{H}^{(n_\mathrm{r})} \mathbf{w}_m + \sigma_\mathrm{n}^2}, \tag{1.13}$$

with σ_n^2 defining the variance of the i.i.d. white Gaussian noise. The UE thus has to find the optimum precoding vector for each stream by solving the SINR-related problem

$$\mathbf{w}_n^\mathrm{opt} = \arg\max_{\mathbf{w}_n} \sum_{m=1}^{N_\mathrm{S}} \sum_{n_\mathrm{r}=1}^{N_\mathrm{R}} \rho^{(n,n_\mathrm{r})}, \tag{1.14}$$

which optimizes the sum SINR over all receive antennas and streams. This is a particularly difficult problem to solve, in particular because (as in the 3GPP precoding codebook) the precoding vectors can have dependencies; thus,the problem in Equation (1.14) cannot be

[11] We utilize the notation $[a, b]$ do denote all integers between a and b; for example, $[3, 9] = 3, 4, \ldots, 9$.

decoupled for the individual streams. Accordingly, a low-complexity approach would be to optimize

$$\mathbf{w}_n^{\text{opt}} = \arg\max_{\mathbf{w}_n} \sum_{n_\text{r}=1}^{N_\text{R}} \mathbf{w}_n^\text{H} \left(\mathbf{H}^{(n_\text{r})}\right)^\text{H} \mathbf{H}^{(n_\text{r})} \mathbf{w}_n = \arg\max_{\mathbf{w}_n} \mathbf{w}_n^\text{H} \underbrace{\left[\sum_{n_\text{r}=1}^{N_\text{R}} \left(\mathbf{H}^{(n_\text{r})}\right)^\text{H} \mathbf{H}^{(n_\text{r})}\right]}_{\triangleq \mathbf{R}} \mathbf{w}_n \quad (1.15)$$

instead of Equation (1.14), for each stream individually. In the D-TxAA MIMO HSDPA case, only one stream has to be evaluated, because the second stream is directly specified by the codebook, given in Equations (1.4)–(1.6). Thus, here, the precoding was chosen according to the optimization over the precoding weight w_2,

$$w_2^{\text{opt}} = \arg\max_{w_2} \mathbf{w}_1^\text{H} \mathbf{R} \mathbf{w}_1, \quad (1.16)$$

which of course favors stream one, thus explaining, for example, the gap between the empirical cumulative density functions (cdfs) of the equivalent fading parameters in Figure 12.3. To overcome the problem of favoring stream one, the optimization problem can be altered to

$$w_2^{\text{opt}} = \arg\max_{w_2} \mathbf{w}_1^\text{H} \mathbf{R} \mathbf{w}_1 + \mathbf{w}_2^\text{H} \mathbf{R} \mathbf{w}_2, \quad (1.17)$$

thus searching for the best combination of the two precoding vectors.

In order to assess the performance of the optimization approaches in Equations (1.16) and (1.17), a simulation was conducted, comparing them with the "optimum precoding" vector choices, given by the eigenvectors corresponding to the largest eigenvalues of $\mathbf{R}$,[12]

$$\mathbf{w}_n^{\text{opt}} = \mathbf{u}_n, \quad \mathbf{u}_n \triangleq \mathbf{U}_{(:,n)} : \mathbf{R} = \mathbf{U} \mathbf{\Lambda} \mathbf{U}^{-1}, \quad (1.18)$$

with $\mathbf{U}$ and $\mathbf{\Lambda}$ denoting the unitary matrix of eigenvectors and the matrix of eigenvalues, both sorted according to the magnitude of the eigenvalue. The simulation parameters for the simulation are given in Table 1.3. The chip-level SINR was chosen to be $E_\text{c}/N_0 = 0\,\text{dB}$, which implies that $\sigma_\text{n}^2 = 1$.

The simulation results are given in Figure 1.10, which shows the sum SINR over all streams and receive antennas, $\sum_{n=1}^{N_\text{S}} \sum_{n_\text{r}=1}^{N_\text{R}} \rho^{(n,n_\text{r})}$. It can be observed that both algorithms perform worse when compared with the nonquantized precoding choice from Equation (1.18). Interestingly, the algorithm greedily aiming for the optimization of stream one from Equation (1.16) outperforms the algorithm considering both streams in Equation (1.17). This is due to the fact that both algorithms are only an approximation to the problem in Equation (1.14).

[12] Note that **R** is a Hermitian positive semi-definite square matrix; thus, Singular Value Decomposition (SVD) and eigenvalue decomposition deliver equal results.

Table 1.3 Simulation parameters for the performance comparison of different precoding-choice algorithms in a 2 × 2 PedA MIMO channel, with $N_S = 2$ streams active

Parameter	Value
Fading model	Improved Zheng model [47, 49]
Antennas	$N_T \times N_R = 2 \times 2$
Precoding codebook	3GPP [3]
Precoding delay	11 slots
Transmitter frequency	2 GHz
Mean equalizer E_c/N_0	0 dB
UE speed	3 km/h
Channel profile	Pedestrian A (PedA)
Simulated slots	10 000, each 2/3 ms

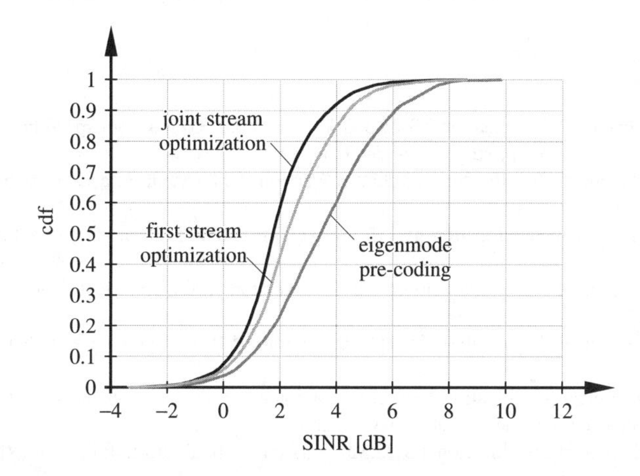

Figure 1.10 Performance of the precoding choice algorithms Equations (1.16) and (1.17) compared with the “best” nonquantized precoding.

1.3.3 MAC Layer Changes for MIMO

The MAC layer of UMTS also needed some enhancements to support the multi-stream operation of D-TxAA. In particular, the MAC-hs now has to deal with:

1. a more complicated scheduling that has to distinguish between single-stream and double-stream transmissions;
2. a more complex resource allocation problem, because a decision regarding the number of utilized streams and the power allocation of those has to be solved; and
3. the HARQ process handling has to be conducted for both streams in the case of a double-stream transmission.

Most of the questions arising in this context are not covered by the standard [2], but are rather left open for vendor-specific implementation, which also leaves much room for research in terms of the RRM opportunities offered by the MIMO enhancements.

1.3.4 Simplifications of the Core Network

As part of Rel'7, 3GPP also tried to simplify the core-network architecture of HSDPA. In particular, 3GPP networks will be increasingly used for IP-based packet services. In Rel'5, the network elements in the user and the Control Plane (C-Plane) are (i) NodeB, (ii) RNC, and (iii) Serving-General packet radio service Support Node (SGSN) and Gateway-General packet radio service Support Node (GGSN); see also Figure 1.2. A flat network architecture, saving unnecessary network elements, however, is expected to reduce latency and thus improve the overall performance of IP-based services. Thus, in Rel'7, the U-Plane can tunnel the SGSN, effectively reducing the number of network elements that have to process the data. This reduced overhead in terms of hardware and delay is important for achieving low cost per bit and enabling competitive flat-rate offerings [31].

References

[1] 3GPP (2001) Technical Specification TS 25.308 Version 5.0.0 'UTRA high speed downlink packet access (HSDPA); overall description; stage 2,' www.3gpp.org.

[2] 3GPP (2006) Technical Specification TS 25.321 Version 7.0.0 'Medium access control (MAC) protocol specification' www.3gpp.org.

[3] 3GPP (2007) Technical Specification TS 25.876 Version 7.0.0 'Multiple-input multiple-output UTRA,' www.3gpp.org.

[4] 3GPP (2007) Technical Specification TS 25.214 Version 7.4.0 'Physical layer procedures (FDD),' www.3gpp.org.

[5] 3GPP (2009) Technical Specification TS 25.212 Version 8.5.0 'Multplexing and channel coding (FDD),' www.3gpp.org.

[6] 3GPP (2009) Technical Specification TS 25.211 Version 8.4.0 'Physical channels and mapping of transport channels onto physical channels (FDD),' www.3gpp.org.

[7] 3GPP (2009) Technical Specification TS 25.322 Version 8.4.0 'Radio link control (RLC) protocol specification,' www.3gpp.org.

[8] 3GPP (2009) Technical Specification TS 25.101 Version 8.6.0 'User equipment (UE) radio transmission and reception (FDD),' www.3gpp.org.

[9] 3GPP (2008) Technical Specification TS 23.107 Version 8.0.0 'Quality of service (QoS) concept and architecture,' www.3gpp.org.

[10] Ahmed, M. H., Yanikomeroglu, H., and Mahmoud, S. (2003) 'Fairness enhancement of link adaptation techniques in wireless access networks,' in *Proceedings of IEEE Vehicular Technology Conference Fall (VTC)*, volume 3, pp. 1554–1557.

[11] Alamouti, S. (1998) 'A simple transmit diversity technique for wireless communications,' *IEEE Journal on Selected Areas in Communications*, **16** (8), 1451–1458.

[12] Belfiore, J.-C., Rekaya, G., and Viterbo, E. (2005) 'The Golden code: a 2 × 2 full-rate space–time code with nonvanishing determinants,' *IEEE Transactions on Information Theory*, **51** (4), 1432–1426.

[13] Betz, S. M. and Poor, H. V. (2008) 'Energy efficient communications in CDMA networks: a game theoretic analysis considering operating costs,' *IEEE Transactions on Signal Processing*, **56** (10), 5181–5190.

[14] Blomeier, S. (2006) *HSDPA – Design Details & System Engineering*, INACON.

[15] Bölcskei, H., Gesbert, D., Papadias, C. B., and Van der Veen, A.-J. (eds.) (2006) *Space–Time Wireless Systems – From Array Processing to MIMO Communications*, Cambridge University Press.

[16] Caire, G., Ramprashad, S. A., Papadopoulos, H. C. *et al.* (2008) 'Multiuser MIMO downlink with limited inter-cell cooperation: approximate interference alignment in time, freuquency and space,' in *Proceedings*

of 46th Annual Allerton Conference on Communication, Control and Computing, pp. 730–737.

[17] Costa, M. H. M. (1983) 'Writing on dirty paper,' *IEEE Transactions on Information Theory*, **29** (3), 439–441.

[18] Dabbagh, A. D. and Love, D. J. (2007) 'Precoding for multiple antenna Gaussian broadcast channels with successive zero-forcing,' *IEEE Transactions on Signal Processing*, **55** (7), 3837–3850.

[19] De Bruin, I., Heijenk, G., El Zarki, M., and Zan, L. (2003) 'Fair channel-dependent scheduling in CDMA systems,' in *Proceedings of IST Mobile & Wireless Communications Summit*, pp. 737–741.

[20] European Telecommunications Standards Institute (ETSI) 'The 3rd generation partnership project (3GPP),' Available from www.3gpp.org.

[21] Foschini, G. J. (1996) 'Layered space-time architecture for wireless communication in a fading environment when using multiple antennas,' *Bell Labs Technical Journal*, **1** (2), 41–59.

[22] Frenger, P., Parkvall, S., and Dahlman, E. (2001) 'Performance comparison of HARQ with chase combining and incremental redundancy for HSDPA,' in *Proceedings of VTS IEEE 54th Vehicular Technology Conference Fall (VTC)*, volume 3, pp. 1829–1833.

[23] Fuchs, M., Del Galdo, G., and Haardt, M. (2007) 'Low-complexity space–time frequency scheduling for MIMO systems with SDMA,' *IEEE Transactions on Vehicular Technology*, **56** (5), 2775–2784.

[24] Gao, J., Vorobyov, S. A., and Jiang, H. (2008) 'Game theoretic solutions for precoding strategies over the interference channel,' in *Proceedings of IEEE Global Telecommunications Conference (GLOBECOM)*.

[25] García, A.-B., Alvarez-Campana, M., Vázquez, E., and Berrocal, J. (2002) 'Quality of service support in the UMTS terrestrial radio access network,' in *Proceedings of 9th HP Openview University Association Conference*.

[26] Gesbert, D., Shafi, M., Shiu, D. *et al.* (2003) 'From theory to practice: an overview of MIMO space–time coded wireless systems,' *IEEE Journal on Selected Areas in Communications*, **21** (3), 281–302.

[27] GSM Association 'HSPA – high speed packet access – mobile broadband today,' Available from hspa.gsmworld.com.

[28] Guo, Y. and Levy, B. C. (2005) 'Worst-case MSE precoder design for imperfectly known MIMO communications channels,' *IEEE Transactions on Signal Processing*, **53** (8), 2918–2930.

[29] Holma, H. and Toskala, A. (2005) *WCDMA for UMTS – Radio Access For Third Generation Mobile Communications*, third edition, John Wiley & Sons, Ltd.

[30] Holma, H. and Toskala, A. (2006) *HSDPA/HSUPA for UMTS: High Speed Radio Access for Mobile Communications*, John Wiley & Sons, Ltd.

[31] Holma, H., Toskala, A., Ranta-Aho, K., and Pirskanen, J. (2007) 'High-speed packet access evolution in 3GPP release 7,' *IEEE Communications Magazine*, **45** (12), 29–35.

[32] Jafar, S. A. and Fakhereddin, M. J. (2007) 'Degrees of freedom for the MIMO interference channel,' *IEEE Transactions on Information Theory*, **53** (7), 2637–2642.

[33] Lozano, A. and Jindal, N. (2010) 'Transmit diversity vs. spatial multiplexing in modern MIMO systems,' *IEEE Transactions on Wireless Communications*, **9** (1), 186–197.

[34] Malkamaki, E., Mathew, D., and Hamalainen, S. (2001) 'Performance of hybrid ARQ techniques for WCDMA high data rates,' in *Proceedings of IEEE 53rd Vehicular Technology Conference Spring (VTC)*, volume 4, pp. 2720–2724.

[35] Mehlführer, C., Caban, S., Wrulich, M., and Rupp, M. (2008) 'Joint throughput optimized CQI and precoding weight calculation for MIMO HSDPA,' in *Proceedings of 42nd Asilomar Conference on Signals, Systems and Computers*, Pacific Grove, USA. Available from http://publik.tuwien.ac.at/files/PubDat_167015.pdf.

[36] Members of TSG-RAN Working Group 4 (2002) 'Revised HSDPA CQI proposal,' Technical Report R4-020612, 3GPP.

[37] Motorola (2006) 'MIMO evaluation proposal,' Technical Report TSG R1 #44 060615, 3rd Generation Partnership Project (3GPP).

[38] Paulraj, A., Nabar, R., and Gore, D. (2003) *Introduction to Space–Time Wireless Communications*, Cambridge University Press.

[39] Pedersen, K. and Michaelsen, P. (2006) 'Algorithms and performance results for dynamic HSDPA resource allocation,' in *Proceedings of IEEE 64th Vehicular Technology Conf. (VTC)*, pp. 1–5.

[40] Rupp, M. (2009) *Video and Multimedia Transmissions over Cellular Networks: Analysis, Modeling and Optimization in Live 3G Mobile Networks*, John Wiley & Sons, Ltd.

[41] Sharma, V., Wajid, I., Gershman, A. B. *et al.* (2008) 'Robust downlink beamforming using positive semi-definite covariance constraints,' in *Proceedings of ITG International Workshop on Smart Antennas (WSA)*.

[42] Tarokh, V., Seshadri, N., and Calderbank, A. R. (1998) 'Space–time codes for high data rate wireless communication: performance criterion and code construction,' *IEEE Transactions on Information Theory*, **44** (2), 744–765.

[43] Telatar, I. E. (1999) 'Capacity of multi-antenna Gaussian channels,' *European Transactions on Telecommunications*, **10**, 585–595.

[44] Wengerter, C., Ohlhorst, J., and Golitschek Edler von Elbwart, A. (2005) 'Fairness and throughput analysis for generalized proportional fair frequency scheduling in OFDMA,' in *Proceedings of IEEE 61st Vehicular Technology Conference (VTC) Spring*, volume 3, pp. 1903–1907.

[45] Wrulich, M. (2006) 'Capacity analysis of MIMO systems,' Master's thesis, Vienna University of Technology. Available from http://publik.tuwien.ac.at/files/pub-et_11276.pdf.

[46] Wu, P. and Jindal, N. (2010) 'Performance of hybrid-ARQ in block-fading channels: a fixed outage probability analysis,' *IEEE Transactions on Communications*, **58** (4), 1129–1141.

[47] Zemen, T. and Mecklenbräuker, C. (2005) 'Time-variant channel estimation using discrete prolate spheroidal sequences,' *IEEE Transactions on Signal Processing*, **53** (9), 3597–3607.

[48] Zheng, L. and Tse, D. N. C. (2003) 'Diversity and multiplexing: a fundamental tradeoff in multiple-antenna channels,' *IEEE Transactions on Information Theory*, **49** (5), 1073–1096.

[49] Zheng, Y. and Xiao, C. (2003) 'Simulation models with correct statistical properties for Rayleigh fading channels,' *IEEE Transactions on Communications*, **51** (6), 920–928.

2

UMTS Long-Term Evolution

Contributed by Josep Colom Ikuno
Vienna University of Technology (TU Wien), Austria

This chapter provides an overall picture of the Long-Term Evolution (LTE) architecture, as well as a more detailed description of the Physical (PHY) and Medium Access Control (MAC) layers, of which a good knowledge is needed in order to understand the results presented in the following chapters of this book.

2.1 LTE Overview

2.1.1 Requirements

In its Rel'8, LTE was standardized by the 3rd Generation Partnership Project (3GPP) as the successor of the Universal Mobile Telecommunications System (UMTS). LTE was designed from the start with the assumption that all of the services would be packet switched rather than circuit switched, thus continuing the trend set from the evolution of Global System for Mobile communications (GSM), to General Packet Radio Service (GPRS), Enhanced Data Rates for Global system for mobile communications Evolution (EDGE), UMTS, and High-Speed Packet Access (HSPA). During this evolution, the focus has been seen to be moving towards providing ubiquitous availability of broadband communications, as well as the classical voice/text communication capabilities. From the early mobile packet services, not only has throughput been dramatically increased, but also latency has been greatly decreased. Early Second-Generation (2G)-based systems, such as GPRS, were able to offer data transfer rates in the order of 10 kbit/s. In its latest current iteration, HSPA could theoretically reach 80 Mbit/s, combining multiple carriers and Multiple-Input Multiple-Output (MIMO) techniques [27, 40, 3]. The combination of higher throughput requirements, lower latency, and affordability, as data traffic does not

Evaluation of HSDPA and LTE: From Testbed Measurements to System Level Performance, First Edition.
Sebastian Caban, Christian Mehlführer, Markus Rupp and Martin Wrulich.

Table 2.1 3GPP requirements for E-UTRAN [8]

			Requirements	Comments
DL	UE throughput	Peak data rate	100 Mbit/s	2 TX antennas
		5 % point of cdf	2–3 × Rel'6 HSDPA	(eNodeB),
		Avg. throughput	3–4 × Rel'6 HSDPA	2 RX antennas (UE),
	Spectral efficiency		3–4 × Rel'6 HSDPA	20 MHz DL
UL	UE throughput	Peak data rate	50 Mbit/s	1 TX antenna (UE),
		5 % point of cdf	2–3 × Rel'6 HSDPA	2 RX antennas
		Avg. throughput	2–3 × Rel'6 HSDPA	(eNodeB),
	Spectral efficiency		2–3 × Rel'6 HSDPA	20 MHz UL
Spectrum allocation			1.4, 3, 5, 10, 15, 20 MHz possible	

scale linearly with revenue, contributed to the requirements specified for LTE by 3GPP, which can be summarized as follows [8]:

- increased peak data rate (100 Mbit/s in the Downlink (DL), 50 Mbit/s in the Uplink (UL)), cell edge bit rate, and spectrum efficiency;
- scalable bandwidth;
- interworking and cost-effective migration with/from existing 3GPP systems;
- reduced Capital Expenditure (CAPEX) and Operational Expenditure (OPEX);
- simplified network architecture; and
- support for high mobile speed.

Table 2.1 shows the 3GPP requirements for the Evolved Universal Terrestrial Radio Access Network (E-UTRAN) of LTE. The final capabilities of LTE, however, go beyond those of the defined target requirements. For instance, although, the targets for DL and UL peak data rate were set to 100 Mbit/s and 50 Mbit/s [18], LTE User Equipment (UE) supports up to 300 Mbit/s and 75 Mbit/s DL/UL peak data rates.

While the first performance evaluations show that the spectral efficiencies of the LTE physical layer and MIMO-enhanced Wideband Code-Division Multiple Access (WCDMA) [28] are approximately the same [17, 21, 37, 43], LTE incorporates several other benefits. The LTE DL transmission scheme is based on Orthogonal Frequency-Division Multiple Access (OFDMA), which converts the wide-band frequency-selective channel into a set of many flat-fading subchannels. The flat-fading subchannels have the advantage that, even in the case of MIMO transmission, optimum receivers can be implemented with reasonable complexity, in contrast to WCDMA systems. OFDMA additionally allows for frequency-domain scheduling, typically trying to assign physical resources to users with optimum channel conditions. This offers large throughput gains in the DL due to multi-user diversity [25, 42]. LTE also includes an inter-eNodeB interface, termed an X2-interface, which can be used for interference management aiming at decreasing inter-cell interference.

Regardless of the network capabilities, the system is nevertheless constrained by the actual capabilities of the receiver mobile equipment; that is, the UE capabilities. LTE defines five UE radio capability categories, to which a given UE has to conform to

Table 2.2 LTE UE categories [7]. Each UE category constrains the maximum throughput and SM capabilities supported in DL and UL

UE Capabilities		UE Category				
		1	2	3	4	5
DL	Peak throughput [Mbit/s]	10.3	51	102	150.8	302.8
	Max. number of supported layers for SM in DL	1	2	2	2	4
	Max. number of supported streams for SM in DL	1	2	2	2	2
UL	Peak throughput [Mbit/s]	5.2	25.5	51	51	75.4
	Support for 64-QAM in UL	No	No	No	No	Yes

[7]. These range from a non-Spatial Multiplexing (SM)-capable UE with a maximum upload/download of 10/5 Mbit/s to a 4×4 MIMO-capable terminal capable of 300/70 Mbit/s. Table 2.2 depicts the maximum supported throughput for both UL and DL, as well as the SM capabilities.

2.2 Network Architecture

Although a detailed view of the LTE network architecture is not in the scope of this book, this section will provide readers with some basic know-how on the architecture of the LTE core network, termed the System Architecture Evolution (SAE). SAE promises a flat architecture with low delay and good scalability, as well as being simpler than in previous 3GPP technologies.

SAE can be divided into the Evolved Packet Core (EPC), representing the core network, and E-UTRAN. The core network provides access to external packet networks based on Internet Protocol (IP) and performs a number of functions for idle and active terminals. In contrast, the Radio Access Network (RAN) performs all radio interface-related functions for terminals in active mode [24].

The overall architecture and a brief description of the functionalities of each element are shown in Figure 2.1. E-UTRAN comprises the Evolved base stations (eNodeBs), which are connected to the EPC via S1 interfaces. The X2 interface provides inter-eNodeB communication and enables a meshed RAN architecture. In LTE, 3G-RNC-functionality has been integrated into the eNodeB, hence the elimination of the Radio Network Controller (RNC) in the SAE architecture. The eNodeB hosts the following functions:

- all PHY and MAC layer procedures, including link adaptation, Hybrid Automatic Repeat reQuest (HARQ), and cell search;
- Radio Resource Management (RRM) – radio bearer control, radio admission control, connection mobility control, and UL/DL scheduling;
- Packet Data Convergence Protocol (PDCP) – IP header compression and encryption of the user data stream;
- selection of a Mobility Management Entity (MME) at UE attachment;
- routing of the User Plane (U-Plane) data towards the Serving Gateway (S-GW).

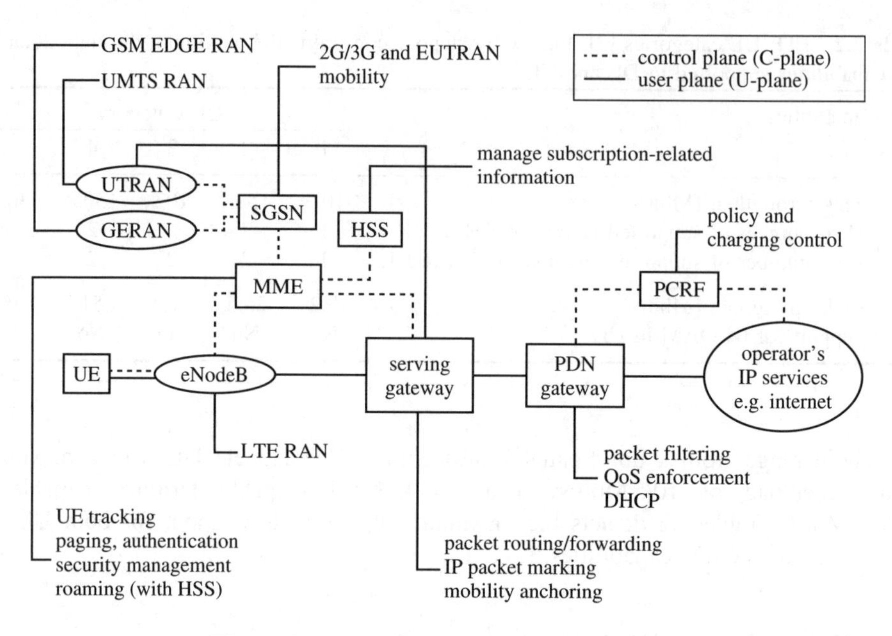

Figure 2.1 Overall LTE architecture [22, 13].

The MME, S-GW, and PDN Gateway (P-GW) elements of the EPC host the following functionalities [33, 12].

- Mobility Management Entity (MME) – Control Plane (C-Plane) functions related to subscriber and session management:
  - ciphering and integrity protection of Non-Access Stratum (NAS) signaling;
  - distribution of paging messages to the eNodeBs;
  - S-GW and P-GW selection;
  - inter-core network node signaling for mobility between 3GPP access networks, including Serving-General packet radio service Support Node (SGSN) selection for handovers to 2G or 3G 3GPP networks;
  - security control linked to the Home Subscriber Server (HSS), which supports the database containing all the user subscription information;
  - idle-state mobility control; and
  - roaming.
- Serving Gateway (S-GW) – termination point towards Universal mobile telecommunications system Terrestrial Radio Access Network (UTRAN):
  - termination of U-Plane packets and switching of U-Plane for support of UE mobility; and
  - packet routing and forwarding.
- PDN Gateway (P-GW) – anchor point for sessions towards external Packet Data Networks (PDNs):
  - packet filtering and/or marking;

– service level charging and rate enforcing, together with the Policy and Charging Rules Function (PCRF); and
– Dynamic Host Configuration Protocol (DHCP); that is, UE IP address allocation.

This functional split of the SAE elements allows for a more specialized implementation of the MME, S-GW, and P-GW. Thus, the MME is optimized for C-Plane processing, while the S-GW is optimized to process high-throughput U-Plane data, and the P-GW works as an exit edge router from the LTE core network, analogous to the DiffServ case [15]. A layer description of the functionalities described above is shown in Figure 2.2, comprising the PHY, MAC, Radio Link Control (RLC), PDCP, and core network. The rest of the chapter offers a more detailed bottom-up description of the PHY and MAC layers of the LTE DL, as well as the mapping between the physical, transport, and logical channels defined in each layer.

2.3 LTE Physical Layer

The LTE DL transmission scheme is based on OFDMA, which converts the wide-band frequency-selective channel into a set of many flat-fading subchannels, while Single-carrier FDMA (SC-FDMA) is applied on the UL due to its lower Peak-to-Average Power Ratio (PAPR). The flat-fading subchannels have the advantage that (even in the case of MIMO transmission) optimum receivers can be implemented with reasonable complexity, in contrast to WCDMA systems. OFDMA additionally allows for easy frequency-domain scheduling, typically trying to assign only "good" subchannels to the individual users. This offers large throughput gains in the DL due to multi-user diversity.

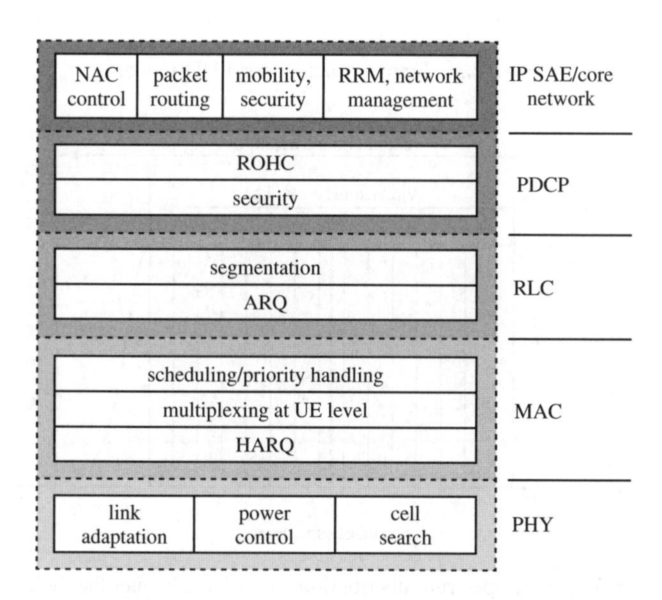

Figure 2.2 Layer structure for LTE [1, 13].

2.3.1 LTE Frame Structure

DL and UL transmissions are organized into radio frames with a duration of 10 ms, both for the Time Division Duplex (TDD) and Frequency Division Duplex (FDD) modes. Focusing on the FDD DL, and as shown in Figure 2.3, the LTE radio frame is subdivided into 10 subframes of 1 ms duration each, subsequently divided into two slots [5]. In the frequency domain, the available bandwidth is divided into equally spaced orthogonal subcarriers. A typical subcarrier spacing is 15 kHz, although a smaller 7.5 kHz spacing is also possible. Subcarriers are organized in groups of N_{sc}^{RB} consecutive subcarriers, which is 12 for the normal-length Cyclic Prefix (CP) and 24 when employing 7.5 kHz subcarrier spacing. Each subcarrier grouping is termed a Resource Block (RB).

In the center of the allocated bandwidth, a zero-DC subcarrier is allocated, which is shown in Figure 2.4. The figure shows the spectrum and RB allocation for a 3 MHz bandwidth. Table 2.3 lists the available bandwidth allocations for LTE when using the normal-length CP, as well as the number of available RBs on each, referred to as N_{RB}^{DL}

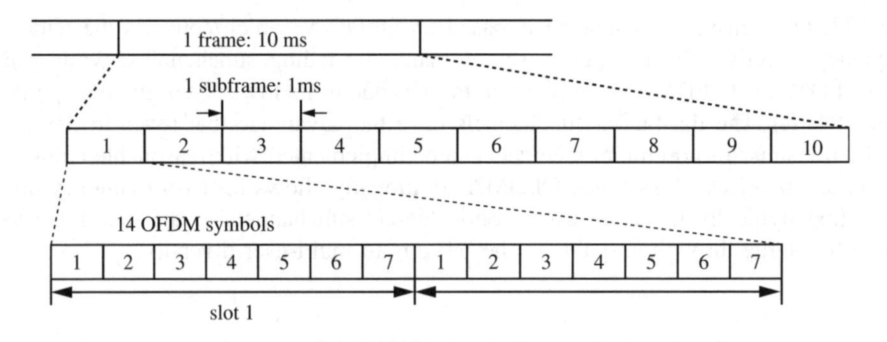

Figure 2.3 Frame structure for FDD mode.

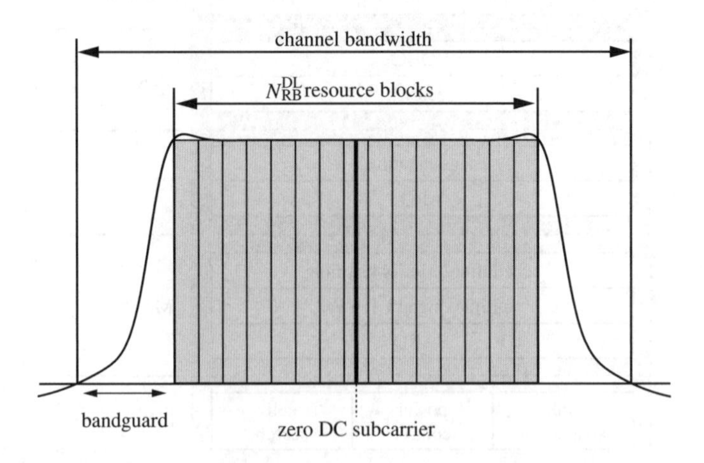

Figure 2.4 OFDM frequency spectrum distribution. A 3 MHz channel bandwidth is shown as an example, encompassing 15 RBs ($N_{RB}^{DL} = 15$).

Table 2.3 Available LTE system bandwidths and available RBs [2]

Channel bandwidth B_{channel} [MHz]	1.4	3	5	10	15	20
Transmission bandwidth configuration $N_{\text{RB}}^{\text{DL}}$	6	15	25	50	75	100
Number of data subcarriers	72	180	300	600	900	1200
Bandguard size [% of B_{channel}]	23	10	10	10	10	10

Table 2.4 OFDM symbols in the time domain for each subcarrier for the several available cyclic prefix lengths and subcarrier spacing configurations [5]. The number of available RBs for the given CP and subcarrier spacing allocation is denoted by $N_{\text{sc}}^{\text{RB}}$

Configuration		$N_{\text{sc}}^{\text{RB}}$	$N_{\text{symb}}^{\text{DL}}$	CP length (μs)
Normal cyclic prefix	$\Delta f = 15\,\text{kHz}$	12	7	4.69
Extended cyclic prefix	$\Delta f = 15\,\text{kHz}$	12	6	16.67
	$\Delta f = 7.5\,\text{kHz}$	24	3	33.33

[2], and the bandguard size relative to the system bandwidth. Thus, depending on the chosen CP length and subcarrier spacing, the number of Orthogonal Frequency-Division Multiplexing (OFDM) symbols in the time domain in one Transmission Time Interval (TTI) may vary [5]. The available configurations are shown in Table 2.4. Not noted in the table is the fact that when a normal-length CP is employed, the first OFDM symbol in a slot has a slightly longer length (5.21 μs instead of 4.69 μs) than the remaining six symbols. This has been implemented in order to preserve the 0.5 ms slot timing.

The whole time–frequency grid structure is depicted in Figure 2.5. It is shown for the 15 kHz subcarrier spacing and normal CP length; thus, it comprises 12 subcarriers per RBs and 7 OFDM symbols per slot (0.5 ms). It depicts the frequency division in RBs, as well as the time division in subframe, slot, and OFDM symbols. Not shown is the CP protecting each OFDM symbol (5.21 μs for the first OFDM symbol in each slot and

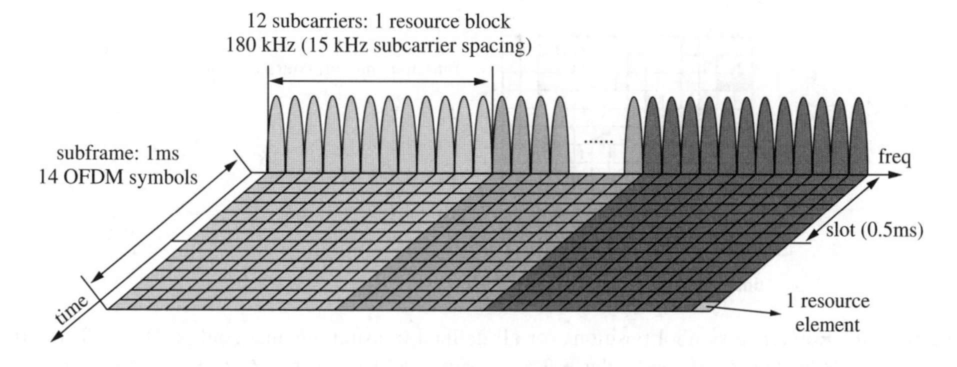

Figure 2.5 LTE time–frequency grid.

4.69 μs for the rest). Each element resulting from this time–frequency separation is a Resource Element (RE), and defines the unit to position the transmitted symbols, be it reference symbols or data/control channels in the DL frame.

2.3.2 Reference and Synchronization Symbols

Two kinds of training symbols are specified (mentioned as physical signals) in the DL: the synchronization signal and the reference signal. These signals do not carry information originating from higher layers and can be utilized for channel estimation or frequency synchronization at the receiver (see Chapter 9 for details):

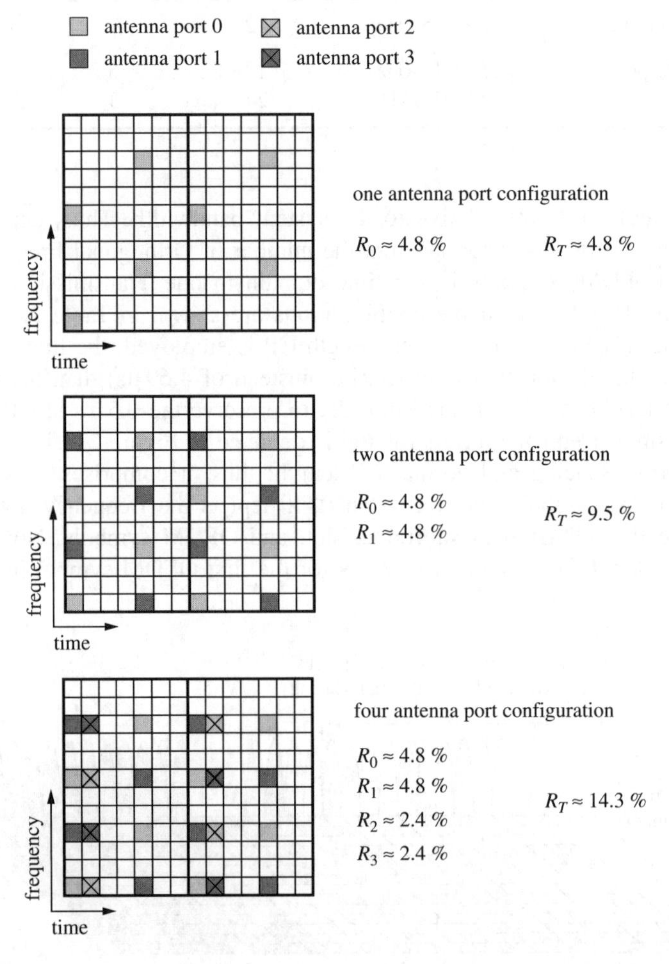

Figure 2.6 Reference symbol positions for all defined transmit antenna configurations. The ratio of symbols dedicated to one particular reference signal is denoted by $R_i; i = 1, 2, 3, 4$, while R_T refers to the ratio over the sum of all the reference symbols.

Synchronization signals: Two synchronization signals are defined, namely the primary synchronization signal and the secondary synchronization signal. In FDD mode, these are located on the 62 subcarriers symmetrically arranged around the DC-carrier in the first slot in the sixth and seventh OFDM symbols of the first and the sixth subframes.

Reference signals: The reference signals are cell-specific and are modulated with 4-QAM. For all the possible transmit antenna configurations defined in LTE (one, two, and four antenna ports) Figure 2.6 shows the location of the pilot symbols in the OFDM time–frequency grid, as well as the ratio of the total number of REs they occupy. Whenever one antenna port is transmitting a reference symbol on one RE, all the other antenna ports transmit a 'zero' symbol at this position, avoiding interference of the reference symbols transmitted from different antennas.

2.3.3 *MIMO Transmission*

MIMO techniques are one of the main enablers to improve the spectral efficiency in LTE with respect to Rel'6. The LTE standard defines support for one, two, and four transmit antennas. As explained in the previous section, employing more transmit antennas causes a penalty in the form of an increased pilot overhead. The supported multi-antenna transmit modes employ Transmit Diversity (TxD) or Spatial Multiplexing (SM) transmission in order to increase diversity, data rate, or both. This is described in detail for the two-transmit-antenna case to illustrate the concept, as the four-transmit-antenna case is an extension of this case. SM can be operated in two modes: Open-Loop Spatial Multiplexing (OLSM) and Closed-Loop Spatial Multiplexing (CLSM). In OLSM, no precoding matrix feedback is employed, while in CLSM, the optimum precoding matrix information is fed back by the UE.

2.3.3.1 Transmit Diversity

The TxD mode utilizes an Alamouti Space–Time Block Code (STBC) [14]. For the two-transmit-antenna case, the two transmitted symbols x_0, x_1, are mapped to the output symbols from each antenna y_0^0, y_1^0 (first antenna) and y_0^1, y_1^1 (second antenna) as follows [5]:

$$\begin{bmatrix} y_0^0 \\ y_1^0 \\ y_0^1 \\ y_1^1 \end{bmatrix} = \frac{1}{\sqrt{2}} \begin{bmatrix} 1 & 0 & \mathrm{i} & 0 \\ 0 & -1 & 0 & \mathrm{i} \\ 0 & 1 & 0 & \mathrm{i} \\ 1 & 0 & -\mathrm{i} & 0 \end{bmatrix} \begin{bmatrix} \Re\{x_0\} \\ \Re\{x_1\} \\ \Im\{x_0\} \\ \Im\{x_1\} \end{bmatrix}. \tag{2.1}$$

2.3.3.2 Open-Loop Spatial Multiplexing

In an SM scheme, the process can be described by a transmit vector $\mathbf{x}$, a precoding matrix $\mathbf{W}$, and an output vector $\mathbf{y}$. Thus,

$$\mathbf{y} = \mathbf{W}\mathbf{x} \tag{2.2}$$

In LTE the length of the vector $\mathbf{x}$ is referred to as the number of layers, and expressed by ν. It is the number of symbols simultaneously transmitted over the available two or four transmit antennas. Thus, the precoding matrix $\mathbf{W}$ is an $N_T \times \nu$ matrix that maps the ν transmit symbols to the N_T antennas. Each layer can also be analogously interpreted as each of the beams produced by $\mathbf{W}$.

With appropriate Channel State Information (CSI), the transmitter could choose the optimum precoding matrix so as to maximize data transfer. However, in the OLSM mode, only Rank Indicator (RI) feedback information is available; that is, how many layers should be employed. As a solution, OLSM incorporates Cyclic Delay Diversity (CDD). CDD shifts the transmit signal in the time direction and transmits these modified signal copies over separate transmit antennas. The time shifts are inserted in cyclically (hence the name), so that there is no additional Inter-Symbol Interference (ISI). This results in increasing the number of resolvable channel propagation paths and, hence, increased diversity with no additional receiver complexity [19]. For the two-transmit-antenna case, and at a time instant k, the transmission of a vector $\mathbf{x}_k$ of length ν symbols can be formulated as

$$\mathbf{y}_k = \mathbf{W}\mathbf{D}_k\mathbf{U}\mathbf{x}_k, \tag{2.3}$$

where $\mathbf{D}_k$ cyclically shifts the delay depending of the time index k:

$$\mathbf{U} = \frac{1}{\sqrt{2}}\begin{bmatrix} 1 & 1 \\ 1 & \mathrm{e}^{-\mathrm{i}2\pi/2} \end{bmatrix}, \tag{2.4}$$

$$\mathbf{D}_k = \begin{bmatrix} 1 & 0 \\ 0 & \mathrm{e}^{-\mathrm{i}2\pi k/2} \end{bmatrix} \tag{2.5}$$

In practice, the CDD matrix cycles with a period of two (that is, odd/even time indexes), allowing expression of the CDD matrix as $\mathbf{D}_{k \bmod 2}$ for the two-transmit-antenna case. Depending on whether the chosen number of layers ν is one or two, the precoding matrix $\mathbf{W}$ is fixed to either

$$\mathbf{W} = \frac{1}{\sqrt{2}}\begin{bmatrix} 1 \\ 1 \end{bmatrix}, \quad \nu = 1, \tag{2.6}$$

or

$$\mathbf{W} = \frac{1}{\sqrt{2}}\begin{bmatrix} 1 & 0 \\ 0 & 1 \end{bmatrix}, \quad \nu = 2 \tag{2.7}$$

For the four-transmit-antenna case there is the difference that, instead of a fixed $\mathbf{W}$ matrix, a different precoder is applied after ν vectors, as well as a CDD matrix $\mathbf{D}_k$ with a period of four; the rest of the process remains the same (appropriate $\mathbf{U}$, $\mathbf{D}_k$, and $\mathbf{W}$ matrices are defined for $\nu = 3, 4$ in [5]).

2.3.3.3 Closed-Loop Spatial Multiplexing

Unless the feedback is invalidated by a rapidly changing channel, gains can be obtained compared with an open-loop scheme by signaling the eNodeB an optimum precoding

Table 2.5 LTE codebook for CLSM mode and two transmit antennas [5]

Codebook index	Number of layers ν	
	1	2
0	$\frac{1}{\sqrt{2}}\begin{bmatrix}1\\1\end{bmatrix}$	–
1	$\frac{1}{\sqrt{2}}\begin{bmatrix}1\\-1\end{bmatrix}$	$\frac{1}{2}\begin{bmatrix}1 & 1\\1 & -1\end{bmatrix}$
2	$\frac{1}{\sqrt{2}}\begin{bmatrix}1\\\mathrm{i}\end{bmatrix}$	$\frac{1}{2}\begin{bmatrix}1 & 1\\\mathrm{i} & -\mathrm{i}\end{bmatrix}$
3	$\frac{1}{\sqrt{2}}\begin{bmatrix}1\\-\mathrm{i}\end{bmatrix}$	–

matrix via Precoding Matrix Indicator (PMI) feedback in addition to RI feedback. In order to simplify signaling, the precoding matrix is chosen from a codebook, which for the two-transmit-antenna is comprised of four ($\nu = 1$) and two ($\nu = 2$) precoder options. For four transmit antennas, the codebook spans 15 precoding matrices. In CLSM, the output vector $\mathbf{y}_k$ can be written as

$$\mathbf{y}_k = \mathbf{W}\mathbf{x}_k, \tag{2.8}$$

where $\mathbf{W}$ is chosen from the precoding codebook, which for the two-transmit-antenna case is shown in Table 2.5.

2.3.4 Modulation and Layer Mapping

LTE allows for up to two parallel data streams, or Codewords (CWs), to be sent on each TTI. As discussed in Section 2.3.3, by means of SM, up to four transmit symbols can be sent per TTI; hence the need for a mapping between the bits from each CW and the spatial layers which transport the modulated symbols. The modulation order is signaled from upper layers, the decision being taken by the scheduler in the MAC layer and signaled downwards to the PHY layer. As shown in Figure 2.7, the modulation and mapping procedures comprise a scrambling, modulation mapper, and layer mapping, which maps the one or two CWs to the ν available spatial layers.

A bit-level scrambling is applied based on a length-31 Gold sequence which is initialized with a UE- and cell-ID-dependent value. It is applied on a CW-basis and it ensures interference randomization between the cells. After scrambling, data modulation is applied to convert the CWs to blocks of complex-modulated symbols. Possible modulation alphabets are 4-, 16-, and 64-QAM, which allow for 2, 4, or 6 bits to be transmitted per symbol. While the precoding maps the symbols of the ν layers to N_{T} antennas, the layer mapping maps the modulated CWs' symbols to ν layers. Depending on the number of CWs and layers, the mapping can be as simple as a one-to-one mapping between CW and layer for the two layer-two CW case or a CW taking twice as many symbols as the

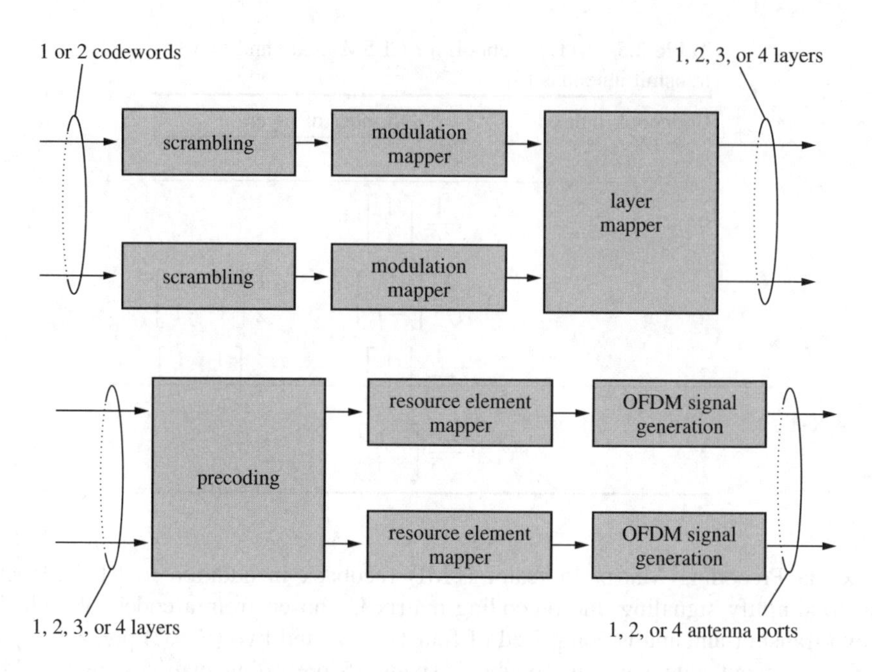

Figure 2.7 Modulation and layer mapping procedures [5].

other when having three layers. The ν spatial layers carry the same number of symbols, and the symbols from each layer come from a single CW. Table 2.6 shows the layer mapping applied for the SM modes. For single-antenna transmission, this is a one-to-one layer mapping, as only one CW is allowed. For TxD, only one CW is allowed and the number of layers equals the number of transmit antennas, so either 1/2 or 1/4 of the CW symbols are placed on each layer. When one CW is mapped to more than one layer, the symbols are alternatively placed on each layer, as in a round-robin fashion.

Table 2.6 Layer mapping for spatial multiplexing [5]. The layer mapping column indicates the codeword, either the first (1) or the second (2), from which the symbols of the νth layer come

Layer number ν	CW number	Layer mapping			
		1	2	3	4
1	1	1	–	–	–
2	2	1	2	–	–
2	1	1	1	–	–
3	2	1	2	2	–
4	2	1	1	2	2

2.3.5 *Channel Coding*

LTE, as well as HSPA, relies on Adaptive Modulation and Coding (AMC) in order to provide adaptability to the channel conditions. In order to match the radio channel capacity and Block Error Ratio (BLER) requirements for each UE, the eNodeB dynamically adjusts both the applied code rate and modulation. The decision about which combination of Effective Code Rate (ECR) and modulation order to use is taken at the MAC layer by the scheduler and signaled down to the PHY layer, where the data is coded at the required code rate and the appropriate modulation scheme is applied. The scheduler decision is aided by a closed feedback loop, which provides the scheduler information on the instantaneous channel conditions, which is explained further in Section 2.3.6.

The channel coding works on a per-CW basis, such that the whole coding procedures are performed either once or twice per TTI and UE, for the one and two Codeword (CW) transmission cases, respectively. Figure 2.8 depicts the channel coding procedures, from the data bits to the coded bits, as in [4]. It provides error detection and correction capabilities to the transmitted bits, as well as generating the different redundancy versions of the Transport Block (TB) that are needed for HARQ operation. Although the channel coding procedures detailed here correspond to the Downlink Shared Channel (DL-SCH), the error detection, error correcting, rate matching, and interleaving procedures applied to the DL-SCH channel are in concept identical to those applied to the other channels. As depicted in Figure 2.8, the channel coding procedures comprise the following functional blocks: transport block CRC attachment, code block segmentation and CRC attachment, channel coding, and rate matching and code block concatenation, which are detailed in the following sections.

2.3.5.1 Transport Block CRC Attachment

Error detection for the whole TB is provided by means of a 24-bit Cyclic Redundancy Check (CRC). The 24 CRC bits are computed and then attached to the end of the TB. If at receive time the CRC checksum does not match that of the TB, then the TB is deemed incorrect and, as such, is signaled to upper layers.

2.3.5.2 Code Block Segmentation and CRC Attachment

While a TB may occupy the PHY resources of the whole 20 MHz maximum LTE bandwidth allocation, implementation issues mean that the maximum block length the channel coder will process is limited to 6144 bits. The channel coding in the DL-SCH is a rate-1/3 Parallel Concatenated Convolutional Code (PCCC) with two 8-state constituent encoders and one turbo code internal interleaver. Before the channel coding, the TB is segmented into C Code Blocks (CBs), so that each CB is smaller than or equal to the maximum turbo interleaver size of 6144 bits. In order to ensure that the CBs fit any of the 188 possible interleaver sizes, filler bits are added. The possible interleaver sizes are shown in Table 2.7. If segmentation is needed, an additional 24-bit CRC with a different cyclic generator polynomial than the TB CRC is attached to ensure CB integrity. The C CBs output after

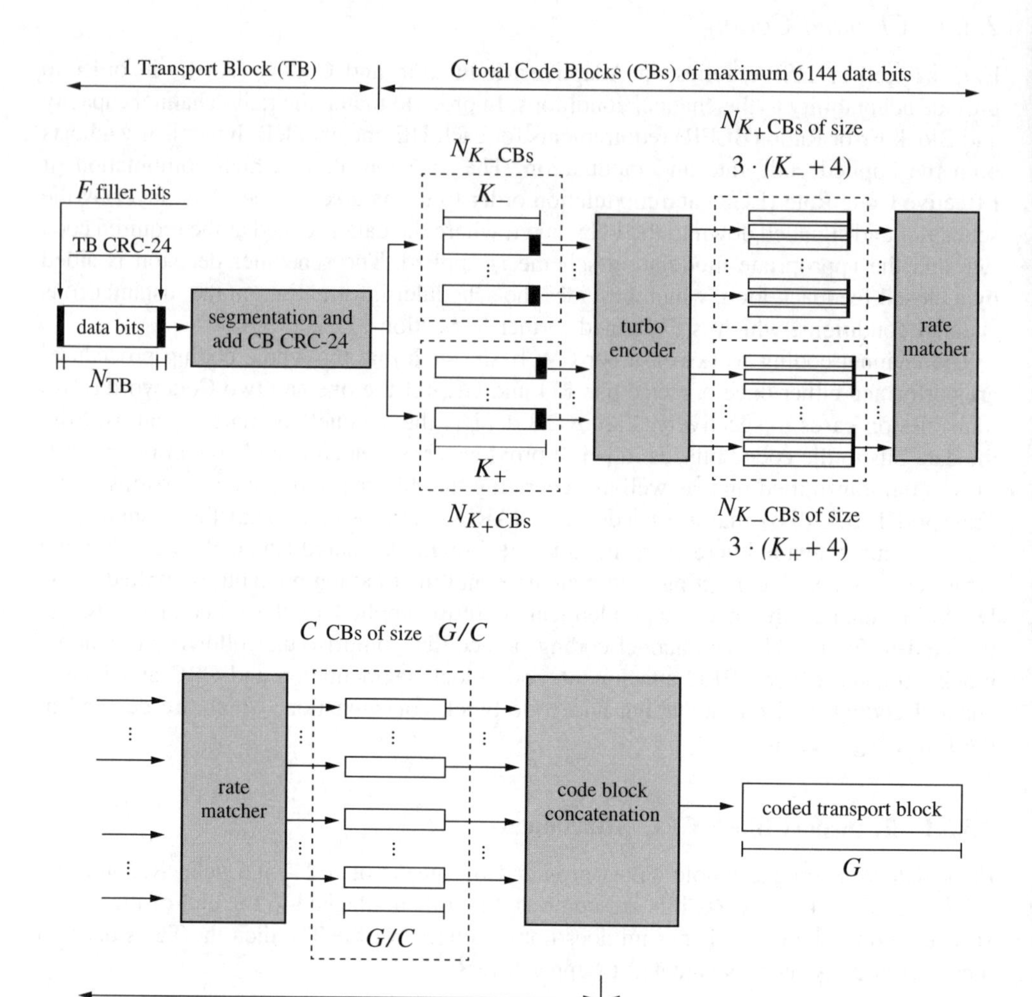

Figure 2.8 LTE channel coding procedures for the DL-SCH [4]. From [30]. Reproduced by permission of © IEEE 2011.

Table 2.7 Possible turbo interleaver sizes. There are 188 sizes defined, ranging from 40 to 6144 bits

$K_i = 40 + 8(i - 1)$	$i = 1, \dots, 60$
$K_i = 512 + 16(i - 60)$	$i = 61, \dots, 92$
$K_i = 1024 + 32(i - 92)$	$i = 93, \dots, 124$
$K_i = 2024 + 64(i - 124)$	$i = 125, \dots, 188$

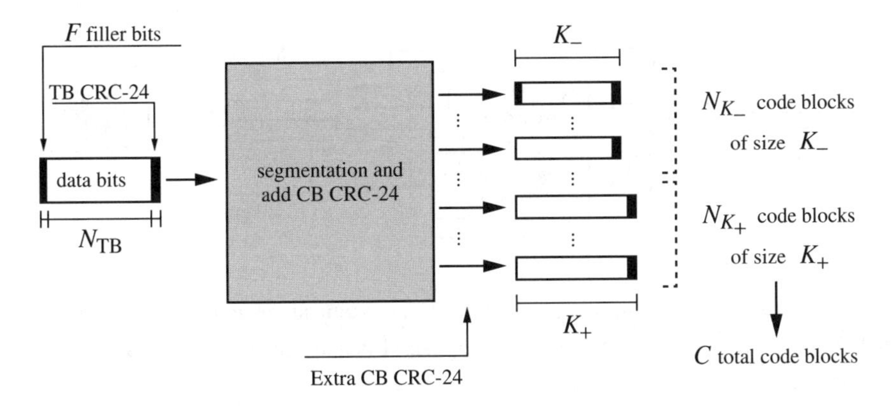

Figure 2.9 Code block segmentation and CRC attachment [4].

the segmentation and CRC attachment are comprised of N_{K_+} CBs of size K_+ and N_{K_-} CBs of size K_-. K_+ and K_- are chosen to be the two consecutive interleaver sizes closest to K/C, where K is the total size of the TB and satisfies $K = N_{K_+}K_+ + N_{K_-}K_- + F$. As each CB is to be independently coded, but all need to be correctly received for the TB to be correct, choosing them to be as close as possible improves the overall performance, as a single excessively short CB would hinder the overall BLER performance. If no segmentation is needed (that is, $K \leq 6144$), no segmentation or extra CRC attachment is performed. Figure 2.9 depicts the segmentation process, from the TB of size N_{TB} to the C CBs of size K_+ and K_-.

2.3.5.3 Channel Coding

With some minor differences, the turbo code in LTE is the same as that in WCDMA [11]. Turbo coding is applied on the Uplink Shared Channel (UL-SCH), DL-SCH, Paging Channel (PCH), and Multicast Channel (MCH). Of the various transport channels, only the Broadcast Channel (BCH) uses a tail biting convolutional code. Control channels use only convolutional, repetition, or block coding. As mentioned previously, the channel coding for the DL-SCH consists of a rate-1/3 PCCC with two 8-state constituent encoders and one turbo code internal interleaver. Rate adaption is obtained by rate-matching the rate 1/3 coded CBs in the rate-matching block. For the turbo interleaver, owing to the performance limitations of those used in WCDMA, a new interleaver based on a Quadratic Permutation Polynomial (QPP) has been employed [36]. As in the WCDMA standard, the generator polynomial for the encoder is

$$G(D) = \left[1, \ \frac{1 + D + D^3}{1 + D^2 + D^3}\right] \tag{2.9}$$

Figure 2.10 depicts the turbo coding procedure for the DL-SCH, which outputs $3K + 12$ bits and includes the 12 tail bits, which are distributed across the three outputs.

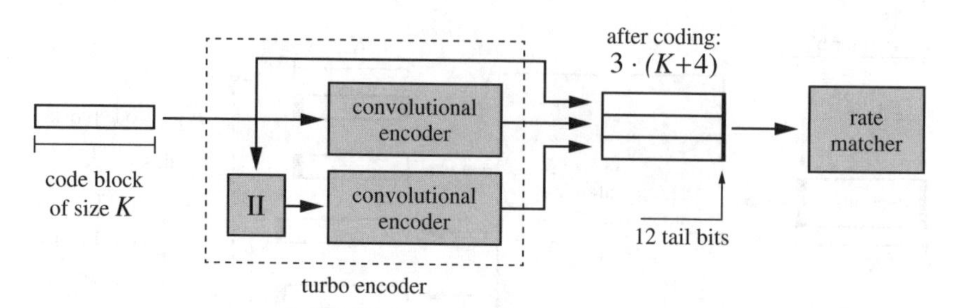

Figure 2.10 DL-SCH turbo encoding procedure. The constituent convolutional encoders are constructed with the $\left[1, \frac{1+D+D^3}{1+D^2+D^3}\right]$ generator polynomial; Π denotes the turbo interleaver [4].

2.3.5.4 Rate Matching and Code Block Concatenation

The rate-matching process allows any arbitrary code rate to be obtained from the mother-rate-1/3 code by a process of puncturing and/or repetition of the coded bits. In addition, as all coded bits are obtained from the original 1/3-encoded codeword, rate matching also allows for HARQ combining to be performed [26, 41]. The rate matching, as depicted in Figure 2.11, comprises an independent interleaving of each of the bit outputs of the channel coder (sub-block interleaving) and a rearranging of the bits of the three branches, which comprise the systematic bits $v^{(0)}$ and the parity bits $v^{(1)}$ and $v^{(2)}$ from each of the convolutional encoders. Before the bit collection (the actual puncturing/repetition), the bits are positioned in a virtual circular buffer [16], following the pattern shown in Figure 2.12. The target bit size G of the rate-matched TB is signaled from upper layers, and applied CB-wise, meaning each CB is rate matched to G/C bits, rounded so as the sum of all CBs is G bits. Depending on the redundancy version index (that is, the HARQ retransmission index of the specific TB), an initial starting value k_0 is chosen. Starting from this position, a contiguous set of bits is taken from the circular buffer until the target number of bits for the given CB has been reached, as seen also in Figure 2.12. For low

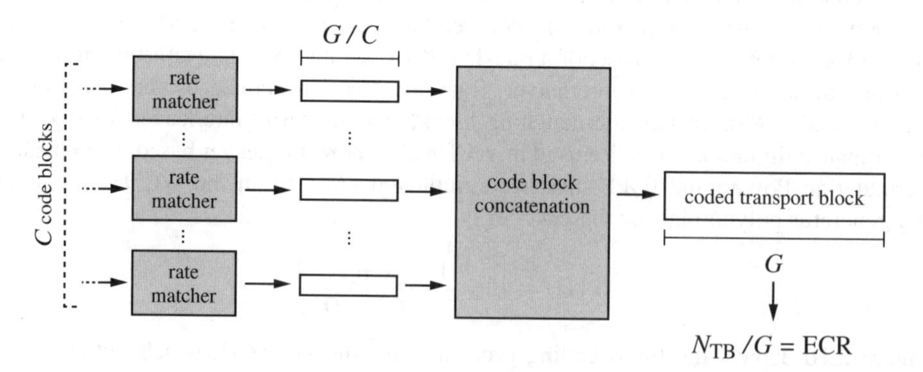

Figure 2.11 DL-SCH rate-matching procedure. Rate matching is performed on a per-CB basis and adjusts each coded CB to G/C bits, rounded so as to ensure that G total bits are output [4].

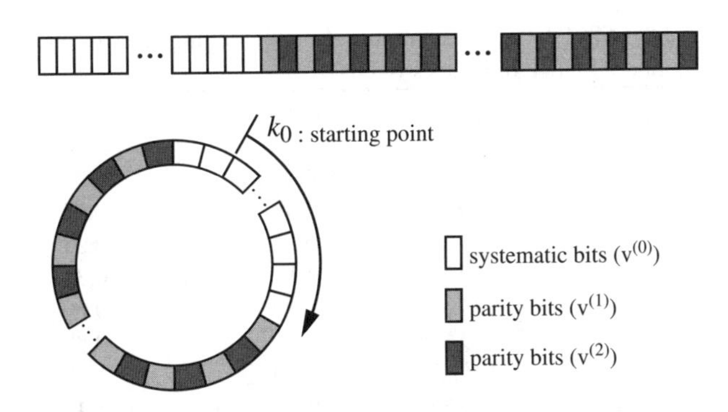

Figure 2.12 *Top*: positioning of the systematic and parity bits in the circular buffer after sub-block interleaving (referred to as $v^{(0)}$, $v^{(1)}$, and $v^{(2)}$ in [5]). *Bottom*: bit collection of the same bits when positioned in the circular buffer [4].

ECRs, where the rate matcher acts as an outer repetition code, no practical difference exists between each retransmission [29]. For higher ECRs, the redundancy version index dependence of k_0 will originate retransmissions, containing different bits from the original subset for each retransmission.

After rate matching is performed on each individual CB, the whole set of CBs are then sequentially concatenated, the output of which is the coded TB.

2.3.6 Channel Adaptive Feedback

LTE implements AMC, as well as closed-loop MIMO in order to adapt the transmission rate based on instantaneous channel knowledge. LTE requires the calculation of up to three different feedback values at the receiver: Channel Quality Indicator (CQI), Rank Indicator (RI), and Precoding Matrix Indicator (PMI) [6], which aim at maximizing the obtainable throughput while maintaining the BLER below a threshold, set to 10 % [34, 39]. By means of the 4-bit CQI, the UE signals on a per-codeword basis which one of the 15 Modulation and Coding Schemes (MCSs) specified in Table 2.8 ensures a BLER lower or equal to 10 % (a zero-CQI represents out of range). The CQIs specify code rates between 0.08 and 0.92, and 4-, 16-, and 64-QAM modulations. As such, it can be interpreted as a mapping from the receiver SINR 10 % BLER points from the BLER curve from each of the 15 MCSs. It should be noted, however, that such a Signal to Interference and Noise Ratio (SINR)-to-CQI mapping depends on the type of receiver. In the same channel conditions, a better receiver (for example, one implementing interference cancelation) would be able to report higher CQIs than another simpler or poorly implemented one.

For the SM modes, the RI informs the transmitter about the number of useful transmission layers for the current MIMO channel. In the current LTE standard it will range between one and four (for the four-transmit-antenna case), while LTE-Advanced (LTE-A) defines even up to eight spatial layers with up to eight transmit antennas [9]. The PMI signals in CLSM mode the optimum index of the precoding matrix from the codebook which should be applied by the eNodeB. The standard [5] defines 4 and 16 possible precoding matrices for two and four-transmit-antenna configurations, respectively.

Table 2.8 Modulation scheme and effective coding rate for each of the Channel Quality Indicators (CQIs)

CQI Index	Modulation	ECR	Data [bit/symbol]
0		Out of range	
1	4-QAM	0.08	0.15
2	4-QAM	0.12	0.23
3	4-QAM	0.19	0.38
4	4-QAM	0.30	0.60
5	4-QAM	0.44	0.88
6	4-QAM	0.59	1.18
7	16-QAM	0.37	1.48
8	16-QAM	0.48	1.91
9	16-QAM	0.60	2.41
10	64-QAM	0.46	2.73
11	64-QAM	0.55	3.32
12	64-QAM	0.65	3.90
13	64-QAM	0.75	4.52
14	64-QAM	0.85	5.12
15	64-QAM	0.93	5.55

2.4 MAC Layer

The MAC layer, as its name implies, controls the access to the transmission medium. It provides data transfer and radio resource allocation services to the upper layers, while the physical layer provides them with lower level data transfer services, signaling (HARQ feedback and scheduling requests), and channel measurements such as PMI, RI, or CQI reports [10]. The PHY, MAC, and upper layer are connected via the use of channels, which are detailed in Section 2.5.

2.4.1 Hybrid Automatic Repeat Request

HARQ with soft combining is a fast retransmission scheme utilized in the MAC layer in LTE. A transmission scheme based on HARQ combines detection and Forward Error Correction (FEC) plus a retransmission of the erroneous packet. From the two types of HARQ – Chase Combining (CC) and Incremental Redundancy (IR) – LTE uses IR, in which each retransmission comprises a different subset of the bits output from the channel coder (whenever that is possible, of course). Figure 2.13 illustrates the CC and IR differences. While CC has a lower implementation complexity, IR offers higher performance gains, especially at higher MCSs. In these cases, IR offers high coding gains, as well as repetition gain, while CC only gains by means of repetition [23, 29]. Employing soft combining and IR, in which a given received packet is combined with the previously received packets, results in a more powerful FEC code for the retransmissions [35].

In order for IR to work, the HARQ functionality must be able to create appropriate redundancy versions from a given coded block and prevent terminal buffer overflow.

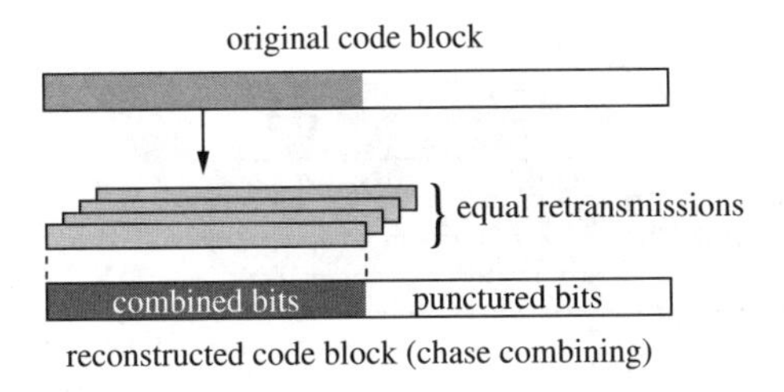

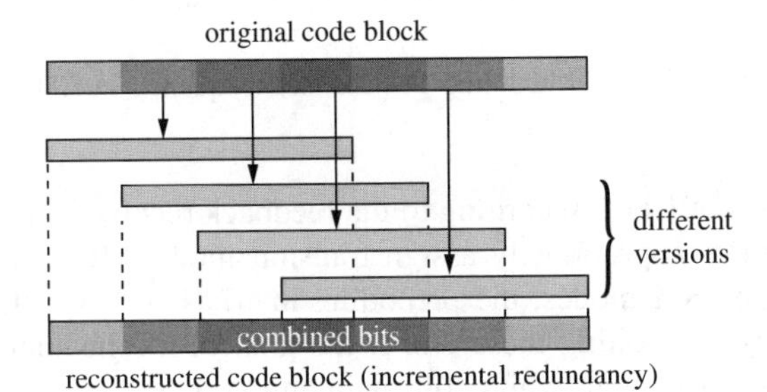

Figure 2.13 In chase combining (top), each retransmission consists of the same bits from the output of the mother code. In incremental redundancy (bottom), a different subset of bits is sent in each retransmission.

As detailed in Section 2.3.5.4, this is achieved through the rate matcher located after the fixed-rate channel encoder. LTE implements this functionality in a one-step rate matching [4], as opposed to HSDPA, where it is accomplished in a two-step process [20, 11]. For a specific target ECR, the rate-matching process is capable of producing different coded packets for each retransmission. HARQ transmissions are indexed differently by a $\mathrm{rv}_{\mathrm{idx}}$ value, which tells the receiver whether the currently transmitted TB is new (zero), or the nth retransmission, up to a maximum of three ($\mathrm{rv}_{\mathrm{idx}} = 0, 1, 2, 3$). For a given target TB size G, the rate matcher can, therefore, produce four differently punctured versions of the original coded TB, depending on the value of $\mathrm{rv}_{\mathrm{idx}}$. The rate-matching process defined for LTE is shown in Figure 2.11, as defined in [4].

2.4.2 Scheduling

In LTE, multi-user diversity is exploited in both the time and frequency domains. UEs are assigned physical resources in the time–frequency grid over time, thus exploiting both Degrees of Freedom (DoFs). Owing to signaling limitations, subcarriers are not individually allocated, but they are aggregated and scheduled on a per-RB basis, each of which spanning 180 kHz, as explained in Section 2.3.1. While the exact RB allocation mechanism can vary between different modes [6], the procedure comprises that each codeword which spans one or more RBs and/or spatial layers is then coded employing a common MCS [32]. In the time domain, a scheduling granularity of 1 ms, which is the time duration of one subframe, is applied. Figure 2.14 depicts the time–frequency

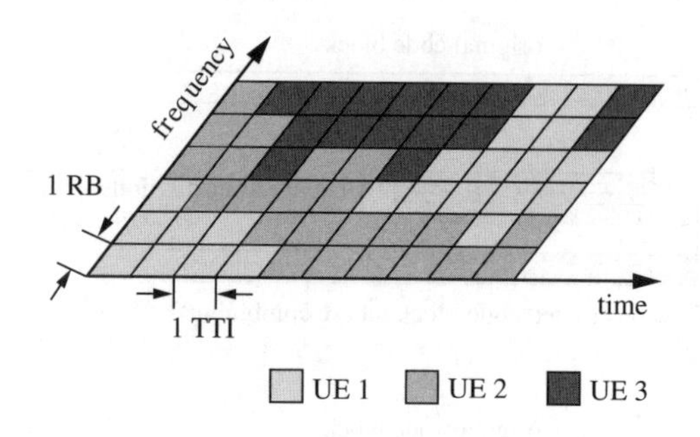

Figure 2.14 Scheduling in time and frequency in LTE.

scheduling mechanism of LTE. According to the feedback received from the UEs (Section 2.3.6), a scheduler must appropriately assign transmit mode, MCS, time, and frequency allocation. Also, in the SM modes, the precoding matrix has to be chosen based on the RI and PMI feedback. Exploiting these DoFs, the goal of a scheduler is typically to try to achieve maximum throughput while still maintaining a certain degree of fairness [31, 32, 38, 44].

2.5 Physical, Transport, and Logical Channels

The physical layer procedures described in the previous sections provide basic functionalities, which are mapped to higher layer channels. In this context, analog to High-Speed Downlink Packet Access (HSDPA), the physical channels for the communication to the MAC layer are combined into so-called transport and logical channels; see Chapter 1.

Data from the radio bearers sent to the RLC layer traverses to the MAC layer via logical channels. Correspondingly, the connection between MAC and PHY is carried out through transport channels by physical channels, transporting the encoded transport channel data over the radio interface. The logical channels can be interpreted as defining "what" information is being transmitted, be it from the radio bearers or RLC-layer signaling, while the transport channels define "how" and with what characteristics the data is transferred. Finally, the over-the-air transmission is carried out via the physical channels. Figure 2.15 illustrates how the radio bearers and logical, transport, and physical channels relate to each other. As the traffic channels define the characteristics, but not the content of the transmitted data, no absolute separation between traffic (U-Plane) and control (C-Plane) can be performed channel-wise. As long as the transport requirements for two logical channels (logical channels are defined based on "what" they are transporting) are equivalent, they can be transmitted by the same transmit channel (which is defined by "how" the data should be transported) [10]. The logical channels of LTE can be divided in traffic and control channels. In UL, only the DTCH is available; the DL supports a multicast channel. Control, common, and dedicated channels are available in both UL and DL, as well as in a Random Access Channel (RACH) for UL. A full list of all the DL channels is provided in Table 2.9, as well as a description of the information being transported. As implied by

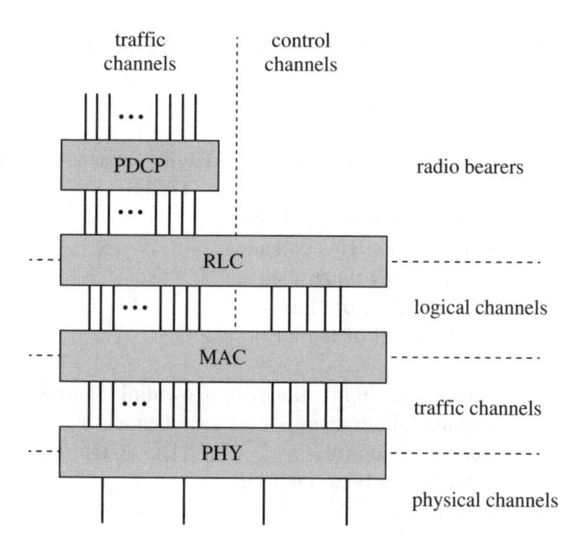

Figure 2.15 Layer and channel structure for LTE.

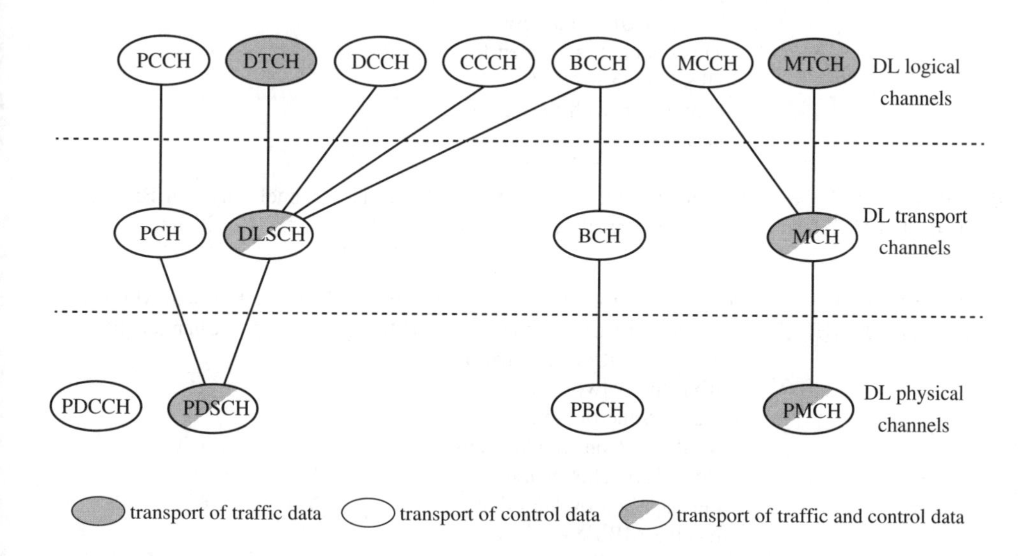

Figure 2.16 LTE DL channel mapping.

Figure 2.15, one-to-one mappings do not exist between logical, transport, and physical channels. In DL, only four physical channels are defined: the PDSCH and the PMCH transmit over the dedicated traffic and multicast respectively, as well as related control information; PBCH and PDCCH are used exclusively for control data. The mapping of the DL logical channels to the transport and physical channels is shown in Figure 2.16 and Table 2.9.

Table 2.9 DL channels description and mapping to traffic and physical channels [1]

Logical channel	Description	Traffic channel	Physical channel
Control			
Paging Control Channel (PCCH)	DL channel that transfers paging information and system information change notifications. This channel is for paging when the network does not know the location cell of the UE	Paging Channel (PCH)	Physical Downlink Shared Channel (PDSCH)
Dedicated Control Channel (DCCH)	Point-to-point bidirectional channel that transmits dedicated control information between a UE and the network	Downlink Shared Channel (DL-SCH)	PDSCH
Common Control Channel (CCCH)	Channel for transmitting control information between UEs and network. This channel is applied for UEs having no RRC connection with the network	DL-SCH	PDSCH
Broadcast Control Channel (BCCH)	DL channel for broadcasting system control information	Broadcast Channel (BCH), DL-SCH	Physical Broadcast Channel (PBCH), PDSCH
Multicast Control Channel (MCCH)	Point-to-multipoint DL channel for transmitting MBMS control information from the network to the UE, for one or several Multicast Control Channels (MTCHs). This channel is only applied by UEs that receive MBMS	Multicast Channel (MCH)	Physical Multicast Channel (PMCH)
–	Informs the UE about the resource allocation of PCH and DL-SCH, and HARQ information related to DL-SCH. Also carries the UL scheduling grant	–	Physical Downlink Control Channel (PDCCH)
Traffic			
Dedicated Traffic Channel (DTCH)	Point-to-point channel, dedicated to one UE, for the transfer of user information	DL-SCH	PDSCH

Table 2.9 (*Continued*)

Logical channel	Description	Traffic channel	Physical channel
MTCH	Point-to-multipoint DL channel for transmitting traffic data from the network to the UE. This channel is only applied by UEs that receive MBMS	MCH	Physical Multicast Channel (PMCH)

References

[1] 3GPP (2009) Technical Specification TS 36.306 Version 9.2.0 'Evolved universal terrestrial radio access (E-UTRA) and evolved universal terrestrial radio access network (E-UTRAN); overall description; stage 2,' www.3gpp.org.
[2] 3GPP (2009) Technical Specification TS 36.104 Version 8.5.0 'Evolved universal terrestrial radio access (E-UTRA); base station (BS) radio transmission and reception,' www.3gpp.org.
[3] 3GPP (2009) Technical Specification TS 36.201 Version 8.3.0 'Evolved universal terrestrial radio access (E-UTRA); LTE physical layer – general description,' www.3gpp.org.
[4] 3GPP (2009) Technical Specification TS 36.212 'Evolved universal terrestrial radio access (E-UTRA); multiplexing and channel coding,' www.3gpp.org.
[5] 3GPP (2009) Technical Specification TS 36.211 Version 8.7.0 'Evolved universal terrestrial radio access (E-UTRA); physical channels and modulation; (release 8),' http://www.3gpp.org/ftp/Specs/html-info/36211.htm.
[6] 3GPP (2009) Technical Specification TS 36.213 'Evolved universal terrestrial radio access (E-UTRA); physical layer procedures,' www.3gpp.org.
[7] 3GPP (2009) Technical Specification TS 36.306 Version 8.3.0 'Evolved universal terrestrial radio access (E-UTRA) user equipment (UE) radio access capabilities,' www.3gpp.org.
[8] 3GPP (2009) Technical Specification TS 25.913 Version 9.0.0 'Requirements for evolved UTRA (E-UTRA) and evolved UTRAN (E-UTRAN),' www.3gpp.org.
[9] 3GPP (2010) Technical Specification TS 36.814 'E-UTRA; further advancements for E-UTRA physical layer aspects,' www.3gpp.org.
[10] 3GPP (2010) Technical Specification TS 36.321 'E-UTRA; medium access control (MAC) protocol specification,' www.3gpp.org.
[11] 3GPP (2010) Technical Specification TS 25.212 'Multiplexing and channel coding (FDD),' www.3gpp.org.
[12] 3GPP (2010) Technical Specification TS 23.402 Version 10.0.0 'Architecture enhancements for non 3GPP accesses,' www.3gpp.org.
[13] 3GPP (2010) Technical Specification TS 23.402 Version 10.0.0 'General packet radio service (GPRS) enhancements for evolved universal terrestrial radio access network (E-UTRAN) access,' www.3gpp.org.
[14] Alamouti, S. (1998) 'A simple transmit diversity technique for wireless communications,' *IEEE Journal on Selected Areas in Communications*, **16** (8), 1451–1458.
[15] Blake, S., Black, D., Carlson, M. *et al.* (1998) 'An architecture for differentiated service,' Technical Report RFC 2475, IETF.
[16] Cheng, J.-F., Nimbalker, A., Blankenship, Y. *et al.* (2008) 'Analysis of circular buffer rate matching for LTE turbo code,' in *IEEE 68th Vehicular Technology Conference (VTC 2008-Fall).*
[17] Dahlman, E., Ekstrom, H., Furuskar, A. *et al.* (2006) 'The 3G long-term evolution – radio interface concepts and performance evaluation,' in *Proceedings of 63rd IEEE Vehicular Technology Conference 2006 (VTC2006-Spring)*, volume 1, pp. 137–141, doi: 10.1109/VETECS.2006.1682791. Available from http://ieeexplore.ieee.org/stamp/stamp.jsp?arnumber=1682791.
[18] Dahlman, E., Parkvall, S., Skold, J., and Beming, P. (2007) *3G Evolution: HSDPA and LTE for Mobile Broadband*, Academic Press.

[19] Dammann, A. and Plass, S. (2006) 'Cyclic delay diversity: effective channel properties and applications,' in *Proceedings of 17th WWRF Meeting*, Heidelberg, Germany.

[20] Dottling, M., Michel, J., and Raaf, B. (2002) 'Hybrid ARQ and adaptive modulation and coding schemes for high speed downlink packet access,' in *The 13th IEEE International Symposium on Personal, Indoor and Mobile Radio Communications*.

[21] Ekstrom, H., Furuskar, A., Karlsson, J. *et al.* (2006) 'Technical solutions for the 3G long-term evolution,' *IEEE Communications Magazine*, **44** (3), 38–45, doi: 10.1109/MCOM.2006.1607864. Available from http://ieeexplore.ieee.org/stamp/stamp.jsp?arnumber=1607864.

[22] Ericsson (2009) 'LTE – an introduction. 284 23-3124 Uen Rev B,' White paper. Available from http://www.ericsson.com/res/docs/whitepapers/lte_overview.pdf.

[23] Frenger, P., Parkvall, S., and Dahlman, E. (2001) 'Performance comparison of HARQ with chase combining and incremental redundancy for HSDPA,' in *IEEE 54th Vehicular Technology Conference (VTC 2001-Fall)*.

[24] Fritze, G. (2008), 'SAE – the core network for LTE,' Available from http://www.3g4g.co.uk/Lte/SAE_Pres_0804_Ericsson.pdf.

[25] Gyasi-Agyei, A. (2005) 'Multiuser diversity based opportunistic scheduling for wireless data networks,' *IEEE Communications Letters*, **9** (7), 670–672, doi: 10.1109/LCOMM.2005.1461700. Available from http://ieeexplore.ieee.org/stamp/stamp.jsp?tp=&arnumber=1461700.

[26] Hagenauer, J. (1988) 'Rate-compatible punctured convolutional codes (RCPC codes) and their applications,' *IEEE Transactions on Communications*, **36** (4), 389–400.

[27] Holma, H. and Toskala, A. (2007) *WCDMA for UMTS: HSPA Evolution and LTE*, John Wiley & Sons.

[28] Holma, H., Toskala, A., Ranta-aho, K., and Pirskanen, J. (2007) 'High-speed packet access evolution in 3GPP release 7,' *IEEE Communications Magazine*, **45** (12), 29–35, doi: 10.1109/MCOM.2007.4395362. Available from http://ieeexplore.ieee.org/stamp/stamp.jsp?arnumber=4395362.

[29] Ikuno, J. C., Mehlführer, C., and M. Rupp (2011) 'A novel LEP model for OFDM systems with HARQ,' in *Proc. IEEE International Conference on Communications (ICC 2011)*.

[30] Ikuno, J. C., Schwarz, S., and Šimko, M. (2011) 'LTE rate matching performance with code block balancing,' in *Proceedings of 17th European Wireless Conference*, Vienna, Austria.

[31] Jain, R. K., Chiu, D.-M. W., and Hawe, W. R. (1984) 'A quantitative measure of fairness and discrimination for resource allocation in shared computer systems,' Technical report, Digital Equipment Corporation.

[32] Kwan, R., Leung, C., and Zhang, J. (2008) 'Multiuser scheduling on the downlink of an LTE cellular system,' *Research Letters in Communications*, **2008** (January), article ID 323048, doi: 10.1155/2008/323048.

[33] Lescuyer, P. and Lucidarme, T. (2008) *Evolved Packet System (EPS)*, John Wiley & Sons, Inc.

[34] Love, D. and Heath, R. (2003) 'Limited feedback precoding for spatial multiplexing systems using linear receivers,' in *IEEE Military Communications Conference*.

[35] Malkamaki, E., Mathew, D., and Hamalainen, S. (2001) 'Performance of hybrid ARQ techniques for WCDMA high data rates,' in *IEEE 53rd Vehicular Technology Conference (VTC 2001)*.

[36] Nimbalker, A., Blankenship, Y., Classon, B., and Blankenship, T. (2008) 'ARP and QPP interleavers for LTE turbo coding,' in *IEEE Wireless Communications and Networking Conference*.

[37] Sánchez, J. J., Morales-Jiménez, D., Gómez, G., and Enbrambasaguas, J. T. (2007) 'Physical layer performance of long term evolution cellular technology,' in *Proceedings of 16th IST Mobile and Wireless Communications Summit 2007*, doi: 10.1109/ISTMWC.2007.4299090. Available from http://ieeexplore.ieee.org/stamp/stamp.jsp?arnumber=4299090.

[38] Schwarz, S., Mehlführer, C., and Rupp, M. (2010) 'Low complexity approximate maximum throughput scheduling for LTE,' in *Conference Record of the 44th Asilomar Conference on Signals, Systems and Computers*. Available from http://publik.tuwien.ac.at/files/PubDat_187402.pdf.

[39] Schwarz, S., Wrulich, M., and Rupp, M. (2010) 'Mutual information based calculation of the precoding matrix indicator for 3GPP UMTS/LTE,' in *IEEE Proceedings of Workshop on Smart Antennas (WSA 2010)*, Bremen. Available from http://publik.tuwien.ac.at/files/PubDat_184424.pdf.

[40] Sesia, S., Toufik, I., and Baker, M. (2009) *LTE, The UMTS Long Term Evolution: From Theory to Practice*, John Wiley & Sons.

[41] Sohn, I. and Bang, S. C. (2000) 'Performance studies of rate matching for WCDMA mobile receiver,' in *IEEE 52nd Vehicular Technology Conference (VTC 2000-Fall)*.

[42] Tang, T. and Heath, R. (2005) 'Opportunistic feedback for downlink multiuser diversity,' *IEEE Communications Letters*, **9** (10), 948–950, doi: 10.1109/LCOMM.2005.10002. Available from http://ieeexplore.ieee.org/stamp/stamp.jsp?tp=&arnumber=1515679.
[43] Tanno, M., Kishiyama, Y., Miki, N. *et al.* (2007) 'Evolved UTRA – physical layer overview,' in *Proceedings of IEEE 8th Workshop on Signal Processing Advances in Wireless Communications 2007 (SPAWC 2007)*, doi: 10.1109/SPAWC.2007.4401427. Available from http://ieeexplore.ieee.org/stamp/stamp.jsp?arnumber=4401427.
[44] Viswanath, P., Tse, D., and Laroia, R. (2002) 'Opportunistic beamforming using dumb antennas,' *IEEE Transactions on Information Theory*, **48** (6), 1277–1294.

Part Two

Testbeds for Measurements

Introduction

In Part Two we explain the concept of testbeds, as opposed to prototypes or demonstrators. While prototypes typically contain a lot of detailed implementation issues and, thus, are tedious in construction, demonstrators are typically used as sales vehicle to provide the costumer a look and feel of a future product (for more details, see also [1]). Our approach is to use testbeds instead, as they are relatively fast to build and rather flexible compared with prototypes. The basic idea is to use MATLAB® or some other high-level language, such as C, for programming, precompute blocks of data for transmission, and then simply emit the blocks to the true physical channel and capture them at the other end. A cluster of back-end PCs then evaluates the received data in offline mode, typically using a high-level language (C and/or MATLAB®) again. The advantages are obvious: the simulation software can basically be reused for generating the data blocks and for receiver processing, as well as the visual front end to display performance curves such as Bit Error Ratio (BER) or throughput. Moreover, the signaling and receiver processing part utilizes identical code, providing consistency in simulation as well as measurements. The testbed thus has the same kind of flexibility as the simulator as long as it is about the transmission system. However, the great benefit is that all transmissions are over real physical channels.

In Chapters 3 and 4 we provide the challenges in building such testbeds and explain what can be expected from testbed measurements. Essentially, it turns out that simply connecting subsystems does not make a testbed. All subsystems require particular care in selection and offer many different pitfalls. A major challenge turned out to be synchronicity of the numerous Multiple-Input Multiple-Output (MIMO) chains, which most vendors claim this property to provide but hardly anybody does. This feature, however, is of utmost importance if beamforming is to be applied. A further important issue is the synchronicity of transmit blocks to ensure the receiver is capturing exactly the transmitted data blocks from their beginning to their end. We explain here our technique of using Global Positioning System (GPS) with a rubidium standard to ensure the highest precision even for remote transceiver pairs.

In Chapter 4 the focus is on the measurement techniques, including evaluation of the measurement quality by statistical inference techniques and measurement automation, as only this step assures high-quality results in a moderate time-frame. Although inference techniques by now are standard in measurements, such techniques are typically absent in wireless measurement (and simulation!) reports in the literature. We thus introduce all the terms required for understanding and explain how to optimally report measurement results by statistical inference techniques such as bootstrapping. Very important is our approach for ensemble averaging by *XY* positioning tables. If just one point-to-point connection is measured, repeating the experiment the next day will hardly achieve identical results. Owing to the time-varying nature of wireless channels, the reproducibility of the experiment can only be ensured in terms of ensemble averages, providing the mean over a small-scale fading environment, say of $3\lambda \times 3\lambda$.

As modern cellular standards rely heavily on feedback information we had to come up with solutions to provide feedback control mechanisms in our testbed in spite of the fact that testbeds are typically designed for unidirectional links. As the channels change over time, simply relying on a static channel, as is commonly found in simulations, or refeeding channel sounding data into a simulation is not realistic. Such feedback techniques are also explained in this chapter.[1]

Reference

[1] Rupp, M., Burg, A., and Beck, E. (2003) 'Rapid prototyping for wireless designs: the five-ones approach,' *Signal Processing*, **83**, 1427–1444, doi: 10.1016/S0165-1684(03)00090-2. Available from http://dl.acm.org/citation.cfm?id=860195.

[1] The chapters of this book part are based on the Ph.D. thesis of Sebastian Caban. The full thesis can be downloaded from EURASIP's open Ph.D. library at http://www.arehna.di.uoa.gr/thesis or directly from http://publik.tuwien.ac.at/files/PubDat_181156.pdf.

3

On Building Testbeds

Setting up a testbed for wireless multi-antenna measurements can be considered a straightforward task. Several companies promise off-the-shelf working baseband hardware and software, whilst other companies offer the high-frequency hardware also required.[1] To put the elements together, "nothing more" than a skilled engineer is required.

An error-free testbed that is able to "reliably" carry out novel "outdoor" wireless measurements 24 hours a day, seven days a week, is another story. It is a story of ongoing redesign, spending five times as much on accessories than on the actual testbed hardware, always searching for bugs and faulty hardware, and never having enough computing, let alone manpower.

This chapter will first introduce our testbed design before reporting on possible pitfalls regarding testbed design in general.

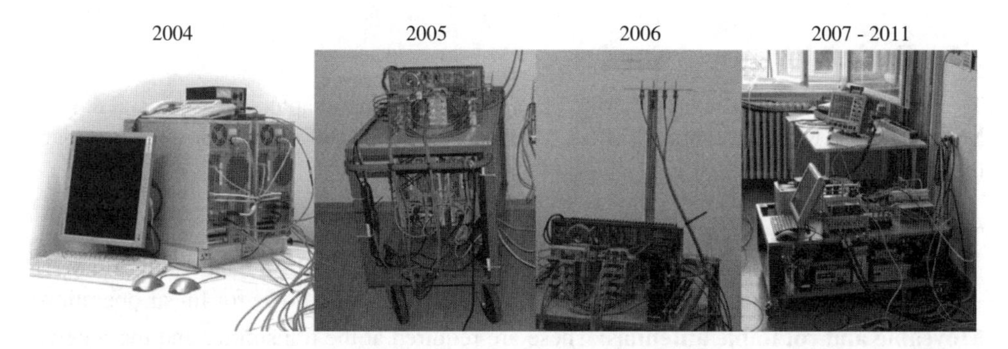

Figure 3.1 Development of our testbed. It took about 4 years from the first publication in 2004 to actually carry out realistic real-world outdoor-to-indoor throughput measurements at the end of 2007.

[1] Examples include the following: Lyrtech (www.lyrtech.com), GE Fanuc (www.gefanucembedded.com), Signalion (www.signalion.com), Hunt-Engineering (hunteng.co.uk), Nallatech (www.nallatech.com), Pentek (www.pentek.com), The Dini Group (www.dinigroup.com), gbm (dsp.gbm.de), 4DSP (www.4dsp.com), GV & Associates Inc. (www.gvassociates.com), Alpha Data Ltd (www.alpha-data.com), Ettus Research LLC (www.ettus.com), Mangocomm (www.mangocomm.com), Sundance Multiprocessor (www.sundance.com), Berkeley (bee2.eecs.berkeley.edu), OpenAir Interface (www.openairinterface.org), Silvus Technologies (www.silvustechnologies.com), Innovative Integration (www.innovative-dsp.com), Digilent (www.digilentinc.com), Spectrum Signal (www.spectrumsignal.com), and MIMOON (www.mimoon.de).

Evaluation of HSDPA and LTE: From Testbed Measurements to System Level Performance, First Edition.
Sebastian Caban, Christian Mehlführer, Markus Rupp and Martin Wrulich.

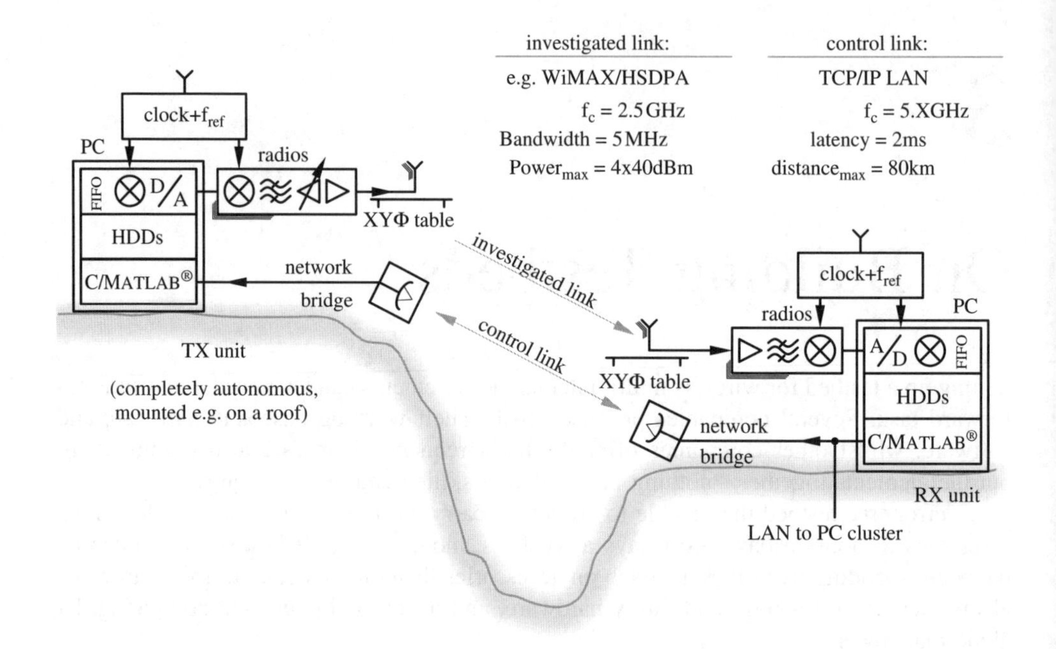

Figure 3.2 The Vienna MIMO testbed.

3.1 Basic Idea

To carry out measurements as described in all the following chapters we designed a 4 × 4 Multiple-Input Multiple-Output (MIMO) testbed that consists of the following key parts (see Figure 3.2):

TX unit: The TX unit is capable of transmitting pre-generated, complex-valued baseband data samples on four antennas at a center frequency of 2.5 GHz, a bandwidth of 5 MHz, and a maximum transmit power of 4 × 40 dBm (which still allows for linear operation).

Moveable and rotatable antennas: These are required at the transmitter and the receiver site (see Figure 4.9).

RX unit: The RX unit is capable of receiving and storing the transmitted data in real time on hard-disk arrays, as well as controlling the overall measurement procedure via a single, fully automated, script (see Section 4.7, p. 95).

Network bridge: A network bridge allows the RX unit to actually control the TX unit via a wireless point-to-point Local Area Network (LAN) connection over distances of up to 80 km with a maximum latency of 2 ms [17] (this being fast enough for quasi-real-time feedback; see Section 4.8, p. 96).

Cluster of PCs: A PC cluster evaluates the data collected by the RX unit offline in MATLAB® effectively[2] (see Section 4.5, p. 88).

[2] Efficiently in terms of total time and manpower needed to write the programs for evaluation and to evaluate the data.

Global Positioning System (GPS) units and rubidium frequency standards: These are located at the transmitter and receiver site to allow for accurate timing and frequency synchronization if necessary [19, 20].

Cabling: A large number of cables are required to connect and control the required PCs, webcams, antennas, linear guides, lights, GPS units, and Radio Frequency (RF) switches.

Host of additional accessories: This is material is shown in Figure 3.3 and the many boxes therein.

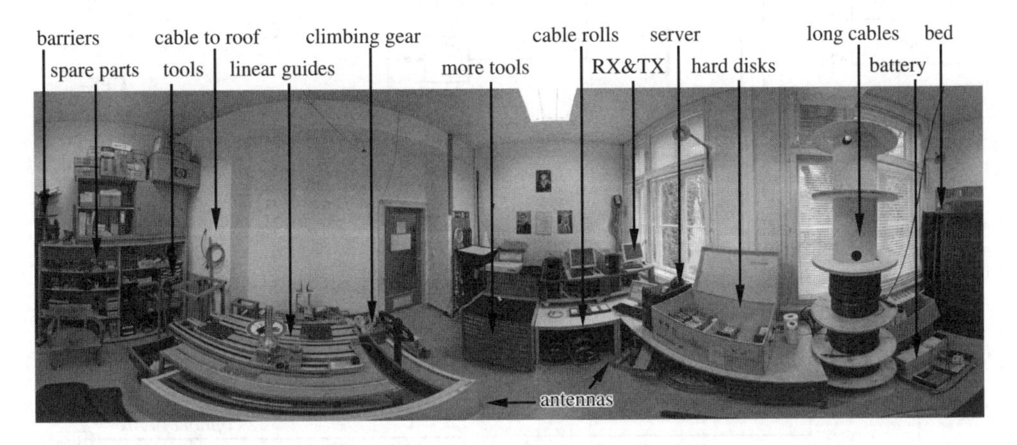

Figure 3.3 The complete Vienna MIMO testbed stored in a room.

Figure 3.4 For carrying out a measurement campaign, the testbed and a small PC cluster are transported in a van.

3.2 Transmitter

As shown in Figure 3.5, the transmitter consists of a PC (A1), a Peripheral Component Interconnect (PCI) plug-in board for digital to analog conversion (B1:6), RF front ends (C1:7), antennas (C9), accessories such as webcams (A12), devices for synchronization (S1:10) (see Section 3.4, p. 65), power supplies (A14), an uninterruptible power supply (A15), a power set (A16), and a lot of cabling (C6, A8, A9, A13, S2, S9, A17:19).

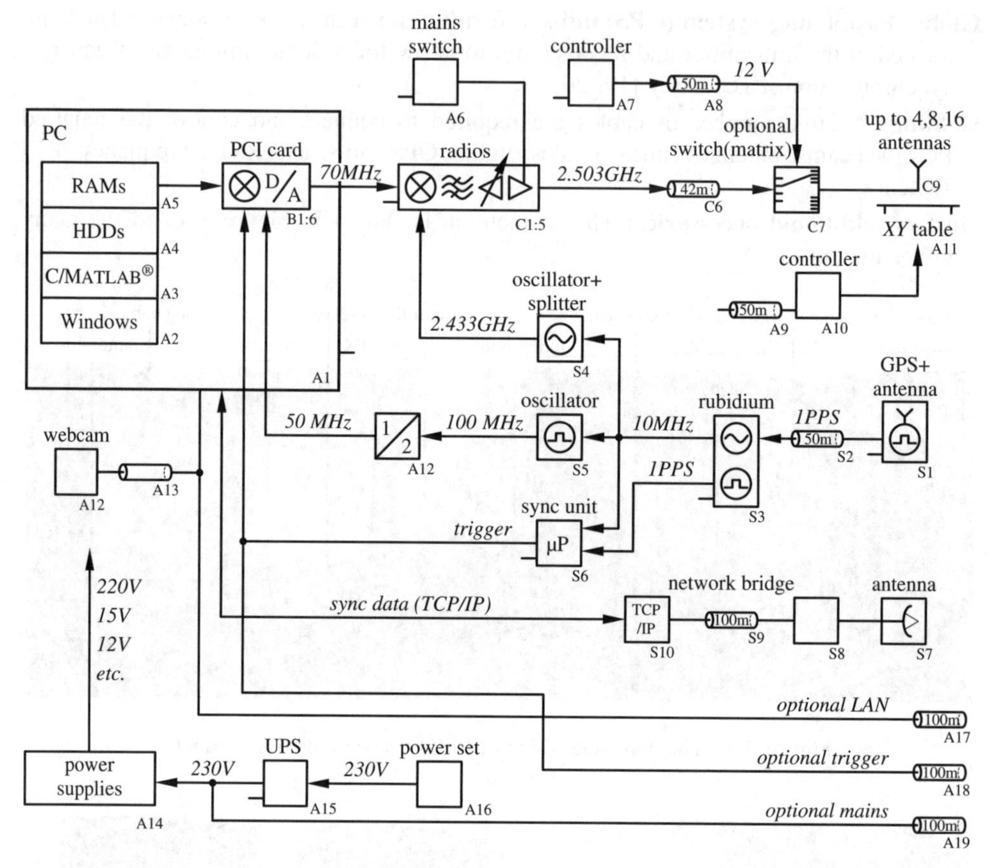

Figure 3.5 TX unit. Open connections are USB or RS232 control connections to the transmit PC.

In the transmit PC, operating Windows 2003 Server® (A2), baseband data samples are pre-generated by MATLAB® (A3) to be stored on Redundant Array of Independent Disks flash hard disks (A4) and/or the internal memory for fast access (A5) by the PCI plug-in board (B1:6).

As pointed out in Figure 3.6, when data is to be transmitted, the previously stored samples are copied to First In, First Out (FIFO) buffers (B1), real-time interpolated to 200 Msample/s (B2&B3), digitally upconverted to a center frequency of 70 MHz (B4&B5), and finally converted to the analog domain (B6). See [2, 3] for further details.

Designing the RF hardware for upconverting the 70 MHz intermediate frequency signal to 2.5 GHz is a straightforward task (see [1, 9]). As shown in Figure 3.7, the signal is attenuated (C1), mixed to 2.5 GHz (C2), variably attenuated to be able to generate different transmit powers (C3), filtered (C4), power amplified (C5), guided by cables to the roof (C6), optionally switched to different antennas (C7), guided to the antennas (C8), and finally transmitted (C9).

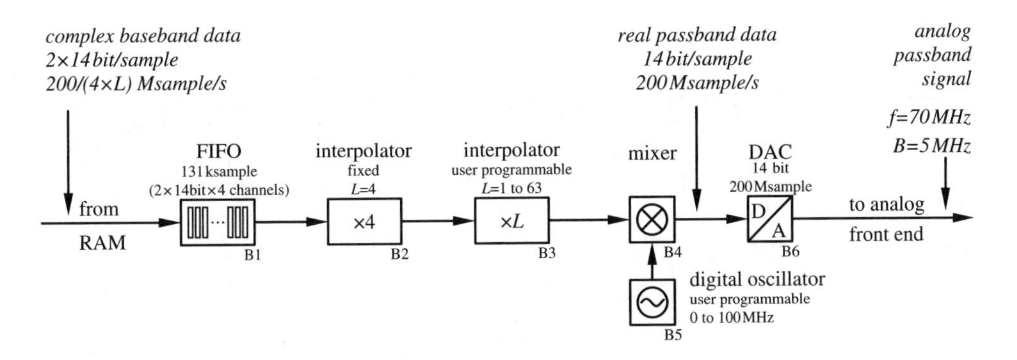

Figure 3.6 The TX unit: digital signal processing.

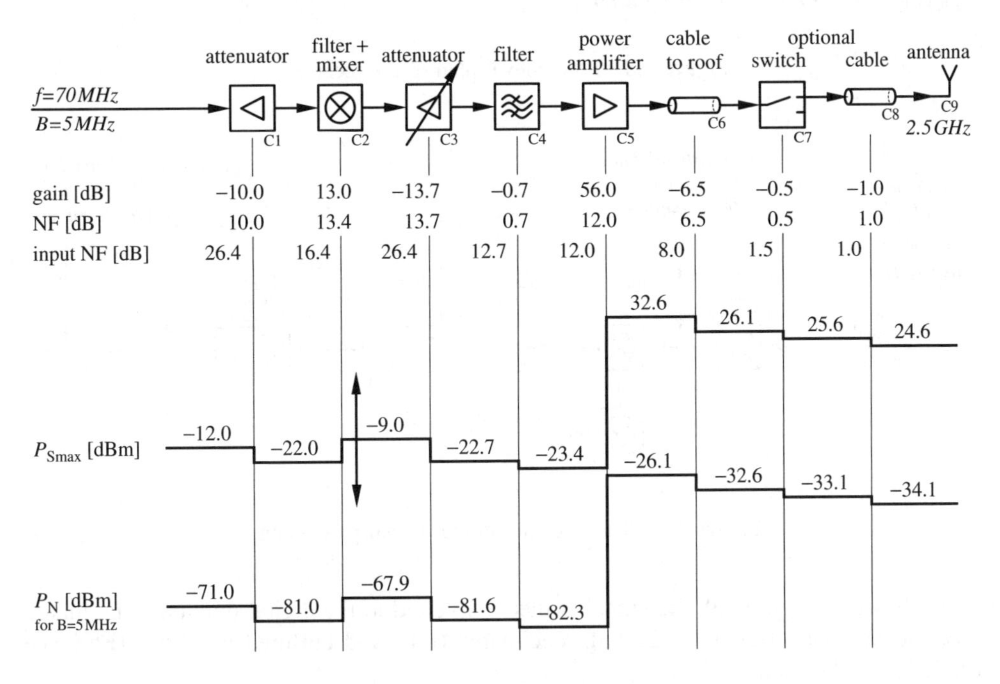

Figure 3.7 Power-level plan of the transmitter.

3.3 Receiver

At the receiver, see Figure 3.8, the transmitted signal is received by up to four antennas (D1), guided (D2) to the receive filter (D3), low-noise amplified (D4), reattenuated (D5) (or guided over long distances if the previous part (D1:4) is mounted on a mast top), refiltered (D6), mixed down to 70 MHz (D7), refiltered once more (D8), and finally amplified to the signal level required by the analog-to-digital converters (D9). See [1, 9] for further details.

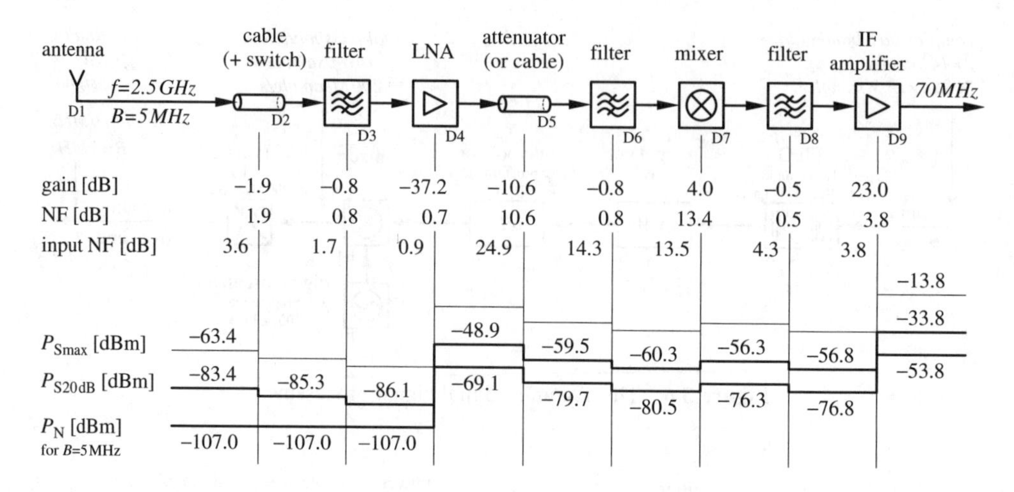

Figure 3.8 Power-level plan of the receiver.

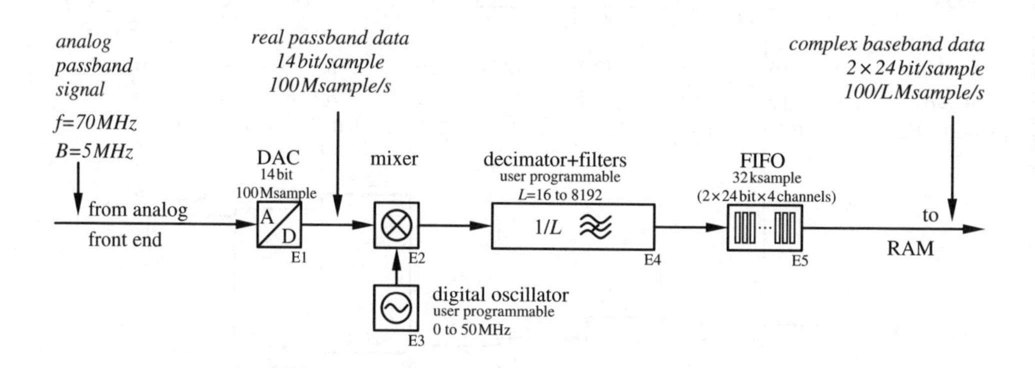

Figure 3.9 The RX unit: digital signal processing.

As shown in Figure 3.9, the signal is then converted to the digital domain (E1), mixed down to baseband (E2, E3), digitally decimated (E4), and buffered in FIFOs (E5). See [2, 3] for further details.

The RX unit, see Figure 3.10, is set up in a very similar fashion to the transmitter. It consists of a PC (F2), the parts previously described for receiving a signal (D1:9, E1:5), accessories such as webcams (F18), devices for synchronization (S1:10) (see Section 3.4, p. 65), power supplies (F16), an uninterruptible power supply (F15), a battery (F13), a DC/AC converter (F14) for mobile use, and, again, extensive cabling (D2, S2, S9, F10:12, F17).

In the PC, the received data samples stored in the FIFOs (E5) of the PCI plug-in board are either directly copied to the hard disks (F4) or copied to MATLAB® (F5) for fast feedback calculation. In the latter case, the resulting optimal block to be transmitted is fed back to the transmitter via a network bridge (S7:S10). This network bridge is also used for remote control of the transmitter operation in general.

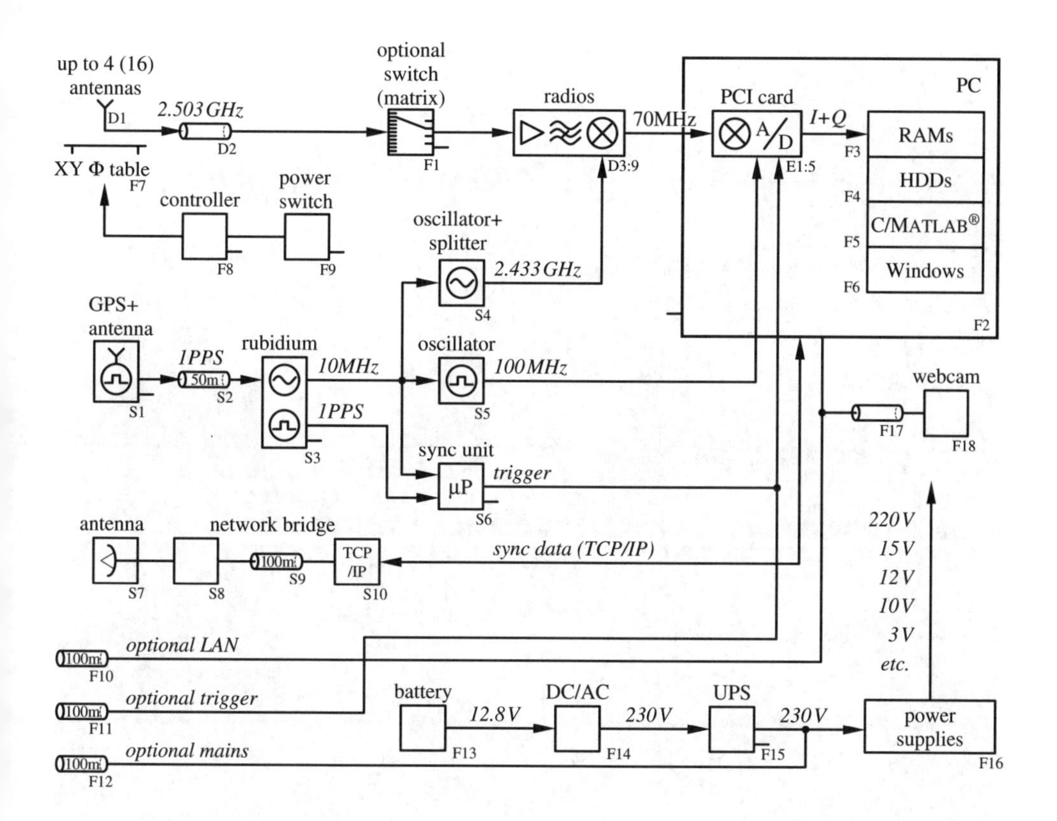

Figure 3.10 The RX unit.

The measured power consumption of the receiver is about 250 W; in other words, a decent 47 kg heavy battery (F13) can supply the whole setup with the energy required for approximately 5 h.

3.4 Synchronization

As there is (in general) no cable connection between the TX and the RX unit, some sort of synchronization is needed to ensure that the blocks of data are transmitted (at the transmit site) and received (at the receive site) at the same time. Furthermore, it should optionally be possible to synchronize the frequency of the internal clocks of the TX unit to those of the RX unit. This is all achieved as follows (see Figure 3.12):

- Locked to geostationary satellites, a GPS (S1) at the transmit site outputs one Pulse Per Second (PPS), as does a second GPS (S1) at the receive site. The starting time of these pulses differs by up to ± 20 ns.
- In the next step, the long-term-stable PPS pulses are short-term stabilized by a rubidium frequency standard (S3).

Figure 3.11 The transmitter (top) and the receiver (bottom).

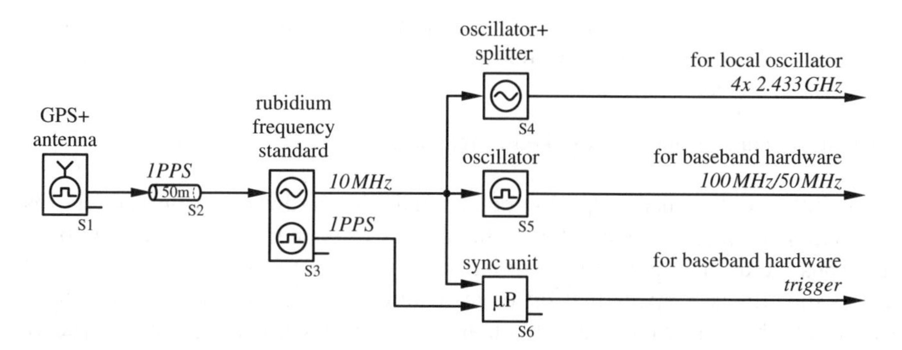

Figure 3.12 Synchronization units at the transmit and receive site (both equal). The open connections are USB or RS232 control connections to the transmit PC.

- The resulting PPS signal is forwarded to a self-built synchronization unit (S6). This unit basically consists of a counter which runs at a frequency of 10 MHz (provided by the rubidium). At the beginning of each measurement, this counter is reset at the transmit and the receive site at exactly the same time by the use of the PPS signal and handshaking via the network bridge (S7:S10) in Figure 3.5 (p. 62). During the measurement, the transmit site then just has to tell the receive site at which counter-value it should receive the transmitted data. See [4] for further details.
- The 10 MHz output of the rubidium is also used to lock a 2.433 GHz low phase-noise oscillator (S4) that is used to mix the 70 MHz intermediate frequency signal to 2.503 GHz (C2) in Figure 3.7 (p. 63) and back (D7) in Figure 3.8 (p. 64). Another 100 MHz oscillator (S5) is used to provide the PCI plug-in boards (B1:6) in Figure 3.5 (p. 62) and (E1:5) in Figure 3.10 (p. 65) with a stable reference clock.
- Since the rubidium (S3), the 2.433 GHz oscillator (S4), and the 100 MHz oscillator (S5) can be used without a reference input as free running devices, all measurements can be carried out with/without external synchronization (or optionally by a tunable oscillator with a given frequency offset).

3.5 Possible Pitfalls

3.5.1 Digital Baseband Hardware

Our experience over the years has shown that it takes a considerable amount of time and manpower to get even the most basic demo programs to work. Reasons for this are:

- missing, misleading, or carelessly written and not updated manuals, pinout documentation, and sample programs coming with off-the-shelf products;
- a lack of support or extremely long support-answer times, not only when buying from low-cost vendors;
- compatibility problems between different versions of hardware and software; and

- stability problems – even a demo program provided by the manufacturer that initially appears to run smoothly may reveal astonishing and unforeseen instabilities if executed over several weeks.

One has to be careful that, for marketing reasons:

- Bandwidths and transfer speeds proclaimed by hardware manufacturers usually represent pure marketing numbers. Busses are often not able to transfer bus-clock times bus-width bits on a permanent basis, only in burst mode.
- It also sometimes happens that different features proclaimed for a product cannot be achieved at the same time, but only on an either–or basis. For example, a synchronous transfer is only possibly at a low speed; the proclaimed high speed only works in asynchronous mode.

Another problem that arises with digital baseband hardware is that products are sometimes sold *not* including the possibility to flexibly reprogram them as advertised – and this fact is not obvious to the customer:

- Typically, firmware can be used in its delivered form but cannot be modified. In order to modify it, additional licenses and tools are required, costing considerable additional amounts of money.
- The software included with the product may also be limited "on purpose" in its functionalities (for example, it will work on one Field-Programmable Gate Array (FPGA) but not on two). The unlock keys required imply additional expenditure often overlooked at the date of purchase.
- And even if the software keys are not limited, one has to be careful that all the documentation needed to modify the firmware is included and does not have to be purchased additionally. In such situations one may find oneself performing the programming and development work that has already been paid for.

3.5.2 Tool and Component Selection

One also has to bear in mind that the "specific" software tools sold with a product require other "specific" software packages – in a "specific" version – to operate:

- It often happens that errors found are corrected in newer software versions, but this also implies that all the other specific "software packages have to be updated" in order to work correctly – a vicious circle that consumes a lot of time and money, since it is common for other software packages from other vendors to also be affected.
- It has been reported that companies "charge for maintenance," which means that one has to pay special attention to whether charge-free bug fixes are included in the delivered software.
- In addition, highly specialized software tools used to program hardware do not, unfortunately, often work reliably.
- Owing to poor documentation and the endless number of possibilities as to why bugs could occur, the tracking of these errors, reporting them, and getting them fixed is usually an endless challenge. Therefore, one should consider only employing

extensively tested "standard tools" (for example, Texas Instruments Development Environment or Xilinx ISE), instead of spending money on tools that seem to save time at first glance but afterwards turn out to be error prone.

Baseband hardware is often sold on a module basis in complete packages. The hardware develops so rapidly that hardly any company has the time to extensively test its hardware and prepare well-written manuals and source codes. The best way to deal with these problems is to buy from the company with the best support. A fast and competent support often makes the difference between achieving anticipated goals or not.

In multiple antenna architectures, special emphasis has to be put on 'synchronous' operation:

- It must be possible to synchronize the digital signal paths used for different antennas, even if they are spread among several chips or modules. Memories, interpolators, filters, and digital mixers may not allow for this, especially if dedicated hardware is used instead of FPGAs.
- Because they are just scaled Single-Input Single-Output solutions, many MIMO solutions offered do not allow for a "synchronous RF oscillator" for analog up/down mixing – even if advertised.
- One has to make sure that the word "synchronous" really means "equal in phase and frequency without any jitter" if used in product advertising brochures. Failure in synchronicity may result in dramatic performance loss compared with perfectly synchronous transmission.

3.5.3 Analog RF Front Ends

When it comes to analog front ends (for example, analog upconverters, filters, amplifiers), the opposite is true. On the one hand, it is extremely difficult to buy the hardware needed, since very few products exist (most only as components rather than complete). One has to buy everything on a per module basis, sometimes even from different suppliers. On the other hand, once the hardware has been obtained, components can be set up and got working relatively quickly. Some very expensive measurement equipment is required, however, in order to check if the hardware is working within the desired specifications or not. There is usually no hidden cost afterwards, no software tools needed, and no carelessly written manuals and demo applications. On the downside, analog high-frequency hardware is hardly ever flexible. Once one changes the center frequency, one has to rebuy most of the hardware – in the digital domain, this only requires the modification of some bits.

Therefore, when buying a new analog RF front end, it is very important to choose the "proper" center frequency:

- It is a lot easier and cheaper to obtain hardware for free Industrial, Scientific and Medical (ISM) bands (for example, 2.4 GHz or 5.2 GHz). Other bands (1.5 GHz, 2 GHz, 3.5 GHz, 5.8 GHz) are sufficiently close to draw the right conclusions for the experiments.

- Choosing a frequency within a free ISM band implies that other interferers, such as cordless computer peripherals, Wireless Local Area Network (WLAN), Bluetooth, and microwave ovens, may inherently influence every single measurement. If this unpredictable interference is accepted (or even desired), a transmit frequency within an ISM band should be chosen.

3.5.4 Cost

As pointed out, setting up a testbed requires a considerable amount of money, manpower, and (most important) time, but in many cases this may be still more economical than, for example, buying an extremely expensive but rather inflexible channel sounder.

A high-quality channel sounder costs typically between €300 000 and €1 000,000, the hardware for a good testbed (€100 000) plus four person-years for setting it up may add up to €250 000 – still considerably cheaper. Furthermore, it allows one to perform more research than just extracting channel coefficients, and also to test realistic transmissions over the air with the signals that will be applied in the final product.

The main downside of a testbed, however, is the time needed to set everything up and get it working.[3] This makes testbeds very suitable for basic research, where time to market is usually not the primary factor, but often inappropriate for other purposes. Companies, on the other hand, are well recommended to continuously put effort into testbeds in order to constantly have them available and not to start from scratch every time a new product design cycle is started. Note also that, if a testbed is set up and working, it may allow for measurements to be carried out within minutes, especially when the data is processed offline in tools such as MATLAB®. Intelligent consideration of similar experiments, which can utilize the testbed without time-consuming hardware modifications, can enable "lost" time to be made up easily.

3.5.5 MATLAB® Code and Testbeds

Once a MATLAB® code has been proven to work well in simulation, a testbed can clarify whether the algorithms are suitable for real over-the-air communications. Unfortunately, running MATLAB® code with a testbed is not a simple process. There are many things that have to be taken into account. For the simple case of offline processing in MATLAB®, these are, for example:

- MATLAB® simulations often operate in the discrete baseband only. Therefore, transmit and receive filters and interpolators have to be added.
- Interpolating to a fixed, given hardware sample rate may also introduce impractical interpolation factors (for example, 3.84 MHz Universal Mobile Telecommunications System to 100 MHz sampling), requiring interpolation filters with extreme length. Alternatively, decreasing and optimizing the filter complexity may result in a lengthy project in its own right.

[3] As a rule of thumb, this time cannot be reduced to less than a year because of delivery times and unforeseeable problems.

- MATLAB® simulations often assume perfectly synchronized signals. In measurements, one now has the choice to:
  - synchronize transmitter and receiver perfectly (typically by cables);
  - nearly perfectly synchronize them by rubidium frequency standards and GPS receivers (which may be required if, for example, the receiver is mounted in a car);
  - use special training sequences prior to the transmitted data (which may only be possible in static scenarios); or
  - implement proper synchronization algorithms.

  Several of these options may even be used together for all required synchronizations (for example, local oscillator frequency, timing, and block start).

  In some cases, perfect synchronization may be the method of choice to avoid all undesired effects (for example, for reference purposes to test the performance loss of proper synchronization algorithms). Even the first famous MIMO experiments carried out used cables for synchronizing transmitter and receiver clocks [22]. In other cases, implementing proper synchronization algorithms in the receiver may deliver a better view of the reality. Unfortunately, this is not always possible; for example, if only a limited number of blocks is available and synchronization requires averaging over long periods of time.
- The channel is never known to the receiver. Therefore, channel estimation cannot be omitted, as is often the case in MATLAB® simulations. Fortunately, in quasi-static scenarios, long training sequences can be used to nearly perfectly estimate the channel (for reference purposes).
- For many (optimal) receiver algorithms the noise variance also has to be estimated at the receiver. However, the simple trick of measuring the noise variance in the absence of transmitting signals may often save coding time and provides accurate estimates regardless of the modulation scheme used.

Once it is working properly, a testbed (plus subsequent offline processing of the received data in MATLAB®) is a very powerful and swift method for evaluating algorithms by realistic over-the-air transmissions. One has the choice of measuring the absolute performance of a transmission scheme or of comparing two transmission schemes relative to each other. It is particularly easy to measure the relative difference between two types of receiver because:

- the same stored receive data can be evaluated, thus making the comparison fair;
- debugging is also made easier, because the received data remains equal;
- the number of channel realizations can be significantly reduced, since, for measuring relative performance (compared with absolute performance), a much smaller number of measured realizations is sufficient;
- systematic errors in relative performance measurements play a less dramatic role than in absolute performance measurements.

3.6 Summary

The testbed presented in this chapter perfectly meets the specifications required by the quasi-real-time measurement procedure described in Chapter 4. It consists of a TX unit and an RX unit comprising moveable and rotatable antennas, a network bridge, GPS units, rubidium frequency standards, extensive cabling, and a host of accessories, as well as a cluster of evaluation PCs.

Of course, one could argue that our testbed is just another approach to building a piece of "publishable" MIMO measurement hardware. Therefore, the engineering part of this chapter was kept short, whilst still describing all the parts of our testbed before listing possible issues regarding testbed design in general.

On the other hand, only very few groups nowadays are continuously producing results based on testbeds owing to the extremely time-consuming efforts involved and the required educational profile ranging from computer engineering through telecommunication engineering and electrical engineering to even mechanical engineering in many cases. What actually differentiates our approach from others is the quantity of efficiently produced measurement results [14–8, 10–16, 18, 21, 23] showing more than scatter plots, estimated transfer functions, or estimated mutual information.

References

[1] Behzad, A. (2007) *Wireless LAN Radios*, 1st edition, John Wiley & Sons, Inc., Hoboken, NJ.

[2] Caban, S. (2005) 'Development and setting up of a 4 × 4 real-time MIMO testbed,' Master's thesis, Institut für Nachrichtentechnik und Hochfrequenztechnik. Available from http://publik.tuwien.ac.at/files/pub-et_9754.pdf.

[3] Caban, S., Mehlführer, C., Langwieser, R. *et al.* (2006) 'Vienna MIMO Testbed,' *EURASIP Journal on Applied Signal Processing*, **2006**, article ID 54868, doi: 10.1155/ASP/2006/54868.

[4] Caban, S., Mehlführer, C., Lechner, G., and Rupp, M. (2009) 'Testbedding MIMO HSDPA and WiMAX,' in *Proceedings of VTC 2009 Fall*, Anchorage, AK. Available from http://publik.tuwien.ac.at/files/PubDat_176574.pdf.

[5] Caban, S., Mehlführer, C., Mayer, L. W., and Rupp, M. (2008) '2 × 2 MIMO at variable antenna distances,' in *Proceedings of VTC 2008 Spring*, Singapore, doi: 10.1109/VETECS.2008.276. Available from http://publik.tuwien.ac.at/files/PubDat_167444.pdf.

[6] Caban, S., Mehlführer, C., Scholtz, A. L., and Rupp, M. (2005) 'Indoor MIMO transmissions with Alamouti space–time block codes,' in *Proceedings of the 8th International Symposium on DSP and Communication Systems, DSPCS 2005*, Noosa Heads, Australia. Available from http://publik.tuwien.ac.at/files/pub-et_9815.pdf.

[7] Caban, S. and Rupp, M. (2007) 'Impact of transmit antenna spacing on 2 × 1 Alamouti radio transmission,' *Electronics Letters*, **43** (4), 198199, doi: 10.1049/el:20073153.

[8] García-Naya, J. A., Mehlführer, C., Caban, S. *et al.* (2009) 'Throughput-based antenna selection measurements,' in *Proceedings of the 70th IEEE Vehicular Technology Conference (VTC2009-Fall)*, Anchorage, AK, doi: 10.1109/VETECF.2009.5378992. Available from http://publik.tuwien.ac.at/files/PubDat_176573.pdf.

[9] Luzatto, A. and Shirazi, G. (2007) *Wireless Transceiver Design*, 1st edition, John Wiley & Sons, Ltd.

[10] Mehlführer, C., Caban, S., and Rupp, M. (2008) 'An accurate and low complex channel estimator for OFDM WiMAX,' in *3rd International Symposium on Communications, Control and Signal Processing (ISCCSP 2008)*, St. Julians, Malta, pp. 922–926. Available from http://publik.tuwien.ac.at/files/pub-et_13650.pdf.

[11] Mehlführer, C., Caban, S., and Rupp, M. (2008) 'Experimental evaluation of adaptive modulation and coding in MIMO WiMAX with limited feedback,' *EURASIP Journal on Advances in Signal Processing*, **2008**, article ID 837102, doi: 10.1155/2008/837102. Available from http://publik.tuwien.ac.at/files/pub-et_13762.pdf.

[12] Mehlführer, C., Caban, S., and Rupp, M. (2008) 'Measurement based evaluation of low complexity receivers for D-TxAA HSDPA,' in *Proceedings of the 16th European Signal Processing Conference*, Lausanne, Switzerland. Available from http://publik.tuwien.ac.at/files/PubDat_166132.pdf.

[13] Mehlführer, C., Caban, S., and Rupp, M. (2009) 'MIMO HSDPA throughput measurement results in an urban scenario,' in *Proceedings of VTC 2009 Fall*, Anchorage, AK. Available from http://publik.tuwien.ac.at/files/PubDat_176321.pdf.
[14] Mehlführer, C., Caban, S., and Rupp, M. (2010) 'MIMO HSDPA throughput measurement results', in *HSDPA/HSUPA Handbook* (eds. B. Furht and S. A. Ahson), CRC Press, Boca Raton, FL, pp. 357–377.
[15] Mehlführer, C., Caban, S., Rupp, M., and Scholtz, A. L. (2005) 'Effect of transmit and receive antenna configuration on the throughput of MIMO UMTS downlink,' in *Proceedings of the 8th International Symposium on DSP and Communication Systems, DSPCS 2005*, Noosa Heads, Australia. Available from http://publik.tuwien.ac.at/files/pub-et_10269.pdf.
[16] Mehlführer, C., Caban, S., Wrulich, M., and Rupp, M. (2008) *Joint Throughput Optimized CQI and Precoding Weight Calculation for MIMO HSDPA*, IEEE, Piscataway, NJ. Available from http://publik.tuwien.ac.at/files/PubDat_167015.pdf.
[17] Pulse Supply 'Airmux backhaul wireless IP and E1/T1 radio.' Available from http://www.airmux.com/.
[18] Rupp, M., Caban, S., and Mehlführer, C. (2007) 'Challenges in building MIMO testbeds,' in *Proceedings of the 13th European Signal Processing Conference (EUSIPCO 2007)*, Poznan, Poland. Available from http://publik.tuwien.ac.at/files/PubDat_112138.pdf.
[19] Stanford Research Systems, Inc. 'FS725 – Benchtop rubidium frequency standard.' Available from http://www.thinksrs.com/products/FS725.htm.
[20] Trimble 'Acutime Gold GPS smart antenna.' Available from http://www.trimble.com/timing/acutime-gold-gps-antenna.aspx?dtID=overview.
[21] Wang, Q., Caban, S., Mehlführer, C., and Rupp, M. (2009) 'Measurement based throughput evaluation of residual frequency offset compensation in WiMAX,' in *Proceedings of 51st International Symposium ELMAR-2009*, Zadar, Croatia. Available from http://publik.tuwien.ac.at/files/PubDat_176679.pdf.
[22] Wolniansky, P., Foschini, G., Golden, G., and Valenzuela, R. (1998) 'V-BLAST: an architecture for realizing very high data rates over the rich-scattering wireless channel,' in *Proceedings of URSI International Symposium on Signals, Systems, and Electronics (ISSSE 98)*, pp. 295–300, doi: 10.1109/ISSSE.1998.738086.
[23] Wrulich, M., Caban, S., and Rupp, M. (2007) 'Testbed measurements of optimized linear dispersion codes,' in *Proceedings of the ITG Workshop on Smart Antennas*. Available from http://publik.tuwien.ac.at/files/pub-et_12313.pdf.

4

Quasi-Real-Time Testbedding

When it comes to evaluating the performance of mobile radio communication systems, the most common method by far is computer simulations. Fortunately, much of the back-breaking work has been eliminated by the incredible amount of computing power and predefined toolboxes available nowadays. Although certainly convenient, such simplicity is bewitching.

As we will show in this chapter, measuring the performance of the physical layer of a mobile communication system can be remarkably similar to simulating it. The main difference is that a measurement has to obey the laws of nature. For example, one cannot just assume perfect channel knowledge, perfect frequency and timing synchronization, known noise variance, double-precision feedback, and so on. One also cannot measure a million independent channel realizations within a small-scale fading scenario. Compared with simulations, all this may seem troublesome; on the other hand, reality is like this.

In other words, simulations reflect the simulation environment (and may reflect reality), whereas measurements do reflect reality.

4.1 Basic Idea

"Simulation is always a form of sampling experiment whenever the model contains one or more stochastic variables (although it is a very special type of sampling experiment since simulations are performed on abstract models instead of real-life objects)" [17, p. xi]. Figure 4.1 shows such an abstract model, which will henceforth be used to obtain the average throughput performance of a radio transmission.

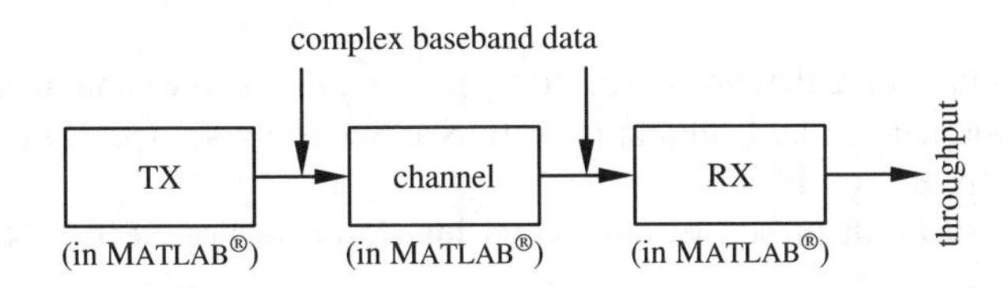

Figure 4.1 Simulating a sample for a throughput simulation.

Evaluation of HSDPA and LTE: From Testbed Measurements to System Level Performance, First Edition.
Sebastian Caban, Christian Mehlführer, Markus Rupp and Martin Wrulich.

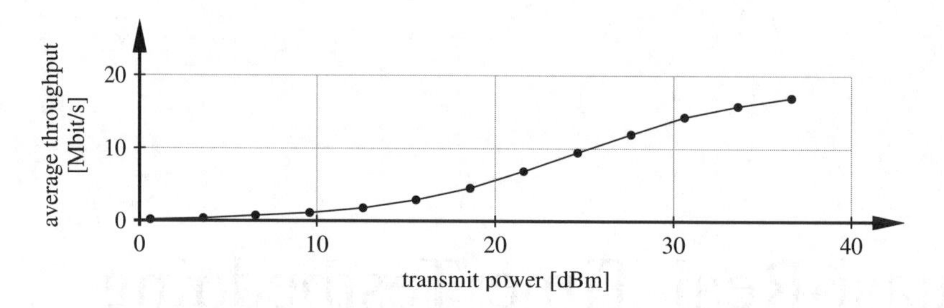

Figure 4.2 Average throughput versus transmit power.

The Monte Carlo simulation now works as follows:

1. "Random" complex-valued baseband data is created in the block named "TX." Next, this data is sent through a "random" channel realization disrupted by "random," usually Gaussian, noise. Finally, the resulting baseband data is decoded in the block named "RX" to calculate the throughput of one transmitted block.
2. The above transmission is repeated to receive an average throughput.
3. The above procedure is repeated to receive a set of average throughputs over the factor of interest (for example, different transmit power levels; see Figure 4.2).

Measuring the performance of a mobile communication system can be carried out in a very similar way by "simply" replacing the channel with some real hardware (in our case, a testbed); that is, transmitting the data in a realistic scenario over a wireless channel (see Figure 4.3):

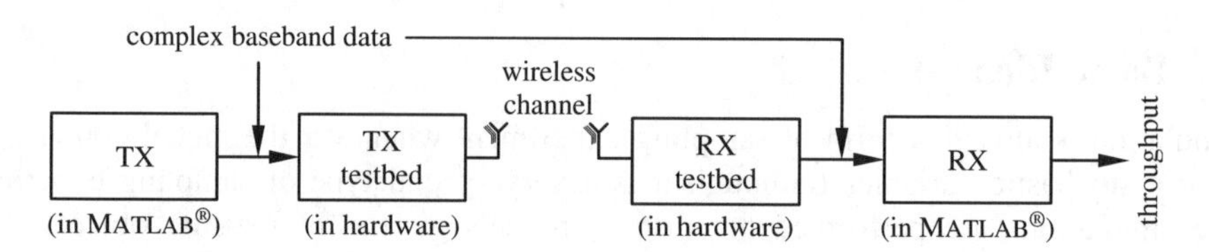

Figure 4.3 Measuring a sample for a throughput measurement.

However, the following differences to simulation arise:

- The different channel realizations created by moving the receive antennas of the testbed may not be independent and subject to drift. See Section 4.4.1 (p. 81) on sampling and Section 4.4.3 (p. 85) on bias.
- The channel can be disturbed by undesired interference. See Section 4.4.4 (p. 86) on outliers.
- The receiver cannot have genie-driven knowledge of the channel and the timing. See Section 4.4.5 (p. 87) on parameter estimation.
- The channel cannot be controlled as it can be in simulations. In particular, it is impossible to hold the channel constant over a given period of time or even to switch it back to a previous state. See Section 4.6.3 (p. 92) on reproducibility and repeatability and see Section 4.8 (p. 96) on feedback and retransmissions.

- Measurement speed is limited owing to the fact that a measurement cannot be parallelized on lots of computers. Neither is it possible to transmit faster than the actual transmission. Furthermore, changing the transmit power (by a programmable attenuator) or changing the receive antenna positions (by a linear guide) takes a considerable amount of time. As a result, a thorough statistical design and analysis is needed, especially regarding the sample size required. See Section 4.6 (p. 90) on statistical inference.
- The quasi-real-time testbed design employed by us requires the transmitted data blocks to be pregenerated in MATLAB®, stored on flash hard-disks for fast access, and then transferred to small but real-time-capable First In, First Out buffers. Therefore, to speed up and simplify things, the transmit blocks used may not be random, but constant during one complete experiment. See Chapter 3 (p. 59) for more details on our testbed.
- Starting a new experiment, the testbed may simply refuse to work as expected because, for example, a cable is loose. See Section 4.3 (p. 78) on pretesting.

4.2 Problem Formulation

Compared with pure simulations, measurements usually require a frequently underestimated amount of money, manpower, and time. Unfortunately, this amount can never be reduced to the level required for simulation, as the simulation itself is still required as "part" of the measurement to validate the measurement. Nevertheless, we still try to minimize our research expenses by simplifying our measurements compared with "commercial systems" in the following way:

- We only analyze the physical layer of radio communication systems; namely, we do not investigate multi-user scheduling and other higher layer issues.
- We only employ one TX and one RX unit with a maximum of four antennas each (16 each by employing switches in the Radio Frequency (RF) front end). Therefore, we can only consider the up-/down-link scenario from a single Base station (NodeB) to a single user. By measuring in an unused band we can later choose between having no interference at all (noise is only thermal) or adding interference from other previously made measurements with different transmit locations in the digital domain (assuming that the receiver is linear).
- We only consider such scenarios on a block-by-block basis in real time; that is, the time between blocks is short (some milliseconds), but typically longer than in most real systems.
- We do not put constraints on the size of our TX and the RX unit. Currently, they are as big as a small table.
- We use more sophisticated RF hardware than a low-cost commercial product would. We do so for three reasons: (i) to carry out the very precise measurements required by some experiments; (ii) to be prepared to meet the specifications of future communication systems; and (iii) to easily evaluate the throughput-impact of a commercial component by comparing it with our "reference" (for example, the impact of a low-cost, noisy oscillator).
- We (optionally) employ external synchronization in frequency and time (see Section 3.4, p. 65). This gives us the choice to measure with no, a constant, or a randomly selected frequency offset.

- We implement all necessary algorithms in MATLAB®/C in floating point. In addition to being "feasible," this gives us all the flexibility needed to quickly change the code if new ideas arise that need to be tried out (for example, Low Density Parity Check instead of turbo channel coding).
- We measure systems employing feedback only in sufficiently static scenarios in order to have enough time for calculating the feedback and, as a result, the required new transmit signal (see Section 4.8, p. 96).

> Given the aforementioned simplifications, our goal is to "scientifically" compare the "performance" of given "wireless communication schemes" in specific, "realistic scenarios."

In the aforementioned goal:

Scientifically "suggests a process of objective[1] investigation that ensures that valid conclusions can be drawn from the experimental study" [19, p. 3]. Furthermore, we want this process not only to be effective, but also to be efficient.[2]

Performance may, for example, be expressed in terms of average throughput or median Bit Error Ratio (BER).

Wireless communication schemes can, for example, be simple single carrier uncoded transmissions or more complex systems such as Worldwide Inter-operability for Microwave Access (WiMAX) or High-Speed Downlink Packet Access (HSDPA). Their actual implementation, however, is not of importance for our measurements as long as a MATLAB® code for the transmitter and receiver exists.

Compare means that we have to be able to measure different communication systems employing the same hardware over channels that are as similar as possible; that is, to transmit the block with scheme two immediately after the one with scheme one.

Realistic scenarios are, for example, indoor-to-indoor, outdoor-to-indoor, and outdoor-to-outdoor urban/suburban/alpine scenarios. For us, a realistic scenario also includes the usage of realistic antennas, such as, for example, 60° flat-panel X-polarized antennas at the NodeB. Monopole antennas at the NodeB are not realistic in our opinion.

4.3 Employing the Basic Idea

To reach the above stated goal and produce, for example, Figure 4.6, we perform the following steps:

1. We design a testbed that enables us to transmit baseband data pre-generated in MATLAB® over realistic channels (see Chapter 3, p. 59).
   In the case of Figures 4.4–4.6, the testbed has to transmit blocks of 2×2 HSDPA and 2×2 WiMAX that are already stored in its memory. Details of the 2×2 HSDPA (see Chapter 1 (p. 75), Chapter 5 (p. 103), and [23, 24]) and 2×2 WiMAX (see [21, 22]) systems, as well as the scenario in which we measured this curve (see Section 7.3

[1] "Independent of individual thought and perceptible by all observers" [25].

[2] The difference between effectiveness and efficiency is easily visualized by the following example: killing a fly with a sledgehammer is effective, but not efficient.

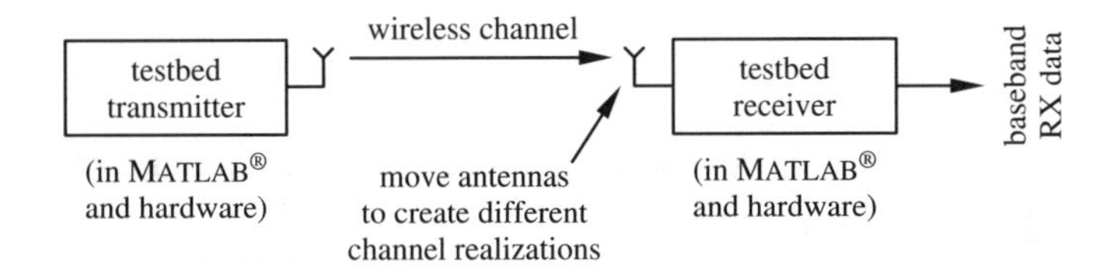

Figure 4.4 Measuring a sample of baseband data that in a subsequent step is automatically converted by a cluster of PCs into the corresponding throughput values.

(p. 155), Figure 7.3 (p. 155), and Figure 7.4 (p. 155)) are not important to understand the following sections. The techniques presented from here onwards work with virtually any MATLAB® code representing the transmitter and receiver of a communication system.

2. We carry out a pretest [4, p. 7]; that is, we collect a very few observations of the experiment outcome (See Figure 4.5).
   We do so to find and solve problems before spending a lot of time on a complete measurement, only to discover afterwards that, for example, a cable was not mounted correctly and the throughput is just zero all the time.

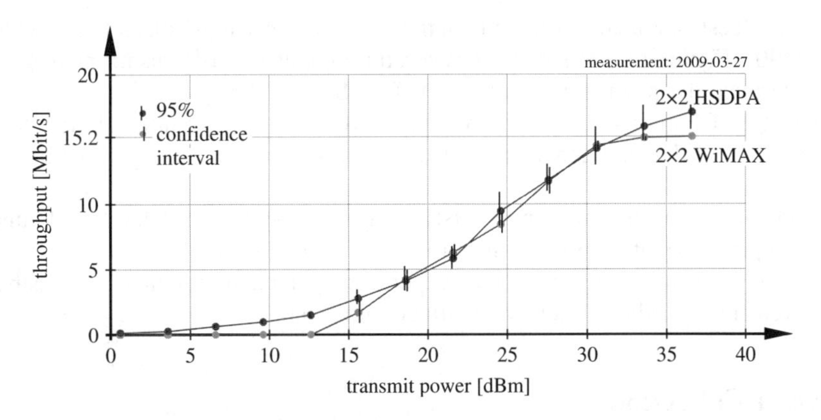

Figure 4.5 Inferred mean throughput performance of a scenario using only 20 throughput observations. By visual inspection we conclude that the measurement works.

3. We continue collecting measurement data, that is, we collect a representative sample of the population of interest (see Section 4.4, p. 80).
   In the case of Figures 4.5 and 4.6, the target population is the aggregate of all possible throughputs in a specific urban scenario.
4. After evaluating the measured baseband data by a PC cluster, we summarize and analyze the data obtained; that is, we draw inferences from the collected sample to plot graphs or tables showing estimated parameters of the target population (see Section 4.5, p. 88).
   In the case of Figure 4.6, the average throughput of the scenario is inferred from the measured sample. Having no other information than the throughputs of our sample at

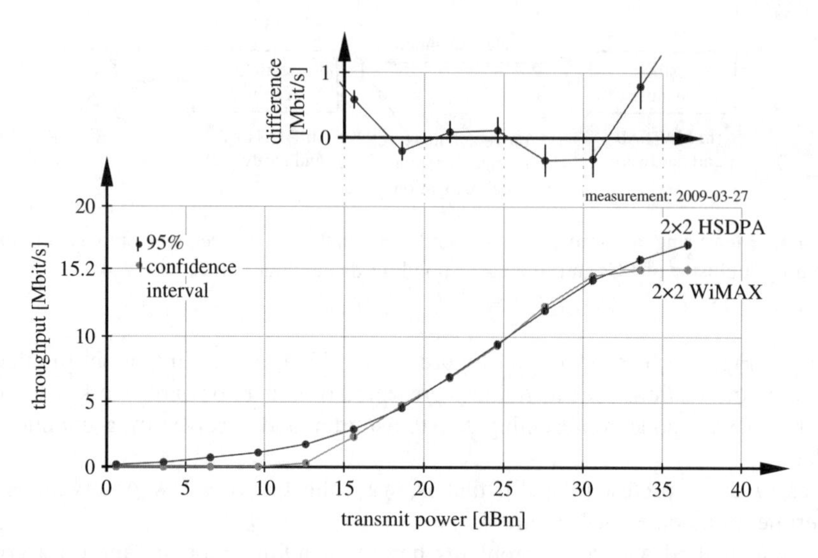

Figure 4.6 Inferred mean throughput performance using all 434 throughput observations. Note that the size of the confidence intervals is considerably smaller than in Figure 4.5.

hand, the "best" estimate for the population mean is the sample mean [14] (see Section 4.6, p. 90). Figure 4.6 plots this estimated throughput versus transmit power.

5. We determine the precision of our results (see Section 4.6, p. 88).
   In the case of Figure 4.6, we draw the 95 % BCa[3] confidence intervals for the mean; that is, 95 % of the confidence intervals are supposed to cover the population mean of interest.
6. We compare our results with previously measured results and theoretical bounds (as, for example, the mutual information) to check for plausibility.
7. If possible, we use the results and experience gained to upgrade our testbed and measurement procedure. Then we continue with Step 2.

4.4 Data Collection

Having a testbed that can transmit a given block of data, we identify the following three sources of variation in our experiment set-up, as is shown in Figure 4.7.

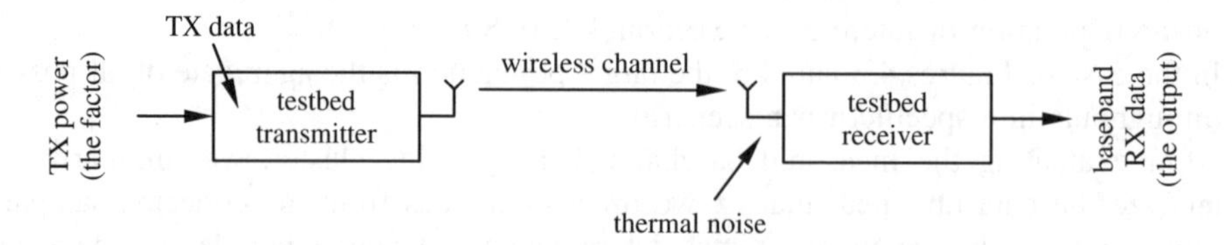

Figure 4.7 Measuring a sample.

[3] BCa stands for the bootstrap Bias-Corrected and accelerated (BCa) algorithm [7, p. 176].

TX data: the TX data to be transmitted consists of many different bits; we are interested in the performance for all combinations.

Wireless channel: the wireless channel is determined by the measurement scenario as well as the transmit and receive antenna positions; we are interested in the performance for a small area, let us say 3×3 wavelengths (i.e. $36\,\text{cm} \times 36\,\text{cm}$).[4]

Thermal noise: it disturbs the received signal; we measure in unused frequency bands to ensure no expected interference from other sources.

The output[5] of a single transmission carried out by the testbed consists of baseband data samples that can be evaluated to obtain, for example, a throughput. The input[6] to our experiment is, for example, the transmit power for which we want to infer the average throughput of a scenario.

In order to obtain parameters of the target population, one method would be to conduct a census; that is, measuring "all" possible combinations of input signals, channel realizations, and noise – or in other words, "all" items of the target population. Obviously, this is not possible. Fortunately, collecting a "representative" sample for a given transmit power is easy. Fix the transmit power and observe the throughput for:

1. a block of random transmit data (by a pseudo-random generator);
2. a random channel realization (by moving the receive antennas to a pseudo-random position and rotation within the area of interest[7]); and
3. random noise (by ensuring that there is no interference).

When collecting such a "simple random sample,"[8] the precision can then be doubled by quadrupling the sample size[9] (see Section 4.6, p. 90).

4.4.1 More Sophisticated Sampling Techniques

"The art of sampling consists in making the most efficient use of available resources so as to afford the best possible estimate concerning the quality of a population under consideration as is consistent with the ever-present limitation in time and funds" [27, p. 15)]. "The purpose of sampling theory is to make sampling more efficient" [4, p. 8].

[4] By considering only such a small area, we exclude large-scale fading effects from our investigations.

[5] The literature knows many names for the output of an experiment; for example, dependent variable, the observed values, the explained variable, the response variable, or the outcome variable [2, 4–13, 17–19, 26–28].

[6] The literature also knows many names for the input of an experiment; for example, the independent variable, the controlled variable, the explanatory variable, or the manipulated variable [2, 4–13, 17–19, 26–28].

[7] Note that we define the scenario to be investigated by the area and orientations our receive antennas can cover. Otherwise, we would not be able to sample the target population.

[8] In simple random sampling, at each draw, every possible item of the population should have the same probability of being chosen [4, p. 18].

[9] "For a random sample of size n with variance σ^2 from an infinite population, it is well known that the variance of the mean is σ^2/n" [4, p. 24].

4.4.1.1 Sampling Homogeneous Subpopulations

The first goal is stratification; that is, dividing the heterogeneous population of interest into subpopulations (so-called strata) that by themselves experience a smaller variation than does the entire population as a whole.[10] Having such homogeneous subpopulations at hand, their mean can be precisely estimated by employing only a few samples. A precise estimate for the whole population can then be obtained by combining the subpopulation means. "If intelligently used, stratification nearly always results in a smaller variance for the estimated mean or total than is given by comparable simple random sampling" [4, p. 99].

In our example, the variation of the throughput introduced by different transmit bits is much lower than the variance introduced by the different channel realizations. In the case of uncoded single carrier 4-QAM (Quadrature Amplitude Modulation) transmissions, we can even conduct a census in our subpopulation; that is, measuring all four symbols of the 4-QAM. In the case of HSDPA, simulations and experience have shown that measuring many channel realizations at the cost of only transmitting one deterministic block[11] of data does indeed reduce the variance of the sample collected. The same holds true for WiMAX.

4.4.1.2 Sampling Spatially Autocorrelated Populations

Regarding the wireless channel, simple random sampling is also not the best method. As shown in Figure 4.8, we have to deal with spatial data where the observations are correlated due to their positions in space. Simply speaking, sampling at distances below 0.2 wavelengths (2.4 cm) ceases to work efficiently because it just repeats the same values. For this case of autocorrelated populations experiencing negative exponential correlation functions (see the spatial correlograms[12] in Figure 4.8), the literature suggests the use of "systematic sampling" (that is, sampling a grid of fixed size), thus minimizing correlation [10, p. 180]. However, systematic sampling is not only an efficient procedure but also a dangerous one, especially when periodic variations exist in the area to be sampled. In the end, the choice of sampling procedure is a tradeoff between the loss of precision due to correlation and possible errors introduced by the systematic sampling approach [4, p. 221]. See [10, p. 180 *et seq.*] for a detailed analysis of this topic.

To measure, for example, 324 different throughputs in an area of 3×3 wavelengths ($36 \times 36\,\text{cm}^2$), we utilize a fully automated XYΦ positioning table that can move the receive antennas along its x and y axes, as well as rotate them (see Figure 4.9).

- To reduce vibrations created by very fast moving/accelerating antennas, we employ a very heavy, low-profile positioning table.
- To minimize the effects of possible periodic variations, we make sure that a wavelength is not an integral multiple of the x and y position increments. In our case this is 0.17

[10] Stratification may be used for other reasons, variance reduction being just one of them.

[11] Measuring less than a complete block is not possible because of the coding that requires a complete block to be transmitted over a constant channel.

[12] A correlogram plots the correlation of items some distance (x-axis) apart over this distance.

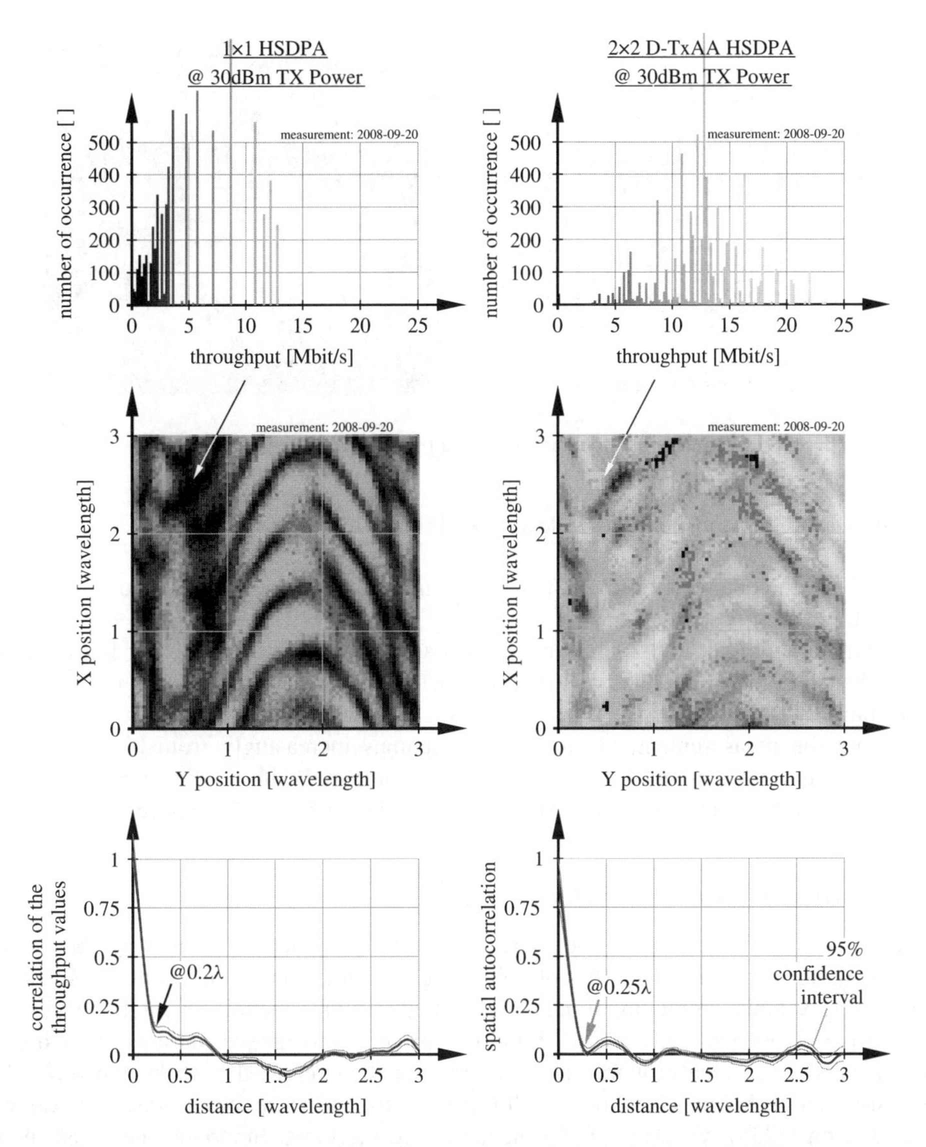

Figure 4.8 To visualize the spatial correlation observed in 1 × 1 and 2 × 2 HSDPA, we systematically sampled (see also Figure 4.9 p. 84) an area of 3 × 3 wavelengths to obtain 7360 throughput observations. The top of the figure shows two histograms, the middle two *XY*-plots, and the bottom spatial correlograms (see [3] on how they are constructed) for these measured throughput values. Note that, for distances larger than a quarter of a wavelength, the throughputs turn out to be uncorrelated ($|\rho| < 0.2$).

wavelengths, where the 0.17 is arbitarily chosen. Therefore, to measure in an area of 3.06 × 3.06 wavelengths, we need 18 × 18 measurements. Then we position the table such that its axes do not run parallel with the walls of the room. This helps to determine later whether the metal of the table has influenced the measurement.

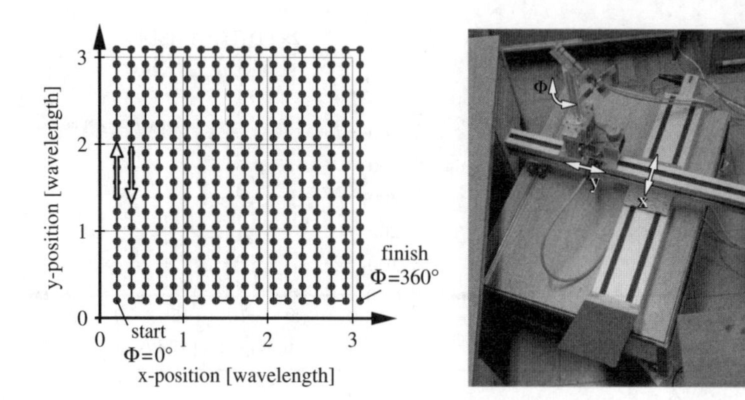

Figure 4.9 Systematically sampling an area of $3 \times 3\,\lambda$.

We measure 324 positions as follows (see Figure 4.9):

- We move the antennas on the XYΦ positioning table to the reset position of $x = 0\,\lambda$, $y = 0\,\lambda$, and $\Phi = 0$.
- Starting at an offset of $0.2\,\lambda$ (owing to mechanical reasons specific to our linear guide), we first move the antennas in the direction of the linear guide on the top to reduce vibrations.
- During the measurement, we rotate the antennas increasingly from 0° to 360°. As the rotation unit employed by us is very slow compared with the linear guides, we are forced to only utilize such small rotation increments to keep the measurement time low.

4.4.2 Variance Reduction Techniques

When inferring, for example, the mean of a target population, increasing the sample size is by no means the only way of enhancing the precision of the estimate. As shown above, for example, employing more sophisticated sampling techniques might be an easier alternative. More generally, D. C. Handscomb calls a technique variance reducing "if it reduces the variance proportionately more than it increases the work involved" [26]. Because such techniques have been well known in the literature for decades (for example, see [8] from 1935), we only briefly mention here the most important ones used in our experiments.

Comparison: As it is very hard to reproduce measurement results exactly (see Section 4.4.3, p. 85, on bias), comparing different communication schemes is far more accurate [5, p. 23].
For example, we compare 2×2 HSDPA with 2×2 WiMAX in Figure 4.6 (p. 80). Note that the confidence intervals shown in this figure only account for the precision of the estimates, not for their accuracy.

Local control: This includes transmitting the schemes to be compared ("grouping") immediately after each other over the same channels ("blocking"), equally often ("balancing") [19, p. 316].

For comparing 2 × 2 HSDPA with 2 × 2 WiMAX in Figure 4.6 this means that we transmit WiMAX and HSDPA for different transmit powers over the same different channels, and not in some other order. The goal of local control is to keep the measurement conditions as similar as possible for the different schemes to be compared over time, since, for example, snow may fall during the measurement and change the channel (see Section 4.6.3, p. 92, on reproducibility).

Randomization: At first, randomizing the order of the transmitted blocks allows for "time trends to average out" [13, p. 5]. In addition, randomization ensures that all transmitted blocks face (on average) the same measurement conditions (for example, regarding training, antenna swing,[13] and interference from adjacent blocks [5, p. 19]). The third reason for randomization is to avoid the measurement fitting some pattern in the uncontrolled variation and producing systematic errors [5, p. 74)].
For comparing 2 × 2 HSDPA with 2 × 2 WiMAX in Figure 4.6, p. 80, this means that we randomly shuffle the order of the adjacent HSDPA and WiMAX blocks. For the comparison of different single carrier schemes, this means that we also shuffle the blocks (possibly around a training sequence) in a random way (see Figure 4.11, p. 87). Nevertheless, even if in principle the best would be to shuffle everything, we did not always do so in order to keep things simple and avoid errors (which are really hard to detect).

Factorial experiments: When evaluating the influence of several factors on a transmission, the use of one-factor-at-a-time experiments[14] is not advisable [19, p. 316], as the factors may jointly influence the response.
See Chapter 7, p. 153, on experiments carried out jointly investigating transmit power, antenna distance, and antenna polarization.

Concomitant observations: A supplementary (concomitant) observation that is correlated with the observations of interest can be used to increase precision [5, p. 48].
For example, see Section 7.4.1 (p. 157) on how we corrected for the average path loss of Multiple-Input Multiple-Output (MIMO) transmissions by the corresponding Single-Input Multiple-Output transmissions.

4.4.3 Bias

Systematic errors, or so-called bias, "affect all measurements in the same way, pushing them in the same direction" [9, p. 103]. This fundamental property is, on the one hand, a real godsend, as it does not introduce any error when looking at the difference between two measured communication systems – and that is what we do.[15] On the other hand, a bias is hard to detect. The only way is to compare the measurement result with some external standard, such as a measured throughput curve with a mutual information curve, a simulation result, or a result obtained by different measurement equipment. In any case,

[13] For example, when moving 25 kg antennas quickly from one measurement position to the next one, they will never be absolutely static.
[14] For example, only investigating antenna distance while keeping the transmit power and polarization constant.
[15] The absolute position of the curves in the figures plotted is never really of interest for us. What we are interested in is their relative positions. A bias would shift all curves alike, thus not changing their relative positions.

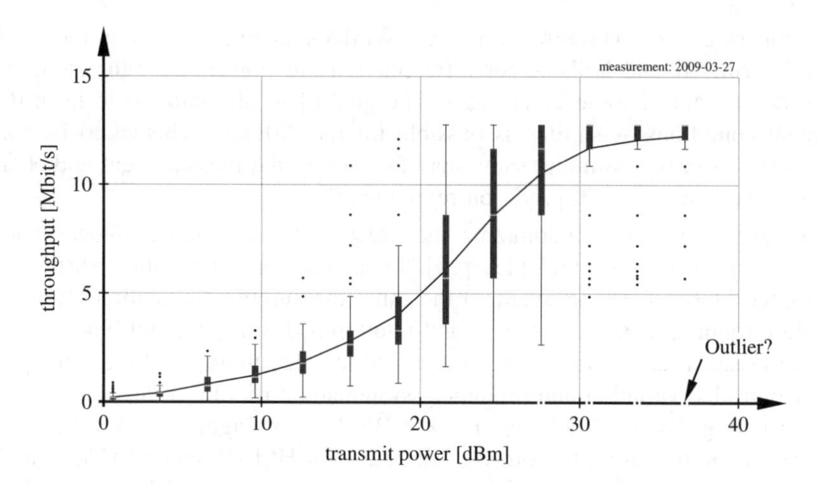

Figure 4.10 Boxplot of 434 samples taken at different transmit power levels. Note the marked zero-throughput observation.

detecting biases is a very demanding, yet necessary endless task.[16] Thus, it is better to start measuring uncoded single-carrier Single-Input Single-Output (SISO) schemes over a coaxial cable instead of wireless channels.

4.4.4 Outliers

An outlier is "an observation which derivates so much from the other observations as to arouse suspicions that it was generated by a different mechanism" [12, p. 1]. For example, the marked observation in Figure 4.10 is more than 10 standard deviations away from the mean. Such an extreme of a measurement result should, at least, attract our attention.

Outliers can arise due to errors in the measurement or its execution [2, p. 28]. Possible causes may be:

- Equipment malfunction.
- Unexpected, typically nonrepeating events, such as a door slamming, people moving through the room, or window washers climbing on the facade in front of the antennas. Such errors can be detected by examining webcam pictures or prevented by blocking access to the measurement site.
- Interference from other transmitters in the same frequency band.
  The way we detect undesired interference is remarkably simple, yet effective, and not dependent on the data transmitted. As shown in Figure 4.11, we do not transmit anything immediately adjacent to the data frames. These "noise gaps" can then be used in the evaluation to detect deviations from the known[17] noise power, thus identifying interference that is usually not shorter than a transmitted block.

[16] A piece of advice for those wishing to build a testbed by themselves: do not underestimate the work needed to test and validate a testbed; it takes "at least" a year.

[17] Since noise is thermal, its power is expected to remain constant during the course of a measurement.

- Errors in the MATLAB® code.
  Since the received data samples are stored for offline evaluation, it is easy to step through the MATLAB® code of the receiver in order to find possible errors.[18]

On the other hand, outliers may also arise due to inherent variability in the measurement result [2, p. 27]; that is, the population measured is simply not as homogeneous as we might believe.

In the case of Figure 4.10, the marked outlier may at first glance look like measurement error. We searched for its cause to find possible flaws in our measurement setup but could not find one; therefore, we did not reject the zero throughput observation, as it might be a legitimate part of the measurement result.[19]

4.4.5 Parameter Estimation

In order to obtain realistic estimates we need training that is separate from our data. Ideally, this training would come in the form of some new data from the same population that produced the original samples. If the transmission scheme itself does not include a preamble, we additionally transmit a known training sequence (see Figure 4.11). Employing the data to be received in a genie-driven procedure to estimate the channel, which is then used for maximum likelihood decoding, is a dangerous procedure (because the genie-driven knowledge can lead to unrealistic results).

In order to estimate the Signal to Noise Ratio (SNR) at the receiver independently of the transmission scheme (nonparametric) we obtain two intermediate estimates for each block received (index i):

- $P_{i,\text{signal+noise}}$, the average power of the signal received; that is, also including noise.
- $P_{i,\text{noise}}$, the power of the noise estimated in the noise gaps before and after the signal (see Figure 4.11).

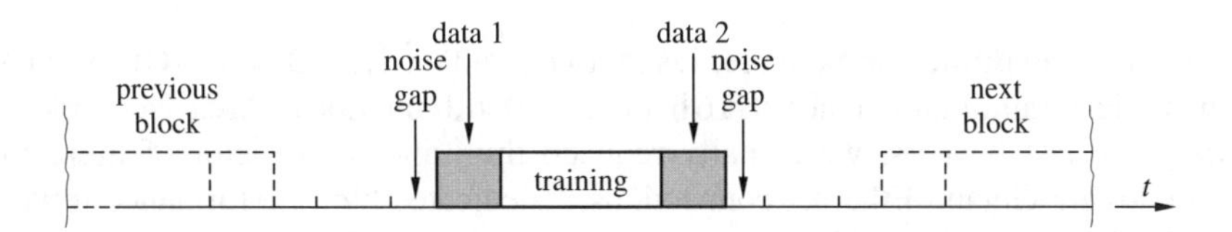

Figure 4.11 Obeying the principles outlined in Section 4.4.2 (p. 84), we use the data transmitted to compare the throughput of “data 1” and “data 2” (comparison) by transmitting them immediately next to each other (blocking), randomly shuffling their order (randomization).

[18] Some errors do not affect pure simulations, especially those arising from unanticipated real-world channel conditions.

[19] A possible explanation. In the scenario we measured 1×1 HSDPA. This communication system is designed to have a Block Error Ratio of 10 %. One block may be retransmitted twice, resulting in about 1 ‰ of blocks with zero throughput. At 434 blocks measured, the probability of observing an “outlier” is in the order of 50 %.

These intermediate estimates are then averaged over all blocks transmitted. Since noise and data are uncorrelated, the estimated average SNR can be easily calculated by

$$\widehat{\mathrm{SNR}} = \frac{\widehat{P}_{\mathrm{signal}}}{\widehat{P}_{\mathrm{noise}}} = \frac{\mathrm{average}(\widehat{P}_{i,\mathrm{signal+noise}}) - \mathrm{average}(\widehat{P}_{i,\mathrm{noise}})}{\mathrm{average}(\widehat{P}_{i,\mathrm{noise}})}. \tag{4.1}$$

Note that the averaging has to be carried out before the nonlinear division (see J. L. W. V. Jensen from 1906 [16]).

4.5 Evaluating and Summarizing the Data

To evaluate our measurements, we employ a self-written cluster software that is set up for the parallel processing of large data sets on standard PCs. The basic idea is simple:

- During the measurement, the complex-valued received data is collected at the receiver site in the form of one large file per receive antenna position (for example, 434 files requiring 1.5 GB each). In addition, a central server is informed each time data for a new receive position becomes available.
- Employees leaving their workplace at night start a simple client program for every processor core they want to make available for carrying out our evaluations. This client program then contacts the server, checks for the availability of a new receive antenna position, copies the data required for the evaluation, evaluates it, and finally writes the results of the evaluation back to the server.
- The progress of the current evaluation can easily be checked at the central server. Furthermore, the results of the previously calculated receive antenna positions can be gathered at any time during the calculation, so the data can be easily pre-tested, pre-analyzed, and pre-summarized.

During the last 3 years, we have identified two critical bottlenecks in this procedure. Depending on the type of evaluation required, their influence varies greatly (for example, many different channel estimators using the same received data or very simple receivers each using different receive data):

Time to copy the data: For example, just copying 500 × 1.5 GB = 750 GB of data takes a theoretical minimum of about 16 h on a 100 Mbit/s Local Area Network (LAN). During the last 3 years, we actually reduced this time by a factor of more than 10 by employing Gigabit Ethernet connections,[20] dedicated high-performance servers, and optimized server-to-client data transfer.

Calculation time: Here, things are straightforward, as the evaluation can be split over several computers on a position-by-position basis. Doubling the number of clients available halves the time required for evaluation. In typical numbers this means that 50 clients need 1 day for an evaluation that would otherwise require about 50 days on a single core.

[20] Note that employing Gigabit Ethernet instead of 100 Mbit/s Ethernet only allows for an approximate fivefold increase in speed in practice, due to excessively slow client computer components. This limitation was overcome by transferring data to several clients simultaneously.

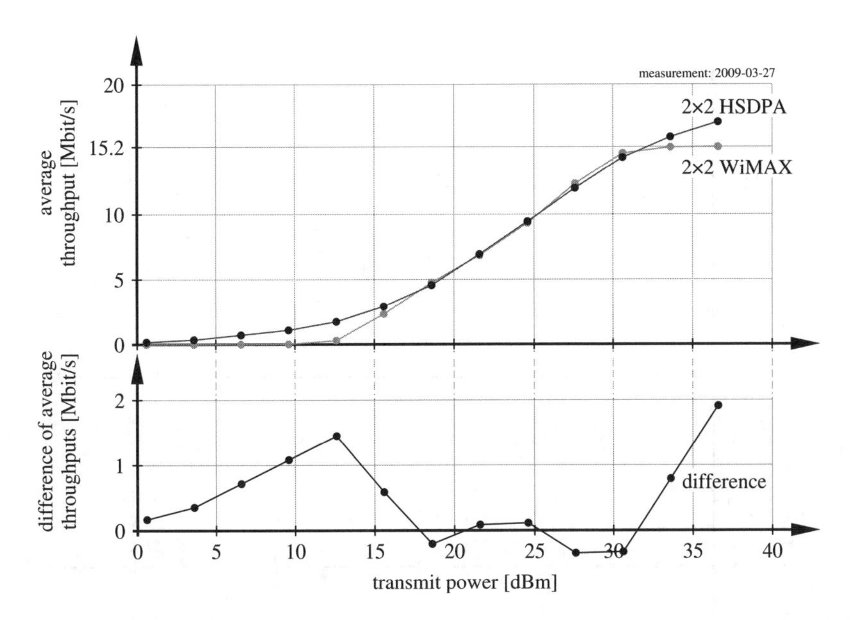

Figure 4.12 Average throughput of the measured HSDPA and WiMAX sample (and the corresponding difference) over transmit power.

The evaluation results stored on the server are then automatically combined to calculate the desired result. For example, gathering the calculated throughputs at each receive antenna position or (in other words) channel realization and transmit power produces, averaged over the 434 receive antenna positions measured, the graph shown in Figure 4.12. As average values tell us only little about the distribution of the sample, cumulative probability functions (Figure 4.13) can be drawn to provide further insight.

Since HSDPA offers 256 [1, Table 7I, p. 54] different modulation and coding schemes in contrast to seven [15, Section 8.3.3.2.3] in the case of WiMAX, the HSDPA curve shows a much finer granularity. The corresponding boxplot is shown in Figure 4.14.

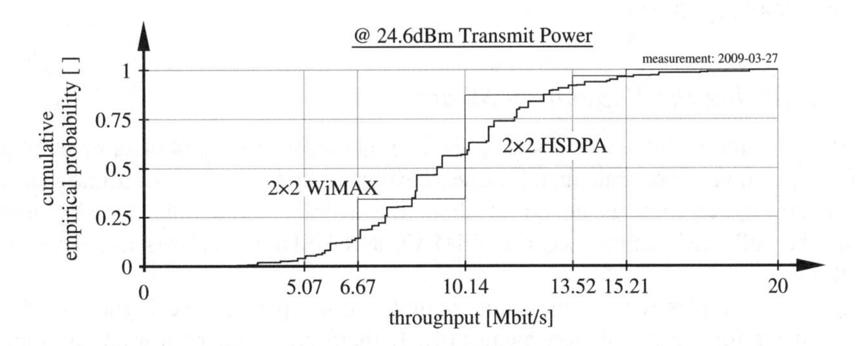

Figure 4.13 Cumulative empirical probability of the 434 HSDPA and WiMAX samples measured at a transmit power of 24.6 dBm.

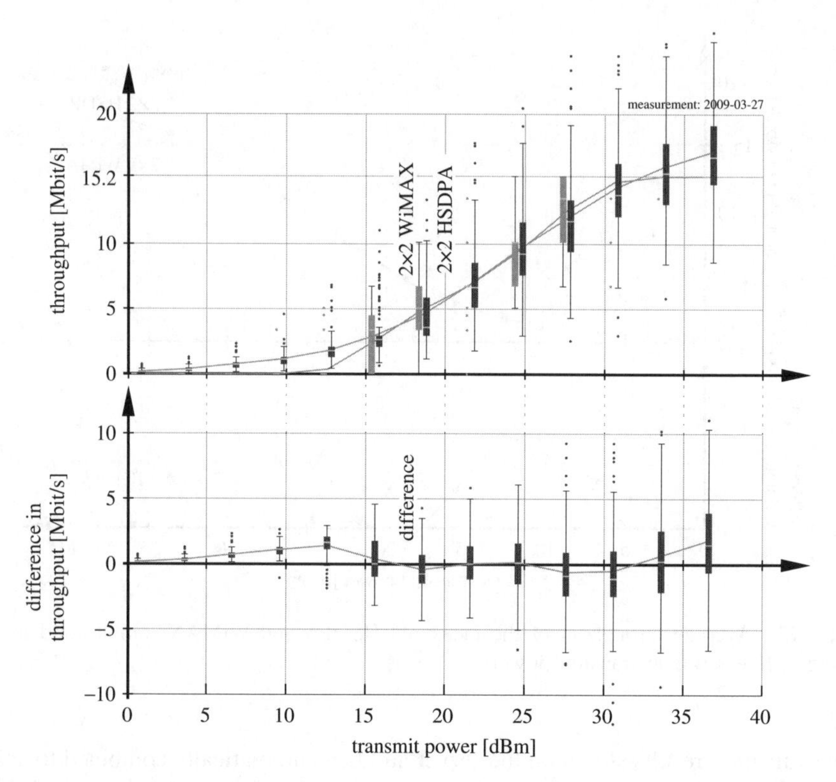

Figure 4.14 Boxplot of the HSDPA and WiMAX throughputs measured for the same transmit powers (the boxplots have deliberately been shifted to avoid overlap).

4.6 Statistical Inference

Once a representative sample of the population of interest has been collected, we are not only interested in summarizing this sample, but also in inferring population parameters. That is the goal of statistical inference – to say what we have learned about the population from the sample [7, p. 18].

4.6.1 Inferring the Population Mean

The left-hand side of Figure 4.15 shows the average throughput of 434 samples obtained per transmit power. The calculated value is exact (which is why no confidence intervals are drawn), as there is no doubt about the average throughput actually achieved "given" the collected sample because WiMAX and HSDPA receivers use deterministic algorithms.

If, now, all samples receive the same weight, the sample average is the "only" unbiased estimator for the population mean [14]. If there is no other information about the population than the sample obtained, this estimated mean also "cannot be improved on" (see plug-in principle [7, p. 37]).

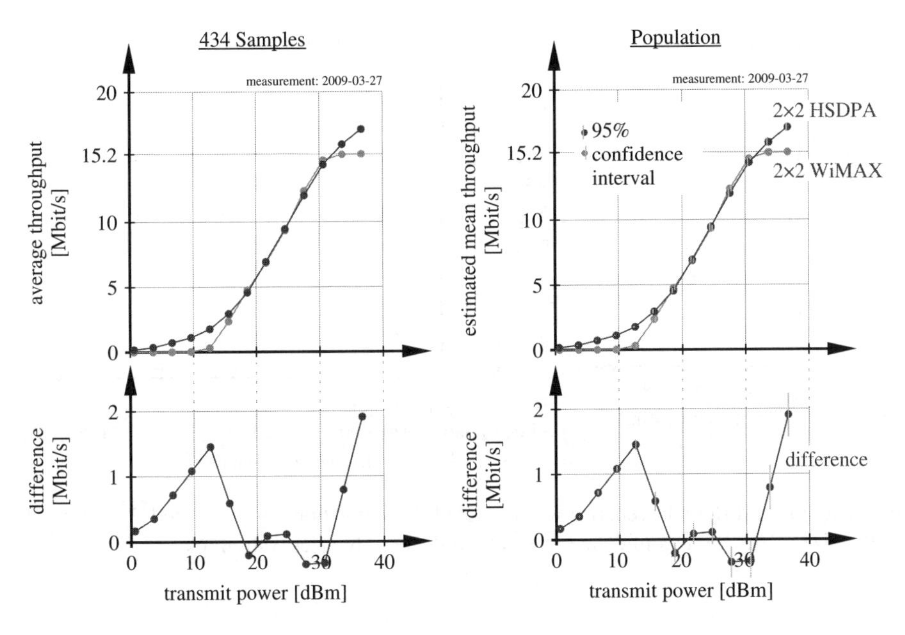

Figure 4.15 Sample average = estimated population mean.

4.6.2 *Precision and Sample Size*

In complicated situations, gauging the uncertainty of an estimate based on an assumed probability model is "tedious and difficult." In addition, inappropriate simplifications and assumptions may lead to potentially misleading results [6, p. 1]. Bootstrapping is a recently developed technique for avoiding such calculations. It is "a computer-based method for assigning measures of accuracy to statistical estimates," thereby "aiming to carry out familiar statistical calculations such as confidence intervals in an unfamiliar way: by purely computational means, rather than through the use of mathematical formulae" [7, p. 10, p. 392]. "The nonparametric bootstrap[21] is a fairly crude form of inference that can be used when the data analyst is either unable or unwilling to carry out extensive modeling. Nonparametric bootstrap inferences are asymptotically efficient – that is, for large samples they give accurate answers no matter what the underlying population" [7, p. 395]. Nevertheless, the real power of the bootstrap lies in the analysis of small data sets from an unknown population – and that is what we do here.

Simply speaking, the bootstrap (through resampling of the few observations obtained) does what we "would do in practice if it were possible: repeat the experiment" [28, p. 4]. The new sets of observations thus created are then used in a subsequent step to calculate, for example, confidence intervals. The inferences of the thus-used percentile-t, BCa, or ABC algorithm [7] "are not perfect, but they are substantially better than most in the other traditional inference methods" [7, p. 393]. As programs such as MATLAB® are already

[21] Experience has shown that the use of a nonparametric bootstrap approach is more accurate than a parametric bootstrap approach with incorrect model assumptions [28, p. 28].

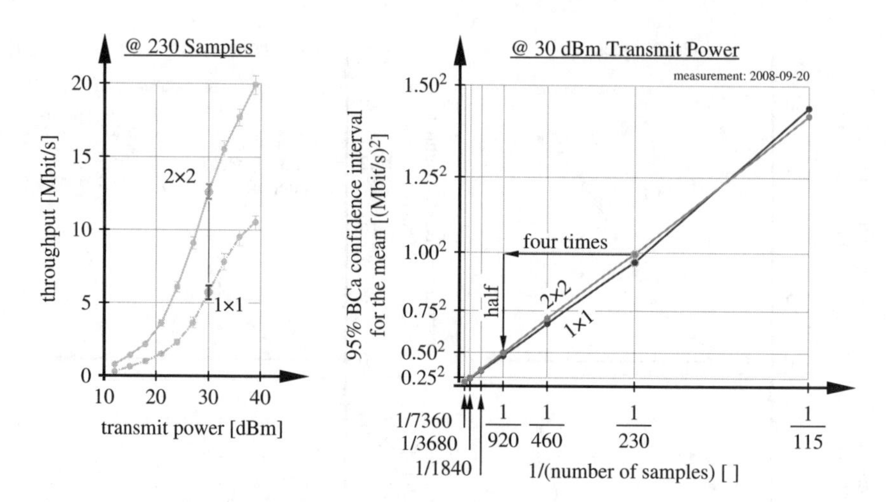

Figure 4.16 In a different scenario, we measured 7360 positions of 1×1 and 2×2 HSDPA to relate the size of the 95 % confidence intervals to differently sized subsamples.

equipped with ready-to-use functions,[22] we will skip the details of the algorithms and refer the interested reader to [7] for additional theoretical background.

Being able to gauge the precision of the results is important in order to determine the number of measurements needed to reach the desired degree of precision. Looking back at the 434 observations measured in Figure 4.15, the right-hand side also shows next to the estimated population mean (that is, the sample average) the corresponding 95 % bootstrap BCa confidence intervals (that is, 95 % of the confidence intervals are supposed to cover the true, for us unknown, population mean[23]).

As the variance of the mean decreases as the reciprocal to the number of samples, so does the square root of the confidence interval size[24] (see Figure 4.16). For the 230 samples measured, we obtain the precision shown in Figure 4.17. In the case of a narrowband single-carrier 4-QAM transmission, about 8610 receive antenna positions are needed to obtain the same relative precision of about 10 % (see Figure 4.18).

4.6.3 Reproducibility and Repeatability

To judge the reproducibility[25] of our measurements, consider the following outdoor-to-indoor experiment carried out in the inner city of Vienna, Austria:

[22] The MATLAB® function for the BCa algorithm is `bootci()`, whilst for the accelerated ABC algorithm a built-in function does not exist but can be downloaded from the internet. The R function for the BCa algorithm is `boot.ci{boot}`. For the accelerated ABC algorithm it is `abc.ci{boot}`.

[23] Because "the chances are in the sampling procedure, not in the parameter," it would be wrong to say that the true population mean is covered by a given confidence interval with a probability of 95 % [9, p. 384].

[24] The size of the range that the confidence interval covers.

[25] Reproducibility: "The closeness of agreement between independent results obtained with the same method on identical test material but under different conditions (different operators, different apparatus, different laboratories and/or after different intervals of time)" [20].

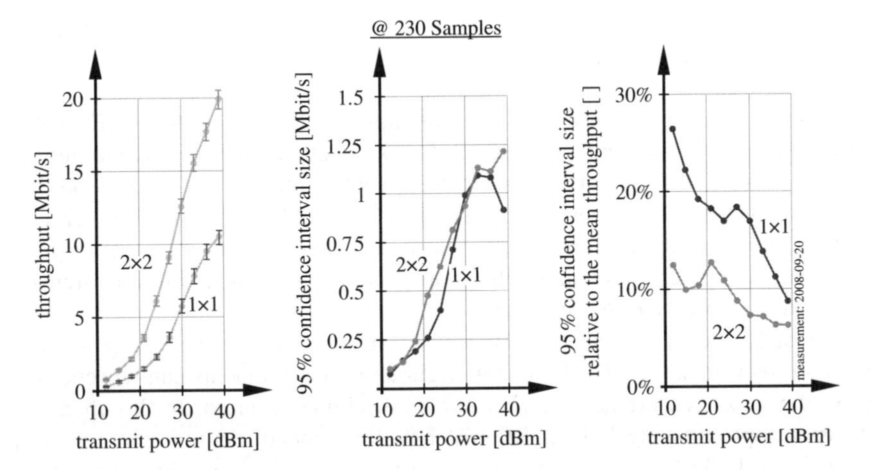

Figure 4.17 For the scenario shown in Figure 4.16, p. 92, we plot the absolute and relative sizes of the 95 % confidence intervals over transmit power.

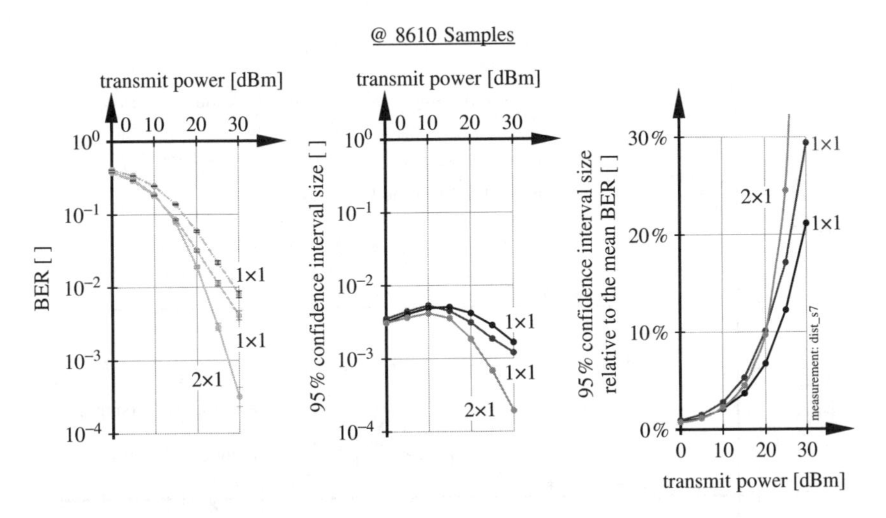

Figure 4.18 For narrowband indoor single-carrier 4-QAM SISO on transmit antenna 1 and 2 as well as 2 × 1 Alamouti we evaluate the absolute and relative size of the 95 % confidence intervals for the mean BER over transmit power.

- Three equal measurements, each lasting 12 h, evaluating many different HSDPA schemes – we only look at a single result, namely the throughput of 2 × 2 Double Transmit Antenna Array (D-TxAA) HSDPA.
- The first measurement was commenced in good weather conditions. After about 10 h, towards the end of the first measurement, a snowstorm covered the dry roofs and streets with a 3 cm thick layer of wet snow.
- The second measurement was carried out when the snowstorm had already ended, although the wet snow remained.

- The third measurement was carried out another 12 h later, under weather conditions similar to measurement number two.

Figure 4.19 shows the resulting mean throughput of 2 × 2 D-TxAA HSDPA obtained from these three measurements. The lower two subplots show the absolute and relative differences between the two later measurements carried out under similar conditions.

Variance reduction techniques, as explained in Section 4.4.2 (p. 84), are therefore necessary to combat undesired influences such as drifts and changes in the environment. Simply speaking, if repeated randomly shortly after each other, different transmission schemes can be more precisely compared.

Unfortunately, we did not measure two equal transmission schemes in any of the above-mentioned urban scenarios. To show how accurately measurements can be repeated[26] we have to refer to a measurement carried out in an alpine scenario. In this measurement, which lasted approximately 12 h, an RF switch malfunctioned and, therefore, we measured the same 2 × 2 D-TxAA HSDPA transmission twice (see Figure 4.20). Note that as the observations of the two measured curves are highly correlated, the confidence intervals for their difference are much smaller than suggested by the confidence intervals for the absolute values.

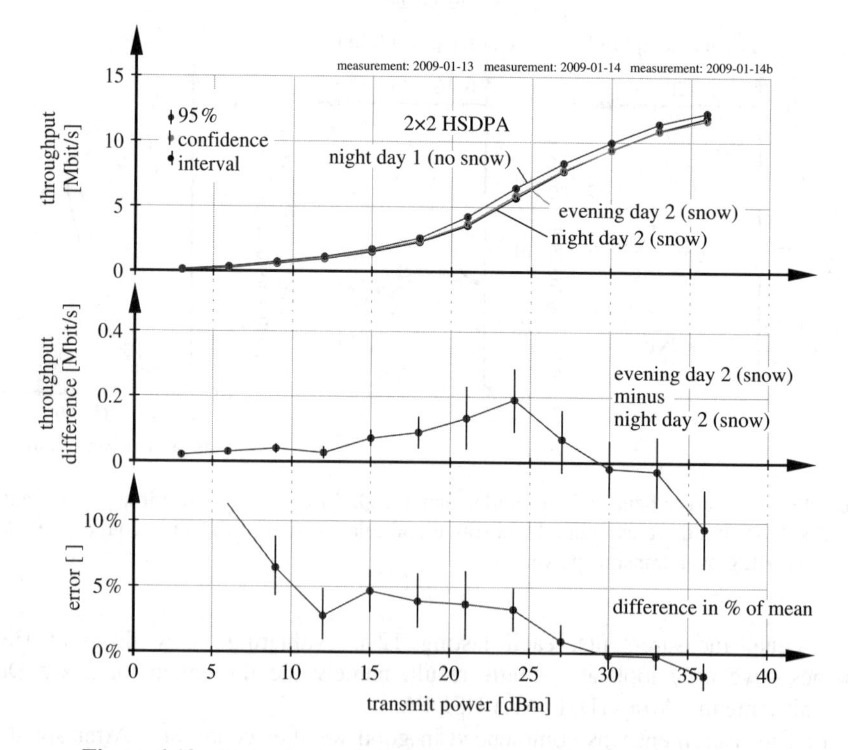

Figure 4.19 Reproducing the same measurements 12 and 24 h later.

[26] Repeatability: "The closeness of agreement between independent results obtained with the same method on identical test material, under the same conditions (same operator, same apparatus, same laboratory and after short intervals of time)" [20].

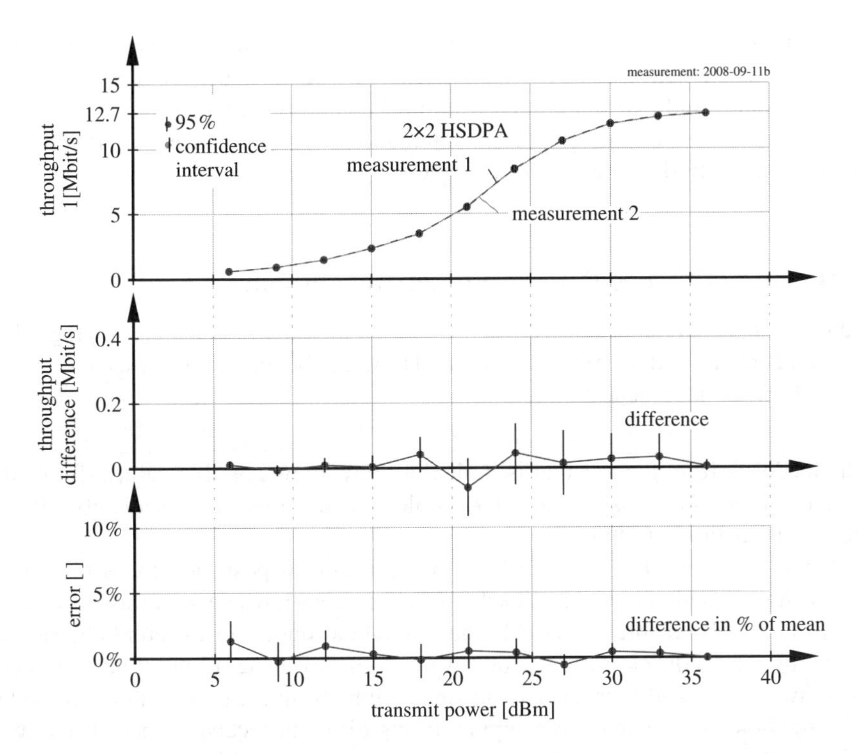

Figure 4.20 Repeating the same measurement at the same receive positions.

4.7 Measurement Automation

As already noted, the whole measurement, evaluation, and graph-plotting procedure is carried out by a single, fully automated, script. This also includes elements whose value is sometimes underestimated, such as:

- taking webcam pictures at the beginning and during the measurement;
- constantly monitoring temperature and humidity at the transmit and receive sites;
- texting the operator in charge if there is an important issue, such as a linear guide getting stuck; and
- archiving the complete source codes used for this particular measurement and for every single evaluation carried out using the measurement data.

At first sight, automating absolutely everything seems to be a waste of resources, particularly time. In addition, automation by itself also offers no protection from errors in the measurement procedure and setup. On the other hand, scripts execute the programmed steps with 100 % certainty. Even if this does not imply that these steps are correct, the scripts can be gradually improved over time to cover every possible flaw in the measurement procedure.

Humans executing steps manually (at best from a checklist) behave differently. In particular, they tend to take shortcuts, either because they are lazy or try to be smart. For example, on one occasion we did not archive the source code of a measurement to save time because we thought that all the parameters were equal to the previous measurement anyway. Now, we find that there is no back-up of some important parameters.

To sum up: in our opinion, "complete" automation is the only means of creating repeatable and consistently documented measurements.

4.8 Dealing with Feedback and Retransmissions

As radio channels stay constant over short periods of time, systems employing adaptive channel adjustment, such as WiMAX and HSDPA, can be efficiently measured by one of the two following approaches:

- If the feedback from the receiver is limited to only a few possible values (for example, seven in WiMAX), transmitting all possible transmit blocks immediately after each other is the method of choice.
  As shown in Figure 4.21 for WiMAX, we transmit all possible blocks one after the other without calculating any feedback information. An advantage of this method is that we will later still be able to evaluate the throughput impact of different algorithms for calculating the feedback. Furthermore, we can also evaluate all possible combinations of one, two, three, and four receive antennas offline from the same set of recorded data. The same holds true when, for example, trying different receiver types. The downside of this method is that the received data increases linearly with the number of possible transmit blocks.
- If the number of possible transmit blocks becomes too large (for example, a few thousand in MIMO HSDPA because of precoding and feedback), transmitting all of them may no longer be feasible. The method of choice is then to employ the feedback in a quasi-real-time fashion. Therefore, we first pre-generate all possible transmit blocks at the transmitter before the measurement is started. At every single transmission, a "previous block" is transmitted first. Next, the receiver calculates "just" the feedback bits in MATLAB®, which requires considerably less computational effort than evaluating the complete received block. The resulting transmit block number is then used to trigger the transmitter (see Figure 4.22). Note that the throughput-evaluation of the data blocks is still carried out later offline.

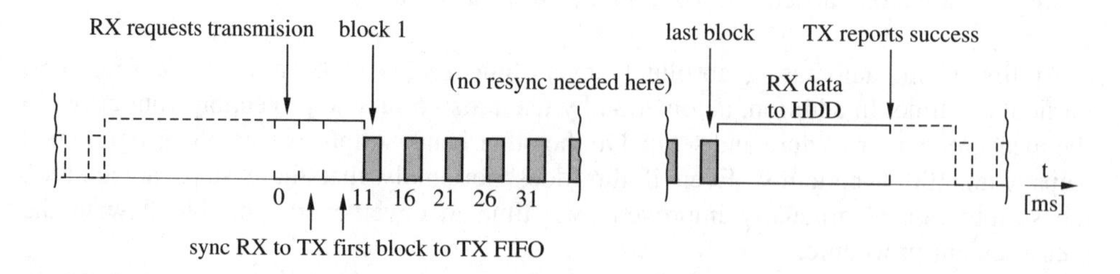

Figure 4.21 Measuring all possible transmit blocks.

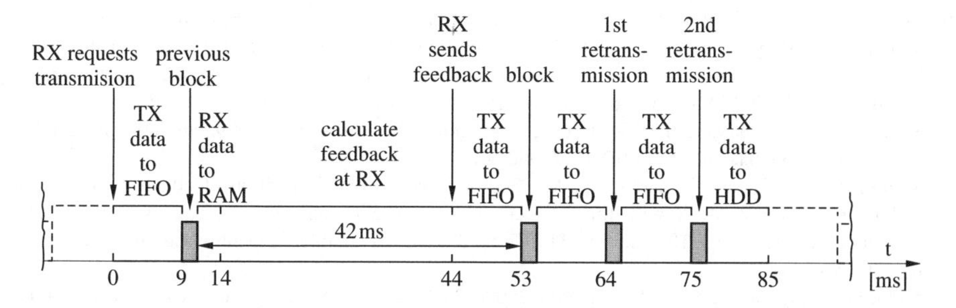

Figure 4.22 Calculating the feedback in MATLAB® to transmit pre-generated transmit blocks and the possibly required feedback via an LAN connection. If the feedback is carried out via a wireless LAN bridge or similar, about 4 ms has to be added for the feedback time.

If a communication system requires the retransmission of data blocks in case of an erroneous detection at the receiver (as is also the case in HSDPA), we always transmit all possible retransmissions (see Figure 4.22). If required, we evaluate them later offline.

References

[1] 3GPP (2007) Technical Specification TS 25.214 version 7.7.0 'Physical layer procedures (FDD),' www.3gpp.org.
[2] Barnett, V. and Lewis, T. (1980) *Outliers in Statistical Data*, first edition, John Wiley & Sons, Ltd.
[3] Bjørnstad, O. and Falck, W. (2001) 'Nonparametric spatial covariance functions: estimation and testing,' *Environmental and Ecological Statistics*, **8** (1), 53–70. Available from http://www.springerlink.com/content/j1436865g74041r2/.
[4] Cochran, W. G. (1977) *Sampling Techniques*, third edition, John Wiley & Sons, Inc.
[5] Cox, D. (1958) *Planning of Experiments*, John Wiley & Sons.
[6] Davison, A. C. and Tibshirani, R. J. (1999) *Bootstrap Methods and Their Application*, first edition, Cambridge Series in Statistical and Probabilistic Mathematics, No. 1, Cambridge University Press.
[7] Efron, B. and Hinkley, D. V. (1994) *An Introduction to the Bootstrap*, first edition, CRC Monographs on Statistics & Applied Probability 57, Chapman & Hall.
[8] Fisher, R. (1935) *The Design of Experiments*, John Wiley & Sons, Inc., New York.
[9] Freedman, D., Pisani, R., and Purves, R. (2007) *Statistics*, fourth edition, Norton.
[10] Haining, R. P. (1993) *Spatial Data Analysis in the Social and Environmental Sciences*, first edition, Cambridge University Press.
[11] Hammersley, J. M. and Handscomb, D. C. (1979) *Monte Carlo Methods*, first edition, Chapman and Hall.
[12] Hawkins, D. (1980) *Identification of Outliers*, first edition, Springer.
[13] Hicks, C. R. (1993) *Fundamental Concepts in the Design of Experiments*, fourth edition, John Wiley & Sons.
[14] Horvitz, D. G. and Thompson, D. J. (1952) 'A generalization of sampling without replacement from a finite universe,' *Journal of the American Statistical Association*, **47** (260), 663–685. Available from http://www.jstor.org/stable/2280784.
[15] IEEE (2004), 'IEEE standard for local and metropolitan area networks; part 16: air interface for fixed broadband wireless access systems', IEEE Std. 802.16-2004. Available from http://standards.ieee.org/getieee802/download/802.16-2004.pdf.
[16] Jensen, J. L. W. V. (1906) 'Sur les fonctions convexes et les inégalités entre les valeurs moyennes,' *Acta Mathematica*, **30** (1), 175–193, doi: 10.1007/BF02418571.
[17] Kleijnen, J. P. C. (1974) *Statistical Techniques in Simulation*, first edition, Dekker.
[18] Lindner, A. (1953) *Planen und Auswerten von Versuchen*, first edition, Birkhauser.
[19] Mason, R. L., Gunst, R. F., and Hess, J. L. (2003) *Statistical Design and Analysis of Experiments*, second edition, Wiley-Interscience.

[20] McNaught, A. D. and Wilkinson, A. (1997) *Compendium of Chemical Terminology: IUPAC Recommendations: Gold Book*, second edition, IUPAC International Union of Pure and Applied Chemistry.
[21] Mehlführer, C., Caban, S., García-Naya, J. A., and Rupp, M. (2009) 'Throughput and capacity of MIMO WiMAX,' in *Conference Record of the 43rd Asilomar Conference on Signals, Systems and Computers*, Pacific Grove, CA, doi: 10.1109/ACSSC.2009.5469848. Available from http://publik.tuwien.ac.at/files/PubDat_178050.pdf.
[22] Mehlführer, C., Caban, S., and Rupp, M. (2008) 'Experimental evaluation of adaptive modulation and coding in MIMO WiMAX with limited feedback,' *EURASIP Journal on Advances in Signal Processing*, **2008**, article ID 837102, doi: 10.1155/2008/837102. Available from http://publik.tuwien.ac.at/files/pub-et_13762.pdf.
[23] Mehlführer, C., Caban, S., and Rupp, M. (2009) 'MIMO HSDPA throughput measurement results in an urban scenario,' in *Proceedings of VTC 2009 Fall*, Anchorage, AK. Available from http://publik.tuwien.ac.at/files/PubDat_176321.pdf.
[24] Mehlführer, C., Caban, S., and Rupp, M. (2010) 'Measurement-based performance evaluation of MIMO HSDPA,' *IEEE Transactions on Vehicular Technology*, **59** (9), 4354–4367, doi: 10.1109/TVT.2010.2066996. Available from http://publik.tuwien.ac.at/files/PubDat_187112.pdf.
[25] Merriam-Webster (2005) *The Merriam-Webster Dictionary*, first edition, Merriam Webster, Inc.
[26] Naylor, T. H. (1968) *The Design of Computer Simulation Experiments*, first edition, Symposium on the Design of Computer Simulation Experiments, Duke University.
[27] Schumacher, F. X. (1942) *Sampling Methods in Forestry and Range Management*, first edition, Duke University.
[28] Zoubir, A. M. and Iskander, D. R. (2004) *Bootstrap Techniques for Signal Processing*, first edition, Cambridge University Press.

Part Three

Experimental Link-Level Evaluation

Introduction

Part Three reports in six chapters on the experimental and simulative evaluation of cellular systems. We show various results, mainly utilizing High-Speed Downlink Packet Access (HSDPA) and Long-Term Evolution (LTE) standards, but also to a certain extent a comparison with Worldwide Inter-operability for Microwave Access (WiMAX). Ideally, the performance of communication systems such as HSDPA should be evaluated for general wireless channels. This is typically accomplished by two approaches: simulation and measurement. Both approaches have their individual pros and cons that are briefly discussed below.

Simulations always rely on models of wireless channels and do not (at least not directly) require costly measurement campaigns. In order to obtain reasonable simulation results, realistic channel models have to be employed. In general, such models are highly complex and have many parameters which have to be carefully set to meaningful values. One very recent example is the Winner Phase II+ channel model. It includes almost all known physical wave propagation effects. However, two significant effects that are currently (July 2010) not well modeled are antenna polarization effects and multi-user correlation. The problem of these (and possibly also other, currently unknown) missing physical effects in channel models can be circumvented by performing measurements. By performing measurements, all physical effects no matter if known or unknown, (well) modeled or not, are included, thus achieving realistic results. By choosing setups that network providers have chosen (near to existing base stations) we ensured that the scenarios selected are typical and the results obtained are meaningful. Our measurement results include antenna polarization effects and possibly numerous other effects currently not known and thus also not modeled.

In simulations, however, one has to know all physical effects beforehand and in a second step to model them accurately. Both steps are difficult to carry out, and no model up to the present day can guarantee this. Performing measurements is thus a method to obtain realistic results with correct quantitative measures. The conclusions drawn from measurements can directly be applied to modify systems, while the conclusions from simulations depend on the details of the model and how the numerous model parameters were selected. However, measurements are much more time and cost intensive than simulations and, therefore, can only be performed in a limited number of scenarios. Thus, measurements are ideally suited as a supplement to simulations, and in particular to obtain quantitative results.

We start in Chapter 5 with HSDPA measurements, where we explain in detail how we estimate Multiple-Input Multiple-Output (MIMO) channels in this context and how we define the Channel Quality Indicator (CQI) as well as the Precoding Control Indicator

(PCI) as an estimate of estimated post-equalization Signal to Interference and Noise Ratio (SINR). Such estimations turn out to be crucial for the optimal performance of the transmission system. We analyze and finally list in detail the implementation losses of the system as we observed them.

Based on our experimental observations, we report on HSDPA antenna selection experiments in Chapter 6. The most surprising part is that, in real physical channels, simple antenna switching techniques offer strong improvements, much stronger than expected theoretically, the reason being insufficient channel modeling. Most channel models nowadays assume a perfectly symmetric location of scattering objects, whereas this is not given in reality. In all our measurements we found significant differences to this assumption.

In Chapter 7 we provide HSDPA antenna spacing measurements for HSDPA in indoor and outdoor scenarios. There is a general myth that the radiating elements of MIMO base-station antennas need to be mounted at a distance of several wavelengths, or at least in half a wavelength. We show by these measurements that, indeed, much shorter distances are possible and still result in a significant MIMO gain. Furthermore, cross-polarized antennas offer the potential to have at least two antennas being located on top of each other, reducing potential distances to their minimum. We present measurements in this chapter when applying such cross-polarized antennas.

In Chapter 8 we show a direct comparison of the various cellular technologies, mainly HSDPA and WiMAX, not only in terms of throughput, but also from an information theoretical approach, by comparing the performance with mutual information and channel capacity. Our comparisons are purely based on consistent testbed measurements that have been carried out in several measurement campaigns in alpine as well as urban environments. Moreover, we investigated thoroughly the reasons for the differences found in terms of their particular losses, feedback channel loss, design loss, and implementation loss and report them here in detail. While some systems are better designed than others (that is, they have less design loss), they typically turn out to exhibit larger implementation loss. At the end, what counts is the sum of both parts.[1]

Chapter 9 provides insight in synchronization issues in particular for Orthogonal Frequency-Division Multiplexing (OFDM) based systems such as LTE. Based on analytical investigations, we show the different behavior of frequency offset estimation techniques and their performance when applied to optimal linear filters and ML techniques. All results are backed up by LTE standard compliant simulation examples.

We conclude this part of the book with Chapter 10, where first we present LTE performance evaluations based on physical-layer simulations. Similar to Chapter 5, we have to define the CQI as well as the Precoding Matrix Indicator (PMI) as an estimate of estimated post-equalization SINR. New, compared with the HSDPA standard, is the so-called Rank Indicator (RI), which allows for several spatially separated data streams, enriching the transmission possibilities even further.

[1] Various chapters of this book part are based on the Ph.D. theses of Christian Mehlführer and José A. García-Naya. The full theses can be downloaded from EURASIP's open Ph.D. library at http://www.arehna.di.uoa.gr/thesis/ or directly from http://publik.tuwien.ac.at/files/PubDat_181154.pdf and http://gtec.des.udc.es/web/images/diss/garcianaya2010.pdf respectively.

5

HSDPA Performance Measurements

Most[1] of the work published on High-Speed Downlink Packed Access (HSDPA) during recent years concentrates on system-level simulations [27, 28, 32, 33, 45, 48, 49, 51, 57] in which an analytical model [2, 3] of the physical layer is employed. Other theoretical studies focusing entirely on simulations look into specific details of the HSDPA physical layer such as, for example, Hybrid Automatic Repeat reQuest (HARQ) [6], receive antenna diversity [29], equalizer architectures [18, 34], radio-frequency hardware impairments [58], or link adaptation [47]. A meaningful assessment of the performance, however, requires the evaluation of the complete physical layer rather than the evaluation of individual physical layer parts. Such an evaluation was, for example, carried out in [36], which presents simulation results of a complete Single-Input Single-Output (SISO) HSDPA system including link adaptation. Furthermore, [36] shows a comparison between the simulated throughput and the Shannon capacity of the Additive White Gaussian Noise (AWGN) channel. Since HSDPA is typically operated in frequency-selective fading channels, however, the AWGN channel capacity, which only considers the mean Signal to Noise Ratio (SNR), is not a good performance bound. A much better, more realistic, and tighter bound can be obtained by calculating the channel capacity based on the channel realizations of the simulation.

Apart from simulations, HSDPA has been evaluated in a few experimental studies. For example, in [31] the measured pilot power strength, collected in so-called drive test measurements, is utilized in simulations to predict the throughput performance of a real SISO HSDPA system. In [56], measurement results of the average throughput performance of Multiple-Input Multiple-Output (MIMO) HSDPA multi-user detectors at 30 indoor receiver locations are presented. The impact of distributed antenna systems on the HSDPA performance in indoor environments is studied by [22]. In [21], the same authors develop guidelines for indoor HSDPA network planning and optimization based on extensive measurements with available SISO HSDPA hardware. The quality of service in a live HSDPA network is investigated by [24]. Throughput measurement results of an SISO HSDPA system are presented in [19] and those of

[1] Reproduced from [20], C. Mehlführer, S. Caban, M. Rupp "Measurement-based performance evaluation of MIMO HSDP" *IEEE Transactions on Vehicular Technology*, **59** (2010), 9; 4354–4367, by permission of © 2010 IEEE.

Evaluation of HSDPA and LTE: From Testbed Measurements to System Level Performance, First Edition.
Sebastian Caban, Christian Mehlführer, Markus Rupp and Martin Wrulich.

an MIMO HSDPA system in [54]. Unfortunately, the results in [54] were obtained by employing a non-standard-compliant MIMO scheme. Thus, the results are not representative for next-generation HSDPA systems. Also, none of the papers cited above compares the actual data throughput of a standard-compliant MIMO HSDPA system with the Mutual Information (MI) and/or the capacity of the wireless channel. Such a comparison, however, is necessary to determine how far HSDPA operates from the channel capacity. Furthermore, previous publications typically do not employ a systematic averaging approach, and thus a mean scenario performance cannot be determined, with the result that such measurements are not repeatable.

In this chapter we present the results of physical-layer MIMO HSDPA throughput measurements. The results were obtained in two extensive measurement campaigns carried out in an alpine valley in Austria and in the inner city of Vienna, Austria (see Figure 6.2, p. 145, and Figure 6.3, p. 145). In all measurements we utilized cross-polarized antennas at the transmitter and the receiver site. We measured not only the standard-compliant 1×1, 1×2, 2×1, and 2×2 HSDPA schemes, but also an advanced four-transmit-antenna HSDPA scheme. The throughput measurement results are compared with a so-called "achievable mutual information" that is calculated on the basis of the MI of the channel and the restrictions of the HSDPA standard.

The remainder of this chapter is organized as follows. In Section 5.1 we provide a mathematical model of the MIMO HSDPA physical layer. This model is the basis for the derivation of a low-complexity channel estimation scheme and the derivation of an equalizer in Section 5.2. Section 5.3 briefly explains the standard-compliant quantization of the HSDPA precoding and, furthermore, introduces precoding matrices for four transmit antennas. The feedback calculation method applied in our measurements is explained in Section 5.4. The term "achievable mutual information," which serves as a performance bound for the measured throughput, is introduced in Section 5.5. Finally, the measurement results are presented in Section 5.6 and a short summary of this chapter is given in Secion 5.7.

5.1 Mathematical Model of the Physical Layer

In this section we describe the (MIMO) HSDPA physical layer. In particular, we elaborate on the precoding at the transmitter, the mathematical representation of the channel, and the equalization at the receiver. The system model introduced describes a general (MIMO) HSDPA system and does not imply any restrictions on the maximum number of data streams, the number of antennas, and the precoding matrices. Only two restrictive assumptions are made in this section. First, we restrict our analysis to slow fading; that is, we assume that the channel remains approximately constant during the transmission of one subframe (2 ms). Second, we assume that only one user per subframe is scheduled by the Base station (NodeB). This assumption is necessary because of the immense hardware effort required for multi-user measurements. It should also be noted that multi-user scheduling for HSDPA is currently a topic of research [5, 30, 46] and that how to implement such scheduling optimally remains an open question.

In MIMO HSDPA a user can receive several spatially multiplexed data streams simultaneously. The number of data streams is selected by the scheduler in the NodeB and is denoted as N_{s} henceforth. Each data bit stream is rate 1/3 turbo encoded, rate matched,

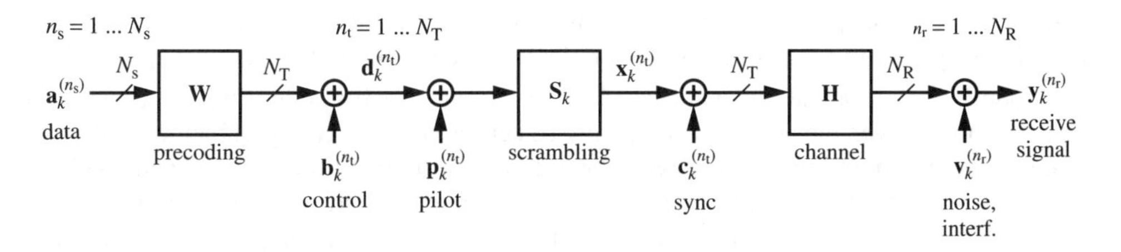

Figure 5.1 Generalized system model of the HSDPA physical layer.

symbol mapped, and spread using a specific number of spreading sequences of length 16. The resulting code rate at the output of the rate matching, the size of the symbol alphabet, and the number of spreading sequences is determined by the scheduler depending on the Channel Quality Indicator (CQI) feedback obtained from the User Equipment (UE). The n_sth spread data chip sequence at time instant k is denoted by the length L_c column vector

$$\mathbf{a}_k^{(n_s)} = \left[a_k^{(n_s)}, \dots, a_{k+L_c-1}^{(n_s)}\right]^T \textit{ with } n_s = 1 \dots N_s$$

(see Figure 5.1). These N_s sequences can be stacked into a matrix $\mathbf{A}_k$ and a vector $\mathbf{a}_k$ to form a compact representation:

$$\mathbf{A}_k = \left[\mathbf{a}_k^{(1)}, \dots, \mathbf{a}_k^{(N_s)}\right] \quad \text{and} \quad \mathbf{a}_k = \text{vec}(\mathbf{A}_k) \tag{5.1}$$

The data chip streams are spatially precoded by multiplying with an $N_T \times N_s$ dimensional precoding matrix:

$$\mathbf{W} = \begin{bmatrix} w^{(1,1)} & \dots & w^{(1,N_s)} \\ \vdots & \ddots & \vdots \\ w^{(N_T,1)} & \dots & w^{(N_T,N_s)} \end{bmatrix} \tag{5.2}$$

Depending on the Precoding Control Indicator (PCI) feedback obtained from the UE, the precoding matrix $\mathbf{W}$ is chosen from a predefined set of matrices (see Section 5.3) by the scheduler. More information about the calculation and implementation of the PCI feedback is provided in Section 5.4.

After precoding, the spread control channel chip streams $\mathbf{b}_k^{(1)} \dots \mathbf{b}_k^{(N_T)}$ are added to the precoded data chip streams on all N_T transmit antennas. The sum of data and control channel chip streams is denoted as $\mathbf{d}_k^{(1)} \dots \mathbf{d}_k^{(N_T)}$. Furthermore, the spread pilot channel chip streams $\mathbf{p}_k^{(1)} \dots \mathbf{p}_k^{(N_T)}$, which are required for channel estimation at the receiver, are added. Analogous to Equation (5.1), the symbols $\mathbf{B}_k$, $\mathbf{b}_k$, $\mathbf{D}_k$, $\mathbf{d}_k$, $\mathbf{P}_k$, and $\mathbf{p}_k$ are defined:

$$\begin{aligned} \mathbf{B}_k &= \left[\mathbf{b}_k^{(1)}, \dots, \mathbf{b}_k^{(N_T)}\right] \quad \text{and} \quad \mathbf{b}_k = \text{vec}(\mathbf{B}_k); \\ \mathbf{D}_k &= \left[\mathbf{d}_k^{(1)}, \dots, \mathbf{d}_k^{(N_T)}\right] \quad \text{and} \quad \mathbf{d}_k = \text{vec}(\mathbf{D}_k); \\ \mathbf{P}_k &= \left[\mathbf{p}_k^{(1)}, \dots, \mathbf{p}_k^{(N_T)}\right] \quad \text{and} \quad \mathbf{p}_k = \text{vec}(\mathbf{P}_k) \end{aligned} \tag{5.3}$$

After adding the pilot sequences $\mathbf{p}_k$, the chip streams of all transmit antennas are scrambled by multiplying with the scrambling sequence $\mathbf{S}_k = \mathrm{diag}\left[s_k, \ldots, s_{k+L_c-1}\right]$ of the NodeB. Synchronization channels $\mathbf{c}_k^{(1)} \ldots \mathbf{c}_k^{(N_T)}$ are finally added before the signals are transmitted over the wireless frequency-selective channel $\mathbf{H}$. The channel matrix $\mathbf{H}$ contains the complex impulse responses between each transmit and receive antenna.

At the receiver, the signals at all N_R receive antennas are additively impaired by $\mathbf{v}_k^{(1)} \ldots \mathbf{v}_k^{(N_R)}$, denoting interference of other base stations and thermal noise. Finally, the received signal that is further processed in the digital baseband receiver is $\mathbf{y}_k^{(1)} \ldots \mathbf{y}_k^{(N_R)}$. The symbols $\mathbf{C}_k$, $\mathbf{c}_k$, $\mathbf{V}_k$, $\mathbf{v}_k$, $\mathbf{Y}_k$, and $\mathbf{y}_k$ are defined in the same way as in Equation (5.1):

$$\begin{aligned} \mathbf{C}_k &= \left[\mathbf{c}_k^{(1)}, \ldots, \mathbf{c}_k^{(N_T)}\right] \quad \text{and} \quad \mathbf{c}_k = \mathrm{vec}(\mathbf{C}_k); \\ \mathbf{V}_k &= \left[\mathbf{v}_k^{(1)}, \ldots, \mathbf{v}_k^{(N_R)}\right] \quad \text{and} \quad \mathbf{v}_k = \mathrm{vec}(\mathbf{V}_k); \\ \mathbf{Y}_k &= \left[\mathbf{y}_k^{(1)}, \ldots, \mathbf{y}_k^{(N_R)}\right] \quad \text{and} \quad \mathbf{y}_k = \mathrm{vec}(\mathbf{Y}_k) \end{aligned} \tag{5.4}$$

We next define two different system models, each requiring a different composition of the channel matrix $\mathbf{H}$.

5.1.1 System Model for the Channel Estimation

For the derivation of the channel estimator it is convenient to define an individual MIMO channel matrix for each tap of the length L_h channel. In this model, the $N_T \times N_R$ dimensional channel matrix at delay m ($m = 0, \ldots, L_h - 1$) is defined as

$$\tilde{\mathbf{H}}_m = \begin{bmatrix} h_m^{(1,1)} & \ldots & h_m^{(1,N_R)} \\ \vdots & \ddots & \vdots \\ h_m^{(N_T,1)} & \ldots & h_m^{(N_T,N_R)} \end{bmatrix}$$

With such a definition of the channel matrix and the matrix descriptions of the HSDPA signals in Equations (5.1), (5.3), and (5.4), the receive signal can be compactly written as

$$\mathbf{Y}_k = \sum_{m=0}^{L_h-1} [\overbrace{\mathbf{S}_{k-m}(\underbrace{\mathbf{A}_{k-m}\mathbf{W}^{\mathrm{T}} + \mathbf{B}_{k-m}}_{\mathbf{D}_{k-m}} + \mathbf{P}_{k-m})}^{\mathbf{X}_{k-m}} + \mathbf{C}_{k-m}]\tilde{\mathbf{H}}_m + \mathbf{V}_k \tag{5.5}$$

5.1.2 System Model for the Equalizer Calculation

For the purpose of the equalizer calculation it is better to define a channel matrix that exhibits a block-Toeplitz structure. Thus, the frequency-selective link between the n_tth transmit and the n_rth receive antenna is modeled by the $L_f \times L_c$ dimensional Toeplitz matrix

$$\mathbf{H}^{(n_r,n_t)} = \begin{bmatrix} h_0^{(n_r,n_t)} & \ldots & h_{L_h-1}^{(n_r,n_t)} & & 0 \\ & \ddots & & \ddots & \\ 0 & & h_0^{(n_r,n_t)} & \ldots & h_{L_h-1}^{(n_r,n_t)} \end{bmatrix}, \quad 1 \le n_r \le N_R, 1 \le n_t \le N_T, \tag{5.6}$$

where $h_m^{(n_r,n_t)}$, $(m = 0, \ldots, L_h - 1)$ represent the channel impulse response between the n_tth transmit and the n_rth receive antenna. The entire frequency-selective MIMO channel is modeled by a block matrix $\mathbf{H}$ consisting of $N_R \times N_T$ Toeplitz matrices:

$$\mathbf{H} = \begin{bmatrix} \mathbf{H}^{(1,1)} & \ldots & \mathbf{H}^{(1,N_T)} \\ \vdots & \ddots & \vdots \\ \mathbf{H}^{(N_R,1)} & \ldots & \mathbf{H}^{(N_R,N_T)} \end{bmatrix} \tag{5.7}$$

With this definition of the channel matrix, the identity matrix $\mathbf{I}_{N_T}$ of dimension $N_T \times N_T$, and the vector descriptions of the HSDPA signals in Equations (5.1), (5.3), and (5.4), the receive signal can be compactly written as

$$\mathbf{y}_k = \mathbf{H} \cdot \{\overbrace{(\mathbf{I}_{N_T} \otimes \mathbf{S}_k)[\underbrace{(\mathbf{W} \otimes \mathbf{I}_{L_c})\mathbf{a}_k + \mathbf{b}_k}_{\mathbf{d}_k} + \mathbf{p}_k]}^{\mathbf{x}_k} + \mathbf{c}_k\} + \mathbf{v}_k \tag{5.8}$$

5.2 Receiver

In this section we explain the specific implementation of the digital baseband HSDPA receiver, that we utilized in our measurements. In particular, a tap-wise Linear Minimum Mean Square Error (LMMSE) channel estimator originally presented in [42] and an equalizer with interference cancelation is derived and explained.

5.2.1 Channel Estimation

Channel estimation in Wideband Code-Division Multiple Access (WCDMA)-based networks is strongly affected by interference, which can be divided into intra-cell and inter-cell interference. At the NodeB usually only a small amount of power (approximately 10 %) is dedicated to the pilot channels which are transmitted to enable channel estimation at the receiver. The remaining power is dedicated to all other channels and is thus considered as intra-cell interference for the channel estimator. The amount of intra-cell interference increases with the number of users simultaneously receiving data, since more spreading codes have to be used by the NodeB. Furthermore, the intra-cell interference also increases with the number of transmit antennas at the NodeB, since the same spreading codes are reused at each transmit antenna. Inter-cell interference, on the contrary, only becomes crucial at the cell edges, where the received power of the desired NodeB is in the order of the received power from other base stations.

A good channel estimator performance can only be achieved if the estimator takes all types of interference into account. This can be obtained for example by an LMMSE channel estimator. However, it turns out that the resulting matrices that have to be inverted become very large (5120 × 5120 for a 2 × 2 MIMO system) [61], prohibiting real-time implementations. Another problem of LMMSE channel estimation is that the autocorrelation matrix of the receive signal (which depends on the autocorrelation matrices of the channel and the noise) is unknown and has to be estimated. Especially in time dispersive MIMO channels, the full autocorrelation matrix becomes huge and very hard to estimate with high accuracy.

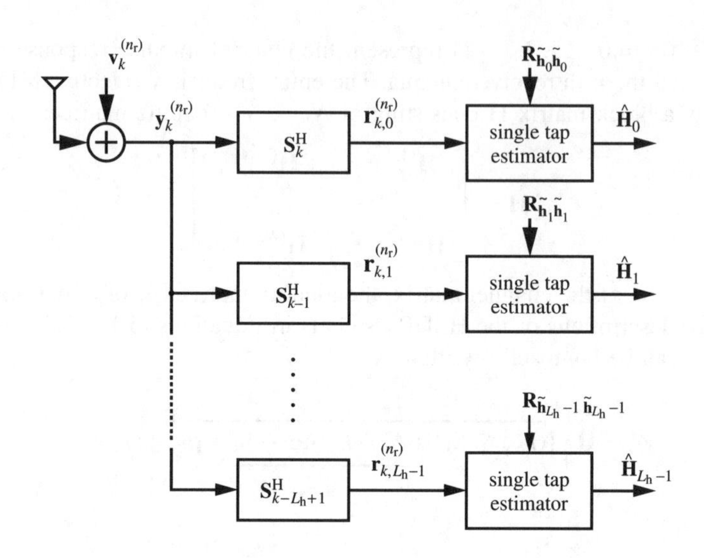

Figure 5.2 Tap-wise LMMSE channel estimator.

Both problems are tackled with the estimator structure illustrated in Figure 5.2. Here, the receive signal is first descrambled by multiplying with conjugated, delayed versions of the scrambling sequence $\mathbf{S}_k$ of the NodeB. The descrambling operations act as whitening or decorrelation filters. The output signals of the descramblers are thus only spatially correlated and the full autocorrelation matrix can be broken up into several, smaller autocorrelation matrices. This structure allows for the estimation of the MIMO channel matrix at every delay individually. Hence, this estimator is called the tap-wise LMMSE channel estimator.

The derivation of the tap-wise LMMSE channel estimator is based on the system model in Equation (5.5). The receive signal $\mathbf{Y}_k$ in this model is given by

$$\mathbf{Y}_k = \sum_{m=0}^{L_h-1} [\underbrace{\mathbf{S}_{k-m}(\mathbf{D}_{k-m} + \mathbf{P}_{k-m})}_{\mathbf{X}_{k-m}} + \mathbf{C}_{k-m}]\tilde{\mathbf{H}}_m + \mathbf{V}_k$$

With $\mathbf{S}_{k-n}^{\mathrm{H}}\mathbf{S}_{k-n} = \mathbf{I}$, the signal $\mathbf{R}_{k,n} = [\mathbf{r}_{k,n}^{(1)}, \ldots, \mathbf{r}_{k,n}^{(N_\mathrm{R})}]$ after the descrambling operation in the nth tap ($n = 0, \ldots, L_\mathrm{h} - 1$) of Figure 5.2 is given by

$$\begin{aligned}\mathbf{R}_{k,n} &= \mathbf{S}_{k-n}^{\mathrm{H}}\mathbf{Y}_k = \sum_{m=0}^{L_h-1} \mathbf{S}_{k-n}^{\mathrm{H}}[\mathbf{S}_{k-m}(\mathbf{D}_{k-m} + \mathbf{P}_{k-m}) + \mathbf{C}_{k-m}]\tilde{\mathbf{H}}_m + \mathbf{S}_{k-n}^{\mathrm{H}}\mathbf{V}_k \\ &= \mathbf{D}_{k-n}\tilde{\mathbf{H}}_n + \mathbf{P}_{k-n}\tilde{\mathbf{H}}_n + \mathbf{S}_{k-n}^{\mathrm{H}}\left[\mathbf{C}_{k-n}\tilde{\mathbf{H}}_n + \sum_{m\neq n}(\mathbf{X}_{k-m} + \mathbf{C}_{k-m})\tilde{\mathbf{H}}_m + \mathbf{V}_k\right]\end{aligned}$$

This description can be reformulated by applying the vec(.) operator and the Kronecker product $\otimes$:

$$\begin{aligned}\mathbf{r}_{k,n} &= \left[\mathbf{r}_{k,n}^{(1)\mathrm{T}}, \ldots, \mathbf{r}_{k,n}^{(N_\mathrm{R})\mathrm{T}}\right]^\mathrm{T} = \mathrm{vec}(\mathbf{R}_{k,n}) = \mathrm{vec}(\mathbf{S}_{k-n}^\mathrm{H}\mathbf{Y}_k)\\
&= \underbrace{(\mathbf{I}_{N_\mathrm{R}} \otimes \mathbf{P}_{k-n})}_{\mathbf{T}_{k-n}} \underbrace{\mathrm{vec}(\tilde{\mathbf{H}}_n)}_{\tilde{\mathbf{h}}_n} + \underbrace{[\mathbf{I}_{N_\mathrm{R}} \otimes (\mathbf{D}_{k-n} + \mathbf{S}_{k-n}^\mathrm{H}\mathbf{C}_{k-n})]}_{\tilde{\mathbf{D}}_{k-n}} \underbrace{\mathrm{vec}(\tilde{\mathbf{H}}_n)}_{\tilde{\mathbf{h}}_n}\\
&+ \sum_{m \neq n} \underbrace{\{\mathbf{I}_{N_\mathrm{R}} \otimes [\mathbf{S}_{k-n}^\mathrm{H}(\mathbf{X}_{k-m} + \mathbf{C}_{k-m})]\}}_{\tilde{\mathbf{X}}_{k-n,k-m}} \underbrace{\mathrm{vec}(\tilde{\mathbf{H}}_m)}_{\tilde{\mathbf{h}}_m} + \underbrace{(\mathbf{I}_{N_\mathrm{R}} \otimes \mathbf{S}_{k-n}^\mathrm{H})\mathrm{vec}(\mathbf{V}_k)}_{\mathbf{w}_k}\end{aligned} \tag{5.9}$$

The LMMSE estimator of the MIMO channel matrix at time delay n is given by

$$\hat{\mathbf{h}}_n = \mathbf{R}_{\tilde{\mathbf{h}}_n\mathbf{r}}\mathbf{R}_{\mathbf{rr}}^{-1}\mathbf{r}_{k,n} \tag{5.10}$$

The cross-correlation $\mathbf{R}_{\tilde{\mathbf{h}}_n\mathbf{r}}$ in Equation (5.10) can be computed to

$$\mathbf{R}_{\tilde{\mathbf{h}}_n\mathbf{r}} = \mathbb{E}\left\{\tilde{\mathbf{h}}_n\mathbf{r}_{k,n}^\mathrm{H}\right\} = \mathbb{E}\left\{\tilde{\mathbf{h}}_n\tilde{\mathbf{h}}_n^\mathrm{H}\mathbf{T}_{k-n}^\mathrm{H}\right\} = \mathbf{R}_{\tilde{\mathbf{h}}_n\tilde{\mathbf{h}}_n}\mathbf{T}_{k-n}^\mathrm{H},$$

where the zero mean property of the data+control channel chip stream $\mathbb{E}\{\mathbf{D}_{k-n}\} = \mathbf{0}$ and the scrambling sequence $\mathbb{E}\{\mathbf{S}_{k-n}\} = \mathbf{0}$ was used. Assuming that the time correlation of the channel coefficients is zero between different delays m–that is, $\mathbb{E}\left\{\mathbf{h}_m\mathbf{h}_n^\mathrm{H}\right\} = \mathbf{0}$ for $m \neq n$–the autocorrelation matrix of the receive signal $\mathbf{R}_{\mathbf{rr}}$ is given by

$$\begin{aligned}\mathbf{R}_{\mathbf{rr}} = \mathbb{E}\left\{\mathbf{r}_{k,n}\mathbf{r}_{k,n}^\mathrm{H}\right\} &= \mathbf{T}_{k-n}\mathbf{R}_{\tilde{\mathbf{h}}_n\tilde{\mathbf{h}}_n}\mathbf{T}_{k-n}^\mathrm{H} + \mathbb{E}\left\{\tilde{\mathbf{D}}_{k-n}\tilde{\mathbf{h}}_n\tilde{\mathbf{h}}_n^\mathrm{H}\tilde{\mathbf{D}}_{k-n}^\mathrm{H}\right\}\\
&+ \sum_{m \neq n}\mathbb{E}\left\{\tilde{\mathbf{X}}_{k-n,k-m}\mathbf{h}_m\mathbf{h}_m^\mathrm{H}\tilde{\mathbf{X}}_{k-n,k-m}^\mathrm{H}\right\} + \mathbf{R}_{\mathbf{ww}}\end{aligned} \tag{5.11}$$

Equation (5.11) can be simplified by the lemma defined below.

Lemma 5.1 *Given a matrix* **A** *and a vector* **b** *with the structure*

$$\mathbf{A} = \begin{bmatrix} a^{(1,1)} & \ldots & a^{(1,N_\mathrm{b})}\\ \vdots & \ddots & \vdots\\ a^{(N_\mathrm{a},1)} & \ldots & a^{(N_\mathrm{a},N_\mathrm{b})}\end{bmatrix}, \quad \mathbf{b} = \begin{bmatrix} b^{(1)}\\ \vdots\\ b^{(N_\mathrm{b})}\end{bmatrix},$$

and assuming that the elements of **A** *are uncorrelated and identically distributed with zero mean and variance* σ_a^2*, the elements of* **b** *are identically distributed (not necessarily uncorrelated) with zero mean and variance* σ_b^2*, and the elements of* **A** *and* **b** *are statistically independent, we have*

$$\mathbb{E}\left\{\left(\mathbf{I}_{N_\mathrm{d}} \otimes \mathbf{A}\right)\mathbf{b}\mathbf{b}^\mathrm{H}\left(\mathbf{I}_{N_\mathrm{d}} \otimes \mathbf{A}^\mathrm{H}\right)^\mathrm{H}\right\} = N_\mathrm{b}\sigma_a^2\sigma_b^2\mathbf{I}_{N_\mathrm{a}N_\mathrm{d}}$$

The proof of this expression is carried out by first showing

$$\mathbb{E}\left\{\mathbf{A}\mathbf{b}\mathbf{b}^\mathrm{H}\mathbf{A}^\mathrm{H}\right\} = N_\mathrm{b}\sigma_a^2\sigma_b^2\mathbf{I}_{N_\mathrm{a}},$$

which itself can be shown by simple inspection of the matrix-vector multiplication.

Lemma 5.1 simplifies Equation (5.11) if it is assumed that the transmit sequences $\tilde{\mathbf{D}}_k$ and $\tilde{\mathbf{X}}_k$ are spatially and temporally uncorrelated. This, in fact, is only partially fulfilled in the measurements since the synchronization signals $\mathbf{C}_k$ and the control channel signals $\mathbf{B}_k$ are identical at all transmit antennas. However, since only a small fraction of the total transmit power is allocated to $\mathbf{C}_k$ and $\mathbf{B}_k$, this assumption will hold approximately. Furthermore, it has to be assumed that the additive noise-and-interference term $\mathbf{w}_k$ is white with variance σ_w^2. This assumption is always fulfilled owing to the descrambling operation that whitens the input signal and thus also the noise-and-interference term. The autocorrelation of the receive signal is thus obtained as

$$\mathbf{R}_{\mathbf{rr}} = \mathbf{T}_{k-n}\mathbf{R}_{\mathbf{h}_n\mathbf{h}_n}\mathbf{T}_{k-n}^{\mathrm{H}} + \underbrace{N_{\mathrm{T}}\sigma_d^2\sigma_{h_n}^2\mathbf{I}_{N_\mathrm{c}N_\mathrm{R}} + N_{\mathrm{T}}\sigma_x^2\sum_{m\neq n}\sigma_{h_m}^2\mathbf{I}_{N_\mathrm{c}N_\mathrm{R}} + \sigma_w^2\mathbf{I}_{N_\mathrm{c}N_\mathrm{R}}}_{\approx\sigma_{rn}^2\mathbf{I}_{N_\mathrm{R}}\otimes\mathbf{I}_{N_\mathrm{c}}} \tag{5.12}$$

The first term in Equation (5.12) corresponds to the received power of the pilot signals arriving at delay n at the receiver. The three remaining terms in Equation (5.12) almost add up to the total receive power. The difference to the total receive power is only given by the pilot power, which is usually low (approximately 10 % of the total NodeB transmit power), thus resulting in a small approximation error. The last three terms in Equation (5.12) can, therefore, be replaced by a diagonal matrix with an estimate of the received signal power on the main diagonal $\hat{\sigma}_{rn}^2 = \frac{1}{N_\mathrm{R}L_\mathrm{c}}\|\mathbf{r}_{k,n}\|_2^2$:

$$\mathbf{R}_{\mathbf{rr}} \approx \mathbf{T}_{k-n}\mathbf{R}_{\mathbf{h}_n\mathbf{h}_n}\mathbf{T}_{k-n}^{\mathrm{H}} + \underbrace{\hat{\sigma}_{rn}^2\mathbf{I}_{N_\mathrm{R}}}_{=\mathbf{R}_{\sigma rn}}\otimes\mathbf{I}_{N_\mathrm{c}} \tag{5.13}$$

The first term in Equation (5.13) is composed of the pilot signals and the channel correlation matrix at time delay n. The second term is approximated by the total received signal power. With such an approximation, the LMMSE estimator for the channel coefficients at delay n is given by

$$\hat{\mathbf{h}}_n = \mathbf{R}_{\tilde{\mathbf{h}}_n\mathbf{r}}\mathbf{R}_{\mathbf{rr}}^{-1}\mathbf{r}_{k,n} = \mathbf{R}_{\tilde{\mathbf{h}}_n\tilde{\mathbf{h}}_n}\mathbf{T}_{k-n}^{\mathrm{H}}\cdot(\mathbf{T}_{k-n}\mathbf{R}_{\tilde{\mathbf{h}}_n\tilde{\mathbf{h}}_n}\mathbf{T}_{k-n}^{\mathrm{H}} + \mathbf{R}_{\sigma_r}\otimes\mathbf{I}_{N_\mathrm{c}})^{-1}\mathbf{r}_{k,n}$$

Applying the Woodbury identity [44, 52] for positive definite matrices Φ and Ω

$$\Phi\Theta^{\mathrm{H}}(\Theta\Phi\Theta^{\mathrm{H}} + \Omega)^{-1} = (\Phi^{-1} + \Theta^{\mathrm{H}}\Omega^{-1}\Theta)^{-1}\Theta^{\mathrm{H}}\Omega^{-1},$$

the definition $\mathbf{T}_{k-n} = \mathbf{I}_{N_\mathrm{R}}\otimes\mathbf{P}_{k-n}$ from Equation (5.9) in a second step, and the relation $(\mathbf{A}\otimes\mathbf{B})(\mathbf{C}\otimes\mathbf{D}) = \mathbf{AC}\otimes\mathbf{BD}$ in a third step, the LMMSE estimator can be reformulated to

$$\begin{aligned}
\hat{\mathbf{h}}_n &= [\mathbf{R}_{\tilde{\mathbf{h}}_n\tilde{\mathbf{h}}_n}^{-1} + \mathbf{T}_{k-n}^{\mathrm{H}}(\mathbf{R}_{\sigma_r}^{-1}\otimes\mathbf{I}_{N_\mathrm{c}})\mathbf{T}_{k-n}]^{-1}\mathbf{T}_{k-n}^{\mathrm{H}}(\mathbf{R}_{\sigma_r}^{-1}\otimes\mathbf{I}_{N_\mathrm{c}})\mathbf{r}_{k,n} \\
&= [\mathbf{R}_{\tilde{\mathbf{h}}_n\tilde{\mathbf{h}}_n}^{-1} + (\mathbf{I}_{N_\mathrm{R}}\otimes\mathbf{P}_{k-n}^{\mathrm{H}})(\mathbf{R}_{\sigma_r}^{-1}\otimes\mathbf{I}_{N_\mathrm{c}})(\mathbf{I}_{N_\mathrm{R}}\otimes\mathbf{P}_{k-n})]^{-1} \\
&\qquad\cdot(\mathbf{I}_{N_\mathrm{R}}\otimes\mathbf{P}_{k-n}^{\mathrm{H}})(\mathbf{R}_{\sigma_r}^{-1}\otimes\mathbf{I}_{N_\mathrm{T}})\mathbf{r}_{k,n} \\
&= [\mathbf{R}_{\tilde{\mathbf{h}}_k\tilde{\mathbf{h}}_k}^{-1} + (\mathbf{R}_{\sigma_r}^{-1}\otimes\mathbf{P}_{k-n}^{\mathrm{H}}\mathbf{P}_{k-n})]^{-1}(\mathbf{R}_{\sigma_r}^{-1}\otimes\mathbf{P}_{k-n}^{\mathrm{H}})\mathbf{r}_{k,n}
\end{aligned} \tag{5.14}$$

This LMMSE estimator requires the knowledge of the channel autocorrelation matrix $\mathbf{R}_{\tilde{\mathbf{h}}_n\tilde{\mathbf{h}}_n}$, which is usually not known. Therefore, a so-called "instantaneous" estimate of $\mathbf{R}_{\tilde{\mathbf{h}}_n\tilde{\mathbf{h}}_n}$ is utilized. This instantaneous estimate is obtained similarly as in an approach for Orthogonal Frequency-Division Multiplexing (OFDM) channel estimation presented in [37].

5.2.1.1 Instantaneous Channel Autocorrelation Estimation

The LMMSE channel estimator employing the instantaneous channel autocorrelation can be described by a three-stage process:

1. Perform a low-complexity, correlation-based channel estimation

$$\hat{\mathbf{h}}_k^{(\text{cor})} = \frac{1}{\|\mathbf{p}\|_2^2}\left(\mathbf{I}_{N_\text{R}} \otimes \mathbf{P}_{k-n}^{\text{H}}\right)\mathbf{r}_{k,n} \tag{5.15}$$

   with $\|\mathbf{p}\|_2^2 = \|\mathbf{p}^{(n_\text{t})}\|_2^2$, assuming that the pilot power is equally distributed over the transmit antennas.
2. Calculate the channel gains and assume them to be spatially uncorrelated. Here, we denote the lth element of a vector by $(.)_l$:

$$\hat{\mathbf{R}}_{\tilde{\mathbf{h}}_n\tilde{\mathbf{h}}_n}^{-1} = \text{diag}\left(\frac{1}{|(\hat{\mathbf{h}}_n^{(\text{cor})})_l|^2}\right), \quad l = 1, \ldots, N_\text{R}N_\text{T} \tag{5.16}$$

3. Improve the correlation-based channel estimate by applying the estimate of $\mathbf{R}_{\tilde{\mathbf{h}}_n\tilde{\mathbf{h}}_n}^{-1}$ in Equation (5.14):

$$\hat{\mathbf{h}}_n^{(\text{LMMSE})} = [\hat{\mathbf{R}}_{\tilde{\mathbf{h}}_n\tilde{\mathbf{h}}_n}^{-1} + \mathbf{R}_{\sigma_r}^{-1} \otimes (\mathbf{P}_{k-n}^{\text{H}}\mathbf{P}_{k-n})]^{-1}\|\mathbf{p}\|^2(\mathbf{R}_{\sigma_r}^{-1} \otimes \mathbf{I}_{N_\text{T}})\hat{\mathbf{h}}_n^{(\text{cor})}$$

At the cost of higher complexity, the LMMSE channel estimate obtained in Step 3 can improve the estimate of the channel autocorrelation matrix $\hat{\mathbf{R}}_{\tilde{\mathbf{h}}_n\tilde{\mathbf{h}}_n}$ in Step 2. It was found that more than three iterations no longer yield a significant performance increase. Therefore, in the measurements, the iterative tap-wise LMMSE channel estimator structure was used with three iterations.

5.2.1.2 Complexity Considerations

The computational complexity of the tap-wise LMMSE channel estimator with instantaneous autocorrelation estimation is slightly higher than that of the Least Squares (LS) channel estimator. The first step of both estimators – the calculation of the correlation-based channel estimate in Equation (5.15) – is exactly the same [42]. The calculation of the actual channel estimate requires N_R matrix inverses of dimension $N_\text{T} \times N_\text{T}$ for each channel tap. The increased complexity of the tap-wise LMMSE channel estimator basically comes from the multiple descrambling operations and the estimation of the channel autocorrelation matrix, as in Equation (5.16). If the iterative channel estimator is employed, the complexity obviously increases linearly with the number of iterations.

5.2.2 *Equalizer*

After the channel estimation, as explained in Section 5.2.1, the interference of the deterministic signals (that is, the pilot channels $\mathbf{p}_k$ and synchronization channels $\mathbf{c}_k$) and the interference of the control channels $\mathbf{b}_k$ is canceled. Since the control channels are transmitted with a large spreading factor of 256, error-free reception is assumed. If, furthermore, perfect channel knowledge and, thus, perfect cancelation is assumed, the system model in Equation (5.8) is simplified to

$$\mathbf{y}_k = \mathbf{H}(\mathbf{I}_{N_\mathrm{T}} \otimes \mathbf{S}_k)(\mathbf{W} \otimes \mathbf{I}_{L_\mathrm{c}})\mathbf{a}_k + \mathbf{v}_k \tag{5.17}$$

Note that without interference cancelation (at least of the synchronization channel) the post-equalization Signal to Interference and Noise Ratio (SINR) would saturate at approximately 20 dB. Values of the CQI requiring higher SINR can thus not be selected, leading to a saturation of the throughput [15, 16, 38]. As an alternative to interference cancelation, interference-aware equalization is possible and was shown to achieve high performance [43,63–65].

Since the precoding and the scrambling are both linear operations, their order in Equation (5.17) can be exchanged to obtain

$$\mathbf{y}_k = \mathbf{H}(\mathbf{W} \otimes \mathbf{I}_{L_\mathrm{c}}) \underbrace{(\mathbf{I}_{N_\mathrm{s}} \otimes \mathbf{S}_k)\mathbf{a}_k}_{\tilde{\mathbf{a}}_k} + \mathbf{v}_k = \mathbf{H}(\mathbf{W} \otimes \mathbf{I}_{L_\mathrm{c}})\tilde{\mathbf{a}}_k + \mathbf{v}_k$$

With this system model, the equalizer coefficients for the reconstruction of the n_sth transmitted chip sequence can be calculated by minimizing the quadratic cost function [11, 35]

$$J(\mathbf{f}^{(n_\mathrm{s})}) = \mathbb{E}\left\{\left|\mathbf{f}^{(n_\mathrm{s})\mathrm{H}}\mathbf{y}_k - \tilde{a}_{k-\tau}^{(n_\mathrm{s})}\right|^2\right\}, \tag{5.18}$$

where $\mathbf{f}^{(n_\mathrm{s})}$ is given by a stacked vector of N_R equalization filters

$$\mathbf{f}^{(n_\mathrm{s})} = \left[\mathbf{f}^{(1,n_\mathrm{s})\mathrm{T}}, \ldots, \mathbf{f}^{(N_\mathrm{R},n_\mathrm{s})\mathrm{T}}\right]^\mathrm{T},$$

each of length L_f:

$$\mathbf{f}^{(n_\mathrm{r},n_\mathrm{s})} = \left[f_0^{(n_\mathrm{r},n_\mathrm{s})}, \ldots, f_{L_f-1}^{(n_\mathrm{r},n_\mathrm{s})}\right]^\mathrm{T}$$

Setting the derivative of the cost function (Equation (5.18)) with respect to $\mathbf{f}^{(n_\mathrm{s})*}$ equal to zero,

$$\begin{aligned}\frac{\mathrm{d}J}{\mathrm{d}\mathbf{f}^{(n_\mathrm{s})*}} &= \left[\sigma_a^2\mathbf{H}\left(\mathbf{W}\mathbf{W}^\mathrm{H} \otimes \mathbf{I}_{L_\mathrm{c}}\right)\mathbf{H}^\mathrm{H} + \sigma_a^2\mathbf{I}_{N_\mathrm{R}L_\mathrm{c}}\right]\mathbf{f}^{(n_t)} - \sigma_s^2\mathbf{H}\left(\mathbf{W} \otimes \mathbf{I}_{L_\mathrm{c}}\right)\mathbf{e}_{\tau+(n_\mathrm{s}-1)L_\mathrm{c}}\\ &= 0,\end{aligned}$$

yields the equalizer coefficients for the n_sth transmitted chip sequence:

$$\mathbf{f}^{(n_\mathrm{s})} = \left[\mathbf{H}(\mathbf{W}\mathbf{W}^\mathrm{H} \otimes \mathbf{I}_{L_\mathrm{c}})\mathbf{H}^\mathrm{H} + \frac{\sigma_\mathrm{n}^2}{\sigma_a^2}\mathbf{I}_{N_\mathrm{R}L_\mathrm{c}}\right]^{-1} \cdot \mathbf{H}(\mathbf{W} \otimes \mathbf{I}_{L_\mathrm{c}})\mathbf{e}_{\tau+(n_\mathrm{s}-1)L_\mathrm{c}}$$

Here, $\mathbf{e}_{\tau+(n_s-1)L_c}$ denotes a unit vector with a single "one" at cursor position ($\tau + (n_s - 1)$ L_c) and "zeros" at all other positions. The term σ_n^2 denotes the noise variance. The calculation of the equalizer coefficients can be implemented efficiently by Fast Fourier Transform (FFT)-based algorithms such as [13, 14], the conjugated gradient algorithm [12], or other iterative algorithms, such as [41] for example. This receiver thus represents a low-complexity HSDPA receiver that is feasible for real-time implementation in a chip [10].

5.2.3 Further Receiver Processing

The output chip stream of the equalizer is descrambled, despread, soft-demapped, and soft-decoded in a turbo decoder in eight iterations. Finally, the Cyclic Redundancy Check (CRC) bytes of the resulting data block are verified. If the CRC yields "false," then at a maximum two HARQ retransmissions are performed. If these retransmissions also fail, then the data block is considered as lost. Note that, in the actual testbed implementation (see Chapter 4), the two retransmissions are always carried out. The number of retransmissions required for a correct reception is determined in the data evaluation after the measurement.

5.3 Quantized Precoding

In the HSDPA standard [1], the precoding matrix defined in Equation (5.2) is strongly quantized and chosen from a predefined codebook. For single antenna transmissions, in which obviously no spatial precoding can be performed, the precoding matrix $\mathbf{W}$ is reduced to a scalar equal to "one":

$$\mathbf{W}^{(\mathrm{SISO})} = 1$$

For multiple antenna transmissions, the precoding matrices are composed of the scalars

$$w_0 = \frac{1}{\sqrt{2}}$$

and

$$w_1, w_2 \in \left\{ \frac{1+\mathrm{i}}{2}, \frac{1-\mathrm{i}}{2}, \frac{-1+\mathrm{i}}{2}, \frac{-1-\mathrm{i}}{2} \right\}$$

The Transmit Antenna Array (TxAA) transmission mode utilizes two antennas to transmit a single stream. In this mode, the precoding matrix is defined as

$$\mathbf{W}^{(\mathrm{TxAA})} = \begin{bmatrix} w_0 \\ w_1 \end{bmatrix} \tag{5.19}$$

In TxAA, the signal at the first antenna is thus always weighted by the same scalar constant w_0, whereas the signal at the second antenna is weighted by w_1, which is chosen in order to maximize the received post-equalization SINR [40]. In TxAA, the number of possible precoding matrices is equal to four, corresponding to an amount of 2 bit feedback.

In the case of a Double Transmit Antenna Array (D-TxAA) transmission, the precoding matrix is given by

$$\mathbf{W}^{(\text{D-TxAA})} = \begin{bmatrix} w_0 & w_0 \\ w_1 & -w_1 \end{bmatrix}$$

Note that this precoding matrix is a unitary matrix; that is, the precoding vector of the second stream is always chosen orthogonal to that of the first stream. Although D-TxAA defines four precoding matrices, only the first two of them cause different SINRs at the receiver. In the other two cases, the SINRs of the first and the second stream are exchanged. Since the data rates of both streams can be adjusted individually, the third and the fourth precoding matrices are redundant. Note also that, if the user experiences a low channel quality in D-TxAA, only a single stream is transmitted; that is, the precoding matrix in Equation (5.19) is applied to the data streams at the transmitter. Thus, in TxAA a single stream is always transmitted, whereas in D-TxAA either single- or double-stream transmission (whichever leads to a higher throughput) is performed.

The HSDPA standard does not define spatial precoding for four transmit antennas. In order to explore the benefits of four transmit antennas in HSDPA, a very simple extension of the existing precoding vectors is employed here. We define the precoding matrix for double-stream four-transmit-antenna transmission as

$$\mathbf{W}^{(\text{4Tx-D-TxAA})} = \begin{bmatrix} w_0 & 0 \\ 0 & w_0 \\ w_1 & 0 \\ 0 & w_2 \end{bmatrix}$$

In contrast to the two-antenna D-TxAA system, the four-antenna D-TxAA system now transmits the two data streams on individual antenna pairs. Also, the precoding of both streams is individually adjusted, allowing 16 possible precoding matrices.

In order to exploit polarization diversity at the receiver, we utilized a so-called XX-Pol antenna (Kathrein 800 10629 [26]) consisting of two cross-polarized, spatially separated antenna pairs (A_1, A_2) and (A_3, A_4), as shown in Figure 5.3. In the case of transmissions which required two transmitter outputs, one cross-polarized antenna pair (A_1, A_2) or (A_3, A_4) was utilized. When four transmitter outputs were required, the first spatial data stream was transmitted with precoding on one polarization (A_1, A_3) and the second spatial data stream on the other polarization (A_2, A_4). In other words, each data stream was transmitted on the two equally polarized antennas, separated by 0.6 wavelengths, either (A_1, A_3) or (A_2, A_4). Consequently, according to the above definitions of the precoding

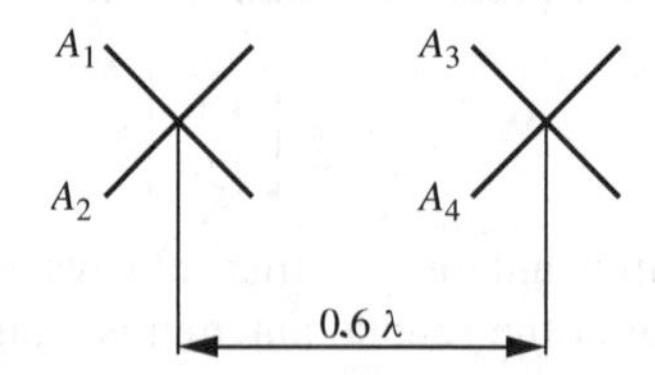

Figure 5.3 Polarization and spatial separation of the transmit antenna elements.

matrices, the first two transmitter outputs were connected to the first antenna pair (A_1, A_2) and the second two transmitter outputs to the second antenna pair (A_3, A_4).

5.4 CQI and PCI Calculation

In this section, analytic expressions for the post-equalization interference terms are derived. By evaluating these expressions, the post-equalization SINR for each possible PCI value can be determined. The estimated SINR values can, furthermore, be mapped to CQI values that correspond to specific Transport Block Sizes (TBSs). The TBS defines the number of data bits to be transmitted within one subframe. Maximizing the TBS of successfully transmitted data blocks over the different PCI values thus maximizes the data throughput [40].

Since the true channel matrix $\mathbf{H}$ is unknown at the receiver, a very important aspect when estimating the post-equalization SINR is the modeling of the channel estimation error. We approximate the true channel matrix $\mathbf{H}$ by the estimated channel matrix $\hat{\mathbf{H}}$ and a matrix $\mathbf{H}_\Delta$ representing the channel estimation error [17]. The matrix $\mathbf{H}_\Delta$ is constructed as the channel matrix $\mathbf{H}$ in Equations (5.6) and (5.7) with zeros at some of the entries. The nonzero elements of $\mathbf{H}_\Delta$ are assumed to be i.i.d. Gaussian with a variance equal to the Mean Square Error (MSE) of the channel estimator.

The SINR expressions derived in this section are verified by simulations and measurements and can be applied not only for implementing the feedback on the physical layer, but also for modeling the physical layer in system-level HSDPA simulations [59, 62, 66–68].

5.4.1 *HS-PDSCH Interference*

The High-Speed Physical Downlink Shared CHannel (HS-PDSCH) consists of the data chip streams in $\mathbf{a}_k$. These code division multiplexed data chip streams generate inter-code interference due to nonperfect equalization at the receiver. The residual interference can be assessed by looking at the total impulse response seen by all scrambled data chip streams in $(\mathbf{I}_{N_\mathrm{s}} \otimes \mathbf{S}_k)\mathbf{a}_k$ at the output of the n_sth equalization filter:

$$\mathbf{h}_{\mathrm{eff},a}^{(n_\mathrm{s})\mathrm{T}} = \mathbf{f}^{(n_\mathrm{s})\mathrm{H}}(\hat{\mathbf{H}} + \mathbf{H}_\Delta)(\mathbf{W} \otimes \mathbf{I}_{L_\mathrm{c}})$$

The interference power at the output of the equalizer can be divided into a deterministic part (caused by $\hat{\mathbf{H}}$) and a stochastic part (caused by $\mathbf{H}_\Delta$). The deterministic interference can be canceled by a decision feedback equalizer [4], whereas the stochastic interference can only be decreased by reducing the MSE of the channel estimator. As will be shown in Section 5.4.4, however, the cancelation of the data channels alone does not yield a significant performance gain.

The deterministic HS-PDSCH interference power is calculated by accumulating the energies of the total impulse response at delays $m \neq \tau$. The remaining interference at delay τ (the chosen delay of the transmitted chip stream after the channel and the equalizer) is irrelevant, since it is perfectly removed by the despreading operation:

$$\gamma_{s,\hat{\mathbf{H}}}^{(n_\mathrm{s})} = \frac{P_{\text{HS-PDSCH}}}{N_\mathrm{s}} \sum_{\substack{m=1 \\ m \neq \tau}}^{L_\mathrm{c}} \left| [\mathbf{f}^{(n_\mathrm{s})\mathrm{T}} \hat{\mathbf{H}} (\mathbf{W} \otimes \mathbf{I}_{L_\mathrm{c}})]_m \right|^2$$

Here, $P_{\text{HS-PDSCH}}$ corresponds to the transmit power available for all N_s data chip streams. The operator $(.)_m$ denotes the mth element of a vector. The stochastic HS-PDSCH interference power is obtained by building the expectation with respect to the unknown channel estimation error:

$$\gamma_{s,\mathbf{H}_\Delta}^{(n_s)} = \frac{P_{\text{HS-PDSCH}}}{N_s} \mathbb{E}_{\mathbf{H}_\Delta} \left\{ \sum_{\substack{m=1 \\ m \neq \tau}}^{L_c} \left| [\mathbf{f}^{(n_s)\mathrm{T}} \mathbf{H}_\Delta (\mathbf{W} \otimes \mathbf{I}_{L_c})]_m \right|^2 \right\}$$

$$\approx \frac{P_{\text{HS-PDSCH}}}{N_s} \cdot \text{MSE} \cdot (L_h - 1) N_T \left\| \mathbf{f}^{(n_s)} \right\|_2^2 \tag{5.20}$$

In this derivation, the last step can be verified by a careful inspection of the matrix-vector multiplications. As shown later in Figure 5.6, the approximation in Equation (5.20) only has a negligible impact on the accuracy of the SINR estimator. The same holds true for similar approximations of the stochastic interference in the next sections.

5.4.2 Pilot Interference

The total impulse response that is experienced by the scrambled Common Pilot CHannels (CPICHs) $(\mathbf{I}_{N_T} \otimes \mathbf{S}_k)\mathbf{p}_k$ at the output of the n_sth equalization filter is given by the $N_T L_c$ length vector:

$$\mathbf{h}_{\text{eff},p}^{(n_s)\mathrm{T}} = \mathbf{f}^{(n_s)\mathrm{T}} \mathbf{H} = \mathbf{f}^{(n_s)\mathrm{T}} (\hat{\mathbf{H}} + \mathbf{H}_\Delta)$$

Again, we can identify a deterministic interference,

$$\gamma_{p,\hat{\mathbf{H}}}^{(n_s)} = \frac{P_{\text{CPICH}}}{N_T} \sum_{n_t=1}^{N_T} \sum_{\substack{m=1 \\ m \neq \tau}}^{L_c} \left| (\mathbf{f}^{(n_s)\mathrm{T}} \hat{\mathbf{H}})_{m+(n_t-1)L_c} \right|^2,$$

and a stochastic interference,

$$\gamma_{p,\mathbf{H}_\Delta}^{(n_s)} = \frac{P_{\text{CPICH}}}{N_T} \mathbb{E}_{\mathbf{H}_\Delta} \left\{ \sum_{n_t=1}^{N_T} \sum_{\substack{m=1 \\ m \neq \tau}}^{L_c} \left| (\mathbf{f}^{(n_s)\mathrm{T}} \mathbf{H}_\Delta)_{m+(n_t-1)L_c} \right|^2 \right\}$$

$$\approx \frac{P_{\text{CPICH}}}{N_T} \cdot \text{MSE} \cdot (L_h - 1) \left\| \mathbf{f}^{(n_s)} \right\|_2^2$$

Here, P_{CPICH} denotes the total pilot channel power for all transmit antennas.

5.4.3 Synchronization and Control Channel Interference

For the calculation of the interference emerging from the synchronization and control channels, it is assumed that these channels are transmitted on all antennas simultaneously;

that is, the power is equally distributed on all transmit antennas. Since the Synchronization Channel (SCH) and the Primary Common Control Physical Channel (PCCPCH) are transmitted time-multiplexed, it is furthermore assumed that both channels have equal power; that is, $P_{\mathrm{SCH}} = P_{\mathrm{CCPCH}}$. The total impulse response experienced by the synchronization channels $\mathbf{c}_k$ and the scrambled control channels $(\mathbf{I}_{N_\mathrm{T}} \otimes \mathbf{S}_k)\mathbf{b}_k$ at the output of the n_sth equalization filter is given by

$$\mathbf{h}_{\mathrm{eff,SCH}}^{(n_\mathrm{s})\mathrm{T}} = \frac{1}{\sqrt{N_\mathrm{T}}}\mathbf{f}^{(n_\mathrm{s})\mathrm{T}}(\hat{\mathbf{H}} + \mathbf{H}_\Delta)(\mathbf{1}_{N_\mathrm{T}} \otimes \mathbf{I}_{L_\mathrm{c}})$$

Here, $\mathbf{1}_{N_\mathrm{T}}$ denotes an $N_\mathrm{T} \times 1$ dimensional vector with all entries equal to one. Note that the multiplication with $(\mathbf{1}_{N_\mathrm{T}} \otimes \mathbf{I}_{L_\mathrm{c}})$ represents the summation of the individual transmit antenna impulse responses. This is required because of the previous assumption that synchronization and control channels are transmitted on all antennas simultaneously. The deterministic part of the interference is calculated as

$$\gamma_{\mathrm{SCH},\hat{\mathbf{H}}}^{(n_\mathrm{s})} = \frac{P_{\mathrm{SCH}}}{N_\mathrm{T}} \sum_{\substack{m=1 \\ m\neq\tau}}^{L_\mathrm{c}} \left| [\mathbf{f}^{(n_\mathrm{s})\mathrm{T}}\hat{\mathbf{H}}(\mathbf{1}_{N_\mathrm{T}} \otimes \mathbf{I}_{L_\mathrm{c}})]_m \right|^2$$

and the stochastic interference as

$$\begin{aligned} \gamma_{\mathrm{SCH},\mathbf{H}_\Delta}^{(n_\mathrm{s})} &= \frac{P_{\mathrm{SCH}}}{N_\mathrm{T}} \mathbb{E}_{\mathbf{H}_\Delta}\left\{ \sum_{\substack{m=1 \\ m\neq\tau}}^{L_\mathrm{c}} \left| [\mathbf{f}^{(n_\mathrm{s})\mathrm{T}}\mathbf{H}_\Delta(\mathbf{1}_{N_\mathrm{T}} \otimes \mathbf{I}_{L_\mathrm{c}})]_m \right|^2 \right\} \\ &\approx \frac{P_{\mathrm{SCH}}}{N_\mathrm{T}} \cdot \mathrm{MSE} \cdot (L_\mathrm{h} - 1) \left\| \mathbf{f}^{(n_\mathrm{s})} \right\|_2^2 \end{aligned}$$

Additional to the two interference terms above, interference at delay lag $m = \tau$ emerges from the SCH, since it is transmitted without spreading and scrambling and is thus not orthogonal to the data channels:

$$\gamma_{\mathrm{SCH},\hat{\mathbf{H}},\tau}^{(n_\mathrm{s})} = \frac{\gamma P_{\mathrm{SCH}}}{N_\mathrm{T}} \left| [\mathbf{f}^{(n_\mathrm{s})\mathrm{T}}\hat{\mathbf{H}}(\mathbf{1}_{N_\mathrm{T}} \otimes \mathbf{I}_{L_\mathrm{c}})]_\tau \right|^2, \tag{5.21}$$

$$\begin{aligned} \gamma_{\mathrm{SCH},\mathbf{H}_\Delta,\tau}^{(n_\mathrm{s})} &= \frac{\gamma P_{\mathrm{SCH}}}{N_\mathrm{T}} \mathbb{E}_{\mathbf{H}_\Delta}\left\{ \left| (\mathbf{f}^{(n_\mathrm{s})\mathrm{T}}\mathbf{H}_\Delta)_\tau \right|^2 \right\} \\ &= \frac{\gamma P_{\mathrm{SCH}}}{N_\mathrm{T}} \cdot \mathrm{MSE} \cdot \sum_{n_\mathrm{r}=1}^{N_\mathrm{R}} \sum_{m=1}^{L_\mathrm{h}} \left| (\mathbf{f}^{(n_\mathrm{s})})_{m+\tau-L_\mathrm{h}+(n_\mathrm{r}-1)L_\mathrm{f}} \right|^2 \end{aligned} \tag{5.22}$$

The constant factor $\gamma = 0.1$ originates from the time-multiplexing of the SCH and the PCCPCH since the SCH occupies only the first 10 % of all chips in every transmitted slot. Note that the PCCPCH does not contribute to the interference terms in Equations (5.21) and (5.22) since it is transmitted with spreading.

5.4.4 Post-equalization Noise and SINR

According to the system model in Equation (5.8), the variance of the post-equalization noise at the output of the n_s-th equalization filter is given by

$$\sigma_{v'}^{(n_s)2} = \|\mathbf{f}^{(n_s)}\|_2^2 \sigma_n^2$$

Note that in this expression it is implicitly assumed that the noise $\mathbf{v}_k$ is white with variance σ_n^2. Given the correlation of the signals received from interfering base stations, it is straightforward to include colored inter-cell interference into the post-equalization noise. Since in the measurements only a single NodeB is considered, colored inter-cell interference is not required here.

The total deterministic interference caused by the pilot, control, and synchronization channels is given by

$$\gamma_{\hat{\mathbf{H}}}^{(n_s)} = \gamma_{p,\hat{\mathbf{H}}}^{(n_s)} + \gamma_{\mathrm{SCH},\hat{\mathbf{H}}}^{(n_s)} + \gamma_{\mathrm{SCH},\hat{\mathbf{H}},\tau}^{(n_s)}$$

Analogously, the stochastic interference is calculated by

$$\gamma_{\mathbf{H}_\Delta}^{(n_s)} = \gamma_{s,\mathbf{H}_\Delta}^{(n_s)} + \gamma_{p,\mathbf{H}_\Delta}^{(n_s)} + \gamma_{\mathrm{SCH},\mathbf{H}_\Delta}^{(n_s)} + \gamma_{\mathrm{SCH},\mathbf{H}_\Delta,\tau}^{(n_s)}$$

Knowing all the above interference terms, the post-equalization SINR [59, 62] of the n_sth data stream at the despreader output can be calculated as

$$\mathrm{SINR}_{\mathrm{est}}{}^{(n_s)} = \frac{\mathrm{SF}\left|(\mathbf{h}_{\mathrm{eff},a}^{(n_s)})_\tau\right|^2 \frac{P_{\text{HS-PDSCH}}}{N_s}}{\gamma_{s,\hat{\mathbf{H}}}^{(n_s)} + \bar{\alpha}_{\mathrm{ic}}\gamma_{\hat{\mathbf{H}}}^{(n_s)} + \gamma_{s,\mathbf{H}_\Delta}^{(n_s)} + \gamma_{\mathbf{H}_\Delta}^{(n_s)} + \sigma_{v'}^{(n_s)2}} \tag{5.23}$$

Here, SF is the spreading factor of the HS-PDSCH (SF = 16 for HSDPA). The factor $\bar{\alpha}_{\mathrm{ic}}$ is equal to zero if the receiver cancels the interference of the synchronization, pilot, and control channels and equal to one if not. In Figure 5.4, the individual post-equalization interference terms of a 2×2 TxAA system are shown over $I_{\mathrm{or}}/I_{\mathrm{oc}}$ (serving NodeB to interfering base stations power ratio). Below $I_{\mathrm{or}}/I_{\mathrm{oc}} \approx 10$ dB, the system performance is mainly dominated by the post-equalization noise. For $I_{\mathrm{or}}/I_{\mathrm{oc}} > 10$ dB, the deterministic interference caused by the pilot and the synchronization channel becomes dominant and has to be canceled to achieve high performance. If interference cancelation is performed, the system performance at high $I_{\mathrm{or}}/I_{\mathrm{oc}} > 10$ is mainly dominated by the channel estimation error. Over the whole $I_{\mathrm{or}}/I_{\mathrm{oc}}$ range, the inter-code interference of the data channels, given by $\gamma_{s,\hat{\mathbf{H}}}^{(n_s)}$, is more than one magnitude smaller than the post-equalization noise. Therefore, interference cancelation of the data channels does not yield additional performance gains.

Figures 5.5 and 5.6 show the estimated SINR and an "observed SINR" for a simulation with perfect channel knowledge and a measurement with channel estimation employed. The "observed SINR" is monitored at the demapper input and is obtained as follows. Consider the transmitted data symbol vector $\mathbf{u}^{(n_s)}$ of the n_sth symbol stream and the corresponding received symbol vector at the demapper input $\hat{\mathbf{u}}^{(n_s)}$. The "observed" or "true" post-equalization SINR is given by

$$\mathrm{SINR} = \frac{\|\mathbf{u}^{(n_s)}\|_2^2}{\|\hat{\mathbf{u}}^{(n_s)} - \mathbf{u}^{(n_s)}\|_2^2}$$

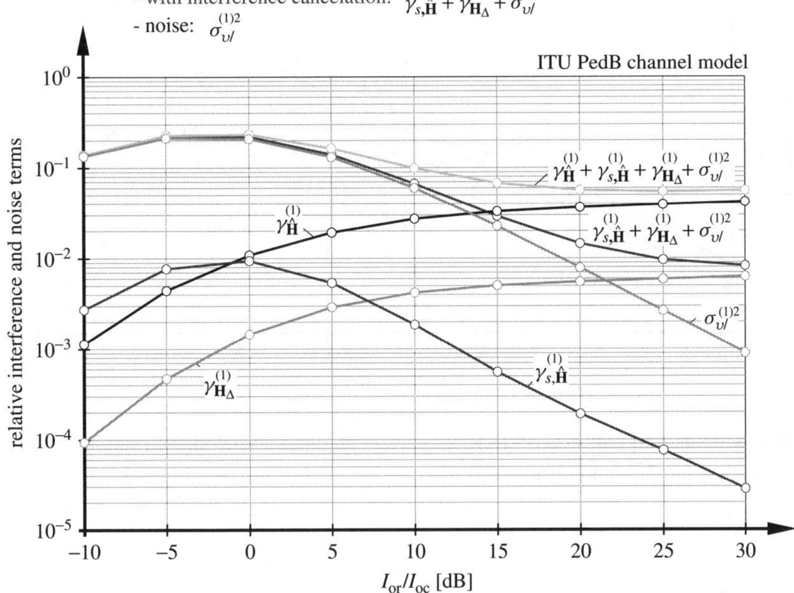

Figure 5.4 Interference terms for a two-receive-antennas TxAA transmission simulated with an uncorrelated ITU Pedestrian B channel model [23].

In the simulations, a very good fit of the estimated SINRs (calculated according to Equation (5.23)) over the full I_{or}/I_{oc} (ratio of the energy of the desired NodeB to the energy of the interfering base stations) range is observed. The SINR increases linearly with increasing I_{or}/I_{oc}. The estimated SINRs in the measurements also show a good fit at all transmit powers. In contrast to the simulations, the SINR saturates at about 30 dB, which is caused by residual interference due to the channel estimation error (which is the dominating factor at high SINR, as shown in Figure 5.4).

5.4.5 SINR to CQI Mapping

Once the SINR is estimated at the receiver, it can be mapped to a CQI value. This is achieved by Table 5.1 for single-stream mode and Table 5.2 for double-stream mode. The SINR-to-CQI mapping tables were obtained by simulating the Block Error Ratio (BLER) performance of a Category 16 UE for all CQI values in an AWGN channel. The SINR values in the table are equal to the AWGN SNRs at 10 % BLER values, as suggested as maximum BLER by the HSDPA standard [1, Section 6A.2]. The mapping is performed

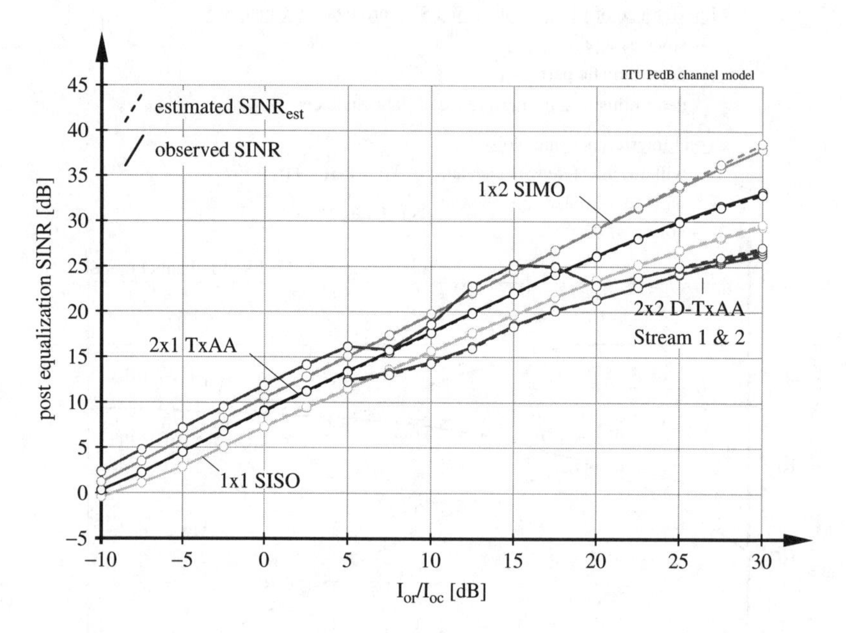

Figure 5.5 Estimated and observed SINR for perfect channel knowledge, simulated with an uncorrelated ITU Pedestrian B channel model [23].

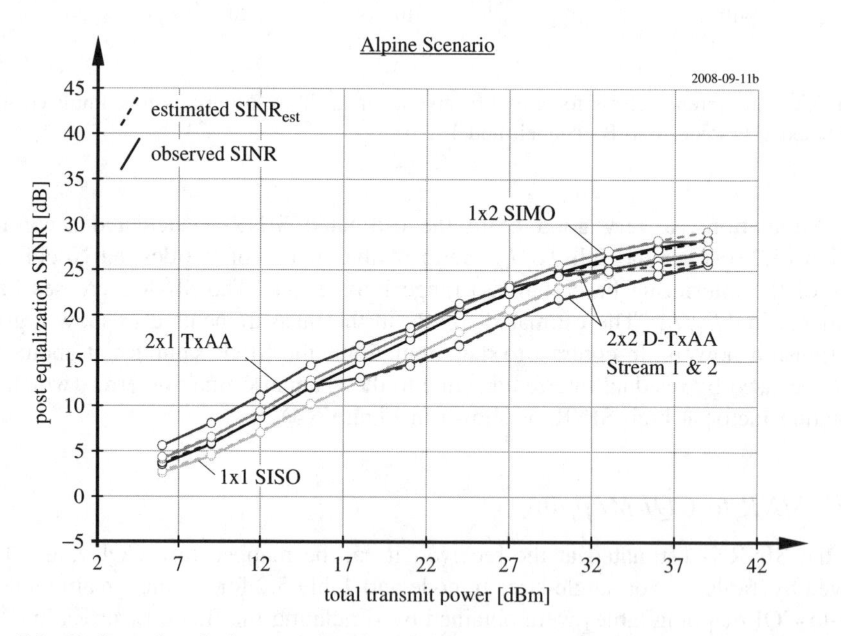

Figure 5.6 Estimated and observed SINR with iterative tap-wise LMMSE channel estimation, measured in the alpine scenario.

Table 5.1 SINR-CQI mapping table for single-stream mode

CQI	1	2	3	4	5	6	7	8
SINR (dB)	−3.5	−2.6	−1.5	−0.3	0.5	1.7	2.5	3.5
CQI	9	10	11	12	13	14	15	16
SINR (dB)	4.4	5.5	6.5	7.5	8.5	9.5	10.7	11.5
CQI	17	18	19	20	21	22	23	24
SINR (dB)	12.6	13.4	14.7	15.7	16.6	17.5	18.6	19.6
CQI	25	26	27	28	29	30		
SINR (dB)	20.6	21.4	22.6	23.5	24.0	24.8		

Table 5.2 SINR-CQI mapping table for double-stream mode

CQI	0	1	2	3	4	5	6	7
SINR (dB)	10.5	10.5	11.2	12.7	14.3	15.7	17.2	18.8
CQI	8	9	10	11	12	13	14	
SINR (dB)	20.4	21.9	23.4	25.3	26.0	26.8	28.3	

as follows. The estimated SINR minus a 1 dB margin (to account for SINR estimation errors) is compared with the SINR values in the tables and the maximum CQI that supports transmission with $<=10\,\%$ BLER is selected. Since this step involves a rounding operation, the actually measured BLER can be expected to be smaller than $10\,\%$.

The CQI values obtained from the mapping tables correspond to TBSs defined in [1, Table 7D, pp. 50] for single-stream mode and in [1, Table 7I, p. 54] for double-stream mode. The maximum TBS in single-stream mode is 25 558 bits that are transmitted in one subframe of length 2 ms. Thus, the maximum data rate of the Category 16 UE in single-stream mode is 12.779 Mbit/s. In double-stream mode, the maximum TBS of one stream is equal to 27 952 bits, allowing for a maximum data rate of 27.952 Mbit/s.

Similar CQI mapping tables to Tables 5.1 and 5.2 are available in documents of the Radio Access Network (RAN) Work Group 1; for example, see [7]. However, a comparison is difficult, since the simulation results in the RAN documents were generated usually before the standard was finalized. Therefore, every company selects different coding rates and/or number of spreading codes in the simulations. For example, [7] employed 15 spreading codes for all coding rates. Nevertheless, Table 5.1 can be compared at the four largest CQI values (since only these also use 15 spreading codes) to the second table in the appendix of [7]. Since [7] provides symbol SNR values in the tables, an offset of $10\log_{10}(15$ dB to the SINR values given in Table 5.1 is present. Besides this offset, the values in Table 5.1 are the same as the values in [7] within a few tenths of a decibel.

5.5 Achievable Mutual Information

In this section, we define a so-called "achievable mutual information." It is a feasible performance bound for the data throughput actually measured, given a specific standard, in this case HSDPA Rel'7. The calculation of the bound is based on the MI between transmit and receive signals. In particular, the achievable mutual information is a function of the wireless channel (that is, the estimated frequency response and the noise variance) and the precoding vectors allowed by the standard. Thus, it incorporates the restrictions imposed by the transmission standard (quantized, frequency-flat precoding), but not the restrictions imposed by the receiver employed.

For the calculation of the achievable mutual information, consider the estimated channel impulse response of length L_h chips between the n_tth transmit and the n_rth receive antenna:

$$\hat{\mathbf{h}}^{(n_\mathrm{r},n_\mathrm{t})} = \left[\hat{h}_0^{(n_\mathrm{r},n_\mathrm{t})} \; \ldots \; \hat{h}_{L_\mathrm{h}-1}^{(n_\mathrm{r},n_\mathrm{t})} \right]^\mathrm{T}$$

Note that the channel coefficients in this vector depend on the receive antenna position as well as the transmit power, as do all the terms defined below. The channel vector $\hat{\mathbf{h}}^{(n_\mathrm{r},n_\mathrm{t})}$ can equivalently be described in the frequency domain as $\hat{\mathbf{g}}^{(n_\mathrm{r},n_\mathrm{t})} = \mathfrak{F}\{\hat{\mathbf{h}}^{(n_\mathrm{r},n_\mathrm{t})}\}$ by applying the N_FFT point Fourier transform $\mathfrak{F}\{.\}$. Thus, the Fourier transform separates the frequency-selective channel into N_FFT frequency flat channels. With the notation $(.)_m$ to extract the mth element of a vector, the estimated $N_\mathrm{R} \times N_\mathrm{T}$ MIMO channel matrix of the mth ($m = 1 \ldots N_\mathrm{FFT}$) frequency bin can then be written as

$$\hat{\mathbf{G}}_m = \begin{bmatrix} \left(\hat{\mathbf{g}}^{(1,1)}\right)_m & \cdots & \left(\hat{\mathbf{g}}^{(1,N_\mathrm{T})}\right)_m \\ \vdots & \ddots & \vdots \\ \left(\hat{\mathbf{g}}^{(N_\mathrm{R},1)}\right)_m & \cdots & \left(\hat{\mathbf{g}}^{(N_\mathrm{R},N_\mathrm{T})}\right)_m \end{bmatrix}$$

With the well-known expressions for the MIMO capacity (for example, see [8, 60]) we obtain the achievable mutual information for the measured channel $\hat{\mathbf{G}}_m$:

$$I_\mathrm{a} = \max_{\mathbf{W} \in \mathcal{W}} \sum_{m=1}^{N_\mathrm{FFT}} \frac{f_\mathrm{s}}{N_\mathrm{FFT}} \log_2 \det \left(\mathbf{I}_{N_\mathrm{R}} + \frac{1}{\sigma_\mathrm{n}^2} \hat{\mathbf{G}}_m \mathbf{W}\mathbf{W}^\mathrm{H} \hat{\mathbf{G}}_m^\mathrm{H} \right) \tag{5.24}$$

Here, σ_n^2 is the variance of the noise $\mathbf{v}_k$ and $f_\mathrm{s} = 3.84\,\mathrm{MHz}$ the equivalent rectangular bandwidth of the Root Raised Cosine (RRC)-filtered HSDPA signal. The maximization in Equation (5.24) is performed over the set $\mathcal{W}$ of all possible precoding matrices. A few remarks should be made about the definition of the achievable mutual information:

- The channel coefficients in Equation (5.24) are the channel coefficients of the data channel that is transmitted at a specific portion (in our case −4 dB) of the NodeB's total transmit power. The remaining power is required for the transmission of the "non-data channels," namely the transmission of the pilot, control, and synchronization channels. By proper scaling of the channel coefficients in Equation (5.24), the inherent loss in spectral efficiency caused by the transmission of the non-data channels is automatically included in the achievable mutual information.
- Equation (5.24) represents neither the MI (obtained without precoding) nor the channel capacity (obtained by optimal water filling), since quantized and frequency flat precoding is utilized in HSDPA. Therefore, we have introduced the new term "achievable mutual information" for the symbol I_a.

- Equation (5.24) only gives the achievable mutual information for a specific channel realization at a specific receive antenna position and at a given transmit power level. In order to obtain the mean achievable mutual information, we perform averaging over all measured receive antenna positions.
- In order to increase the accuracy of the estimated channel coefficients in Equation (5.24), we utilize the channel coefficients at the largest transmit power to calculate the achievable mutual information. At lower transmit powers, the different received SNRs are obtained by scaling the channel coefficients accordingly.

As an alternative to the achievable mutual information, the well-known "channel capacity" C (requiring full channel state information to be available at the transmitter side) and the "mutual information" I can be interpreted as performance bounds. For the case of an SISO transmission in the urban scenario, Figure 5.7 visualizes these three performance bounds, the "measured data throughput" D_{m}, and the relations between them. In this scenario, the channel capacity C that could be theoretically achieved if optimum water-filling [50, 53] is employed at the transmitter side is slightly greater than the mutual information I. In the case of SISO transmissions, the achievable mutual information is the same as the mutual information, shifted by 4 dB in transmit power. This is because, in our measurement, the data channel (HS-PDSCH) is transmitted at an offset of −4 dB against the total transmit power (see Table 5.3). Note that, for any other choice of power offset, the achievable mutual information curve would shift accordingly. The right-hand side of Figure 5.7 shows three losses that we define as follows:

The CSI loss:

$$L_{\mathrm{CSI}} = C - I$$

is the difference between the channel capacity C and the mutual information I. It accounts for a missing full CSI at the transmitter side. An ideal transmission system not having CSI available at the transmitter can only achieve a rate equal to the mutual information. Compared with the other two losses defined below, the CSI loss is almost negligible, especially at higher transmit powers and, thus, higher SNRs [55].

The design loss:

$$L_{\mathrm{d}} = I - I_{\mathrm{a}}$$

is the difference between the mutual information I and the achievable mutual information I_{a}. It accounts for inherent system design losses caused by, for example, the necessary transmission of pilot symbols. In the medium transmit power range of 10–20 dBm, the design loss causes most of the overall system losses.

The implementation loss:

$$L_{\mathrm{i}} = I_{\mathrm{a}} - D_{\mathrm{m}}$$

is the difference between the achievable mutual information I_{a} and the measured data throughput D_{m}. It accounts for losses caused by nonoptimum receivers and channel codes. At higher transmit powers, the implementation loss becomes dominant. A more detailed discussion of the individual parts of the implementation loss is provided in Section 5.6.3.

5.6 Measurement Results

In this section, the throughput measurement results (the solid lines in Figures 5.9–5.12) are presented and compared with the achievable mutual information (the dashed lines in Figures 5.9–5.12). The most important HSDPA parameters are listed in Table 5.3. The E_c/I_{or} values indicate how much of the total transmit power is allocated to the transmission of the HS-PDSCH, the CPICH, the SCH, and the PCCPCH. The user equipment category indicated in Table 5.3 defines the maximum supported data rate. Note that, in a real network, many more physical channels would be transmitted by the NodeB than we did in our experiments. For example, a NodeB needs to transmit the Paging Indication Channel or channels to transport Universal Mobile Telecommunications System (UMTS) voice calls. Although we do not actually transmit these channels in our experiment, we have to take into account their portion of the total transmit power in order to realistically assess the performance losses of HSDPA (as done, for example, in Figure 5.7). We reserve −3.6 dB of the NodeB's total transmit power for these channels in our budget.

The user equipment category indicated in Table 5.3 defines the maximum supported data rate. In our experiment, in which we utilize the CQI mapping table of a Category 16 user equipment, the maximum data rate in single stream mode is 12.779 Mbit/s and in double stream mode is 27.952 Mbit/s.

As opposed to simulations, we plot our measurement results over transmit power. This is necessary for two reasons. First, the adaptive precoding employed in Rel'7 of HSDPA influences the receiver SNR very strongly while leaving the sum transmit power unchanged. Plotting over receive SNR would thus shift the throughput curves of the different HSDPA schemes against each other. For example, in the case of TxAA this shift would be approximately 2 dB compared with SISO. Second, we typically observe large deviations in average receive SNR of 3–5 dB from receive antenna to receive antenna, although they are only spaced half a wavelength apart [9, 39]. Typically, such deviations are not reflected by channel models offering identical average receive SNR for each receive antenna. Identical average receive SNRs would only be obtained in a very symmetrical transmission system in which the scattering objects are homogeneously distributed around the antennas. Nevertheless, to facilitate comparisons of our measurement

Table 5.3 Measurement parameters

	Alpine scenario	Urban scenario
User equipment category	16	16
HS-PDSCH E_c/I_{or} (dB)	−4	−4
CPICH E_c/I_{or} (dB)	−10	−10
SCH/PCCPCH E_c/I_{or} (dB)	−12	−12
E_c/I_{or} of other channels (dB)	−3.6	−3.6
Channel estimator length L_h (chips)	23	48
Equalizer length L_f (chips)	30	60

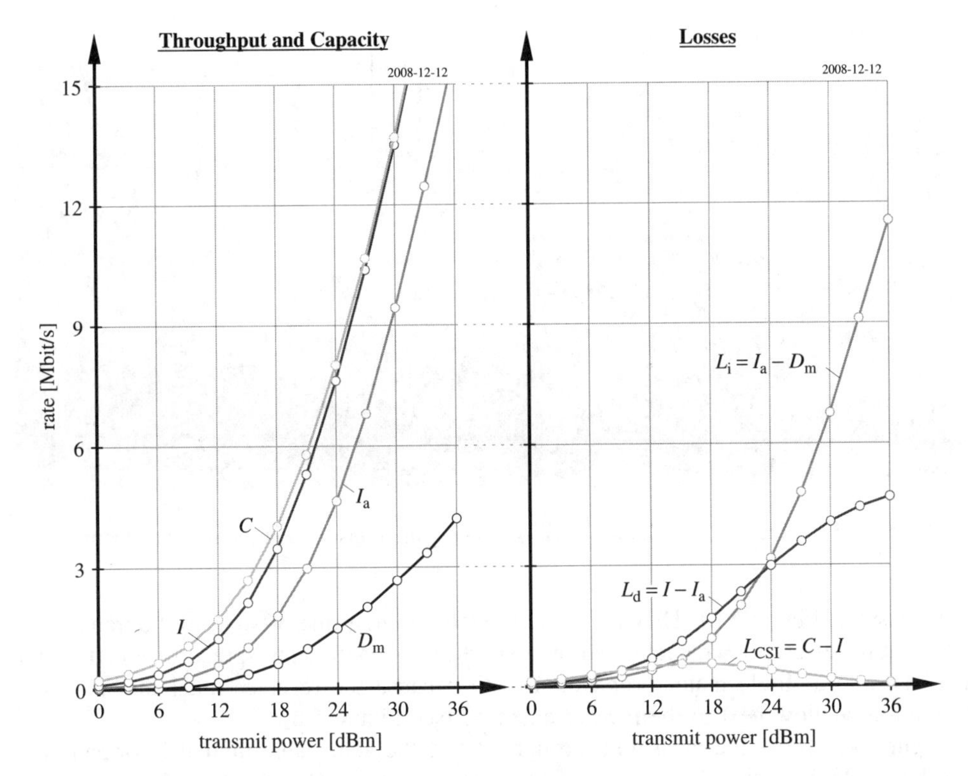

Figure 5.7 Rates of the SISO system in the urban scenario (ID "2008-12-12"). Left-hand side: channel capacity C, mutual information I, achievable mutual information I_a, and measured throughput D_m. Right-hand side: CSI loss L_{CSI}, design loss L_d, and implementation loss L_i.

results with simulation results, we also compute the average SISO receive SNR as well as the average SISO receive power to show them in all result figures below in additional x-axes (see for example Figure 5.9).

5.6.1 Alpine Scenario

Figure 5.8 (p. 126) shows a picture of the alpine scenario in the Drau valley in Austria. In this scenario, the transmit antennas are placed immediately adjacent to existing NodeB antennas on one side of the valley. The receive unit is located at a distance of 5.7 km – without obstacles between transmit and receive units – inside a house on the opposite side of the valley.

At the NodeB, we employ two Kathrein 800 10543 [25] 2X-pol base station panel antennas ($2 \times \pm 45°$ polarization, half-power beam width 60°/6.5°, down tilt 6°); see Figure 5.8. At the receiver we utilize standard (and cheap) Linksys WiFi-router rod

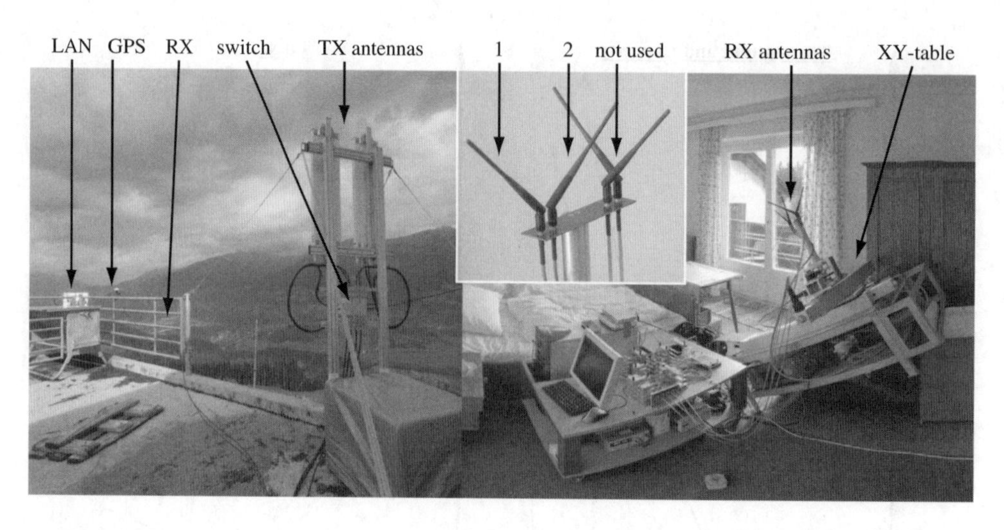

Figure 5.8 Picture of the alpine scenario showing the transmit and the receive antennas.

antennas (see Figure 5.8). They are placed indoors in non-line-of-sight to the transmitter. The complete setup is characterized by a short mean RMS delay spread of about 1 chip (260 ns) and a single major propagation path because the receive signal mainly enters through a window next to the receive antennas (see Figure 5.8).

Figure 5.9 shows the measured throughput and the achievable mutual information of the 1 × 1 SISO, the 1 × 2 Single-Input Multiple-Output (SIMO), the 2 × 1 TxAA, and the 2 × 2 D-TxAA transmission systems in the alpine scenario. Several observations can be made with regard to Figure 5.9:

- The measured throughput of the 2 × 1 TxAA system is significantly (approximately 3 dB) better than the throughput of the SISO system. The scalar precoding of the 2 × 1 TxAA system thus performs very well in this scenario with low delay spread.
- The 2 × 2 D-TxAA system also performs very well, achieving more than twice the throughput of the SISO system. Especially at high SNR, D-TxAA benefits from its optional second data stream.
- The achievable mutual information of the 2 × 1 TxAA system is approximately 3 dB better than that of the SISO system. This is in accordance with the measured data throughput curves, which also show a 3 dB performance increase for the 2 × 1 TxAA system.
- The achievable mutual information of the 2 × 2 D-TxAA system has a steeper slope than that of the other systems. This is a consequence of the higher channel rank of the 2 × 2 D-TxAA system.

The comparison between the measured throughput and the achievable mutual information should be performed at a transmit power of approximately 10–25 dBm or a throughput of approximately 5 Mbit/s to avoid saturation effects of the measured throughput (since outside this range neither smaller nor larger CQI values are available for channel adaptation). At large transmit power levels at which the single-stream transmission

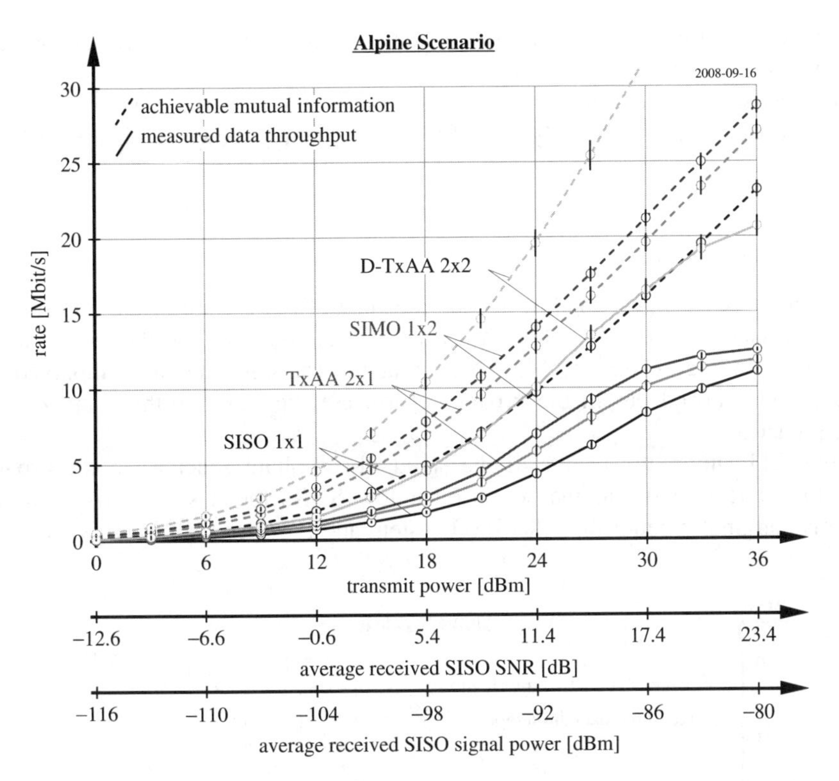

Figure 5.9 Throughput results of the standard compliant schemes in the alpine scenario (ID "2008-09-16"). Averaging was performed over 110 receiver positions. The solid lines represent the measured throughput and the dashed lines the achievable mutual information.

modes saturate, such a comparison would be unfair since the throughput could be easily increased by providing additional modulation and coding schemes. At a throughput of 5 Mbit/s, the measured SISO and 2×2 D-TxAA throughputs lose approximately 7 dB and 6 dB, respectively, compared with their corresponding achievable mutual information.

Figure 5.10 shows the throughput of the four-transmit-antenna HSDPA schemes compared with the standard-compliant 2×2 D-TxAA system in the alpine scenario. We observe:

- The measured throughput of the four-transmit-antenna schemes is increased significantly. For the 4×4 system, the measured throughput is approximately twice that of the 2×2 system. In terms of SNR, the 4×4 system gains approximately 6 dB compared with the 2×2 system.
- The theoretical maximum throughput of all four schemes plotted in Figure 5.10 is 27.952 Mbit/s. However, the measured throughput saturates before reaching this maximum value because of residual interference after the equalization. In the four-receive-antenna schemes, the equalizer is able to reduce the post-equalization interference to a lower level than in the two-receive-antenna schemes. Therefore, the maximum

measured data throughput of the four-receive-antenna schemes is greater than that of the two-receive-antenna schemes.

- A comparison of the achievable mutual information with the measured throughput reveals losses of approximately 6–7 dB for the schemes shown in Figure 5.10.

5.6.2 Urban Scenario

The urban scenario is described in Section 6.3 (p. 144). A picture of the transmit side is shown in Figure 6.2 (p. 145), while the office with the receiver and the receive antennas can be seen in Figure 6.3 (p. 145). Note that the transmit antennas employed in the measurements carried out in the urban scenario are the same as those utilized in the alpine scenario.

Figure 5.11 shows the results of the standard-compliant schemes in the urban scenario. In contrast to the alpine scenario, at low SNRs the 2×1 TxAA system only performs marginally better than the SISO system and is worse than SISO when the SNR

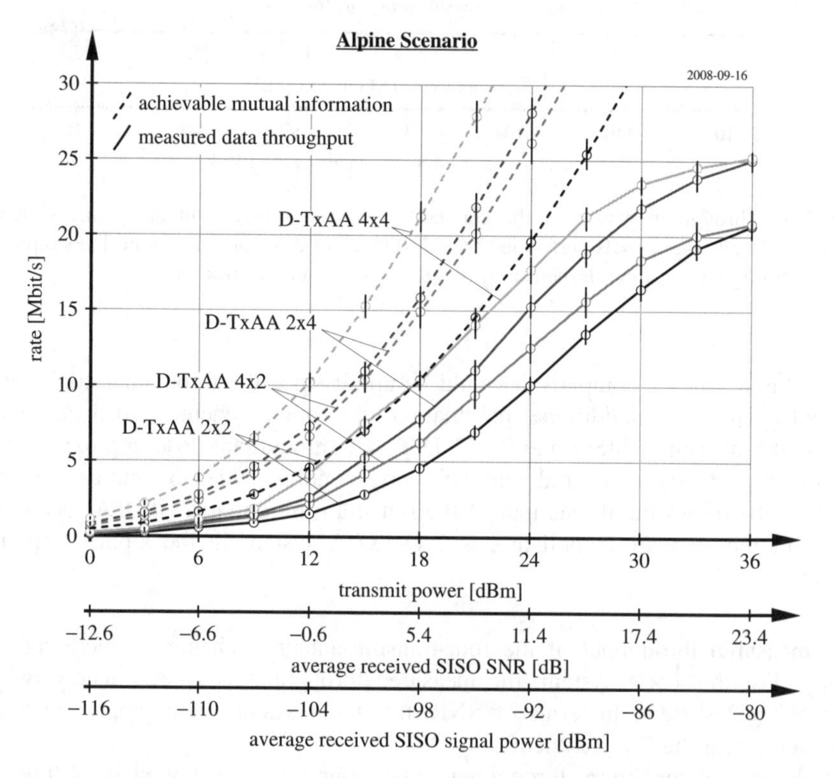

Figure 5.10 Throughput results of the advanced schemes in the alpine scenario (ID "2008-09-16"). Averaging was performed over 110 receiver positions. The solid lines represent the measured throughput and the dashed lines the achievable mutual information.

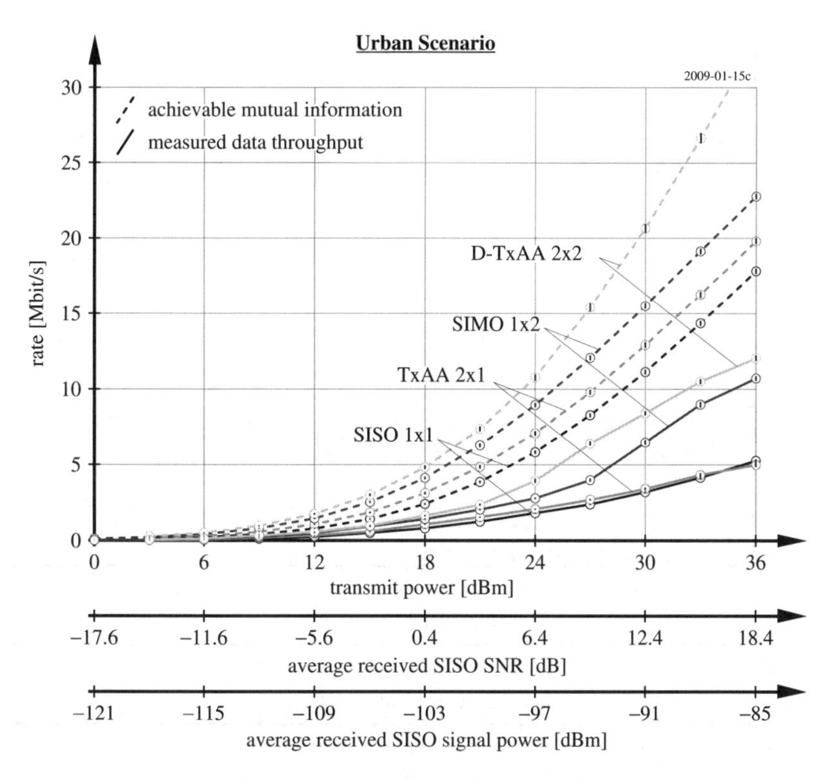

Figure 5.11 Throughput results of the standard-compliant schemes in the urban scenario (ID "2009-01-15c"). Averaging was performed over 484 receiver positions. The solid lines represent the measured throughput and the dashed lines the achievable mutual information.

is large. Thus, it is not favorable to apply the standard-compliant precoding of HSDPA in the urban scenario at high SNR. The reason for this surprising result is the rather large maximum delay spread of approximately 20 chips, which destroys the orthogonality of the spreading codes. As a consequence, the channel estimator performance suffers because the pilots also experience high interference. In the case of transmissions with precoding, the channel estimator performance suffers even more because the available pilot power is divided up among the transmit antennas while the total data power, which acts as interference on the pilots, remains the same. The throughput degradation measured is explained by the fact that at high SNR the HSDPA performance mainly depends on the channel estimator performance, as explained in Section 5.4.4 and [40]. Comparing the measured data throughput with the achievable mutual information yields for the SISO system approximately 9 dB loss in terms of SNR, which is approximately 3 dB more than in the alpine scenario. As in the alpine scenario, the 2×2 D-TxAA system yields approximately twice the throughput of the SISO system.

The results for the four-transmit-antenna schemes in the urban scenario are plotted in Figure 5.12. Here, the gains of the four-transmit-antenna schemes are approximately the

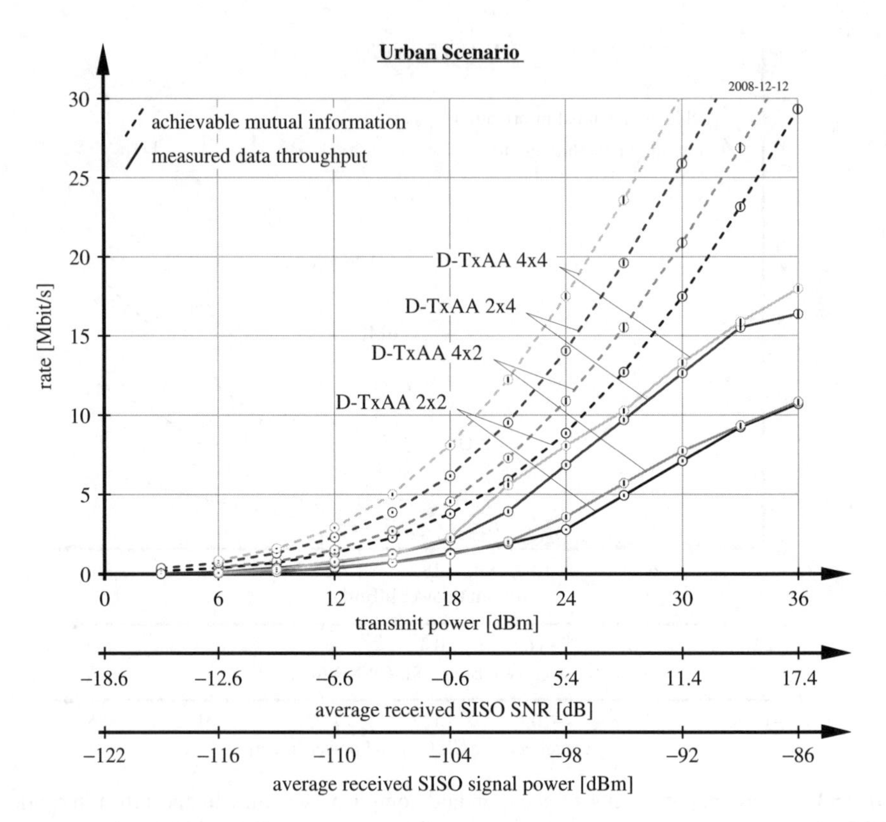

Figure 5.12 Throughput results of the advanced schemes in the urban scenario (ID "2008-12-12"). Averaging was performed over 484 receiver positions. The solid lines represent the measured throughput and the dashed lines the achievable mutual information.

same as in the alpine scenario. The 4×4 system outperforms the 2×2 system by slightly more than 6 dB in SNR or by more than a factor of two in terms of throughput. Figure 5.12 also shows that, in the urban scenario, most of the throughput gains are due to the four receive antennas. Four transmit antennas only yield small throughput gains in this scenario.

5.6.3 Discussion of the Implementation Loss

Although the results of the previous sections show a significant increase in performance of the different MIMO schemes when compared to the SISO transmission, all measured throughput curves are approximately 6 to 9 dB worse than the achievable mutual information. The effects below contribute (perhaps together with other possible causes) to this loss:

- The rate-matched turbo code utilized in HSDPA is good but not optimal. By carrying out a set of comparative Additive White Gaussian Noise (AWGN) simulations, we found that at higher code rates, the rate-matched turbo code loses up to 2 dB when decoded by a max-log-Maximum A-Posteriori (MAP) decoder.

- The equalizer based on an LMMSE channel estimator representing a low-complexity and cost-effective solution is also not optimal. Better receivers, such as the LMMSE-MAP, have the potential to improve the performance by approximately 1 dB [69].
- In the urban scenario, a larger throughput loss was measured than in the alpine scenario because of the larger delay spread and, consequently, the larger inter-code interference. For example, in the alpine scenario the SISO system loses approximately 6 dB to the achievable mutual information, whereas the loss in the urban scenario is approximately 9 dB.
- In addition to the above-mentioned losses, channel estimation errors and overestimation/underestimation of the post-equalization SINR degrade the measured throughput. Exactly quantifying the loss caused by these effects is difficult, because neither perfect channel state information nor perfect post-equalization SINR is available in measurements.

5.7 Summary

This chapter presents MIMO HSDPA throughput measurement results obtained in two extensive measurement campaigns. The campaigns were carried out in an alpine valley in

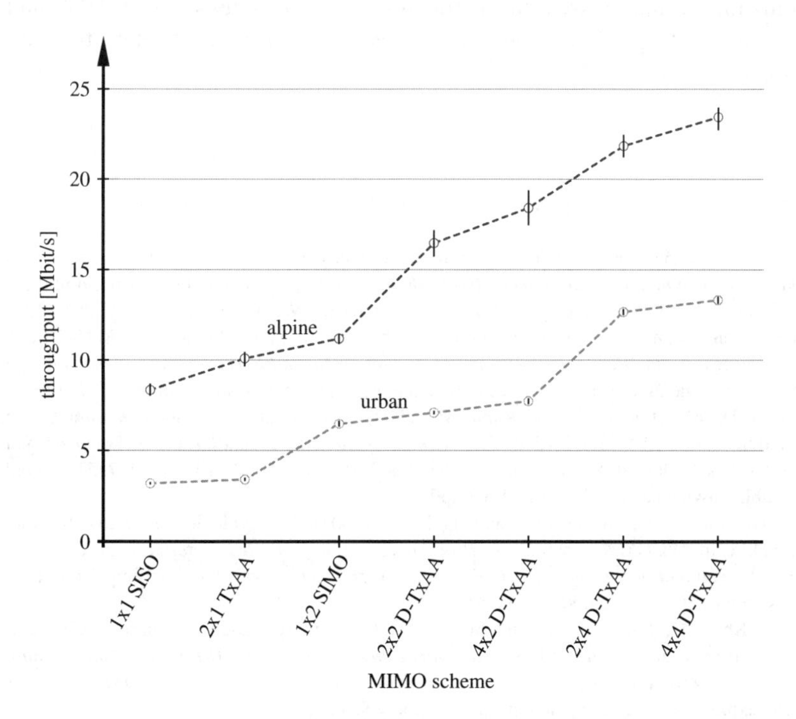

Figure 5.13 Throughput of the different MIMO schemes in the alpine and the urban scenarios for transmit power $P_{\text{TX}} = 30\,\text{dBm}$.

Austria and in the inner city of Vienna, Austria. The scenarios differ significantly in the delay spread of the channel. In both scenarios, multiple antennas considerably increase the physical layer throughput. The standard compliant 2×2 system increases the physical-layer throughput by more than a factor of two compared with the SISO system, while the 4×4 system further increases the throughput by a factor of two. Figure 5.13 provides a comprehensive overview of the MIMO gains in HSDPA.

In order to compare the measured throughput with a performance bound, the achievable mutual information is defined. This achievable mutual information is calculated on the basis of the mutual information of the channel and the precoding employed at the transmitter. Comparison of the measured throughput and the achievable mutual information shows that the measured throughput is far from optimal, losing between 6 and 9 dB in SNR. This loss is caused by the channel coding (approximately 2 dB), the suboptimal equalizer based on LMMSE channel estimation (approximately 1 dB), inter-code interference (3 dB additional loss in the urban scenario compared with the alpine scenario), channel estimation errors, feedback errors (caused by overestimation/underestimation of the post-equalization SINR), and perhaps other effects. Besides this implementation loss, we also define a design loss (mutual information minus achievable mutual information) and a CSI loss (capacity minus mutual information). We find that the CSI loss is almost negligible throughout the entire SNR range. The design loss, however, is dominant in the medium transmit power range that corresponds to a receive SNR of approximately 0 dB.

It is worth reading Chapter 8 and comparing it with this chapter, as it provides more insight into the various losses and compares the same setups for HSDPA and Worldwide Inter-operability for Microwave Access (WiMAX), leading to interesting conclusions when comparing both standards.

References

[1] 3GPP (2007) Technical Specification TS 25.214 Version 7.7.0 ‘Physical layer procedures (FDD),’ www.3gpp.org.

[2] Assaad, M. and Zeghlache, D. (2003) ‘On the capacity of HSDPA,’ in *Proceedings of 46th IEEE Global Telecommunications Conference (GLOBECOM 2003)*, volume 1, pp. 60–64, doi: 10.1109/GLOCOM.2003.1258203. Available from http://ieeexplore.ieee.org/stamp/stamp.jsp?arnumber=1258203.

[3] Assaad, M. and Zeghlache, D. (2004) ‘Comparison between MIMO techniques in UMTS-HSDPA system,’ in *Proceedings of 8th IEEE International Symposium on Spread Spectrum Techniques and Applications*, pp. 874–878. Available from http://ieeexplore.ieee.org/stamp/stamp.jsp?arnumber=1371826.

[4] Bosanska, D., Mehlführer, C., and Rupp, M. (2008) ‘Performance evaluation of intra-cell interference cancelation in D-TxAA HSDPA,’ in *Proceedings of the International ITG Workshop on Smart Antennas (WSA 2008)*, Darmstadt, Germany, pp. 338–342, doi: 10.1109/WSA.2008.4475579. Available from http://publik.tuwien.ac.at/files/pub-et_13677.pdf.

[5] Chao, H., Liang, Z., Wang, Y., and Gui, L. (2004) ‘A dynamic resource allocation method for HSDPA in WCDMA system,’ in *Proceedings of the 5th IEE International Conference on 3G Mobile Communication Technologies (3G 2004)*, pp. 569–573. Available from http://ieeexplore.ieee.org/stamp/stamp.jsp?tp=&arnumber=1434541.

[6] Das, A., Khan, F., Sampath, A., and Su, H.-J. (2001) ‘Performance of hybrid ARQ for high speed downlink packet access in UMTS,’ in *Proceedings of the 54th IEEE Vehicular Technology Conference (VTC2001-Fall)*, volume 4, pp. 2133–2137, doi: 10.1109/VTC.2001.957121. Available from http://ieeexplore.ieee.org/stamp/stamp.jsp?arnumber=957121.

[7] Ericsson (2006) '64-QAM for HSDPA – link level simulation results,' Technical Report TSG-RAN Working Group 1 Meeting #46, R1-062264, 3GPP. Available from http://www.3gpp.org/ftp/tsg_ran/WG1_RL1/TSGR1_46/Docs/R1-062264.zip.

[8] Foschini, G. J. and Gans, M. J. (1998) 'On limits of wireless communications in a fading environment when using multiple antennas,' *Wireless Personal Communications*, **6** (3), 311–335. Available from http://www.springerlink.com/content/h1n7866218781520/fulltext.pdf.

[9] García-Naya, J. A., Mehlführer, C., Caban, S. *et al.* (2009) 'Throughput-based antenna selection measurements,' in *Proceedings of the 70th IEEE Vehicular Technology Conference (VTC2009-Fall)*, Anchorage, AK, doi: 10.1109/VETECF.2009.5378992. Available from http://publik.tuwien.ac.at/files/PubDat_176573.pdf.

[10] Garrett, D., Woodward, G., Davis, L. *et al.* (2004) 'A 28.8Mb/s 4×4 MIMO 3G high-speed downlink packet access receiver with normalized least mean square equalization,' in *Digest of Technical Papers. IEEE International Solid-State Circuits Conference 2004 (ISSCC 2004)*, volume 1, pp. 420–536, doi: 10.1109/ISSCC.2004.1332773. Available from http://ieeexplore.ieee.org/stamp/stamp.jsp?arnumber=1332773.

[11] Geirhofer, S., Mehlführer, C., and Rupp, M. (2005) 'Design and real-time measurement of HSDPA equalizers,' in *Proceedings of the 6th IEEE Workshop on Signal Processing Advances in Wireless Communications (SPAWC 2005)*, New York, NY, pp. 166–170, doi: 10.1109/SPAWC.2005.1505893. Available from http://publik.tuwien.ac.at/files/pub-et_9722.pdf.

[12] Golub, G. H. and van Loan, C. F. (eds.) (1996) *Matrix Computations*, third edition, The Johns Hopkins University Press, Baltimore, MD.

[13] Guo, Y., Zhang, J., McCain, D., and Cavallaro, J. (2004) 'Efficient MIMO equalization for downlink multicode CDMA: complexity optimization and comparative study,' in *Proceedings of the 47th IEEE Global Telecommunications Conference (GLOBECOM 2004)*, volume 4, pp. 2513–2519, doi: 10.1109/GLOCOM.2004.1378459. Available from http://ieeexplore.ieee.org/stamp/stamp.jsp?arnumber=1378459.

[14] Guo, Y., Zhang, J., McCain, D., and Cavallaro, J. R. (2006) 'An efficient circulant MIMO equalizer for CDMA downlink: algorithm and VLSI architecture,' *EURASIP Journal on Applied Signal Processing*, **2006**, article ID 57134, doi: 10.1155/ASP/2006/57134. Available from http://downloads.hindawi.com/journals/asp/2006/057134.pdf.

[15] Harteneck, M., Boloorian, M., Georgoulis, S., and Tanner, R. (2004) 'Practical aspects of an HSDPA 14Mbps terminal,' in *Conference Record of the 38th Asilomar Conference on Signals, Systems and Computers*, volume 1, Pacific Grove, CA, pp. 799–803. Available from http://ieeexplore.ieee.org/stamp/stamp.jsp?arnumber=1399246.

[16] Harteneck, M., Boloorian, M., Georgoulis, S., and Tanner, R. (2005) 'Throughput measurements of HSDPA 14 Mbit/s terminal,' *IEE Electronics Letters*, **41** (7), 425–427, doi: 10.1049/el:20058362. Available from http://ieeexplore.ieee.org/stamp/stamp.jsp?arnumber=1421242.

[17] Hassibi, B. and Hochwald, B. (2003) 'How much training is needed in multiple-antenna wireless links?' *IEEE Transactions on Information Theory*, **49** (4), 951–963, doi: 10.1109/TIT.2003.809594. Available from http://ieeexplore.ieee.org/stamp/stamp.jsp?arnumber=1193803.

[18] Heikkila, M. and Majonen, K. (2004) 'Increasing HSDPA throughput by employing space–time equalization,' in *Proceedings of the 15th IEEE International Symposium on Personal, Indoor and Mobile Radio Communications (PIMRC 2004)*, volume 4, pp. 2328–2332. Available from http://ieeexplore.ieee.org/stamp/stamp.jsp?tp=&arnumber=1368735.

[19] Holma, H. and Reunanen, J. (2006) '3GPP release 5 HSDPA measurements,' in *Proceedings of the 17th IEEE International Symposium on Personal, Indoor and Mobile Radio Communications (PIMRC 2006)*, doi: 10.1109/PIMRC.2006.254116. Available from http://ieeexplore.ieee.org/stamp/stamp.jsp?arnumber=4022310.

[20] Ikuno, J. C., Wrulich, M., and Rupp, M. (2010) 'System level simulation of LTE networks,' in *Proceedings of the 2010 IEEE 71st Vehicular Technology Conference (VTC2010-Spring)*, Taipei, Taiwan, doi: 10.1109/VETECS.2010.5494007. Available from http://publik.tuwien.ac.at/files/PubDat_184908.pdf.

[21] Isotalo, T., Lähdekorpi, P., and Lempiäinen, J. (2008) 'Improving HSDPA indoor coverage and throughput by repeater and dedicated indoor system,' *EURASIP Journal on Wireless Communications and Networking*, **2008**, article ID 951481.

[22] Isotalo, T. and Lempiäinen, J. (2007) 'HSDPA measurements for indoor DAS,' in *Proceedings of the 65th IEEE Vehicular Technology Conference (VTC2007-Spring)*, pp. 1127–1130, doi: 10.1109/VETECS.2007.239. Available from http://ieeexplore.ieee.org/stamp/stamp.jsp?arnumber=4212667.

[23] ITU (1997) 'Recommendation ITU-R M.1225: guidelines for evaluation of radio transmission technologies for IMT-2000,' Technical report, ITU.

[24] Jurvansuu, M., Prokkola, J., Hanski, M., and Perala, P. (2007) 'HSDPA performance in live networks,' in *Proceedings of the IEEE International Conference on Communications (ICC 2007)*, pp. 467–471, doi: 10.1109/ICC.2007.83. Available from http://ieeexplore.ieee.org/stamp/stamp.jsp?arnumber=4288754.

[25] Kathrein 'Technical specification Kathrein antenna type no. 742 211.' Available from http://www.kathrein.de/de/mca/produkte/download/9362108g.pdf.

[26] Kathrein 'Technical specification Kathrein antenna type no. 800 10629.' Available from http://www.nt.tuwien.ac.at/fileadmin/data/testbed/kat-ant.pdf.

[27] Kolding, T., Frederiksen, F., and Mogensen, P. (2002) 'Performance aspects of WCDMA systems with high speed downlink packet access (HSDPA)', in *Proceedings of the 56th IEEE Vehicular Technology Conference (VTC2002-Fall)*, volume 1, pp. 477–481, doi: 10.1109/VETECF.2002.1040389. Available from http://ieeexplore.ieee.org/stamp/stamp.jsp?arnumber=1040389.

[28] Kolding, T. E., Pedersen, K. I., Wigard, J. *et al.* (2003) 'High speed downlink packet access: WCDMA evolution,' *IEEE Vehicular Technology Society News*, (February), 4–10. Available from http://kom.aau.dk/group/05gr943/literature/hsdpa/evolution%20of%20HSDPA.pdf.

[29] Kunze, J., Schmits, C., Bilgic, A., and Hausner, J. (2008) 'Receive antenna diversity architectures for HSDPA,' in *Proceedings of the 67th IEEE Vehicular Technology Conference (VTC2008-Spring)*, pp. 2071–2075, doi: 10.1109/VETECS.2008.465. Available from http://ieeexplore.ieee.org/stamp/stamp.jsp?tp=&arnumber=4526021.

[30] Kwan, R., Aydin, M., Leung, C., and Zhang, J. (2008) 'Multiuser scheduling in HSDPA using simulated annealing,' in *Proceedings of the International Wireless Communications and Mobile Computing Conference (IWCMC 2008)*, pp. 236–241, doi: 10.1109/IWCMC.2008.42. Available from http://ieeexplore.ieee.org/stamp/stamp.jsp?tp=&arnumber=4599941.

[31] Landre, J.-B. and Saadani, A. (2008) 'HSDPA 14,4Mbps mobiles – realistic throughputs evaluation,' in *Proceedings of the 67th IEEE Vehicular Technology Conference (VTC2008-Spring)*, pp. 2086–2090, doi: 10.1109/VETECS.2008.468. Available from http://ieeexplore.ieee.org/stamp/stamp.jsp?arnumber=4526024.

[32] Love, R., Ghosh, A., Nikides, R. *et al.* (2001) 'High speed downlink packet access performance,' in *Proceedings of the 53rd IEEE Vehicular Technology Conference (VTC2001-Spring)*, volume 3, pp. 2234–2238, doi: 10.1109/VETECS.2001.945093. Available from http://ieeexplore.ieee.org/stamp/stamp.jsp?arnumber=945093.

[33] Love, R., Ghosh, A., Xiao, W., and Ratasuk, R. (2004) 'Performance of 3GPP high speed downlink packet access (HSDPA),' in *Proceedings of the IEEE 60th Vehicular Technology Conference (VTC 2004-Fall)*, volume 5, pp. 3359–3363, doi: 10.1109/VETECF.2004.1404686. Available from http://ieeexplore.ieee.org/stamp/stamp.jsp?tp=&arnumber=1404686.

[34] Love, R., Stewart, K., Bachu, R., and Ghosh, A. (2003) 'MMSE equalization for UMTS HSDPA,' in *Proceedings of the IEEE 58th Vehicular Technology Conference (VTC 2003-Fall)*, volume 4, pp. 2416–2420, doi: 10.1109/VETECF.2003.1285963. Available from http://ieeexplore.ieee.org/stamp/stamp.jsp?tp=&arnumber=1285963.

[35] Mailaender, L. (2005) 'Linear MIMO equalization for CDMA downlink signals with code reuse,' *IEEE Transactions on Wireless Communications*, **4** (5), 2423–2434. Available from http://ieeexplore.ieee.org/iel5/7693/32683/01532226.pdf.

[36] Malkowski, M. (2007) 'Link-level comparison of IP-OFDMA (mobile WiMAX) and UMTS HSDPA,' in *Proceedings of the IEEE 18th International Symposium on Personal, Indoor and Mobile Radio Communications (PIMRC 2007)*, doi: 10.1109/PIMRC.2007.4394134. Available from http://ieeexplore.ieee.org/stamp/stamp.jsp?tp=&arnumber=4394134.

[37] Mehlführer, C., Caban, S., and Rupp, M. (2008) 'An accurate and low complex channel estimator for OFDM WiMAX,' in *Proceedings of the 3rd International Symposium on Communications, Control and Signal Processing*, Malta, pp. 922–926, doi: 10.1109/ISCCSP.2008.4537355. Available from http://publik.tuwien.ac.at/files/pub-et_13650.pdf.

[38] Mehlführer, C., Caban, S., and Rupp, M. (2008) 'Measurement based evaluation of low complexity receivers for D-TxAA HSDPA,' in *Proceedings of the 16th European Signal Processing Conference (EUSIPCO 2008)*, Lausanne, Switzerland. Available from http://publik.tuwien.ac.at/files/PubDat_166132.pdf.

[39] Mehlführer, C., Caban, S., and Rupp, M. (2010) 'Measurement-based performance evaluation of MIMO HSDPA,' *IEEE Transactions on Vehicular Technology*, **59** (9), 4354–4367, doi: 10.1109/TVT.2010.2066996. Available from http://publik.tuwien.ac.at/files/PubDat_187112.pdf.

[40] Mehlführer, C., Caban, S., Wrulich, M., and Rupp, M. (2008) 'Joint throughput optimized CQI and precoding weight calculation for MIMO HSDPA,' in *Conference Record of the 42nd Asilomar Conference on Signals, Systems and Computers*, Pacific Grove, CA, pp. 1320–1325, doi: 10.1109/ACSSC.2008.5074632. Available from http://publik.tuwien.ac.at/files/PubDat_167015.pdf.

[41] Mehlführer, C. and Rupp, M. (2006) 'A robust MMSE equalizer for MIMO enhanced HSDPA,' in *Conference Record of the 40th Asilomar Conference on Signals, Systems and Computers*, Pacific Grove, CA, pp. 129–133, doi: 10.1109/ACSSC.2006.356599. Available from http://publik.tuwien.ac.at/files/pub-et_11498.pdf.

[42] Mehlführer, C. and Rupp, M. (2008) 'Novel tap-wise LMMSE channel estimation for MIMO W-CDMA,' in *Proceedings of the 51st IEEE Global Telecommunications Conference (GLOBECOM 2008)*, New Orleans, LA, doi: 10.1109/GLOCOM.2008.ECP.829. Available from http://publik.tuwien.ac.at/files/PubDat_169129.pdf.

[43] Mehlführer, C., Wrulich, M., and Rupp, M. (2008) 'Intra-cell interference aware equalization for TxAA HSDPA,' in *Proceedings of the 3rd IEEE International Symposium on Wireless Pervasive Computing (ISWPC 2008)*, Santorini, Greece, pp. 406–409, doi: 10.1109/ISWPC.2008.4556239. Available from http://publik.tuwien.ac.at/files/pub-et_13749.pdf.

[44] Moon, T. K. and Stirling, W. C. (2000) *Mathematical Methods and Algorithms for Signal Processing*, first edition, Prentice Hall, Upper Saddle River, NJ.

[45] Moulsley, T. (2001) 'Throughput of high speed downlink packet access for UMTS,' in *Proceedings of the 2nd IEEE International Conference on 3G Mobile Communication Technologies (3G 2001)*, pp. 363–367. Available from http://ieeexplore.ieee.org/stamp/stamp.jsp?arnumber=923569.

[46] Naja, R., Claude, J.-P., and Tohme, S. (2008) 'Adaptive multi-user fair packet scheduling in HSDPA network,' in *Proceedings of the International Conference on Innovations in Information Technology (IIT 2008)*, pp. 406–410, doi: 10.1109/INNOVATIONS.2008.4781652. Available from http://ieeexplore.ieee.org/stamp/stamp.jsp?tp=&arnumber=4781652.

[47] Nakamura, M., Awad, Y., and Vadgama, S. (2002) 'Adaptive control of link adaptation for high speed downlink packet access (HSDPA) in W-CDMA,' in *Proceedings of the 5th IEEE International Symposium on Wireless Personal Multimedia Communications*, volume 2, pp. 382–386, doi: 10.1109/WPMC.2002.1088198. Available from http://ieeexplore.ieee.org/stamp/stamp.jsp?arnumber=1088198.

[48] Parkvall, S., Dahlman, E., Frenger, P. *et al.* (2001) 'The evolution of WCDMA towards higher speed downlink packet data access,' in *Proceedings of the 53rd IEEE Vehicular Technology Conference (VTC2001-Spring)*, volume 3, pp. 2287–2291, doi: 10.1109/VETECS.2001.945103. Available from http://ieeexplore.ieee.org/stamp/stamp.jsp?arnumber=945103.

[49] Parkvall, S., Dahlman, E., Frenger, P. *et al.* (2001) 'The high speed packet data evolution of WCDMA,' in *Proceedings of the 12th IEEE International Symposium on Personal, Indoor and Mobile Radio Communications (PIMRC 2001)*, volume 2, doi: 10.1109/PIMRC.2001.965315. Available from http://ieeexplore.ieee.org/stamp/stamp.jsp?arnumber=965315.

[50] Paulraj, A., Nabar, R., and Gore, D. (2003) *Introduction to Space–Time Wireless Communications*, first edition, Cambridge University Press, Cambridge, UK.

[51] Pedersen, K., Lootsma, T., Støttrup, M. *et al.* (2004) 'Network performance of mixed traffic on high speed downlink packet access and dedicated channels in WCDMA,' in *Proceedings of the IEEE 60th Vehicular Technology Conference (VTC 2004-Fall)*, volume 6, pp. 4496–4500, doi: 10.1109/VETECF.2004.1404930. Available from http://ieeexplore.ieee.org/stamp/stamp.jsp?tp=&arnumber=1404930.

[52] Petersen, K. B. and Pedersen, M. S. (eds.) (2008) *The Matrix Cookbook*. Available from http://matrixcookbook.com, November 14.

[53] Raleigh, G. G. and Cioffi, J. M. (1998) 'Spatio-temporal coding for wireless communication,' *IEEE Transactions on Communications*, **46** (3), 357–366. Available from http://www-isl.stanford.edu/cioffi/dsm/wlpap/mimocioffi98.pdf.

[54] Riback, M., Grant, S., Jongren, G. *et al.* (2007) 'MIMO-HSPA testbed performance measurements,' in *Proceedings of the 18th IEEE International Symposium on Personal, Indoor and Mobile Radio Communications (PIMRC 2007)*, doi: 10.1109/PIMRC.2007.4394434. Available from http://ieeexplore.ieee.org/stamp/stamp.jsp?tp=&arnumber=4394434.

[55] Rupp, M., García-Naya, J. A., Mehlführer, C. *et al.* (2010) 'On mutual information and capacity in frequency selective wireless channels,' in *Proceedings of the IEEE International Conference on Communications (ICC 2010)*, Cape Town, South Africa, doi: 10.1109/ICC.2010.5501942. Available from http://publik.tuwien.ac.at/files/PubDat_184660.pdf.

[56] Samardzija, D., Lozano, A., and Papadias, C. (2004) 'Experimental validation of MIMO multiuser detection for UMTS high-speed downlink packet access,' in *Proceedings of the 47th IEEE Global Telecommunications Conference 2004 (GLOBECOM 2004)*, volume 6, pp. 3840–3844, doi: 10.1109/GLOCOM.2004.1379087. Available from http://ieeexplore.ieee.org/stamp/stamp.jsp?arnumber=1379087.

[57] Siomina, I. and Yuan, D. (2008) 'Enhancing HSDPA performance via automated and large-scale optimization of radio base station antenna configuration,' in *Proceedings of the 67th IEEE Vehicular Technology Conference (VTC2008-Spring)*, pp. 2061–2065, doi: 10.1109/VETECS.2008.463. Available from http://ieeexplore.ieee.org/stamp/stamp.jsp?tp=&arnumber=4526019.

[58] Stuhlberger, R., Maurer, L., Hueber, G., and Springer, A. (2006) 'The impact of RF-impairments and automatic gain control on UMTS-HSDPA-throughput performance,' in *Proceedings of the IEEE 64th Vehicular Technology Conference (VTC 2006-Fall)*, doi: 10.1109/VTCF.2006.389. Available from http://ieeexplore.ieee.org/stamp/stamp.jsp?tp=&arnumber=4109654.

[59] Szabo, A., Geng, N., Seeger, A., and Utschick, W. (2003) 'Investigations on link to system level interface for MIMO systems,' in *Proceedings of the 3rd International Symposium on Image and Signal Processing and Analysis 2003 (ISPA2003)*, Rome, Italy, pp. 365–369. Available from http://ieeexplore.ieee.org/iel5/9084/28837/01296924.pdf?tp=&isnumber=&arnumber=1296924.

[60] Telatar, I. E. (1999) 'Capacity of multi-antenna Gaussian channels,' *European Transactions on Telecommunications*, **10** (6), 585–595. Available from http://mars.bell-labs.com/papers/proof/proof.pdf.

[61] Tresch, R., Mehlführer, C., and Guillaud, M. (2008) 'LMMSE channel estimation for MIMO W-CDMA with out-of-cell interference mitigation,' in *Conference Record of the 42nd Asilomar Conference on Signals, Systems and Computers*, Pacific Grove, CA, USA, pp. 331–335, doi: 10.1109/ACSSC.2008.5074419. Available from http://publik.tuwien.ac.at/files/PubDat_167781.pdf.

[62] Wrulich, M., Eder, S., Viering, I., and Rupp, M. (2008) 'Efficient link-to-system level model for MIMO HSDPA,' in *Proceedings of the 4th IEEE Broadband Wireless Access Workshop*. Available from http://publik.tuwien.ac.at/files/PubDat_170334.pdf.

[63] Wrulich, M., Mehlführer, C., and Rupp, M. (2008) 'Interference aware MMSE equalization for MIMO TxAA,' in *Proceedings of the 3rd International Symposium on Communications, Control and Signal Processing*, St. Julians, Malta, pp. 1585–1589, doi: 10.1109/ISCCSP.2008.4537480. Available from http://publik.tuwien.ac.at/files/pub-et_13657.pdf.

[64] Wrulich, M., Mehlführer, C., and Rupp, M. (2010) 'Advanced receivers for MIMO HSDPA,' in *HSDPA/HSUPA Handbook* (eds. B. Furht and S. Ahson), CRC Press, Boca Raton, FL, pp. 89–110.

[65] Wrulich, M., Mehlführer, C., and Rupp, M. (2010) 'Managing the interference structure of MIMO HSDPA: a multi-user interference aware MMSE receiver with moderate complexity,' *IEEE Transactions on Wireless Communications*, **9** (4), 1472–1482, doi: 10.1109/TWC.2010.04.090612. Available from http://publik.tuwien.ac.at/files/PubDat_180743.pdf.

[66] Wrulich, M. and Rupp, M. (2008) 'Efficient link measurement model for system level simulations of Alamouti encoded MIMO HSDPA transmissions,' in *Proceedings of the ITG International Workshop on Smart Antennas (WSA 2008)*, Darmstadt, Germany. Available from http://publik.tuwien.ac.at/files/pub-et_13641.pdf.

[67] Wrulich, M. and Rupp, M. (2009) 'Computationally efficient MIMO HSDPA system-level modeling,' *EURASIP Journal on Wireless Communications and Networking*, **2009**, article ID 382501, doi: 10.1155/2009/382501.

[68] Wrulich, M., Weiler, W., and Rupp, M. (2008) 'HSDPA performance in a mixed traffic network,' in *Proceedings of the 67th IEEE Vehicular Technology Conference Spring (VTC2008-Spring)*, Singapore, pp. 2056–2060. Available from http://publik.tuwien.ac.at/files/pub-et_13769.pdf.
[69] Ylioinas, J., Hooli, K., Kiiskila, K., and Juntti, M. (2004) 'Interference suppression in MIMO HSDPA communication,' in *Proceedings of the of the 6th Nordic Signal Processing Symposium (NORSIG 2004)*, pp. 228–231. Available from http://ieeexplore.ieee.org/stamp/stamp.jsp?arnumber=1344565.

6

HSDPA Antenna Selection Techniques

Contributed by José Antonio García-Naya
University of A Coruña, Spain, and
Vienna University of Technology (TU Wien), Austria

One[1] of the main reasons for the slow introduction of Multiple-Input Multiple-Output (MIMO) technologies in commercial wireless systems is the required increment in complexity and hardware cost with respect to traditional, single-antenna systems. While antenna elements are cheap and usually small, each one requires a complete Radio Frequency (RF) chain (low-noise amplifier, frequency downconverter, analog-to-digital converter, filters). Unfortunately, RF hardware is expensive compared with digital hardware, and it does not follow Moore's law [40]. Additionally, the introduction of new hardware implies more energy consumption, which is very inconvenient for today's handheld mobile devices.

On the other hand, simulations of MIMO systems usually assume that individual single-antenna links comprising a full MIMO system suffer from the same average path loss (for example, see [3, 17, 22, 26, 44, 45]). In practice, however, this is rarely observed because realistic scattering environments are highly asymmetrical, especially when differently polarized or orientated antennas are employed. Therefore, a good strategy to extract the diversity gain offered by MIMO systems – while still maintaining the complexity of the system at a reasonable level – consists in selecting the antenna(s) offering the best performance while keeping the others idle. Such a strategy is known as antenna switching or antenna selection. The performance obtained is mainly determined by the ability of the selection criterion employed to extract all diversity gain offered by the MIMO link and the way implementation issues are solved. Hence, the ideal best selection criterion

[1] Reproduced from [13], Garcia-Naya, J. A., Mehlführer, C., Caban, S., Rupp, M., Castedo, L., "Throughput-based antenna selection measurements," *2009 IEEE 70th Vehicular Technology Conference Fall (VTC 2009-Fall)*, pp. 1–5, 20–23 Sept. 2009, by permission of © IEEE 2009.

Evaluation of HSDPA and LTE: From Testbed Measurements to System Level Performance, First Edition.
Sebastian Caban, Christian Mehlführer, Markus Rupp and Martin Wrulich.

should guarantee that the antenna subset offering the best performance is always selected, without any other kind of penalty.

During recent years, a lot of research has been carried out trying to find the best approximation to the ideal selection criterion. Most of that research is only focused on the channel state information and does not consider the entire system; that is, the transmitter and/or the receiver. On the other hand, the performance of current wireless mobile systems (for example, High-Speed Downlink Packet Access (HSDPA)) is often limited by the interference of other users, and not by noise. Consequently, choosing the antenna(s) receiving the highest signal strength does not guarantee the best performance.

In this chapter we introduce a novel, throughput-based, receive antenna selection criterion that considers the entire system design (including the receiver) rather than only the channel state information. The main idea of such a criterion consists in selecting the antenna(s) leading to the highest post-equalization Signal to Interference and Noise Ratio (SINR) value. Given that the post-equalization SINR is directly related to the physical-layer throughput (see Section 5.4.4, p. 118), maximizing the post-equalization SINR also results in a throughput maximization. The main advantage of the proposed criterion is the enormous impact observed on the measured throughput, which is much higher than that predicted by analytic results and simulations, because real channels are different than those in theoretical analyses. At the same time, the criterion is especially suitable for interference-limited systems, such as, for example, HSDPA.

We restrict our approach to the receiver side of the Downlink (DL) because a selection criterion at the transmitter side of the DL would have to consider all users at the receiver side, resulting in a nonpractical approach. Interestingly, in a great variety of scenarios, multiple-antenna systems based on antenna selection can achieve the same diversity gains as full-complexity systems at a considerably reduced system complexity. On the downside, antenna selection suffers from a loss in array gain (mean receive Signal to Noise Ratio (SNR) gain), but this loss can be mitigated by adequate preprocessing in the RF domain, as it has been suggested in [35, 39, 47, 48, 56].

With the purpose of showing the performance of the proposed antenna selection criterion in real scenarios, we carried out measurements in a realistic urban outdoor scenario with the "Vienna MIMO testbed" (see Chapter 3, p. 59). The results are presented in terms of HSDPA DL physical-layer throughput. As mentioned above, the main difference between our measurement approach (see Chapter 4, p. 75) and typical simulations reported in the literature is that not all individual Single-Input Single-Output (SISO) links of the full MIMO system offer the same average gain. On the contrary, they exhibit significant differences (up to 3.5 dB on average), thus resulting in a great difference in performance depending on the antenna subset selected. This effect is illustrated in Figure 6.1, in which we plot the average SISO HSDPA throughput measured in the alpine scenario (see Section 5.6.1, p. 125, and Figure 5.8) corresponding to each of the 16 SISO channels between each transmit and receive antenna in a 4×4 MIMO system. Additionally, the average measured SISO HSDPA throughput of all 16 SISO channels is also plotted. Note that at a throughput of 6 Mbit/s, the SNR difference between the best and the worst transmit – receive antenna combination is 3.5 dB. This variability is mainly caused by the antenna pattern and by antenna polarization effects and motivates the application of antenna selection.

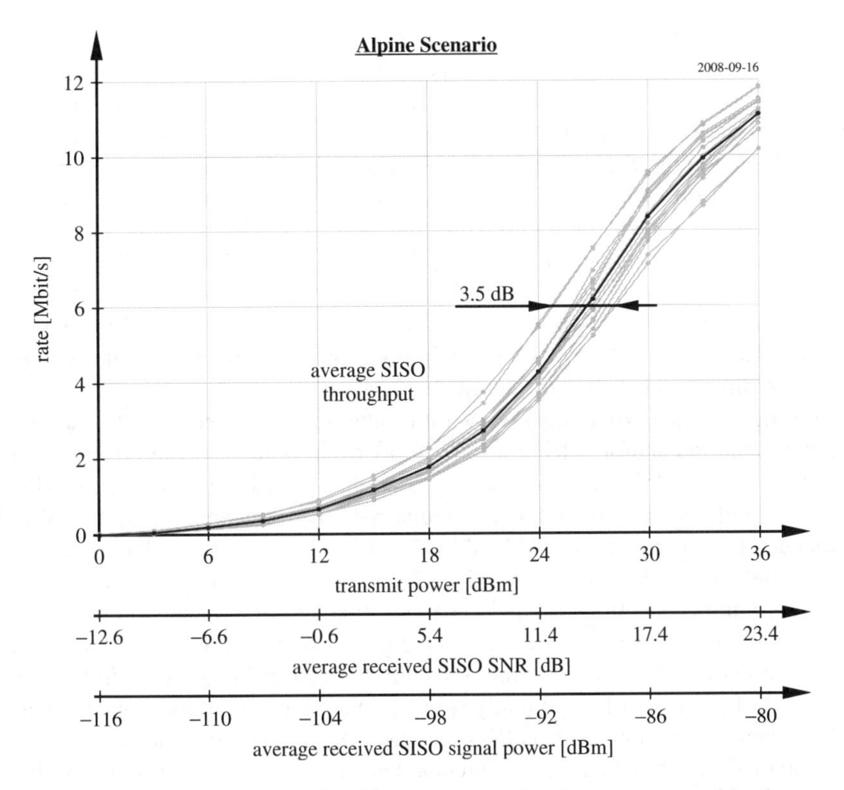

Figure 6.1 SISO HSDPA throughput for different transmit/receive antenna combinations, averaged over 484 different receive antenna positions within an area of $3\,\lambda \times 3\,\lambda$. A picture of the scenario is shown in Figure 5.8 (p. 126).

6.1 Existing Research

In addition to the first publications on antenna selection [30, 31], a considerable amount of research has been reported in the literature with antenna selection as the main topic (for example, see [4, 8, 14, 16, 18–20, 24, 34, 36, 41, 43, 44, 46, 52, 54, 55, 57] and references cited therein). Most of the published work assumes perfect knowledge of the channel at the transmit and at the receive sides, while modeling the channel as i.i.d. flat-fading at each antenna. In [36], the problem of channel frequency selectivity is jointly addressed with some practical issues, such as channel estimation errors and hardware nonidealities. Receive Antenna Selection (RAS) has been extensively studied in [15, 17, 22, 23, 32]. Besides the general antenna selection overview, [21, 53] deal with the problem of the selection criterion for both Transmit Antenna Selection (TAS) and RAS. As mentioned above, most of the available literature addresses the problem from a theoretical point of view and rarely focuses on common practical issues, such as antenna coupling effects, noisy estimates of the channel, or hardware nonidealities. However, some studies are based on more realistic channel models and consider the antenna selection problem for frequency-selective channels when frequency-domain equalizers are used (see [53] and references cited therein), or when only noisy channel estimates are available [1, 2]. Another interesting advantage of antenna selection is given by its potential to reduce the complexity of precoded MIMO Orthogonal Frequency-Division Multiplexing (OFDM) systems [42], including the case of limited feedback [41, 43], as well as its combination with spatial multiplexing [24]. To take advantage of the spatial diversity available in MIMO systems, previous work focused on antenna selection based on the maximization of the unconstrained channel capacity [37, 38]. In recent years, MIMO OFDM systems have become very popular owing to their ability to exploit the benefits from both MIMO and OFDM systems. For this reason, antenna selection has also been studied in such systems. More specifically, research on existing wireless communications systems employed in indoor environments has been carried out [6, 7, 49, 51].

Despite all previous work, only very little attention has been paid to measurements in realistic scenarios and/or selection criteria taking into account the effects of practical implementations. In [25], antenna selection is applied to IEEE 802.11a, and indoor measurements as well as experiments with antenna polarization were carried out. In [49], the performance of TAS and RAS for MIMO OFDM is studied in depth based on measured, correlated, indoor and frequency-selective channels. The results obtained are compared with those computed by the TGn [11] channel model. The comparison reports that the capacity gain predicted from the simulated channels is significantly lower than that achievable with real-world channels. Antenna selection has also been analyzed in the context of the IEEE 802.11n standard [5]. Finally, [35] emphasizes a multitude of practical issues related to antenna selection, such as RF preprocessing, antenna selection training, RF mismatch, nonidealities in selection, or antenna selection in OFDM systems. Additionally, an analysis of capacity in indoor scenarios based on measurements is provided.

6.2 Receive Antenna Selection

Diversity via multiple receive antennas is a direct extension of traditional diversity ideas. The receiver observes several versions of the transmit signal, each one experiencing a different fade and noise. The classic extension of selection combining to RAS consists of choosing the L_R receive antennas that optimize some metric, such as SNR, Bit Error Ratio (BER), or throughput. For that purpose, all channels of the MIMO system need to be known. Typically, this information is obtained by sounding the wireless channel in a time-multiplexed manner. When more than one antenna is selected, Maximum Ratio Combining (MRC) can be performed among L_R out of N_R antennas. Based on this approach, antenna selection can extract most of the diversity of a full MIMO system without requiring all N_R RF chains.

When the wireless channel has sufficient Degrees of Freedom (DoFs), the data streams transmitted from multiple antennas can be separated, hence leading to parallel data paths. The capacity of the radio channel under these conditions (full-rank channel matrix) grows with $\min(N_\mathrm{T}, N_\mathrm{R})$; that is, linearly with the minimum number of antennas [50]. In a multiple-antenna system with N_T transmit and N_R receive antennas, the complex-valued channel matrix $\mathbf{H}$ is of dimension $N_\mathrm{R} \times N_\mathrm{T}$. When RAS is employed, a subset of L_R out of N_R antennas is selected. It is straightforward to see that the equivalent problem consists of selecting a channel matrix $\tilde{\mathbf{H}}$ with dimensions $L_\mathrm{R} \times N_\mathrm{T}$ from the full channel matrix **H**. The selection requires sounding all possible paths using the available L_R RF chains at the receiver, requiring $\lceil N_R/L_R \rceil$ sounding operations. Next, a selection criterion has to be applied by evaluating all $\binom{N_R}{L_R}$ different receive antenna combinations. For example, selecting one out of four receive antennas requires four sounding operations and evaluating the selection criterion four times, whereas selecting two out of four receive antennas involves only two sounding operations but six evaluations of the selection criterion. Consequently, there is a tradeoff between the total number of receive antennas with respect to the number of available RF chains and the latency required to sound the full MIMO channel from the available branches. There is also another tradeoff in terms of costs. The greater the number of RF branches L_R, the more expensive the final system is, whereas greater throughput (as well as array gain) can be achieved.

6.2.1 Antenna Selection Based on System Throughput

In the literature, different selection criteria have been reported based on specific properties of the wireless channel, such as capacity or eigenvalue spread. However, most of these selection methods do not consider the entire wireless system, including the receiver. For example, eigenvalue-based or capacity-based selection methods are based only on the coefficients of the wireless channel. Besides the channel coefficients, an optimum selection criterion for a specific transmission system such as HSDPA has to consider the properties of the transmit signals and the receiver. Therefore, we propose the utilization of analytic expressions of the post-equalization SINR as the criterion to select the receive antennas subset. The SINR is directly obtained from the estimated wireless channel and can be mapped to the physical layer throughput by a look-up table (see Table 5.1 (p. 121), Table 5.2 (p. 121), and [33]). In Section 5.4 (p. 115), it is shown that maximizing the SINR also leads to throughput maximization. Notice that this antenna selection criterion is not only limited to HSDPA. It can be used whenever it is possible to describe the physical-layer throughput by means of analytic expressions that depend on the channel coefficients and the type of receiver implemented.

6.2.2 Hardware Aspects of Antenna Selection

The existing literature about antenna selection mostly considers "ideal" hardware. However, when taking into account the features of actual hardware, the promised gains of antenna selection are smaller. Special attention should be paid to the RF switch. "Ideal" switches do not deteriorate the receiver noise figure and work instantaneously; that is, without any delay. However, such features cannot be guaranteed in practice.

The losses caused by the RF switches at the receive side are difficult to compensate. A possible approach consists in installing the switch after the low-noise amplifier, but this requires as many low-noise amplifiers as receive antennas, hence reducing the benefits provided by RAS in terms of costs and energy consumption. However, it should be noted that current mobile communication systems, such as HSDPA, are interference-limited systems, not noise-limited systems. This means that, although the noise figure of the receiver suffers some deterioration caused by the insertion loss of the switch, the final system throughput will not be significantly degraded.

From the hardware point of view, sequentially sounding the full MIMO channel is a challenging task. It can take a long time for the full MIMO channel to be acquired, especially in systems in which base stations do not continuously broadcast pilot symbols. Until the full MIMO channel state information becomes available, antenna selection cannot be used (at least not optimally). Typically, this leads to a loss in spectral efficiency. Fortunately, base stations in HSDPA are always transmitting pilot channels. As a consequence, the user equipment is able to continuously perform channel sounding over the N_R receive antennas even though no data is being received. As soon as the user equipment is notified by the Base station (NodeB) of a new incoming data block, it can use the previously estimated channel state information to select the optimum antenna subset. This "instantaneous best throughput selection" requires the channel to be approximately constant over several subframes (a few milliseconds). If the channel changes faster, the antennas can only be selected by maximizing the "average" throughput of the last (some hundreds of) subframes. Unfortunately, continuously sounding the wireless channel consumes energy that may be not affordable by a handheld device. Consequently, a tradeoff must be reached between the accuracy of the channel estimate and the energy invested in the channel sounding stage. Investing less energy in sounding the channel leads to a noisy estimation that in turn could lead to a suboptimal subset selection, thus degrading the performance. Fortunately, according to the literature (for example [47]), antenna selection is quite robust to imperfect channel estimations.

6.3 An Exemplary Measurement and its Results

The measurements we present in this chapter were carried out by the "Vienna MIMO testbed" (see Chapter 3, p. 59) in a realistic, urban, outdoor scenario. The details of the steps followed to carry out the measurements, as well as the details of the testbed set-up, can be found in Chapter 4 (p. 75). An example of the wireless communications system exemplarily measured in this work is a laptop with four receive antennas and two RF chains located in an office and connected to the NodeB unit.

6.3.1 Urban Scenario

We estimate the mean HSDPA throughput when applying antenna selection in an urban scenario (see Figures 6.2 and 6.3) and at the carrier frequency of 2.5 GHz. At the NodeB, we employ a Kathrein 800 10543 [29] 2X-pol base station panel antenna ($2 \times \pm 45^\circ$ polarization, half-power beam width $58^\circ/6.2^\circ$, down tilt 6°).

The transmitter is placed on the roof of a tall building in downtown Vienna. The receive antennas are located 460 m away inside an office room (see Figure 6.3). The direct path is blocked by a copper roof and the adjacent buildings, resulting in a non-line-of-sight connection with a very large root-mean-square delay spread of 1 μs.

The receive antennas are moved by a fully automated $XY\Phi$ positioning table (in an area of $3\,\lambda \times 3\,\lambda$) to automatically generate many small-scale-fading channel realizations (see Figure 6.3). The receive antennas (see Figure 6.3) are printed monopole elements [27, 28] that were designed according to the generalized Koch pre-fractal curve with a resonance frequency equal to 2.7 GHz and bandwidth greater than 400 MHz.

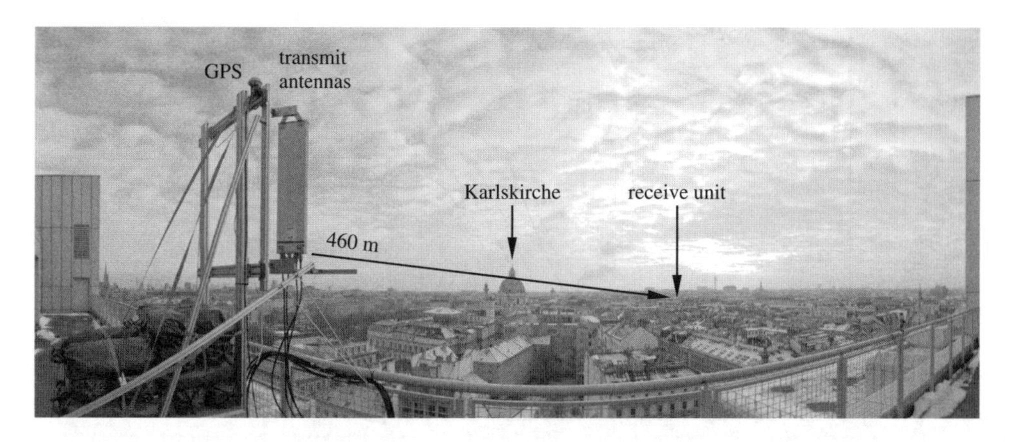

Figure 6.2 Urban scenario showing the antennas at the transmit side.

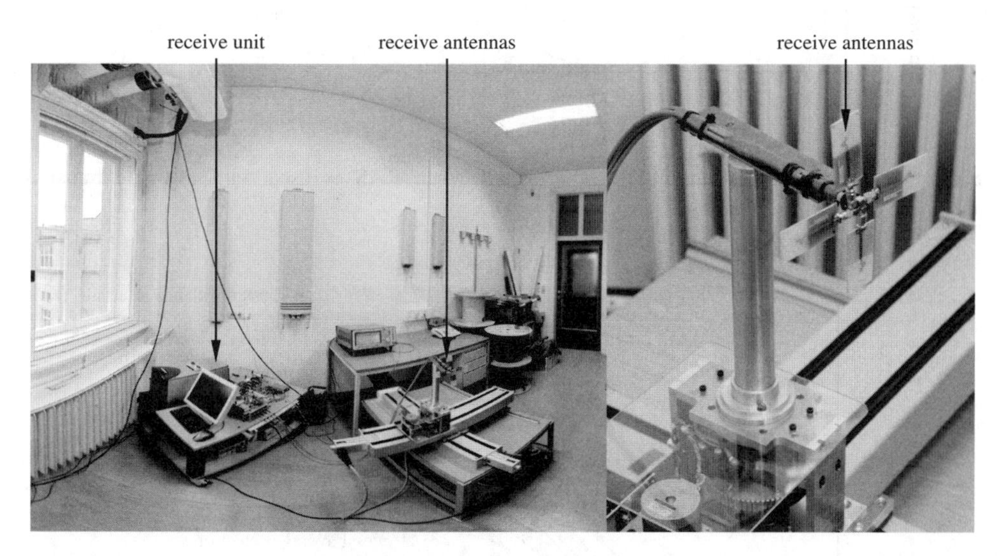

Figure 6.3 Urban scenario at the receive side. The testbed and the four printed monopole receive antennas [27, 28] mounted on the $XY\Phi$ positioning table are shown.

6.3.2 Experimental Assessment of Antenna Selection in HSDPA

The experimental evaluation of antenna selection in HSDPA is carried out by either a single transmit antenna (see Figure 6.4) or two transmit antennas (see Figure 6.5), employing the so-called Transmit Antenna Array (TxAA) HSDPA mode in which a single stream is transmitted regardless of the number of transmit antennas. In both cases, the results are measured by all four receive antennas and they are presented when one, two, or all receive antennas are selected. Obviously, no selection is performed at all when the four receive antennas are utilized. The mean throughput of HSDPA physical-layer DL is

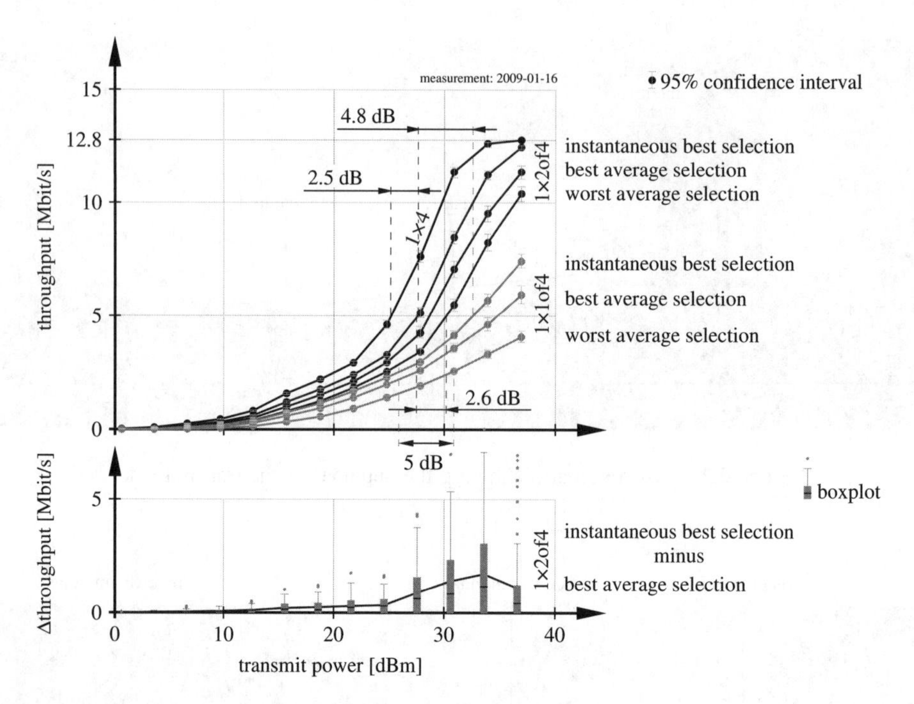

Figure 6.4 Estimated mean HSDPA throughput when RAS is used with a single transmit antenna.

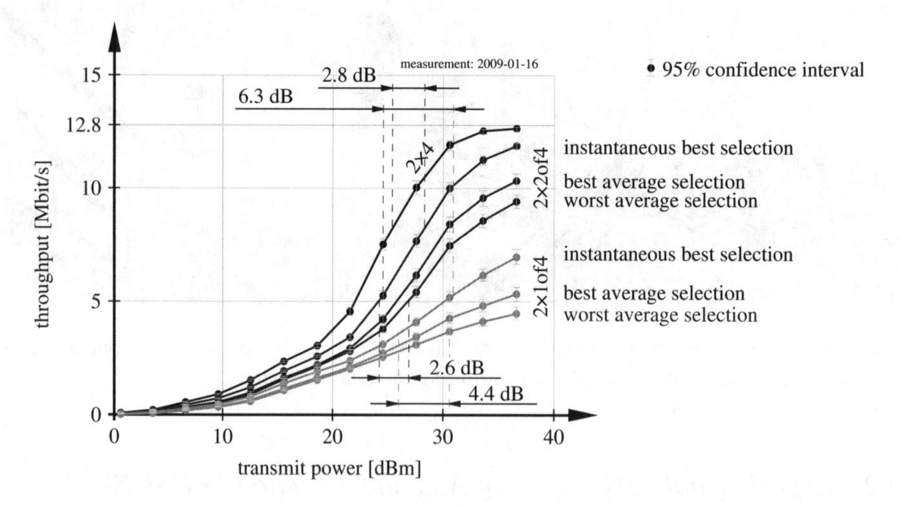

Figure 6.5 Estimated mean HSDPA throughput when RAS is used with two transmit antennas.

evaluated employing the following antenna subset selections:

- All four receive antennas are used, thus no selection is performed at all. This is measured to obtain an upper bound for the throughput. The signals acquired by the four receive antennas are combined by MRC to maximize the receive gain.

- Two receive antennas are selected out of the four receive antennas available.
- A single receive antenna is selected out of the four available.

Therefore, as one or two transmit antennas can be used together with three different subset selections at the receiver, we evaluate six cases. In the four cases in which two receive antennas or a single receive antenna is selected, the following selection criteria are evaluated:

Instantaneous best selection: For each channel realization created by moving the receive antennas, the receive antenna combination (one or two receive antennas out of four) offering the highest throughput is chosen in accordance with the SINR expressions calculated in Equation (5.23) (p. 118). Notice that the measurements are carried out by acquiring the signals simultaneously from all four receive antennas, regardless of the mode being measured. Consequently, during the evaluation stage it is possible to evaluate the receive antenna combination offering the best throughput for each channel realization. This selection criterion is an upper bound for the HSDPA physical-layer throughput when RAS is performed and the channel is known for every realization. Since in HSDPA the channel can be sounded continuously, a performance close to this upper bound can be easily achieved in reality.

Best average selection: For a chosen transmit signal power level, the HSDPA throughput is estimated for every receive antenna combination (one or two receive antennas out of four) and for every channel realization. Next, the estimated mean throughput for every antenna combination over all channel realizations is calculated. Finally, the combination offering the highest throughput is chosen, which corresponds to the mean throughput obtained with this criterion for the selected transmit power level. Notice that, in a real HSDPA system, the average throughput can be calculated by permanently sounding the channel. Note also that this approach is especially suitable when the channel coherence time is short compared with the frame duration.

Worst average selection: Analogous to the best average selection introduced above, after the estimation of the mean throughput for each antenna combination, the antenna combination leading to the worst throughput is chosen. This selection criterion corresponds to the worst possible case, even worse than selecting the antennas randomly, thus constituting a lower bound for the RAS performance.

6.3.3 Measurement Results and Discussion

Measurement results are plotted for the case in which one transmit antenna is used (Figure 6.4) and when two transmit antennas are employed (Figure 6.5). In each figure, three different sets of curves are plotted (from the top down):

- The first set plots a single curve for the case in which all four receive antennas are employed (no RAS is performed at all). This set is labeled as "1 × 4" or "2 × 4."
- The second set corresponds to the case in which two out of four receive antennas are selected. This set is labeled as "1 × 2 of 4" or "2 × 2 of 4."
- Finally, in the third set, a single receive antenna is selected out of four. The resulting curves are labeled as "1 × 1 of 4" or "2 × 1 of 4."

For the second and third sets, three different curves are plotted, corresponding to the three different selection criteria. The topmost curve shows the mean throughput when the instantaneous best selection is performed. The two bottom curves, respectively, plot the mean throughput for the best and for the worst average selection criteria.

Note that the 15 to 30 dBm transmit power region is the most adequate for comparing the different throughput curves. For higher transmit power levels, no higher modulation and coding schemes are available in HSDPA to exploit the channel capacity and, therefore, saturation occurs.

We estimate the precision of the measurement by means of bootstrapping methods [9, 10, 12]. In both throughput graphs (Figures 6.4 and 6.5) the dots represent the inferred mean throughput, the gray vertical lines are the 95 % BCa[2] confidence intervals for the mean, and the corresponding horizontal lines are the 2.5 % and the 97.5 % percentiles. Note that the receive antenna position remains unchanged while measuring the two schemes with one and two transmit antennas at different transmit power levels. This leads to smooth curves and relative positions of the curves that are more accurate than the confidence intervals for the absolute positions suggest. From the results presented for the single transmit antenna case in Figure 6.4 we can obtain the following conclusions:

- The difference between the Single-Input Multiple-Output (SIMO) 1×4 full system curve and the SIMO 1×2 instantaneous best selection – expressed in terms of the transmit power level needed to achieve the same throughput value – is about 2.5 dB. This can be explained by the additional array gain of the four-receive-antenna system.
- The SIMO 1×2 scheme presents a transmit power gain of up to 4.8 dB with respect to the SISO 1×1 system. This is due to the array gain offered by the 1×2 system and, additionally, due to the ability to exploit all available diversity.
- The maximum observed gain offered by selection is about 2.6 dB for the SIMO 1×2 scheme and about 5 dB for the SISO system. Such a gain comes from the difference between the "instantaneous best selection" and the "worst average selection" curves.

The lower part of Figure 6.4 shows the difference between the "instantaneous best selection" and the "best average selection." Average gains of up to 2 dB are observed, while instantaneous gains may be much higher. From the results for the case of two transmit antennas shown in Figure 6.5, the following points should be noticed:

- The difference between the MIMO 2×4 full system and the MIMO 2×2 instantaneous best selection is in the order of 2.8 dB.
- The MIMO 2×2 scheme presents a gain of up to 6.3 dB with respect to the Multiple-Input Single-Output (MISO) 2×1 system.
- Finally, the maximum observed gain offered by antenna selection is about 2.6 dB for the MIMO scheme and about 4.4 dB for the MISO system.

6.4 Summary

In this chapter, a new RAS criterion applicable to HSDPA, as well as other mobile communication systems, was introduced. The criterion takes into account the entire system

[2] BCa stands for the bootstrap Bias-Corrected and accelerated (BCa) algorithm [10, p. 176].

design and it is based on the SINR expressions, describing the behavior of the whole link (including the receiver) instead of employing only channel state information. We stressed the suitability of RAS to HSDPA-style systems because it enables most of the implementation issues exhibited by antenna selection to be overcome.

With the aim of evaluating the performance of such a selection criterion, we carried out measurements in a realistic, urban scenario featuring rich scattering, and with no line of sight between the transmitter and the receiver. The results were presented in terms of HSDPA physical-layer DL throughput and show significant throughput gains when RAS is employed. Such gains are especially remarkable when only one receive antenna is selected out of the four available, in which case they vary from 5 dB to 4.4 dB when one and two transmit antennas are respectively used. It is also worth pointing out the enormous gain obtained when two receive antennas are selected instead of only one. More specifically, such gains are equal to 4.8 dB when one transmit antenna is used and up to 6.3 dB in the case of two transmit antennas. Finally, the maximum achievable gains resulting from antenna selection implementation range from 2.6 dB to 5 dB, depending on the number of transmit antennas and the number of selected receive antennas. Nevertheless, we omitted two important issues in the performance evaluation. First, continuously receiving HSDPA data packets on different antennas in order to obtain knowledge about the channel consumed energy that may not be affordable for handheld devices. Second, switching between different receive antennas takes time. While this delay does not affect the throughput in the case of the "best average selection," the throughput of the "instantaneous best selection" has to be corrected by the switching time needed. Consequently, an optimal switching rate can be defined.

References

[1] Berenguer, I., Wang, X., and Krishnamurthy, V. (2003) 'Adaptive MIMO antenna selection,' in *Conference Record of the Thirty-Seventh Asilomar Conference on Signals, Systems and Computers*, volume 1, pp. 21–26, doi: 10.1109/ACSSC.2003.1291856.

[2] Berenguer, I., Wang, X., and Krishnamurthy, V. (2005) 'Adaptive MIMO antenna selection via discrete stochastic optimization,' *IEEE Transactions on Signal Processing*, **53** (11), 4315–4329, doi: 10.1109/TSP.2005.857056.

[3] Blum, R., Xu, Z., and Sfar, S. (2009) 'A near-optimal joint transmit and receive antenna selection algorithm for MIMO systems,' in *Proceedings of the IEEE Radio and Wireless Symposium (RWS 2009)*, pp. 554–557, doi: 10.1109/RWS.2009.4957411.

[4] Chen, Z., Collings, I. B., Zhou, Z., and Vucetic, B. (2009) 'Transmit antenna selection schemes with reduced feedback rate,' *IEEE Transactions on Wireless Communications*, **8** (2), 1006–1016, doi: 10.1109/TWC.2009.080296.

[5] Chen, Z. and Suzuki, H. (2008) 'Performance of 802.11n WLAN with transmit antenna selection in measured indoor channels,' in *Proceedings of the Australian Communications Theory Workshop (AusCTW)*, pp. 139–143, doi: 10.1109/AUSCTW.2008.4460836.

[6] Collados, M. and Gorokhov, A. (2004) 'Antenna selection for MIMO-OFDM WLAN systems,' in *Proceedings of the 15th IEEE International Symposium on Personal, Indoor and Mobile Radio Communications (PIMRC)*, volume 3, pp. 1802–1806.

[7] Collados, M. and Gorokhov, A. (2005) 'Antenna selection for MIMO-OFDM WLAN systems,' *International Journal of Wireless Information Networks*, **12** (4), 205–213.

[8] Coon, J. P. and Sandell, M. (2010) 'Combined bulk and per-tone transmit antenna selection in OFDM systems,' *IEEE Communications Letters*, **14** (5), 426–428, doi: 10.1109/LCOMM.2010.05.100055.

[9] Cox, D. (1958) *Planning of Experiments*, John Wiley & Sons.

[10] Efron, B. and Hinkley, D. V. (1994) *An Introduction to the Bootstrap*, first edition, CRC Monographs on Statistics & Applied Probability, Chapman & Hall.

[11] Erceg, V., Schumacher, L., Kyritsi, P. *et al.* (2004) 'TGn channel models,' Technical report, IEEE P802.11. Available from http://www.802wirelessworld.com/8802.

[12] Fisher, R. (1935) *The Design of Experiments*, John Wiley & Sons, Inc., New York, NY.

[13] García-Naya, J. A., Mehlführer, C., Caban, S. *et al.* (2009) 'Throughput-based antenna selection measurements,' in *Proceedings of the 70th IEEE Vehicular Technology Conference (VTC2009-Fall)*, Anchorage, AK, doi: 10.1109/VETECF.2009.5378992. Available from http://publik.tuwien.ac.at/files/PubDat_176573.pdf.

[14] Gershman, A. B. and Sidiropoulos, N. D. (2005) *Space–Time Processing for MIMO Communications*, John Wiley & Sons.

[15] Gharavi-Alkhansari, M. and Gershman, A. B. (2004) 'Fast antenna subset selection in MIMO systems,' *IEEE Transactions on Signal Processing*, **52** (2), 339–347.

[16] Ghrayeb, A. (2006) 'A survey on antenna selection for MIMO communication systems,' in *Proceedings of the 2nd IEEE International Conference on Information and Communication Technologies (ICTTA 2006)*, volume 2, pp. 2104–2109, doi: 10.1109/ICTTA.2006.1684727.

[17] Ghrayeb, A. and Duman, T. M. (2003) 'Performance analysis of MIMO systems with antenna selection over quasi-static fading channels,' *IEEE Transactions on Vehicular Technology*, **52** (2), 281–288.

[18] Gorokhov, A. (2002) 'Antenna selection algorithms for MEA transmission systems,' in *Proceedings of the IEEE International Conference on Acoustics, Speech, and Signal Processing (ICASSP)*, volume 3, pp. 2857–2860, doi: 10.1109/ICASSP.2002.1005282.

[19] Gorokhov, A., Gore, D., and Paulraj, A. (2003) 'Performance bounds for antenna selection in MIMO systems,' in *Proceedings of the IEEE International Conference on Communications (ICC)*, volume 5, pp. 3021–3025, doi: 10.1109/ICC.2003.1203962.

[20] Gorokhov, A., Gore, D., and Paulraj, A. (2003) 'Receive antenna selection for MIMO flat-fading channels: theory and algorithms,' *IEEE Transactions on Information Theory*, **49** (10), 2687–2696, doi: 10.1109/TIT.2003.817458.

[21] Gucluoglu, T. and Duman, T. (2008) 'Performance analysis of transmit and receive antenna selection over flat fading channels,' *IEEE Transactions on Wireless Communications*, **7** (8), 3056–3065, doi: 10.1109/TWC.2008.061087.

[22] Gucluoglu, T., Duman, T. M., and Ghrayeb, A. (2004) 'Antenna selection for space time coding over frequency-selective fading channels,' in *Proceedings of the IEEE International Conference on Acoustics, Speech, and Signal Processing (ICASSP)*, volume 4, pp. 709–712.

[23] Hajiaghayi, M. and Tellambura, C. (2008) 'Antenna selection for unitary space–time modulation over correlated Rayleigh channels,' in *Proceedings of the IEEE International Conference on Communications (ICC)*, pp. 3824–3828, doi: 10.1109/ICC.2008.718.

[24] Heath, J., R. W. and Love, D. J. (2005) 'Multimode antenna selection for spatial multiplexing systems with linear receivers,' *IEEE Transactions on Signal Processing*, **53** (8), 3042–3056.

[25] Honda, A., Ida, I., Oishi, Y. *et al.* (2007) 'Experimental evaluation of MIMO antena selection system using RF-MEMS switches on a mobile terminal,' in *Proceedings of the IEEE 18th International Symposium on Personal, Indoor and Mobile Radio Communications (PIMRC)*, pp. 1–5, doi: 10.1109/PIMRC.2007.4394576.

[26] Hwang, K.-S., Ko, Y.-C., Alouini, M.-S., and Lee, W. (2008) 'Slow antenna selection and fast link adaptation MIMO systems,' in *Proceedings of the IEEE/ACS International Conference on Computer Systems and Applications*, pp. 325–332, doi: 10.1109/AICCSA.2008.4493553.

[27] Kakoyiannis, C. G. and Constantinou, P. (2008) 'Co-design of antenna element and ground plane for printed monopoles embedded in wireless sensors,' in *Proceedings of the 2nd International Conference on Sensor Technologies and Applications (SENSORCOMM)*, pp. 413–418.

[28] Kakoyiannis, C. G., Troubouki, S. V., and Constantinou, P. (2008) 'Design and implementation of printed multi-element antennas on wireless sensor nodes,' in *Proceedings of the 3rd International Symposium on Wireless Pervasive Computing (ISWPC)*, pp. 224–228.

[29] Kathrein (2010) 'KATHREIN-Werke KG Antenna No. 80010543', Available from http://www.nt.tuwien.ac.at/fileadmin/data/testbed/kat-ant.pdf.

[30] Lau, C. and Leung, C. (1989) 'Antenna selection in a multi-sector packet radio network,' in *Proceedings of the IEEE International Conference on Communications (ICC)*, volume 1, pp. 188–192.

[31] Lau, C. T. and Leung, C. (1990) 'A slotted ALOHA packet radio system with multiple antennas and receivers,' *IEEE Transactions on Vehicular Technology*, **39** (3), 218–226, doi: 10.1109/25.131003.

[32] Ma, Q. and Tepedelenlioglu, C. (2005) 'Antenna selection for unitary space-time modulation,' *IEEE Transactions on Information Theory*, **51** (10), 3620–3631.

[33] Mehlführer, C., Caban, S., Wrulich, M., and Rupp, M. (2008) 'Joint throughput optimized CQI and precoding weight calculation for MIMO HSDPA,' in *Conference Record of the 42nd Asilomar Conference on Signals, Systems and Computers*, Pacific Grove, CA, USA, pp. 1320–1325, doi: 10.1109/ACSSC.2008.5074632. Available from http://publik.tuwien.ac.at/files/PubDat_167015.pdf.

[34] Molisch, A. F. (2003) 'MIMO systems with antenna selection – an overview,' *Proceedings of the Radio and Wireless Conference (RAWCON)*, 167–170, doi: 10.1109/RAWCON.2003.1227919.

[35] Molisch, A. F., Mehta, N. B., Zhang, H. *et al.* (2006) 'Implementation aspects of antenna selection for MIMO systems,' *Proceedings of the 1st International Conference on Communications and Networking in China (ChinaCom)*, 1–7.

[36] Molisch, A. F. and Win, M. Z. (2004) 'MIMO systems with antenna selection,' *IEEE Microwave Magazine*, **5** (1), 46–56, doi: 10.1109/MMW.2004.1284943.

[37] Molisch, A. F., Win, M. Z., Choi, Y.-S., and Winters, J. H. (2005) 'Capacity of MIMO systems with antenna selection,' *IEEE Transactions on Wireless Communications*, **4** (4), 1759–1772, doi: 10.1109/TWC.2005.850307.

[38] Molisch, A. F., Win, M. Z., and Winters, J. H. (2001) 'Capacity of MIMO systems with antenna selection,' *Proceedings of the IEEE International Conference on Communications (ICC)*, **2**, 570–574.

[39] Molisch, A. F. and Zhang, X. (2004) 'FFT-based hybrid antenna selection schemes for spatially correlated MIMO channels,' *IEEE Communications Letter*, **8** (1), 36–38, doi: 10.1109/LCOMM.2003.822512.

[40] Moore, G. (1998) 'Cramming more components onto integrated circuits,' *Proceedings of the IEEE*, **86** (1), 82–85, doi: 10.1109/JPROC.1998.658762.

[41] Pai, H.-T. (2006) 'Limited feedback for antenna selection in MIMO-OFDM systems,' in *3rd IEEE Consumer Communications and Networking Conference*, volume 2, pp. 1052–1056.

[42] Pande, T., Love, D. J., and Krogmeier, J. V. (2007) 'Reduced feedback MIMO-OFDM precoding and antenna selection,' *IEEE Transactions on Signal Processing*, **55** (5), 2284–2293.

[43] Park, C. S. and Lee, K. B. (2008) 'Statistical multimode transmit antenna selection for limited feedback MIMO systems,' *IEEE Transactions on Wireless Communications*, **7** (11), 4432–4438, doi: 10.1109/T-WC.2008.060213.

[44] Sanayei, S. and Nosratinia, A. (2004) 'Antenna selection in MIMO systems,' *IEEE Communications Magazine*, **42** (10), 68–73, doi: 10.1109/MCOM.2004.1341263.

[45] Sanayei, S. and Nosratinia, A. (2007) 'Capacity of MIMO channels with antenna selection,' *IEEE Transactions on Information Theory*, **53** (11), 4356–4362.

[46] Stüber, G. L., Barry, J. R., McLaughlin, S. W. *et al.* (2004) 'Broadband MIMO-OFDM wireless communications,' *Proceedings of the IEEE*, **92** (2), 271–294, doi: 10.1109/JPROC.2003.821912.

[47] Sudarshan, P., Mehta, N., Molisch, A., and Zhang, J. (2004) 'Antenna selection with RF pre-processing: robustness to RF and selection non-idealities,' in *Radio and Wireless Conference*, pp. 391–394.

[48] Sudarshan, P., Mehta, N. B., Molisch, A. F., and Zhang, J. (2006) 'Channel statistics-based RF pre-processing with antenna selection,' *IEEE Transactions on Wireless Communications*, **5** (12), 3501–3511, doi: 10.1109/TWC.2006.256973.

[49] Tang, Z., Suzuki, H., and Collings, I. B. (2006) 'Performance of antenna selection for MIMO-OFDM systems based on measured indoor correlated frequency selective channels,' *Proceedings of the Australian Telecommunication Networks and Applications Conference (ATNAC)*.

[50] Telatar, I. E. (1999) 'Capacity of multi-antenna Gaussian channels,' *European Transactions on Telecommunications*, **10** (6), 585–595. Available from http://mars.bell-labs.com/papers/proof/proof.pdf.

[51] Tran, Q. T., Hara, S., Honda, A. *et al.* (2006) 'A receiver side antenna selection method for MIMO-OFDM system,' in *Proceedings of the 64th IEEE Vehicular Technology Conference (VTC 2006 Fall)*, pp. 1–5.

[52] Trivedi, Y. N. and Chaturvedi, A. K. (2010) 'Performance analysis of Alamouti scheme with transmit antenna selection in MISO systems,' in *National Conference on Communications (NCC)*, pp. 1–5, doi: 10.1109/NCC.2010.5430237.

[53] Wilzeck, A., Pan, P., and Kaiser, T. (2006) 'Transmit and receive antenna subset selection for MIMO SC-FDE in frequency selective channels,' in *Proceedings of the European Signal Procesing Conference (EUSIPCO)*.

[54] Zeng, X. N. and Ghrayeb, A. (2006) 'Performance bounds for combined channel coding and space–time block coding with receive antenna selection,' *IEEE Transactions on Vehicular Technology*, **55** (4), 1441–1446, doi: 10.1109/TVT.2006.877459.
[55] Zhang, H., Molisch, A. F., and Zhang, J. (2006) 'Applying antenna selection in WLANs for achieving broadband multimedia communications,' *IEEE Transactions on Broadcasting*, **52** (4), 475–482, doi: 10.1109/TBC.2006.884831.
[56] Zhang, X., Molisch, A. F., and Kung, S.-Y. (2005) 'Variable-phase-shift-based RF-baseband codesign for MIMO antenna selection,' *IEEE Transactions on Signal Processing*, **53** (11), 4091–4103, doi: 10.1109/TSP.2005.857024.
[57] Zhu, X. and Yuan, D. (2008) 'Performance analysis of adaptive modulation in MIMO system using transmit antenna selection with Alamouti scheme,' in *Proceedings of the 4th International Conference on Wireless Communications, Networking and Mobile Computing (WiCOM)*, pp. 1–4.

7

HSDPA Antenna Spacing Measurements

In cellular mobile radio communication, not only do users demand small mobile phones, but mobile operators also demand small Base station (NodeB) antennas. Unfortunately, the recently emerged trend towards more antennas at the transmit and receive side of a radio link (referred to as Multiple-Input Multiple-Output (MIMO)) considerably increases the width of NodeB antennas if multiple elements have to be placed immediately adjacent to each other. While several configurations are possible, for the case of two antennas, two setups are usually favored at a mobile phone operator's NodeB (see Figure 7.1).

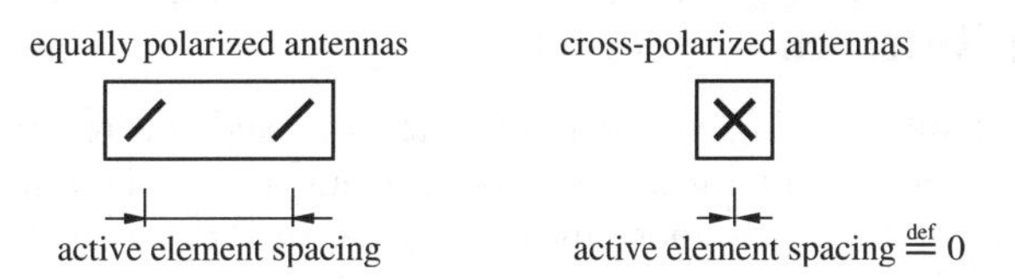

Figure 7.1 Cross-polarized and equally polarized transmit antennas.

7.1 Problem Formulation

Given a specific transmitter and receiver position, we want to directly measure the mean closed-loop physical-layer Downlink (DL) throughput of 2×2 MIMO Universal Mobile Telecommunications System (UMTS) High-Speed Downlink Packet Access (HSDPA) for equally polarized transmit antennas at variable antenna spacings and for cross-polarized transmit antennas. We want to do so by employing realistic transmit antennas (see Figure 7.2), realistic receive antennas (see Figure 7.4, p. 155), and transmitting standard-compliant HSDPA data frames [2] including the HSDPA pilot structure (see Chapter 5, p. 103). In this experiment, we do not want to investigate multi-user and interference effects; that is, we only measure a single, isolated HSDPA DL experiencing

Evaluation of HSDPA and LTE: From Testbed Measurements to System Level Performance, First Edition.
Sebastian Caban, Christian Mehlführer, Markus Rupp and Martin Wrulich.

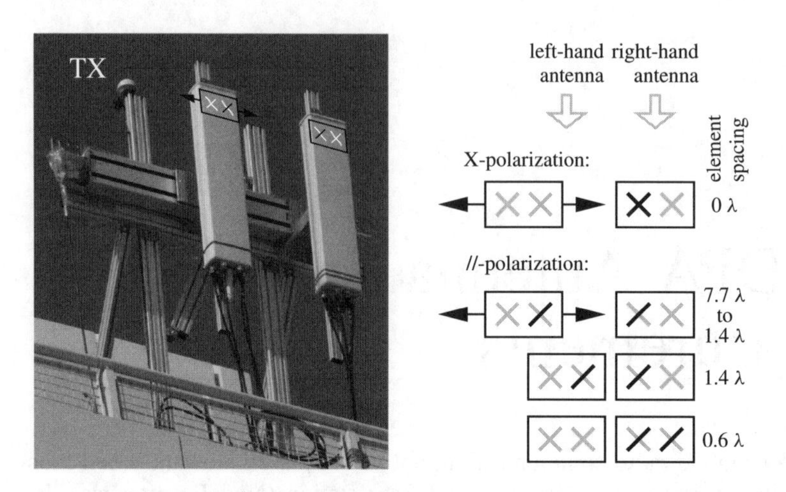

Figure 7.2 A "two-element NodeB antenna" consisting of a moveable 2X-pol (left-hand) antenna and a fixed 2X-pol (right-hand) antenna. In total, only two antenna elements are excited at the same time.

Gaussian noise. Finally, we want to compare the measured throughput with the estimated unconstrained capacity of the channel and explain the difference observed.

7.2 Existing Research

The effects of antenna spacing at the transmitter site and the receiver site have been studied for a long time. The first measurement campaigns carried out in outdoor scenarios date from the 1970s [18, 26]. In these measurements, the "correlation coefficient" of the incoming signals with respect to antenna spacing has been investigated. More recent measurements investigated this effect in indoor-only scenarios [10, 12, 17], in outdoor-to-indoor scenarios [21], and in outdoor-only scenarios [24, 35]. The impact of antenna spacing on "channel capacity" has been measured in a variety of scenarios and conditions, including indoor scenarios [12, 14], outdoor scenarios [6, 14, 30], reverberation chambers [16], and virtual antenna arrays [22]. These measurement results were complemented with ray-tracing techniques [20, 25] and theoretical analyses, such as for example in [7, 29, 32, 34] and the references cited therein. The influence of antenna spacing on the Bit Error Ratio (BER) was also theoretically investigated [9, 19], measured indoors [4, 5], and measured outdoors [11, 33].

Recently, the influence of antenna spacing on the "throughput" of an Orthogonal Frequency-Division Multiplexing (OFDM) transmission was studied in [31] by sounded channel coefficients in a simulation. Similarly, [13] investigates the throughput difference between equally polarized and cross-polarized transmit antennas. Notably, except [11, 31], all the above-cited references do not employ NodeB antennas similar to the ones currently in use in mobile cellular networks. Furthermore, except [31], none of the references found relates transmit antenna spacing to the closed-loop physical-layer throughput of a standard-compliant multi-antenna mobile communication system (such as for example 2×2 MIMO HSDPA in our case; that is, HSDPA with two transmit and two receive antennas).

Figure 7.3 Picture of the urban non-line-of-sight scenario featuring rich-scattering.

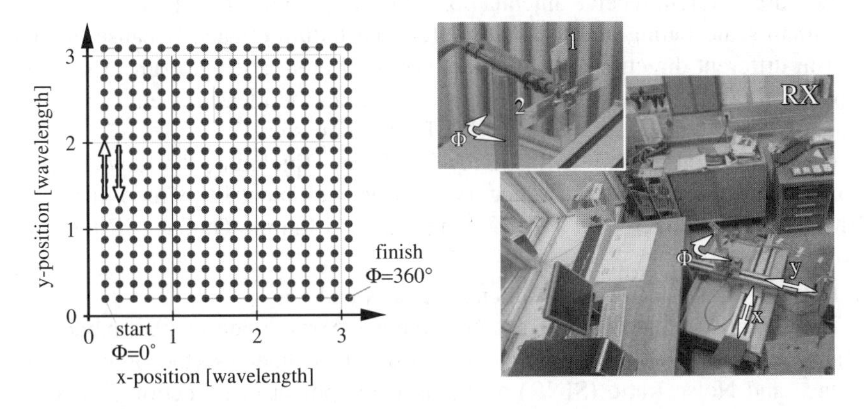

Figure 7.4 The receiver employing two (1, 2) moveable (x, y) and rotatable (Φ) printed monopole antennas. The other two printed monopole antennas shown are not used in this measurement.

7.3 Experimental Setup

The goal of the experimental setup described below is to examine the impact of NodeB antenna spacing on the mean closed-loop physical-layer DL throughput of an isolated 2×2 HSDPA transmission in the specific small-scale fading scenario (see also [3]):

1. The Urban (190 m) outdoor environment – the channel
   We investigate a realistic non-line-of-sight urban scenario denoted by "Urban (190 m)" – in contrast to the urban scenario defined in Section 6.3 (p. 148) (see Figure 5.8, p. 126) – in the inner city of Vienna, Austria (see Figure 6.2 and

Figure 6.3 (p. 145)). The transmit antennas are placed on the roof of a tall building in the vicinity of other operators' NodeB antennas (see Figures 7.2 and 7.3), making the measurement results obtained very realistic and representative for a mobile communication system. The receive antennas are placed in a small office approximately 190 m away (see Figure 7.4, p. 155). The estimated root-mean-square delay spread for this rich-scattering, non-line-of-sight scenario is 0.5 μs.

2. A "two-element" flat-panel antenna at the NodeB
   At the NodeB, we employ two KATHREIN 800 10629 [1] 2X-pol panel antennas with a half-power beam width of 80°/7.5° and a total down tilt of 16°(= 10° mechanical +6° electrical). Each 2X-pol antenna consists of two cross-polarized antennas spaced by 0.6 λ (with λ = 12 cm; namely, the wavelength at our carrier frequency of 2.5 GHz). Only two of the eight possible elements are excited at the same time, to obtain a two-element NodeB antenna with a variable element spacing from 0.6 λ to 7.7 λ for equal polarization and 0 λ for cross-polarization. Using two ordinary X-pol antennas instead of two 2X-pol antennas would have only allowed for measuring down to an element spacing of 1.4 λ (then the two antenna casings touch each other) rather than 0.6 λ.
3. Two printed monopole antennas at the mobile phone
   At the receiver site, we utilize two printed monopole antennas that can be integrated into a mobile handset or a laptop computer [15]. We employ differently polarized antennas to obtain robust and close-to-reality measurement results. As shown in Figure 7.4, we measure different receive antenna positions (x, y) in an area of 3 λ × 3 λ to average over small-scale fading and to avoid large-scale fading effects. Because the antennas point in different directions, they experience a different average path loss. This effect is averaged out by rotating the antennas (Φ) during the measurement.
4. A standard-compliant 2 × 2 MIMO HSDPA transmission
   We transmit standard-compliant HSDPA data frames (including the standard-compliant pilot structure) encoded for a Category 16 user equipment [2]. Three different 2 × 2 MIMO HSDPA modes plus a 1 × 2 Single-Input Multiple-Output (SIMO) mode are measured:
   (i) 1-stream mode = Transmit Antenna Array (TxAA)
   The so-called closed-loop Transmit Antenna Array mode (TxAA mode) employs strongly quantized precoding at the transmitter to increase the Signal to Interference and Noise Ratio (SINR) at the user equipment (see Section 5.3, p. 115). In this mode, the data rate of a Category 16 user equipment can be adjusted between 68.5 kbit/s and 12.779 Mbit/s [2].
   (ii) 1or2-stream mode = D-TxAA
   The so-called Double Transmit Antenna Array (D-TxAA) mode (is downward compatible with TxAA and equals TxAA when the SINR at the user equipment is low. At larger SINR values, D-TxAA switches to transmitting two independently coded and modulated HSDPA data streams. In this mode, the data rate of a Category 16 user equipment can be adjusted between 68.5 kbit/s (lowest rate of single-stream transmission) and 27.952 Mbit/s (the highest rate of double-stream transmission) [2].
   (iii) 2-stream mode
   Additional to the standard-compliant modes defined above, we also implemented a mode (nonexistent in the standard) that behaves as D-TxAA, but consistently

forces the transmitter to use two streams, regardless of the SINR estimated at the receiver. Always enforcing two streams, the data rate of a Category 16 user equipment can be adjusted between 4.581 Mbit/s and 27.952 Mbit/s. Note that this artificial HSDPA mode has been implemented to improve the understanding of the findings reported in Section 7.5 (p. 163).

(iv) SIMO HSDPA
We measure 1×2 SIMO HSDPA as a reference (concomitant observations) in order to enhance the precision of the 2×2 results measured over antenna spacing (see Section 7.4.2, p. 160). For more details about variance reduction techniques, see Section 4.4.2 (p. 88).

Summarizing, in 1-stream mode (TxAA) a single data stream is always transmitted, in the 2-stream mode two data streams are transmitted, and in the 1or2-stream mode (D-TxAA), either single-stream or double-stream transmission – whichever leads to higher physical-layer throughput on a block-by-block basis – is performed (see Section 5.4.5, p. 121).

7.4 Measurement Methodology

The methodology employed to carry out the measurements is explained in detail in Section 4.3 (p. 80). In this measurement approach, all possible transmit data blocks are generated offline in MATLAB®, but only the required data is then transmitted over a wireless channel, which is altered by moving the receive antennas. The feedback calculation mandatory for closed-loop HSDPA is instantly calculated in MATLAB® in approximately 42 ms, which is less than the channel coherence time if the scenario is static (see Figure 4.22, p. 97). The received data itself is not evaluated in real time, but offline by a cluster of Personal Computers (PCs). Results for the scenario measured are automatically obtained by the same program that has already controlled the complete measurement procedure and documentation (see Section 4.5 (p. 89) for a more detailed description).

A different approach would be to use a (more expensive) channel sounder in the same measurement setup and then “simulate” the throughput by the extracted channel coefficients in combination with models for hardware impurities. As our research interest and expertise consists of “directly” measuring the HSDPA throughput actually achieved, we transmit standard-compliant HSDPA frames. Compared with channel sounding, the testbed approach employed is more straightforward while still allowing for extracting the channel coefficients later, although not as precisely as in the case of channel sounding.

7.4.1 Inferring the Mean Scenario Throughput

We use well-known statistical techniques explained in Section 4.4 (p. 88) to infer the mean throughput performance of a specific small-scale fading scenario as described below:

1. We measure the HSDPA modes to be compared (grouping, comparison) in random order (randomization) immediately after each other over the same channels (blocking) equally often (balancing).

2. We measure all the modes of point 1. at 12 different transmit antenna spacings (see Figure 7.2) to draw the abscissas in the resulting figure (see Figure 7.8) (a one-factor-at-a-time experiment).
3. We measure all the above at 11 different transmit power levels (12×11 full factorial design). For simplicity, we draw the result-figure (see Figure 7.8) for a constant transmit power of 24.6 dBm.
4. We measure all the above at 324 different receive antenna positions (see Figure 7.4) (systematic sampling).
5. We evaluate all measured throughput values offline and average them (best estimator for the mean having no other knowledge; "plug-in principle") over the receive antenna positions to obtain the mean scenario throughput.
6. We exploit the correlation between the 2×2 throughput values and the 1×2 throughput values ("concomitant observations" that we define to be constant over the transmit antenna spacing) to enhance the precision of this measured mean scenario throughput (see Figure 7.7).
7. We extract the channel coefficients from the measurements to calculate:
   (i) the unconstrained mean channel capacity assuming Gaussian input signals and full channel knowledge at the transmitter (see Equation (8.1) (p. 169) and [27, Eq. (1)]);
   (ii) the mean Mutual Information (MI) assuming Gaussian input signals and no channel knowledge at the transmit site (see Equation (8.2) (p. 169) and [27, Eq. (1)]); and
   (iii) the so-called mean "achievable mutual information" that we define as the mutual information reduced by the inherent losses of the HSDPA system; namely, the losses caused by the quantization of the precoding as well as losses caused by the transmission of non-data channels (pilot, synchronization, and control channels). For more information, see Equation (8.4) (p. 170) and Section 5.5 (p. 124).
8. Finally, we calculate the 99 % confidence intervals for these mean values (Bias-Corrected and accelerated (BCa) bootstrap algorithm [8]) – the small vertical lines in Figure 7.8 – to gauge the precision of the results. For more information, see Section 4.6.2 (p. 94).

7.4.2 *Issues Requiring Special Attention*

When comparing the throughput of a cross-polarized transmission with the throughput of an equally polarized transmission, the main issue is that these transmissions use different antenna elements, as is shown in Figure 7.5.

As a result, a greatly different (sometimes even several decibels) average path loss leads to a change in throughput that is not caused by equal polarization instead of

Figure 7.5 With two HSDPA blocks transmitted, one cannot excite the same antenna elements (compare with Figure 7.6).

cross-polarization but by $+45^\circ$ versus -45° polarization and/or a different antenna position (see for example 1 Mbit/s between several of the light and dark dots in Figure 7.7). This issue can be overcome by carrying out two additional measurements as it is shown in Figure 7.6.

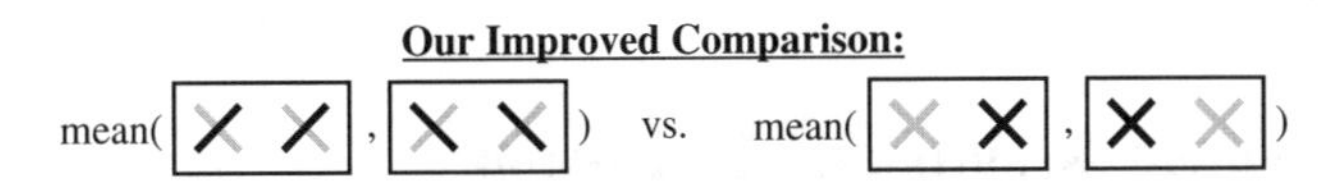

Figure 7.6 With four HSDPA blocks transmitted, one can (at least in total) excite the same antenna elements.

Consequentially, to measure the throughput of the equally polarized mode, we average the throughput of two measurements, one exciting both "/" elements, the other one exciting both "\" elements. When measuring the cross-polarized modes, we average the measurement result of the right-hand transmit antenna with the measurement result of the left-hand transmit antenna. In the end, we have excited each element equally often, giving the receiver the chance to receive (on average) the same signal power.

From the experiment carried out, we conclude that, for a constant transmit power, the 2×2 MIMO and the 1×2 SIMO throughput values are correlated (correlation coefficient 99 % confidence interval [0.87, 0.98], linear regression coefficient 99 % confidence interval [0.79, 1.24]) (see Figure 7.7). However, the 1×2 throughput should not change over

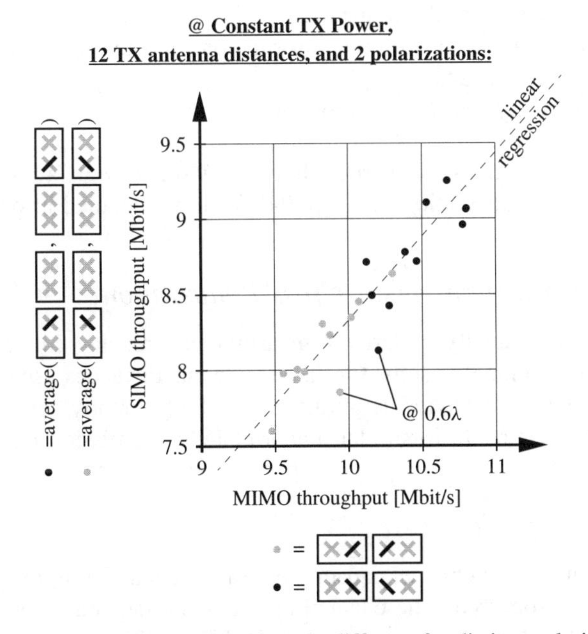

Figure 7.7 The average throughput is significantly different for distinct polarizations, whereas the 1×2 SIMO and 2×2 MIMO throughputs are correlated over antenna spacing.

antenna spacing as there is only one transmit antenna, but it does change because the antenna positions are different. Since we know that the throughput of the 1×2 system has to be constant over the transmit antenna spacing (the concomitant observations), we can use this knowledge in combination with the correlation between the 1×2 and the 2×2 throughput values to enhance the precision of the measured MIMO throughput over transmit antenna spacing.

7.5 Measurement Results and Discussion

The results of our measurements are shown in Figure 7.8. Holding the transmit power constant at 24.6 dBm, the results presented in Figure 7.8 consist of the following.

- Three sets, one for each transmission mode:
  - HSDPA configured in 1-stream mode (=TxAA);
  - HSDPA configured in 2-stream mode; and
  - HSDPA configured in 1or2-stream mode (=D-TxAA).
- Each set contains two graphs showing the performance of:
  - equally polarized transmit antennas on the left-hand side; and
  - cross-polarized transmit antennas on the right-hand side.

Each graph, plotted over the spacing of the active TX antenna elements, shows:

- the estimated mean "unconstrained capacity" of the channel;
- the estimated mean "mutual information" of the channel accounting for the absence of channel state information at the transmitter;
- the calculated mean "achievable mutual information" accounting for the inherent system design losses of HSDPA; and
- the measured mean physical layer "throughput."

For each plotted value, the small vertical lines show the 99% BCa bootstrap confidence intervals for the mean (see Section 4.6.2 (p. 94) for more detailed information).

7.5.1 *Equal Polarization Versus Cross-Polarization*

We observe that the equally polarized measurement results converge towards the cross-polarized measurement results for large antenna distances (see the dotted lines in Figure 7.8). Such a behavior is intuitive, as both greatly spaced antennas and cross-polarized antennas result in similar, uncorrelated transmission paths.

7.5.2 *Channel Capacity*

In theory, the unconstrained channel capacity is exactly equal for all three HSDPA modes because it is only a property of the channel and it is not dependent on the transmission system used (see the channel capacity curves in Figure 7.8). In our measurement, we experience slightly different channels, as:

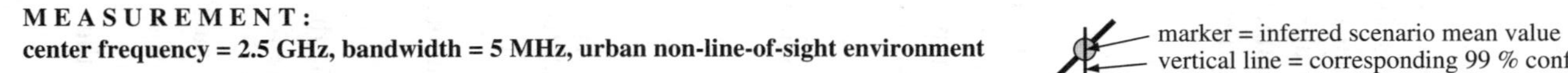

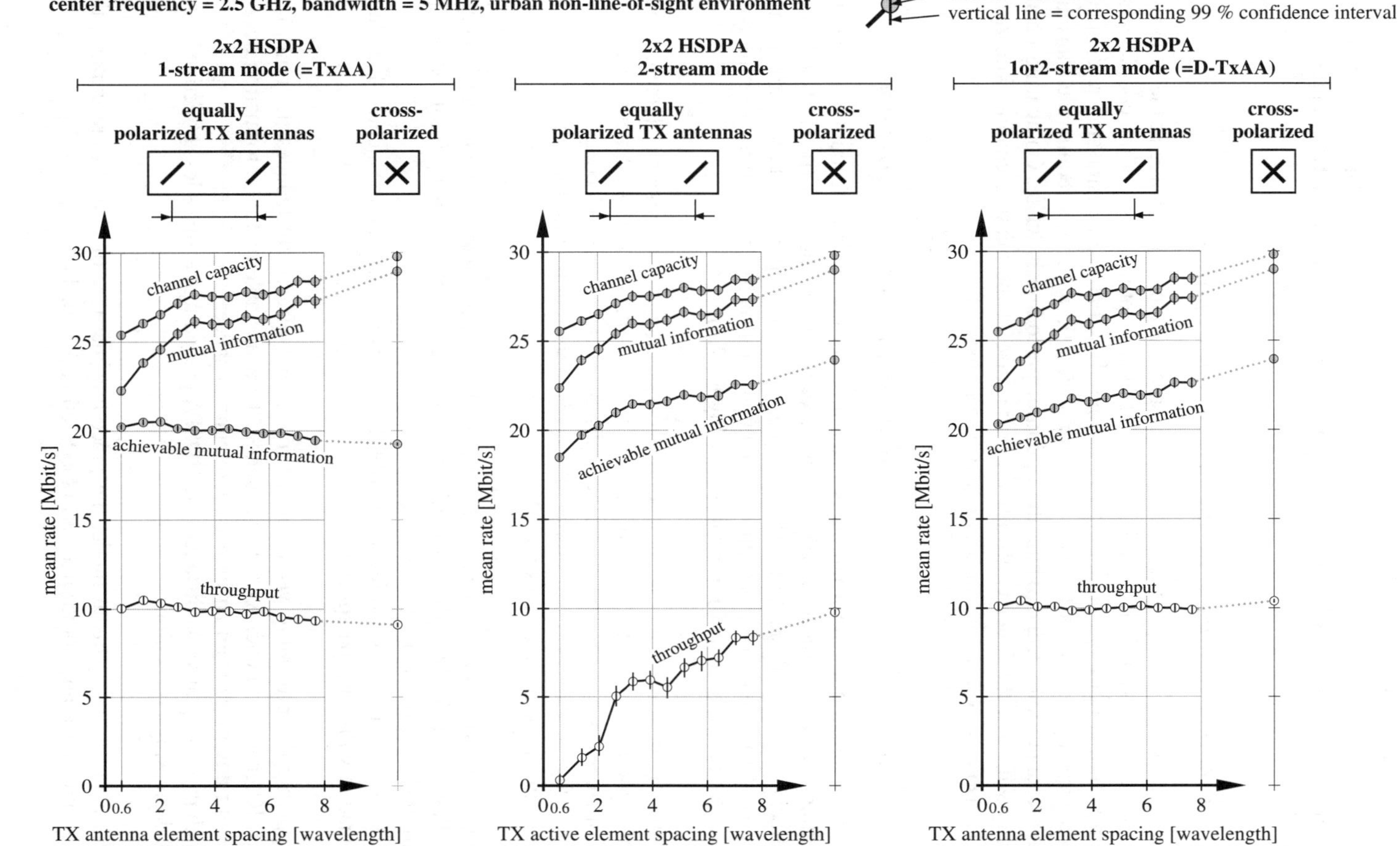

Figure 7.8 Unconstrained capacity, MI, achievable mutual information, and throughput of the closed-loop 2×2 MIMO HSDPA physical-layer DL over transmit antenna spacing.

- the different HSDPA modes are measured time-multiplexed in a not perfectly static environment; and
- the required channel coefficients have to be estimated from the HSDPA signal that is disrupted by additive Gaussian noise with a Signal to Noise Ratio (SNR) of approximately 11 dB.

7.5.3 Channel Capacity Versus Mutual Information

When calculating the channel capacity, we assume channel knowledge at the transmitter side. This channel knowledge can be used to calculate optimum precoding coefficients. A system with this optimum precoding can potentially achieve a transmission rate equal to channel capacity. In contrast, when calculating the MI, we do not require the transmitter to have channel knowledge. Thus, the optimum precoding is white. Comparing channel capacity with MI, the latter is always lower because of the missing channel knowledge. This is also observed in our measurement (see Figure 7.8).

7.5.4 Mutual Information Versus Achievable Mutual Information

The achievable mutual information is calculated by taking into account the quantized precoding and the overhead of HSDPA as described in Section 5.5 (p. 124) and in [23, 28]. Because of the different precoding employed in the 1-stream and the 2-stream transmissions, the behavior of the achievable mutual information also differs (see the curves labeled "achievable mutual information" in Figure 7.8):

- In the "1-stream mode," the precoding works better at low antenna element spacing where the transmit antenna correlation is large. With increasing element spacing, the MI increases while the achievable mutual information slightly decreases.
- Contrarily, in the "2-stream mode," the achievable mutual information increases with the element spacing, as does the MI.

This quite different behavior is caused by taking into account the quantized precoding of HSDPA when calculating the achievable mutual information (see Equation (5.24), p. 110).

In 1-stream mode, two effects contribute to the decrease of the achievable mutual information with antenna spacing:

1. Applying the precoding vector at the transmitter effectively reduces the rank of the MIMO channel to one. As a consequence, the achievable mutual information does not benefit any more from an increased antenna distance.
2. With increasing antenna distance, the HDSPA precoding becomes less effective because of decreasing correlation.

On the contrary, in 2-stream mode, the precoding matrices of HSDPA are unitary; thus, they do not cause any additional loss (see Section 5.3, p. 115).

7.5.5 *Achievable Mutual Information Versus Throughput*

Comparing the measured throughput with the achievable mutual information reveals a rate loss of about 50 % when transmitting with one stream. Since the inherent system losses are already considered in the calculation of the achievable mutual information, this loss is an implementation loss caused by the nonoptimality of the receiver. See Chapter 8 (p. 167) and [23] for a detailed discussion about performance losses in HSDPA.

In the case of 2-stream transmission, the loss between the measured throughput and the achievable mutual information is much larger, especially at small antenna element spacings. There are two reasons for this. First, a small transmit antenna spacing causes a large antenna correlation and thus a low rank of the MIMO channel, degrading the throughput of the 2-stream mode. Second, the lowest transmission rate that can be adaptively selected in the 2-stream mode is 2.29 Mbit/s per stream (4.581 Mbit/s for both streams). If the channel does not support this transmission rate "for both" streams, the throughput is zero.

In the case of 1-stream transmission, there are more transmission rates available to be adaptively selected at the transmitter, the lowest being 68.5 kbit/s [2].

7.5.6 *Throughput*

Looking at the throughput curves in Figure 7.8, we observe for the 1or2-stream mode (D-TxAA) that the measured throughput is only slightly greater than that of the 1-stream mode. Also, the throughput of the 1or2-stream mode is nearly independent of the transmit antenna spacing. This behavior is intuitive, as the 1or2-stream mode can be seen as choosing, for each channel realization, between the 1-stream and the 2-stream mode, whichever leads to a larger throughput. We conclude that the 2-stream mode is only chosen very rarely. Note that, for larger SNR values than the 11 dB chosen in Figure 7.8, the probability of choosing the 2-stream mode increases.

7.6 Different Transmit Power Levels and Scenarios

All results presented in this chapter so far have been extracted from a single measurement in the urban scenario and plotted for a constant transmit power of 24.6 dBm.

- We have investigated 11 transmit power levels (see Item 2 in Section 7.4.1). Besides changes in transmission rates, the conclusions drawn so far do not change. Only at large transmit power levels does the performance of the 1or2-stream mode increase in comparison with the 1-stream mode, as the 2-stream transmissions are less often constrained by the minimum stream rate of 2.29 Mbit/s and the 1-stream mode saturates at its theoretical maximum of 12.779 Mbit/s (Category 16 user equipment [2]).
- We have also investigated different receiver positions in the same room and in rooms on the other side of the building. In all the setups measured, the results obtained showed no significant change except for a variation in the average path loss.
- We have moved the receiver to another building located 460 m instead of 190 m away from the transmitter in the downtown area of Vienna, Austria (see Section 6.3.1 (p. 148),

Figure 6.2 (p. 145), and Figure 6.3 (p. 145)). As a result, the root-mean-square delay spread of the channel increased from 0.5 μs to 1 μs (corresponding to an increase from 1.9 to 3.8 chip durations). Owing to the richer scattering, the correlation at lower antenna spacings is decreased, resulting in a throughput performance nearly constant over antenna spacing.

- Finally, we have also repeated the measurements in an alpine valley. This scenario is characterized by a 5.7 km line-of-sight connection with a root-mean-square delay spread of 260 ns (one HSDPA chip duration). In this scenario, the 2-stream transmission always failed. Thus, the 1or2-stream mode performed as well as the 1-stream transmission. A performance difference for different transmit antenna spacings could not be observed in this scenario.

References

[1] KATHREIN-Werke KG Antenna No. 80010629. Available from http://www.nt.tuwien.ac.at/fileadmin/data/testbed/kat-ant.pdf.

[2] 3GPP (2007) Technical Specification TS 25.214 Version 7.7.0 'Physical layer procedures (FDD),' www.3gpp.org.

[3] Caban, S., García-Naya, J. A., Mehlführer, C. *et al.* (2010) 'Measuring the closed-loop throughput of 2 × 2 HSDPA over TX power and TX antenna spacing,' in *Proceedings of the 2nd International ICST Conference on Mobile Lightweight Wireless Systems*, Barcelona, Spain.

[4] Caban, S., Mehlführer, C., Mayer, L. W., and Rupp, M. (2008) '2 × 2 MIMO at variable antenna distances,' in *Proceedings of the Vehicular Technology Conference 2008 Spring*, Singapore, doi: 10.1109/VETECS.2008.276. Available from http://publik.tuwien.ac.at/files/PubDat_167444.pdf.

[5] Caban, S. and Rupp, M. (2007) 'Impact of transmit antenna spacing on 2 × 1 Alamouti radio transmission,' *Electronics Letters*, **43** (4), 198–199, doi: 10.1049/el:20073153. Available from http://publik.tuwien.ac.at/files/PubDat_167444.pdf.

[6] Chizhik, D., Ling, J., Wolniansky, P. *et al.* (2003) 'Multiple-input–multiple-output measurements and modeling in Manhattan,' *IEEE Journal on Selected Areas in Communication*, **21** (3), 321–331, doi: 10.1109/JSAC.2003.809457.

[7] Chizhik, D., Rashid-Farrokhi, F., Ling, J., and Lozano, A. (2000) 'Effect of antenna separation on the capacity of BLAST in correlated channels,' *IEEE Communications Letters*, **4** (11), 337–339, doi: 10.1109/4234.892194.

[8] Efron, B. and Hinkley, D. V. (1994) *An Introduction to the Bootstrap*, first edition, CRC Monographs on Statistics & Applied Probability 57, Chapman & Hall/CRC, London.

[9] Femenias, G. (2004) 'BER performance of linear STBC from orthogonal designs over MIMO correlated Nakagami-*m* fading channels,' *IEEE Transactions on Vehicular Technology*, **53** (2), 307–317, doi: 10.1109/TVT.2004.823475.

[10] Fernandez, O., Domingo, M., and Torres, R. (2005) 'Empirical analysis of the correlation of MIMO channels in indoor scenarios at 2GHz,' *IEE Proceedings Communications*, **152** (1), 82–88, doi: 10.1049/ip-com:20041019.

[11] Hunukumbure, R. and Beach, M. (2002) 'Outdoor MIMO measurements for UTRA applications,' in *Proceedings of EURO-COST 2002*. Available from http://hdl.handle.net/1983/887.

[12] Intarapanich, A., Kafle, P., Davies, R. *et al.* (2004) 'Spatial correlation measurements for broadband MIMO wireless channels,' in *Proceedings of the Vehicular Technology Conference 2004 Fall*, volume 1, pp. 52–56, doi: 10.1109/VETECF.2004.1399919.

[13] Jungnickel, V., Jaeckel, S., Thiele, L. *et al.* (2006) 'Capacity measurements in a multicell MIMO system,' in *IEEE Global Telecommunications Conference, 2006*, pp. 1–6, doi: 10.1109/GLOCOM.2006.645.

[14] Jungnickel, V., Pohl, V., and von Helmolt, C. (2003) 'Capacity of MIMO systems with closely spaced antennas,' *IEEE Communications Letters*, **7** (8), 361–363, doi: 10.1109/LCOMM.2003.815644.

[15] Kakoyiannis, C., Troubouki, S., and Constantinou, P. (2008) 'Design and implementation of printed multi-element antennas on wireless sensor nodes,' in *Proceedings of the 3rd International Symposium on Wireless Pervasive Computing (ISWPC 2008)*, Santorini, Greece, pp. 224–228, doi: 10.1109/ISWPC.2008.4556202. Available from http://ieeexplore.ieee.org/stamp/stamp.jsp?arnumber = 4556202.

[16] Kildal, P. S. and Rosengren, K. (2004) 'Correlation and capacity of MIMO systems and mutual coupling, radiation efficiency, and diversity gain of their antennas: simulations and measurements in a reverberation chamber,' *IEEE Communications Magazine*, **42** (12), 104–112, doi: 10.1109/MCOM.2004.1367562.

[17] Kivinen, J., Zhao, X., and Vainikainen, P. (2001) 'Empirical characterization of wideband indoor radio channel at 5.3GHz,' *IEEE Transactions on Antennas and Propagation*, **49** (8), 1192–1203, doi: 10.1109/8.943314.

[18] Lee, W. (1973) 'Effects on correlation between two mobile radio base-station antennas,' *IEEE Transactions on Communications*, **21** (11), 1214–1224.

[19] Luo, J., Zeidler, J., and McLaughlin, S. (2001) 'Performance analysis of compact antenna arrays with MRC in correlated Nakagami fading channels,' *IEEE Transactions on Vehicular Technology*, **50** (1), 267–277, doi: 10.1109/25.917940.

[20] Lv, J., Lu, Y., Wang, Y. *et al*. (2005) 'Antenna spacing effect on indoor MIMO channel capacity,' in *Proceedings of Asia-Pacific Microwave Conference*, pp. 1–3.

[21] Medbo, J., Harrysson, F., Asplund, H., and Berger, J.-E. (2004) 'Measurements and analysis of a MIMO macrocell outdoor–indoor scenario at 1947 MHz,' in *Proceedings of the Vehicular Technology Conference 2004 Spring*, volume 1, pp. 261–265.

[22] Medbo, J., Riback, M., and Berg, J.-E. (2006) 'Validation of 3GPP spatial channel model including WINNER wideband extension using measurements,' in *Proceedings of the Vehicular Technology Conference 2006 Fall*, pp. 1–5, doi: 10.1109/VTCF.2006.36.

[23] Mehlführer, C., Caban, S., and Rupp, M. (2010) 'Measurement-based performance evaluation of MIMO HSDPA,' *IEEE Transactions on Vehicular Technology*, **59** (9), 4354–4367, doi: 10.1109/TVT.2010.2066996. Available from http://publik.tuwien.ac.at/files/PubDat_187112.pdf.

[24] Pedersen, K., Mogensen, P., and Fleury, B. (1998) 'Spatial channel characteristics in outdoor environments and their impact on BS antenna system performance,' in *Proceedings of the Vehicular Technology Conference*, volume 2, pp. 719–723, doi: 10.1109/VETEC.1998.683676.

[25] Pohl, V., Jungnickel, V., Haustein, T., and von Helmolt, C. (2002) 'Antenna spacing in MIMO indoor channels,' in *Proceedings of the Vehicular Technology Conference 2002 Spring*, volume 2, pp. 749–753, doi: 10.1109/VTC.2002.1002587.

[26] Rhee, S.-B. and Zysman, G. (1974) 'Results of suburban base station spatial diversity measurements in the UHF band,' *IEEE Transactions on Communications*, **22** (10), 1630–1636.

[27] Rupp, M., García-Naya, J. A., Mehlführer, C. *et al*. (2010) 'On mutual information and capacity in frequency selective wireless channels,' in *Proceedings of the IEEE International Conference on Communications (ICC 2010)*, Cape Town, South Africa, doi: 10.1109/ICC.2010.5501942. Available from http://publik.tuwien.ac.at/files/PubDat_184660.pdf.

[28] Rupp, M., Mehlführer, C., and Caban, S. (2010) 'On achieving the Shannon bound in cellular systems,' in *Proceedings of the 20th International Conference Radioelektronika*, Brno, Czech Republic, doi: 10.1109/RADIOELEK.2010.5478551. Available from http://publik.tuwien.ac.at/files/PubDat_185403.pdf.

[29] Shiu, D.-S., Foschini, G., Gans, M., and Kahn, J. (2000) 'Fading correlation and its effect on the capacity of multielement antenna systems,' *IEEE Transactions on Communications*, **48** (3), 502–513, doi: 10.1109/26.837052.

[30] Skentos, N., Kanatas, A., Pantos, G., and Constantinou, P. (2004) 'Capacity results from short range fixed MIMO measurements at 5.2 GHz in urban propagation environment,' in *Proceedings of the 2004 IEEE International Conference on Communications*, volume 5, pp. 3020–3024, doi: 10.1109/ICC.2004.1313086.

[31] Thomas, T. A., Desai, V., and Kepler, J. F. (2009) 'Experimental MIMO comparisons of a 4-element uniform linear array to an array of two cross polarized antennas at 3.5 GHz,' in *Proceedings of the Vehicular Technology Conference 2009 Fall*.

[32] Thushara, D., Rodney, A., and Jaunty, T. (2002) 'On capacity of multi-antenna wireless channels: effects of antenna separation and spatial correlation,' in *3rd Australian Communications Theory Workshop (AusCTW)*.

[33] Trautwein, U., Schneider, C., and Thomä, R. (2005) 'Measurement-based performance evaluation of advanced MIMO transceiver designs,' *EURASIP Journal on Applied Signal Processing*, **2005** (11), 1712–1724, doi: 10.1155/ASP.2005.1712.

[34] Waldschmidt, C., Kuhnert, C., Schulteis, S., and Wiesbeck, W. (2003) 'Analysis of compact arrays for MIMO based on a complete RF system model,' in *IEEE Topical Conference on Wireless Communication Technology*, pp. 286–287, doi: 10.1109/WCT.2003.1321527.

[35] Zhao, X., Kivinen, J., Vainikainen, P., and Skog, K. (2002) 'Propagation characteristics for wideband outdoor mobile communications at 5.3 GHz,' *IEEE Journal on Selected Areas in Communications*, **20** (3), 507–514, doi: 10.1109/49.995509.

8

Throughput Performance Comparisons

In this chapter[1] we take a closer look at Worldwide Inter-operability for Microwave Access (WiMAX) and High-Speed Downlink Packet Access (HSDPA), two successful cellular systems that are being operated currently in many countries. As researchers claim to be able to design a physical layer very close to the Shannon bound, we believe that it is now the time to check the truly achievable performance of these systems. Therefore, we have measured the physical-layer throughput of WiMAX and HSDPA in various realistic environments (urban and mountainous) with high-quality equipment and different antenna configurations. We furthermore compare the measured throughput with the Shannon bound and many other related bounds, reflecting design as well as implementation aspects.

8.1 Introduction

In 1948, Claude Shannon provided us with a clear channel capacity bound for point-to-point connections [7]. This bound was extended independently in the mid 1990s by Gerry Foschini and Mike Gans [1], as well as by Emre Telatar [8], to the case of multiple transmit and receive antennas; in short, Multiple-Input Multiple-Output (MIMO). It is this bound that we refer to and compare our throughput results with. It is this bound that coding experts compare with their turbo codes, as well as with their iterative receivers. Note, however, that this bound is a single-user bound even if it supports several transmit and receive antennas. Considering multiple users and/or interference is much more difficult. In many situations, the definition of capacity alone is still under discussion. Furthermore, the Shannon bound describes aspects of the physical-layer system, not the scheduling techniques, the network routing issues, or application layer facets. It thus, by far, does not describe the entire 3G system performance, but is merely an important piece.

Nowadays, mathematical models are very common for predicting the performance of systems designed either by mankind or nature. We apply mathematical models to predict tomorrow's weather, our climate in a hundred years from now, and the performance of next-generation wireless systems. Clearly, models reflect the behavior of the system

[1] Portions reprinted, with permission, from [5]. © 2011 IEEE.

Evaluation of HSDPA and LTE: From Testbed Measurements to System Level Performance, First Edition.
Sebastian Caban, Christian Mehlführer, Markus Rupp and Martin Wrulich.

under investigation. Depending on how precisely we know such a modeled system or subsystem, our prediction models can be more or less accurate. Higher accuracy typically comes with the price of parameter and complexity increase, yet the beauty of a model lies in its simplicity. Once models contain hundreds of parameters and require many hours of simulation time for each new parameter setup, they become less and less treatable. Unfortunately, wireless 3G systems are very complex systems with hundreds of parameters. The channel, a significant part of wireless systems, is still a hot topic for investigations. A recent channel model (for example, Winner Phase II+) may contain more than a hundred parameters to support a multitude of situations with extreme flexibility. If not an expert in modeling, such complex channel models can easily be mistreated and wrong results can be generated. We therefore based our investigations on measurements.

Compared with simulations, measurements are tedious. On the other hand, each measurement scenario is a true, realistic, and physically correct scenario. With even more effort, measurements can be made repeatable, although not easily reproducible. For simulations this seems to be a very simple task; still, simulation code is rarely being provided in the signal processing community. In order to minimize the effect of our measurement hardware on the results, we employed rather expensive, self-built measurement equipment with high linearity, a large range of operation, and very low noise figure. In the digital domain, we employed receiver structures with the highest performance still feasible for real-time implementation. Owing to cost limitations, however, such receivers are unlikely to be employed in commercial 3G systems. Therefore, the measured implementation loss can be interpreted as a "truly lower bound" for commercial systems.

Eventually, measurements are more honest, as they exclude assumptions in general. If, for example, you do not know the channel, the frequency offset, or the noise variance, then this is natural for a measurement, but not so for a simulation. More severe is that unknown impairments (of unknown sources) only occur in measurements, never in simulations.

This chapter is organized as follows. In Section 8.2 we provide a brief introduction to MIMO WiMAX and HSDPA. We explain the transmit precoding and the receiver processing, as it is important for understanding and interpreting the results. Section 8.3 briefly explains our measurement setup and methodology; the interested reader can find more details in the references provided. Section 8.4 is the major part of the chapter, in which we present the performance losses observed for WiMAX and HSDPA in Single-Input Single-Output (SISO) and 2×2 MIMO scenarios. For HSDPA we also present some advanced studies employing up to four antennas.

8.2 Cellular Systems Investigated: WiMAX and HSDPA

8.2.1 WiMAX and HSDPA

We implemented, measured, and analyzed two standardized 3G cellular systems:

WiMAX: The WiMAX physical layer as defined in IEEE 802.16–2004, Section 8.3. This standard was developed to provide wireless internet access to stationary and low-mobility users [2]. In our measurements, we employed the Orthogonal Frequency-Division Multiplexing (OFDM) physical layer with 256 narrow-band sub-carriers. By choosing one out of seven Adaptive Modulation and Coding (AMC) schemes at the transmitter, the data rate is adjusted to the current channel conditions, thereby maximizing the data throughput. The standard defines various channel codes, out of which we selected for our evaluations the mandatory Reed–Solomon Convolutional Code (RS-CC) and the optional Convolutional Turbo Code (CTC). We furthermore implemented and measured a regular Low Density Parity Check (LDPC) channel code, which is not defined in the standard. Unless stated otherwise, all our results refer to the CTC. In order to utilize transmit diversity, the standard furthermore foresees a simultaneous transmission on two transmit antennas by Alamouti space–time coding.

HSDPA: The HSDPA mode of the Universal Mobile Telecommunications System (UMTS) [3]. The first version of HSDPA was introduced in Rel'5 of UMTS to provide high data rates to mobile users. This is achieved by several techniques, such as fast link adaptation, fast hybrid automated repeat request, and fast scheduling. In contrast to the pure transmit power adaptation performed in UMTS, fast link adaptation in HSDPA adjusts the data rate and the number of spreading codes depending on a so-called Channel Quality Indicator (CQI) feedback. MIMO HSDPA, standardized in Rel'7 of UMTS, further increases the maximum downlink data rate by spatially multiplexing two independently coded and modulated data streams. Additionally, channel-adaptive spatial precoding is implemented at the base station. This is achieved by a standardized set of precoding vectors of which one vector is chosen based on a so-called Precoding Control Indicator (PCI) feedback obtained from the user equipment.

8.2.2 Throughput Bounds and System Losses

In this section we define bounds for the data throughput. The differences between the bounds can be considered as system losses. In 1998, Foschini and Gans [1] and Telatar [8] extended the Shannon capacity C to MIMO systems. We show the formulas in terms of discrete frequency bins $k = 1 \ldots K$, as this is very natural for OFDM systems. In HSDPA, an equivalent Discrete Fourier Transform (DFT) on the channel impulse responses has been applied to calculate frequency-domain channel coefficients. For a system working on discrete frequencies $k = 1 \ldots K$, the capacity C as a function of the transmit power P_{Tx}, the channel matrix $\mathbf{H}_k$ at the kth frequency bin with bandwidth B/K, the entire channel bandwidth B, the receiver noise variance σ_{n}^2, and the number of transmit antennas N_{T} is

$$C(P_{\mathrm{Tx}}) = \max_{\sum \mathrm{tr}\{\mathbf{R}_k\} \leq K} \frac{B}{K} \sum_{k=1}^{K} \log_2 \det \left(\mathbf{I} + \frac{P_{\mathrm{Tx}}}{\sigma_{\mathrm{n}}^2 N_{\mathrm{T}}} \mathbf{H}_k \mathbf{R}_k \mathbf{H}_k^{\mathrm{H}} \right) \tag{8.1}$$

A transmission system that is designed to achieve the channel capacity $C(P_{\mathrm{Tx}})$ has to perform frequency-selective and spatial precoding according to the matrices $\mathbf{R}_k$. For specific channel realizations, solutions for $\mathbf{R}_k$ are found by the waterfilling algorithm. Such a precoding, of course, only works if a-priori Channel State Information (CSI) is available at the transmitter. If this is not the case, the best strategy is to transmit with equal power over all transmit antennas and over all frequencies. The throughput of a system following this strategy is bounded by the Mutual Information (MI) $I(P_{\mathrm{Tx}})$:

$$I(P_{\mathrm{Tx}}) = \frac{B}{K} \sum_{k=1}^{K} \log_2 \det \left(\mathbf{I} + \frac{P_{\mathrm{Tx}}}{\sigma_{\mathrm{n}}^2 N_{\mathrm{T}}} \mathbf{H}_k \mathbf{H}_k^{\mathrm{H}} \right) \tag{8.2}$$

In Chapter 5 we have already introduced various losses that become very useful now. Since the difference between the channel capacity $C(P_{\mathrm{Tx}})$ and the mutual information $I(P_{\mathrm{Tx}})$ is only caused by the absence of CSI at the transmitter side, we define the difference between them as the "absolute CSI loss" $L_{\mathrm{CSI}}(P_{\mathrm{Tx}})$ and the "relative CSI loss" $L_{\mathrm{CSI\%}}(P_{\mathrm{Tx}})$:

$$L_{\mathrm{CSI}}(P_{\mathrm{Tx}}) = C(P_{\mathrm{Tx}}) - I(P_{\mathrm{Tx}}); \quad L_{\mathrm{CSI\%}}(P_{\mathrm{Tx}}) = 100 \times \frac{C(P_{\mathrm{Tx}}) - I(P_{\mathrm{Tx}})}{C(P_{\mathrm{Tx}})} \tag{8.3}$$

The definitions of capacity and mutual information do not take specific constraints of current cellular systems into account. For example, current systems cannot utilize the whole frequency band and/or the whole transmit power for transmitting information bits. Parts of the spectrum and/or parts of the transmit power have to be allocated to pilot, control, and synchronization information. This is reflected in the measure $I_{\mathrm{a}}(P_{\mathrm{Tx}})$, which we call the "achievable mutual information":

$$I_{\mathrm{a}}(P_{\mathrm{Tx}}) = \max_{\mathbf{W}\in\mathcal{W}} \frac{\beta B}{K} \sum_{k=1}^{K} \log_2 \det \left(\mathbf{I} + \frac{\alpha P_{\mathrm{Tx}}}{\sigma_{\mathrm{n}}^2 N_{\mathrm{T}}} \mathbf{H}_k \mathbf{W} \mathbf{W}^{\mathrm{H}} \mathbf{H}_k^{\mathrm{H}} \right) \tag{8.4}$$

Compare also with Equation (5.24) in Chapter 5. The formula presented here in this chapter offers several additional Degrees of Freedom (DoFs) to tailor it to various transmission schemes. The parameter $\alpha \leq 1$ accounts for transmit power losses and the parameter $\beta \leq 1$ accounts for spectrum and transmission time losses. In our setup, we find $\alpha = 1$ and $\beta \approx 0.65$ for WiMAX and $\alpha \approx 0.4$ and $\beta \approx 0.77$ for HSDPA. The matrix $\mathbf{W}$ accounts for nonoptimal precoding/space–time coding. In the case of HSDPA, the precoding matrix $\mathbf{W}$ is strongly quantized and chosen adaptively out of a predefined codebook $\mathcal{W}$. Depending on the transmission mode, the inclusion of the precoding $\mathbf{W}$ in the calculation of the achievable mutual information may or may not (if, for example, the matrices $\mathbf{W}$ are chosen unitary) impact $I_{\mathrm{a}}(P_{\mathrm{Tx}})$. Note that in HSDPA only one precoding matrix $\mathbf{W}$ is chosen for the entire bandwidth; future standards, such as Long-Term Evolution (LTE), allow for frequency-dependent precoding, thereby increasing the achievable mutual information. In the case of WiMAX, the transmitter employs Alamouti space–time coding, which can be rewritten as a constant precoding. In contrast to the channel capacity and the mutual information, the achievable mutual information, therefore, is a capacity constrained by the specific cellular standard. We define the difference between the mutual information $I(P_{\mathrm{Tx}})$ and the achievable mutual information $I_{\mathrm{a}}(P_{\mathrm{Tx}})$ as the "design loss," as it quantifies the loss imposed by the system design:

$$L_{\mathrm{d}}(P_{\mathrm{Tx}}) = I(P_{\mathrm{Tx}}) - I_{\mathrm{a}}(P_{\mathrm{Tx}}); \quad L_{\mathrm{d\%}}(P_{\mathrm{Tx}}) = 100 \times \frac{I(P_{\mathrm{Tx}}) - I_{\mathrm{a}}(P_{\mathrm{Tx}})}{C(P_{\mathrm{Tx}})} \tag{8.5}$$

The design loss accounts for the inherent system design losses caused by, for example, losses due to the transmission of pilot and synchronization symbols, quantized precoding, or suboptimal space–time coding. By alternating design parameters, which can partially also be achieved adaptively on the transmission scenario, the design loss can be reduced.

Finally, there is the throughput $D_{\mathrm{m}}(P_{\mathrm{Tx}})$ that can be measured in bits per second in a given link allowing one to define the so-called "implementation loss"

$$L_{\mathrm{i}}(P_{\mathrm{Tx}}) = I_{\mathrm{a}}(P_{\mathrm{Tx}}) - D_{\mathrm{m}}(P_{\mathrm{Tx}}); \quad L_{\mathrm{i\%}}(P_{\mathrm{Tx}}) = 100 \times \frac{I_{\mathrm{a}}(P_{\mathrm{Tx}}) - D_{\mathrm{m}}(P_{\mathrm{Tx}})}{C(P_{\mathrm{Tx}})} \tag{8.6}$$

as the difference between the achievable mutual information $I_a(P_{Tx})$ and the measured data throughput $D_m(P_{Tx})$. It accounts for losses caused by nonoptimum receivers and channel codes. A more detailed discussion of the individual parts of the implementation loss in WiMAX or HSDPA is provided further ahead.

In order to complete the picture we also need to define the "relative throughput"

$$D_{m\%}(P_{Tx}) = 100 \cdot \frac{D_m(P_{Tx})}{C(P_{Tx})} \tag{8.7}$$

Summing up the relative losses and the relative throughput results in 100%. We will display the relative losses together with the relative throughput in Section 8.4.

In Figure 8.1, all terms defined above are shown in their absolute and relative measures for an SISO WiMAX transmission. Comparing the CSI loss with the other two losses, it is almost negligible, especially at higher transmit powers and, thus, higher Signal to Noise Ratios (SNRs) [6]. Note that in the plot of relative losses the CSI loss appears as a monotone decreasing function over transmit power with high values (50 %) at the lower end. However, at very low transmit power (approximately −10 dB SNR) the capacity is extremely small and losing 50 % does not make a noticeable performance difference. The initial loss of 50 % varies depending on the selectivity offered in the frequency and spatial domains, as will be shown in Section 8.4. The more diversity that is offered, the larger the CSI loss will be.

In the literature, we find the measures defined above to be a function of the receive SNR. Note, however, that in modern 3G systems such a definition is obsolete. There are three reasons, the first being that we measure in terms of transmit power P_{Tx} and comparing two systems we should compare them at equal transmit power, not at equal receive SNR. Once a precoding matrix is introduced, the main part of the power may

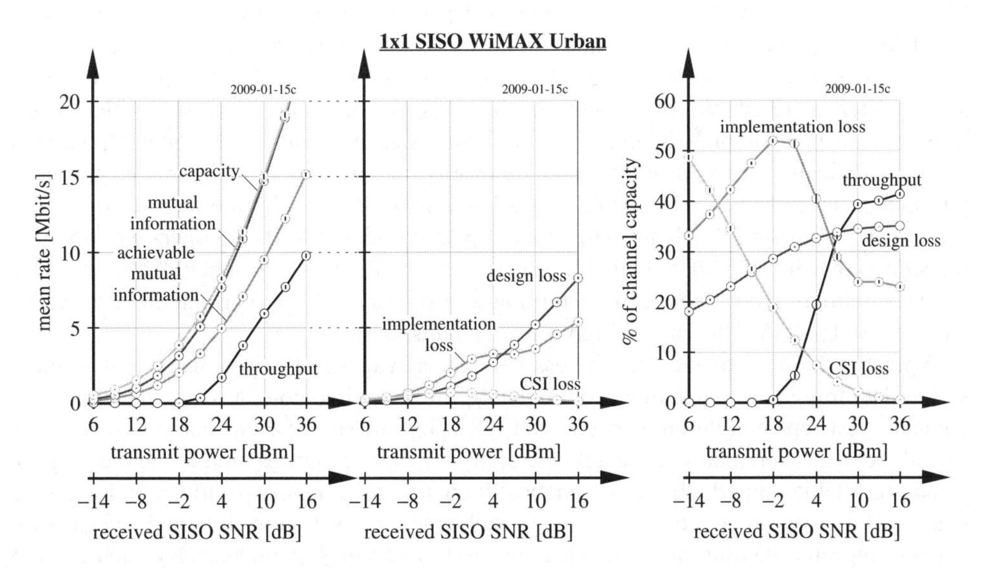

Figure 8.1 Example of measured performance metrics (left) and derived throughput losses, absolute (middle) and relative (right) [5]. Reproduced by permission of © IEEE 2010.

go to a different direction pointing away from the receive antennas, changing the receive SNR without changing P_{Tx}, providing a second reason. Third, depending on the scenario and the construction, one receive antenna will receive on average a different amount of power than another. Such differences can be as large as 6 dB in the mean. The reason for such differences is because the scattering environment is not symmetric in nature, as has been assumed in most earlier MIMO channel models. The Winner Phase II+ model is the first to allow for defining such asymmetries. We thus have to plot over transmit power P_{Tx} rather than SNR. Nevertheless, in order to facilitate comparisons with previous simulation results, we provide two scales in our plots, one being P_{Tx} and the other one a computed equivalent average SISO receiver SNR.

8.3 Measurement Methodology and Setup

Several scenarios have been measured by us in past years, and two of them are compared in this chapter: an alpine (see Section 5.6.1) and an urban scenario (see Section 6.3.1). We picked those two scenarios as they reflect existing cells in a Carinthian alpine valley and in downtown Vienna, Austria, and they reflect the extreme ends of a range of scenarios. In the alpine environment, the transmitter was placed 5.7 km distant to the receiver with an essentially strong Line-Of-Sight (LOS) component and moderate scattering. The Root Mean Square (RMS) delay spread of the channels was 260 ns (corresponding to roughly one chip duration of HSDPA). The other scenario was an urban measurement with only 190 m distance but without an LOS component. The scattering was much richer; the RMS delay spread was 1.1 μs. Other scenarios, not presented in this chapter but measured for reference purposes, had a characteristic performance in between the alpine and the urban scenario.

Although many more situations were investigated, we only present the results here for SISO and 2×2 MIMO with cross-polarized transmit antennas. In addition, we present results of a four-transmit-antenna measurement of advanced HSDPA to provide a flavor of what more antennas can offer. We employed commercially available cross-polarized basestation antennas (Kathrein 800 10543), as they support the highest diversity at the lowest size and are thus of high interest for providers. Measurements with equally polarized antennas typically provided much smaller capacity and correspondingly smaller throughput. Our receiver has a noise figure of 1.9 dB, which increases by the attenuation of the feeder cable (1 dB/m cable length) to a total of approximately 4 dB. We operate our 5 MHz transmit signal in 6.25 MHz available spectrum to facilitate sharp spectral filtering. Our power amplifiers are capable of transmitting at a power of 30 W (per antenna), of which we use less than 15 % to ensure high-linearity transmissions.

Typically, the performance of wireless systems is evaluated by drive tests; that is, measuring the throughput in a vehicle continuously while driving along a road. Unfortunately, the same test repeated on another day can result in entirely different outcomes. In order to make our measurements repeatable we apply a measurement technique, employing *XY* positioning tables on which we automatically change the antenna positions and measure many locations in an area of $3\,\lambda \times 3\,\lambda$. At each antenna position in the same scenario we perform repeated measurements of all transmission schemes of interest. By such a setup we ensure that all schemes are measured under exactly the same conditions. The averaged result of all antenna positions then reflects later an estimated mean value (of, for example, the measured throughput or the channel capacity calculated from the estimated channel coefficients) at a given transmit power P_{Tx}. By measuring hundreds of antenna positions we ensure that we estimate with high precision. In order to evaluate our precision, we compute for each measurement point the 99 % confidence interval by bootstrapping methods and plot the point augmented by a small scale (for example, see Figure 8.1 in the previous section or the following figures) according to the size of the confidence interval achieved. Most of these only appear as points, indicating extremely high precision. See Chapter 4 and Section 5.6 for a more detailed description of the measurement procedure employed.

8.4 Measurement Results

8.4.1 WiMAX Results

Figures 8.2 and 8.3 show our WiMAX measurement results for the selected scenarios. Figure 8.2 plots the absolute values of capacity, MI, achievable mutual information, and throughput and Figure 8.3 shows the relative loss terms defined in Section 8.2.2. We observe the following.

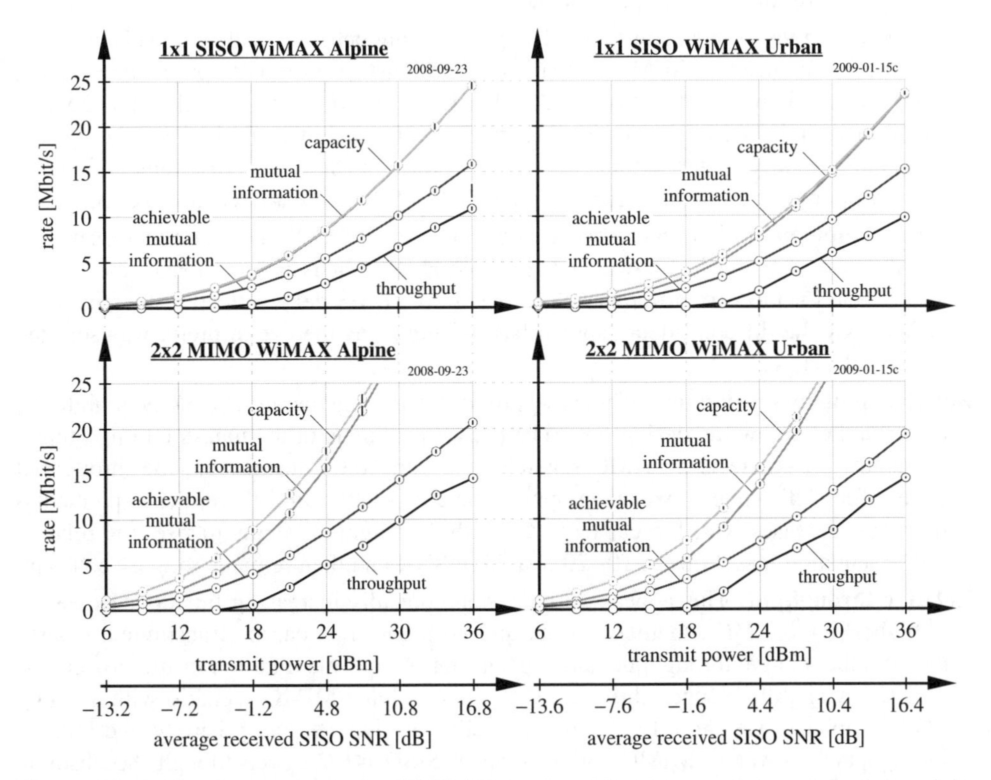

Figure 8.2 Throughput and bounds of WiMAX 1 × 1 and 2 × 2 transmissions: left, alpine; right, urban; top, SISO; bottom, MIMO.

Absolute CSI loss: This is very small, since the curves for capacity and MI are almost on top of each other, as shown in Figure 8.2. In order to combat the CSI loss, a considerable amount of feedback information is required. Therefore, it has to be carefully considered whether combating this loss pays off. For SISO systems the CSI loss can be neglected compared to the other losses. For cross polarized 2×2 MIMO systems the CSI loss is also very small compared with the others. Only in antenna arrays with at least four antennas the CSI loss is worth considering. This has already been pointed out in [6], where, based on this property, a simple channel model was proposed that allows analytical treatment and to match capacity with measurements.

Relative CSI loss: This is a monotonically decreasing function. For moderate to high transmit power values P_{Tx}, the impact of missing CSI at the transmitter can be neglected, questioning why channel information beyond the receiver SNR is useful for the transmitter.

Achievable mutual information and the throughput: These monotonically increase with transmit power P_{Tx}. As Figure 8.2 shows, the throughput curve looks like a shifted version of the achievable mutual information, meaning that the receiver performance does not change over receive SNR. Note the difference later in the HSDPA results in Figure 8.5: here, the gap between throughput and achievable mutual information increases with receive SNR, which is due to increasing influence of inter-code interference on the receiver performance.

Relative design loss: This shows a gradually increasing behavior with P_{Tx}. This behavior is more pronounced in MIMO scenarios, in which the Alamouti space–time coding plays a crucial role. In the case of 2×2 MIMO transmission, the Alamouti code is suboptimal and, hence, the achievable mutual information is limited by the system design. Especially at large transmit power levels, the loss due to Alamouti coding becomes dominant, wasting approximately 50 % of the available channel capacity.

Absolute implementation loss: This is moderately small for WiMAX and increases only slightly with the transmit power, whereas for HSDPA it is monotonously increasing. The reason for the very different behavior is in the RMS delay spread of the channels. In WiMAX the different delay spread has no impact as the cyclic prefix was selected sufficiently large.

Relative implementation loss: This is severe for lower to moderate P_{Tx}. Only at high P_{Tx} does this fall below the design loss. The reason for this is in the measured throughput. As at low P_{Tx} there is no AMC scheme supporting the transmission, the throughput is zero for SNR values lower than approximately −3 dB. Until there, the implementation loss increases with the capacity. Once the throughput begins to rise, the relative implementation loss drops to 10–20 % at high SNR and then remains roughly constant.

Relative throughput: This appears to be a monotonically increasing function. However, at high P_{Tx} the AMC schemes run out and no higher rate can be transmitted. Consequently, the relative throughput starts decreasing. As our maximum transmit power was 36 dBm, we hardly reached this area. Only in the alpine MIMO scenario with a strong LOS component we can observe such a decline at high P_{Tx}. Surprisingly, the relative throughput is lower in MIMO (30 %) than in SISO (40 %) even though the absolute values are much higher in MIMO, showing that this version of WiMAX is not capable of taking advantage of what MIMO offers.

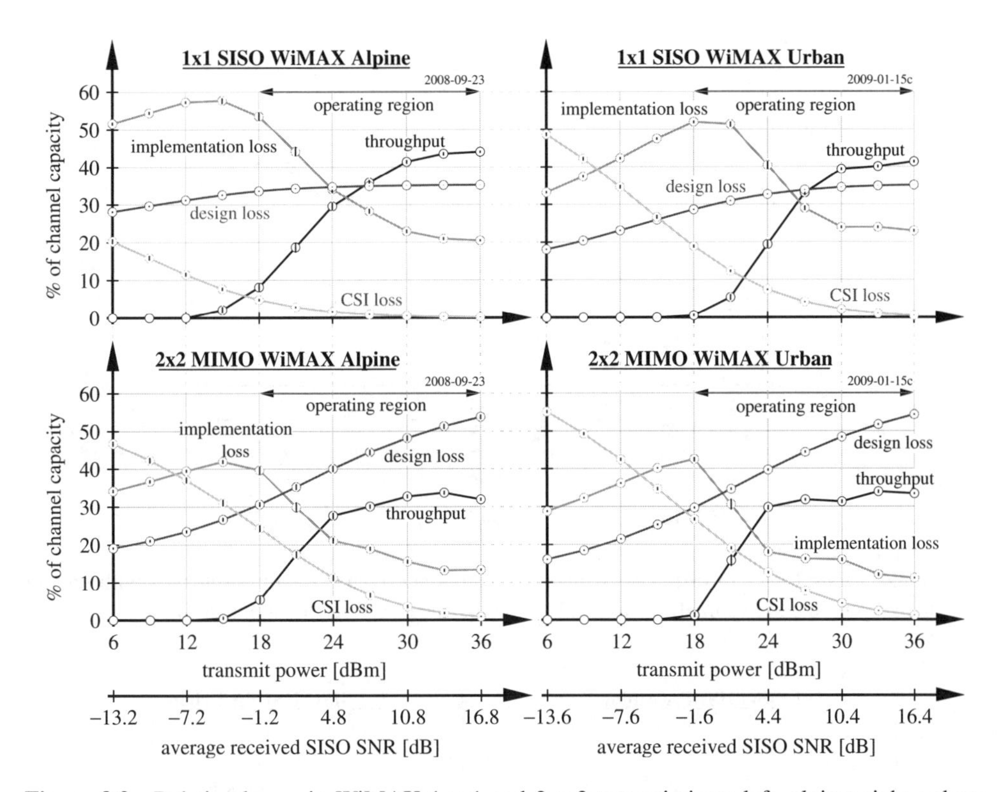

Figure 8.3 Relative losses in WiMAX 1×1 and 2×2 transmissions: left, alpine; right, urban; top, SISO; bottom, MIMO [5]. Reproduced by permission of © IEEE 2010.

In summary, we conclude that WiMAX behaves quite similar in alpine and urban environments as well as in SISO and MIMO. Comparing SISO with MIMO performance, we can only conclude that the various components show larger variations in SISO and less in MIMO.

Let us now take a closer look at the implementation loss. In our displayed measurements we employed the Convolutional Turbo Code (CTC) with an advanced channel estimation scheme (Approximate Linear Minimum Mean Square Error (ALMMSE)). Figure 8.4 depicts the SNR loss of various channel estimation and channel coding schemes at a throughput of 5 Mbit/s when compared with a genie channel estimator and LDPC channel coding. The genie channel estimator knows not only the 200 pilot symbols but also all 9400 transmitted data symbols and uses them to achieve a very high channel estimation quality. Similarly, a regular LDPC code was designed to provide a high-quality channel code for a given SNR. Figure 8.4 shows that there is a 6 dB improvement from poor Least Squares (LS) channel estimation in combination with the mandatory RS-CC channel coding of WiMAX to genie knowledge of the channel in combination with high-quality channel coding. In the case of 2×2 transmissions, when Alamouti coding is employed at the transmitter, we observe a much higher dependence on the type of channel estimator chosen. This can be explained by two facts:

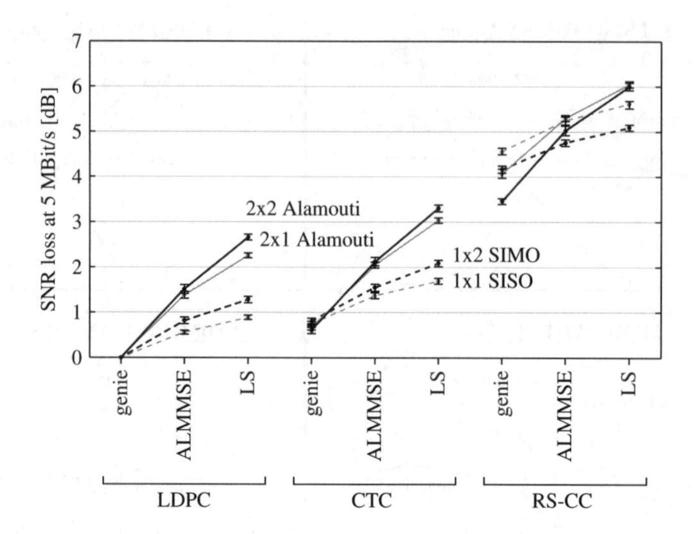

Figure 8.4 SNR losses of different channel estimators and channel coding schemes with respect to genie-driven channel estimation and LDPC coding; measured in the alpine scenario.

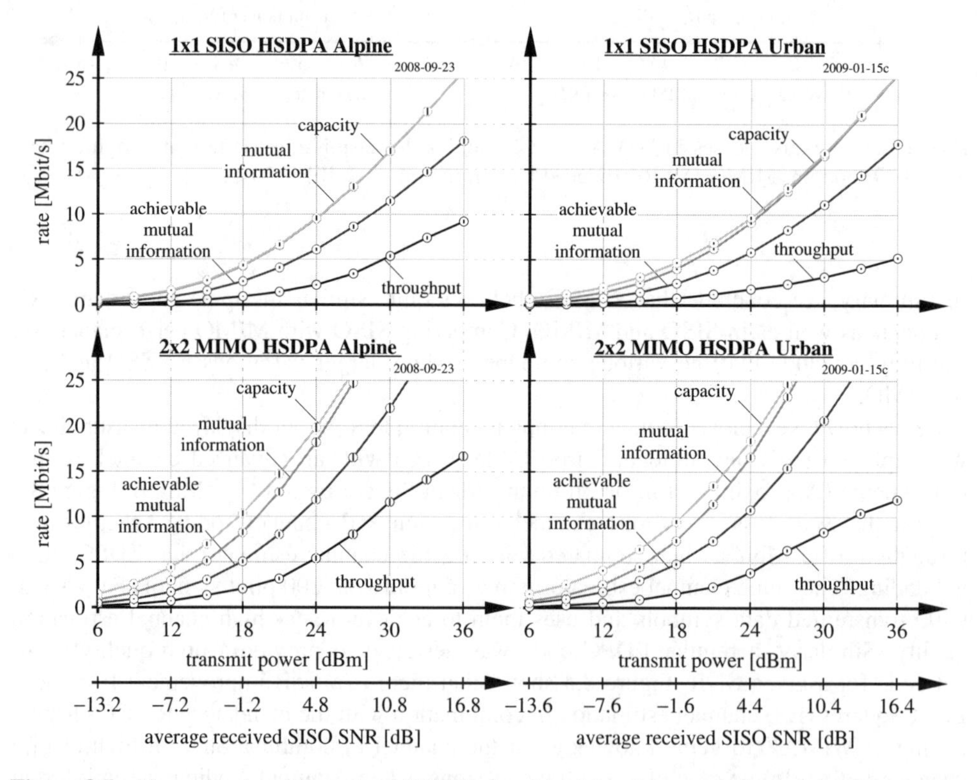

Figure 8.5 Throughput and bounds of HSDPA 1×1 and 2×2 transmissions: left, alpine; right, urban; top, SISO; bottom, MIMO.

1. In the case of Alamouti transmission, the available transmit power and, thus, also the training signal power are equally distributed on the two transmit antennas. Therefore, only half the training signal power is available per channel coefficient to be estimated. As a consequence, the channel estimation performance is poorer than in the one-transmit-antenna case.
2. The comparison in Figure 8.4 is carried out at a constant throughput value of 5 Mbit/s. The Alamouti transmission achieves this throughput at a much lower receive SNR than the SISO transmission. Such lower SNR in turn causes poorer channel estimation quality and, thus, a substantial SNR loss in comparison with the genie channel estimator.

In order to compensate for the above effects, one would have to apply a much better channel estimator for MIMO than in the case of SISO transmissions.

8.4.2 HSDPA Results in Standard-Compliant Setting

All measurement results presented in this section are for a Category 16 HSDPA user equipment. In the MIMO case, a Double Transmit Antenna Array (D-TxAA) with adaptive precoding is applied [4].

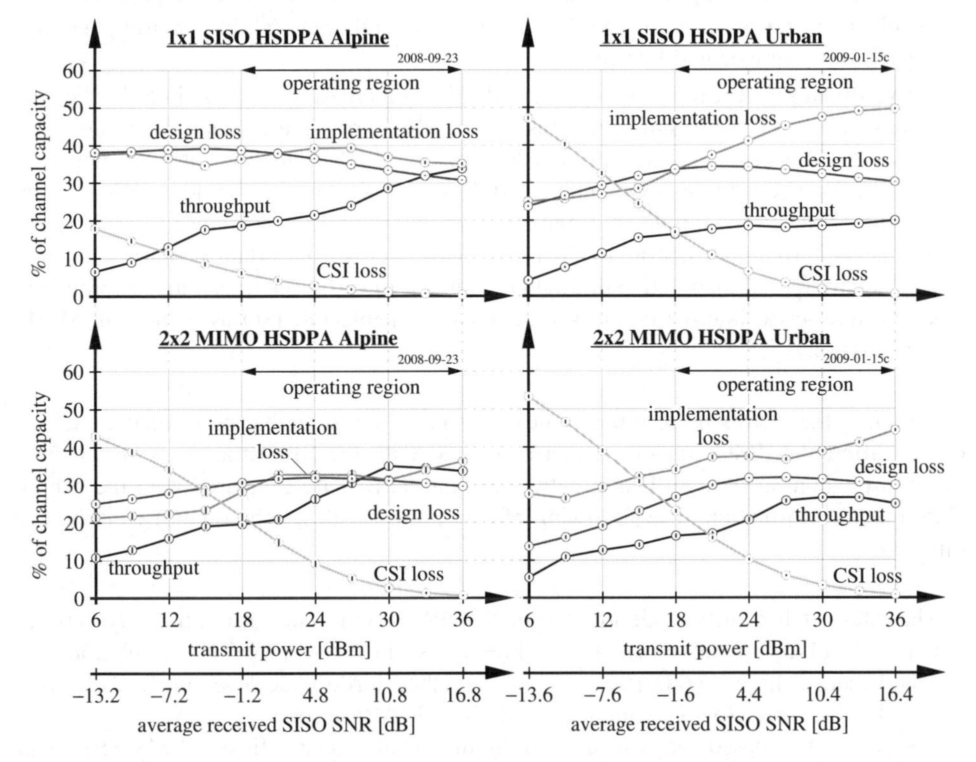

Figure 8.6 Relative losses in HSDPA 1×1 and 2×2 transmissions: left, alpine; right, urban; top, SISO; bottom, MIMO [5]. Reproduced by permission of © IEEE 2010.

The various losses of HSDPA as depicted in Figure 8.5 (absolute values) and Figure 8.6 (relative losses) show a much less lively picture when compared with WiMAX.

Relative CSI loss: As before, this decreases monotonically in exactly the same dimensions as the CSI loss is independent of the transmission standard, proving that both measurements have experienced the same equipment as well as the same wireless conditions. Small differences when compared with the WiMAX results are due to a slightly different occupied bandwidth.

Relative design loss: The other losses are more or less constant functions or slightly increasing with P_{Tx}. The relative design loss with values between 30 and 40 % is a particularly flat curve when compared with WiMAX. Only at low P_{Tx} in the urban MIMO scenario can the values become as small as 15 %.

Relative implementation loss: This is either of the same value as the design loss or larger. The distinct behavior in the urban scenario is of interest. Owing to the larger RMS delay spread in the urban scenario, the inter-code interference is increased and becomes a dominant part at high P_{Tx}. Therefore, we recognize a higher implementation loss and a smaller design loss for high P_{Tx}.

Absolute implementation loss: This behaves very differently for WiMAX and HSDPA, as can be observed in Figure 8.5. In environments with large RMS delay spread (urban environment with 1.1 μs) the implementation loss of HSDPA is extremely large due to self-interference while in small RMS delay spread areas (alpine environment with 260 ns) the behavior is different.

Relative throughput: Similar to WiMAX, this is an increasing function in SNR. For low values of P_{Tx} the spreading functions of HSDPA improve the situation considerably, allowing a transmission even for very low receiver SNR at low bit rates. Furthermore, HSDPA has many more AMC schemes than WiMAX, especially at low SNR. At high P_{Tx} we obtain in the alpine environment with its strong LOS the close to 40 % values in relative throughput just as for WiMAX. However, in the urban scenario with high RMS delay spread the relative throughput is much lower as the transmitter is producing strong inter-code interference in such channels, visible in SISO as well as in MIMO transmissions.

Although the results in absolute values [4] show a considerable performance increase of the different MIMO schemes when compared with the SISO transmission, all measured throughput curves are losing between 3 and 9 dB SNR compared with the achievable mutual information. The following effects (along with maybe others) contribute to this loss:

- The rate-matched turbo code utilized in HSDPA is good but not optimal. By carrying out a set of comparative Additive White Gaussian Noise (AWGN) simulations, we found that, at higher code rates, the rate-matched turbo code loses up to 2 dB when decoded by a max-log-Maximum A Posteriori (MAP) decoder.
- The equalizer based on Linear Minimum Mean Square Error (LMMSE) channel estimation representing a low-complexity and cost-effective solution is also not optimal. Better receivers, such as the Linear Minimum Mean Square Error

Maximum A Posteriori (LMMSE-MAP), have the potential to improve the performance by approximately 1 dB.

- In the urban scenario, a larger throughput loss was measured than in the alpine scenario because of the larger delay spread and, consequently, the larger inter-code interference. For example, in the alpine scenario the SISO system loses approximately 6 dB to the achievable mutual information, whereas the loss in the urban scenario is approximately 9 dB.
- In addition to the above-mentioned losses, channel estimation errors and over-/under-estimation of the post-equalization Signal to Interference and Noise Ratio (SINR) degrade the measured throughput. It is difficult to quantify exactly the loss caused by these effects, because neither perfect channel state information nor perfect post-equalization SINR is available in measurements.

Note that HSDPA supports Hybrid Automatic Repeat reQuest (HARQ) retransmissions, whereas the chosen WiMAX standard does not support this yet. We indeed implemented retransmissions in HSDPA, but the measurement plots shown here do not utilize such retransmissions in order to keep the comparison with WiMAX fair. Note that including up to three retransmissions hardly changed the figures. We conclude that while HARQ offers some advantages in service quality (for example, lower latency), it has no impact on the throughput performance. This may be due to the fact that in our measurements the Block Error Ratio (BLER) is kept at approximately 1 % throughout the whole measurement range.

8.4.3 HSDPA Results in Advanced Setting

By employing a straightforward extension of the two-transmit-antenna precoding, we extended the HSDPA standard towards four antennas at the transmit side and utilized either two or four receive antennas [4]. The results obtained when employing 4×2 and 4×4 MIMO systems are shown in Figure 8.7 (absolute losses) and Figure 8.8 (relative losses). They are very different from the results shown above.

Relative design loss: This is now clearly the major loss, dominating the implementation loss in every scenario, while they were roughly of equal size above. This indicates that implementation of precoding for four antennas should be much more sophisticated than in our experiment.

Relative throughput: This reached 30 % of the capacity at best. We can thus conclude that although advanced settings on HSDPA improve absolute values of throughput by up to 50 %, relatively speaking they utilize less of the now much higher capacity.

Absolute throughput: Here, Figure 8.7 clearly exhibits strong improvements. However, the improvements come at the expense of even less efficiency than before.

8.5 Summary

The measured throughput D_{m} actually observed accounts for only 15–45 % of the available channel capacity. In terms of the individual losses, our findings are summarized as follows.

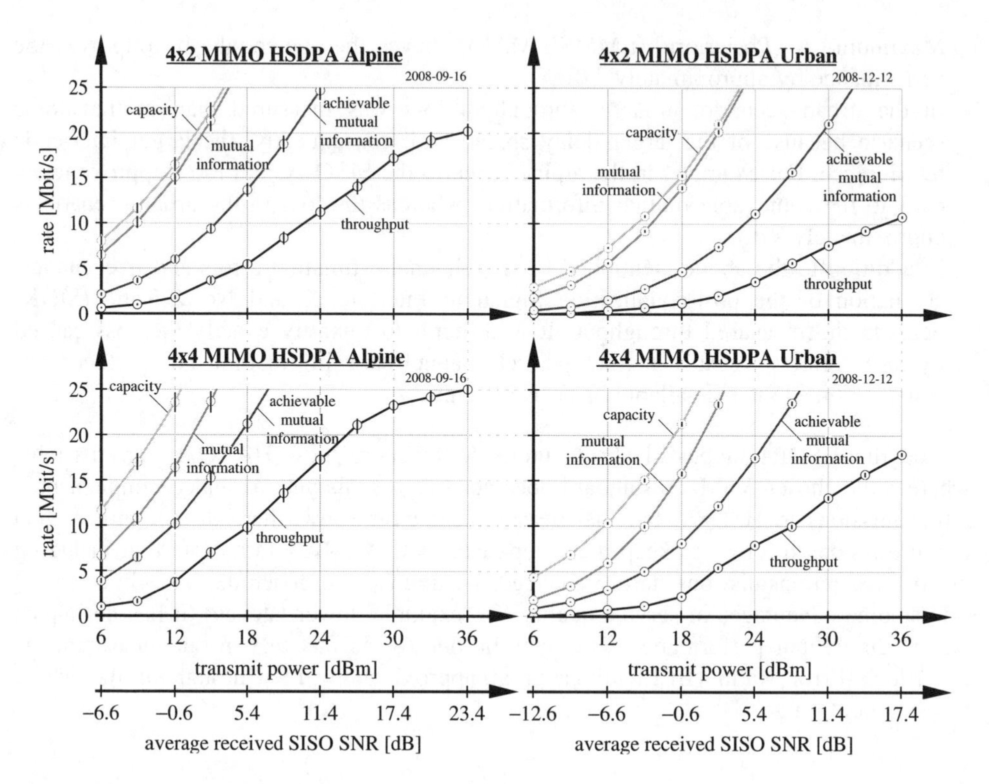

Figure 8.7 Throughput and bounds of HSDPA 4 Tx antenna transmissions: left, alpine; right, urban; top, 4 × 2 MIMO; bottom, 4 × 4 MIMO.

The CSI loss L_{CSI}: This loss is given by absence of full channel state information at the transmitter. The CSI loss is typically small and for SISO and cross-polarized 2 × 2 MIMO systems it can be neglected compared with the other losses. Only when the transmission system employs more than two antennas and is operating at low SNRs it is worth combating the CSI loss by employing detailed CSI feedback. A potential of 50 % improvement in capacity may sound appealing, but note, however, that the absolute values are rather small in the low SNR regime.

The design loss L_d: Here, the differences between HSDPA and WiMAX are not very pronounced. HSDPA is generally better designed, as the design loss is roughly constant over the full SNR range, whereas in WiMAX the design loss is small for low SNR and increasing with SNR. In the case of 2 × 2 MIMO WiMAX this is caused by the suboptimal Alamouti space–time coding. Precoding is not expected to have an impact on the design loss, as optimal precoding matrices are unitary and so are the proposed precoding schemes in the HSDPA and LTE standards. The design loss is "the" vehicle of the future to achieve improvements at design stage. Once the standard is released, the design loss is final.

The implementation loss L_i: In environments with high RMS delay spread (urban environment with 1.1 μs) the implementation loss of HSDPA is extremely large due to self-interference, whereas in small RMS delay spread areas (alpine environment with 260 ns) the behavior is different. In WiMAX, the different delay spread has no impact, as the cyclic prefix was selected sufficiently large. Both WiMAX and HSDPA suffer losses of several decibels due to channel estimation and nonoptimal coding. The much higher number of AMC schemes in HSDPA does not show much of a benefit compared with WiMAX. A good strategy for standardization institutions could be to accept a rather large initial implementation loss, anticipating that manufacturers will figure out how to improve implementation quality by employing more digital signal processing complexity. However, some implementations are close to optimal, not offering much to be gained any more. Once manufacturers are offering close-to-optimal implementations, they cannot differentiate their products between each other.

What counts at the end of the day when designing a system is the throughput actually achieved. Therefore, a system designer should equally take "all losses" into consideration, but note that the design loss imposed by the system design can never be reduced in the future by a clever receiver implementation. Therefore, it may pay off to start with a smaller design loss and a larger implementation loss.

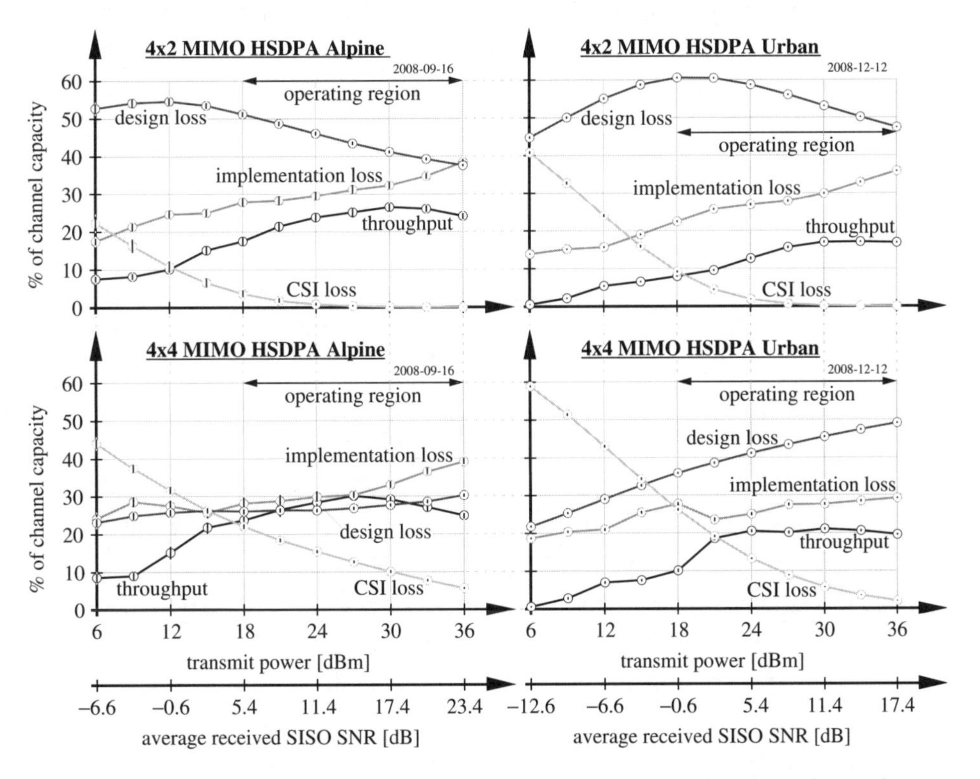

Figure 8.8 Relative losses in HSDPA 4 Tx antenna transmissions: left, alpine; right, urban; top, 4 × 2 MIMO; bottom, 4 × 4 MIMO [5]. Reproduced by permission of © IEEE 2010.

References

[1] Foschini, G. J. and Gans, M. J. (1998) 'On limits of wireless communications in a fading environment when using multiple antennas,' *Wireless Personal Communications*, **6** (3), 311–335. Available from http://www.springerlink.com/content/h1n7866218781520/fulltext.pdf.

[2] Ghosh, A., Wolter, D. R., Andrews, J. G., and Chen, R. (2005) 'Broadband wireless access with WiMax/802.16: current performance benchmarks and future potential,' *IEEE Communications Magazine*, **43** (2), 129–136, doi: 10.1109/MCOM.2005.1391513. Available from http://ieeexplore.ieee.org/iel5/35/30297/01391513.pdf?tp=&arnumber=1391513.

[3] Holma, H., Toskala, A., Ranta-aho, K., and Pirskanen, J. (2007) 'High-speed packet access evolution in 3GPP release 7,' *IEEE Communications Magazine*, **45** (12), 29–35, doi: 10.1109/MCOM.2007.4395362. Available from http://ieeexplore.ieee.org/stamp/stamp.jsp?arnumber=4395362.

[4] Mehlführer, C., Caban, S., and Rupp, M. (2010) 'Measurement-based performance evaluation of MIMO HSDPA,' *IEEE Transactions on Vehicular Technology*, **59** (9), 4354–4367, doi: 10.1109/TVT.2010.2066996. Available from http://publik.tuwien.ac.at/files/PubDat_187112.pdf.

[5] Mehlführer, C., Caban, S., and Rupp, M. (2011) 'Cellular system physical layer throughput: how far off are we from the Shannon bound?' *IEEE Wireless Communications Magazine*, Dec. 2011.

[6] Rupp, M., García-Naya, J. A., Mehlführer, C. et al. (2010) 'On mutual information and capacity in frequency selective wireless channels,' in *Proceedings of the IEEE International Conference on Communications (ICC 2010)*, Cape Town, South Africa, doi: 10.1109/ICC.2010.5501942. Available from http://publik.tuwien.ac.at/files/PubDat_184660.pdf.

[7] Shannon, C. (1948) 'A mathematical theory of communication,' *The Bell System Technical Journal*, **27**, 379–423, 623–656. Available from http://cm.bell-labs.com/cm/ms/what/shannonday/shannon1948.pdf.

[8] Telatar, I. E. (1999) 'Capacity of multi-antenna Gaussian channels,' *European Transactions on Telecommunications*, **10** (6), 585–595. Available from http://mars.bell-labs.com/papers/proof/proof.pdf.

9

Frequency Synchronization in LTE

Contributed by Qi Wang
Vienna University of Technology (TU Wien), Austria

The Long-Term Evolution (LTE) standard[1] employs an Orthogonal Frequency-Division Multiplexing (OFDM) physical layer in the Downlink (DL), making an accurate frequency synchronization become a crucial issue. The mismatch between the two local oscillators at the transmitter and the receiver introduces a Carrier Frequency Offset (CFO). This CFO destroys the orthogonality between subcarriers and degrades the system performance severely. While in simulations the CFO can readily be set to zero, this is never the case in reality. Owing to cost limitations, local oscillators at the User Equipment (UE) side have a typical frequency stability tolerance of ± 10 ppm. Consider an oscillator for LTE at 2.5 GHz, ± 10 ppm results in an offset of ± 25 kHz. With the fixed subcarrier spacing of 15 kHz defined in LTE, a CFO of ± 1.67 subcarrier spacing has to be handled at the receiver.

The literature on CFO estimation falls into two categories: data-aided methods and non-data-aided methods. Data-aided methods estimate the CFO by periodically transmitted reference symbols, where usually a repetitive pattern is required [10, 13, 15, 16, 20, 22]. These methods can be applied to burst transmission protocols such as IEEE 802.11 and Worldwide Inter-operability for Microwave Access (WiMAX). On the other hand, blind or non-data-aided methods [3, 4, 11, 12, 14, 25, 26, 28] do not rely on dedicated training symbols. In [26] and [14], Maximum Likelihood (ML) CFO estimators were proposed for flat-fading and frequency-selective fading channels. These two estimators exploit the redundancy of the Cyclic Prefix (CP) in OFDM. In [11, 12, 25], subspace-based estimators were presented based on the null subcarriers in an OFDM symbol. Other non-data-aided methods exploit either the cyclostationarity of the OFDM signal [3] or a kurtosis-type cost function [28].

[1] Reproduced from Q. Wang, C. Mehlführer, and M. Rupp, "Carrier frequency synchronization in the downlink of 3GPP LTE," in *Proceeding of the 21st Annual IEEE International Symposium on Personal, Indoor and Mobile Radio Communications (PIMRC10)*, Sept. 2010, by permission of © 2010 IEEE.

Evaluation of HSDPA and LTE: From Testbed Measurements to System Level Performance, First Edition.
Sebastian Caban, Christian Mehlführer, Markus Rupp and Martin Wrulich.

In above-mentioned literature, the performance is evaluated in terms of the estimation error; in other words, the Mean Square Error (MSE). On the other hand, the performance of the transmission system is evaluated in terms of Signal to Noise Ratio (SNR), Bit Error Ratio (BER), and most importantly the overall throughput. Therefore, not only the accuracy of the synchronization becomes of interest, but also the effect of frequency synchronization errors on the performance of OFDM systems and in particular in LTE.

9.1 Mathematical Model

In general, the CFO has a twofold impact on OFDM signals. Taking the aforementioned example, a CFO is normalized to the subcarrier spacing for analysis, denoted by $\varepsilon_{\mathrm{CFO}} = \pm 1.67$. The fractional part ± 0.67 of $\varepsilon_{\mathrm{CFO}}$ induces Inter-Carrier Interference (ICI) to the OFDM signal. The integer part ± 1 poses a mismatch of the subcarrier indices between the transmitter and the receiver. In this section, we present a mathematical model in order to show the twofold impact of the CFO on OFDM analytically. We denote the transmitted OFDM signal in the time domain by $x_{l,n}^{(q)}$, the channel impulse response by $h_{l,n}^{(m,q)}$, the additive white Gaussian noise by $v_{l,n}^{(m)}$, and the received time-domain signal by $r_{l,n}^{(m)}$. Here, l is the OFDM symbol index within one subframe, $n \in [-N_{\mathrm{g}}, \ldots, 0, \ldots, N-1]$ is the time index within one OFDM symbol, q is the transmit antenna index, and m is the receive antenna index. The Fast Fourier Transform (FFT) size is denoted by N and the CP length by N_{g}. Applying the above definitions, the transmission with CFO is described in the time domain as

$$r_{l,n}^{(m)} = \left\{ \sum_{q=1}^{N_{\mathrm{T}}} x_{l,n}^{(q)} * h_{l,n}^{(m,q)} + v_{l,n}^{(m)} \right\} \mathrm{e}^{\mathrm{i}\frac{2\pi\varepsilon_{\mathrm{CFO}}[n+l(N+N_{\mathrm{g}})]}{N}}, \tag{9.1}$$

where $*$ denotes a convolution. We assume that all transmitter (receiver) antennas are served by one common local oscillator on the transmitter (receiver) side, leading to a common CFO for all antenna pairs. The normalized CFO between the two sides is denoted by $\varepsilon_{\mathrm{CFO}}$. Correspondingly, in the frequency domain, we define $X_{l,k}^{(q)} = \mathcal{F}\{x_{l,n}^{(q)}\}$ as the transmitted symbol, $H_{l,k}^{(m,q)} = \mathcal{F}\{h_{l,n}^{(m,q)}\}$ as the channel frequency response, $V_{l,k}^{(m)} = \mathcal{F}\{v_{l,n}^{(m)}\}$ as the additive white Gaussian noise, and $R_{l,k}^{(m)} = \mathcal{F}\{r_{l,n}^{(m)}\}$ as the received symbol on the kth subcarrier of the lth OFDM symbol at the mth receive antenna. When the discrete Fourier transform is applied to the received signal $r_{l,n}^{(m)}$ in Equation (9.1), we obtain

$$\begin{aligned} R_{l,k}^{(m)} &= \sum_{n=0}^{N-1} r_{l,n}^{(m)} \mathrm{e}^{-\mathrm{i}\frac{2\pi kn}{N}} \\ &= \sum_{n=0}^{N-1} \left\{ \left\{ \sum_{q=1}^{N_{\mathrm{T}}} x_{l,n}^{(q)} * h_{l,n}^{(m,q)} + v_{l,n}^{(m)} \right\} \mathrm{e}^{\mathrm{i}\frac{2\pi\varepsilon_{\mathrm{CFO}}n}{N}} \mathrm{e}^{\mathrm{i}\frac{2\pi\varepsilon_{\mathrm{CFO}}l(N+N_{\mathrm{g}})}{N}} \right\} \mathrm{e}^{-\mathrm{i}\frac{2\pi kn}{N}} \\ &= \mathrm{e}^{\mathrm{i}\frac{2\pi\varepsilon_{\mathrm{CFO}}l(N+N_{\mathrm{g}})}{N}} \sum_{n=0}^{N-1} \left\{ \sum_{q=1}^{N_{\mathrm{T}}} x_{l,n}^{(q)} * h_{l,n}^{(m,q)} + v_{l,n}^{(m)} \right\} \mathrm{e}^{\mathrm{i}\frac{2\pi\varepsilon_{\mathrm{CFO}}n}{N}} \mathrm{e}^{-\mathrm{i}\frac{2\pi kn}{N}} \end{aligned} \tag{9.2}$$

According to the convolution theorem, this further leads to

$$R_{l,k}^{(m)} = \mathrm{e}^{\mathrm{i}\frac{2\pi\varepsilon_{\mathrm{CFO}}l(N+N_{\mathrm{g}})}{N}} \sum_{q=1}^{N_{\mathrm{T}}} \left\{X_{l,k}^{(q)} H_{l,k}^{(m,q)}\right\} * \left\{\frac{1}{N}\sum_{n=0}^{N-1} \mathrm{e}^{\mathrm{i}\frac{2\pi\varepsilon_{\mathrm{CFO}}n}{N}} \mathrm{e}^{-\mathrm{i}\frac{2\pi kn}{N}}\right\} + \tilde{V}_{l,k}^{(m)} \tag{9.3}$$

$$\begin{aligned} R_{l,k}^{(m)} &= \mathrm{e}^{\mathrm{i}\frac{2\pi\varepsilon_{\mathrm{CFO}}l(N+N_{\mathrm{g}})}{N}} \sum_{q=1}^{N_{\mathrm{T}}} \left\{X_{l,k}^{(q)} H_{l,k}^{(m,q)}\right\} \\ &\quad * \left\{\frac{\sin\left[\pi(\varepsilon_{\mathrm{CFO}} - k)\right]}{N\sin\left[\pi(\varepsilon_{\mathrm{CFO}} - k)/N\right]} \mathrm{e}^{\mathrm{i}\frac{\pi(\varepsilon_{\mathrm{CFO}}-k)(N-1)}{N}}\right\} + \tilde{V}_{l,k}^{(m)} \\ &= \mathrm{e}^{\mathrm{i}\frac{2\pi\varepsilon_{\mathrm{CFO}}l(N+N_{\mathrm{g}})}{N}} \sum_{q=1}^{N_{\mathrm{T}}} \left\{\sum_{p=0}^{N-1} X_{l,p}^{(q)} H_{l,p}^{(m,q)} \frac{\sin\left[\pi(p - k + \varepsilon_{\mathrm{CFO}})\right]}{N\sin\left[\pi(p - k + \varepsilon_{\mathrm{CFO}})/N\right]} \right. \\ &\quad \left. \times\, \mathrm{e}^{\mathrm{i}\frac{\pi(p-k+\varepsilon_{\mathrm{CFO}})(N-1)}{N}}\right\} + \tilde{V}_{l,k}^{(m)} \end{aligned} \tag{9.4}$$

Here, $\tilde{V}_{l,k}^{(m)}$ denotes the noise term $V_{l,k}^{(m)}$ rotationally altered by the CFO; that is, its statistical properties do not change. In order to elaborate the twofold impact, we split the total CFO into a fractional part $\varepsilon_{\mathrm{FFO}}$ and an integer part $\varepsilon_{\mathrm{IFO}}$ with $\varepsilon_{\mathrm{CFO}} = \varepsilon_{\mathrm{FFO}} + \varepsilon_{\mathrm{IFO}}$. This allows the separation of signal and interference terms from Equation (9.4) to

$$R_{l,k}^{(m)} = \underbrace{I(0)\sum_{q=1}^{N_{\mathrm{T}}} X_{l,k'}^{(q)} H_{l,k'}^{(m,q)}}_{\text{signal}} + \underbrace{\sum_{p\neq k'}\sum_{q=1}^{N_{\mathrm{T}}} X_{l,p}^{(q)} H_{l,p}^{(m,q)} \cdot I(p-k')}_{\text{interference}} + \underbrace{\tilde{V}_{l,k}^{(m)}}_{\text{noise}}, \tag{9.5}$$

with

$$I(0) = \frac{\sin(\pi\varepsilon_{\mathrm{FFO}})}{N\sin(\pi\varepsilon_{\mathrm{FFO}}/N)} \mathrm{e}^{\mathrm{i}\frac{\pi\varepsilon_{\mathrm{FFO}}(N-1)}{N}} \mathrm{e}^{\mathrm{i}\Phi(\varepsilon_{\mathrm{CFO}},l)}, \tag{9.6}$$

$$\begin{aligned} I(p-k') &= \frac{\sin[\pi(p - k + \varepsilon_{\mathrm{CFO}})]}{N\sin[\pi(p - k + \varepsilon_{\mathrm{CFO}})/N]} \mathrm{e}^{\mathrm{i}\frac{\pi(p-k+\varepsilon_{\mathrm{CFO}})(N-1)}{N}} \mathrm{e}^{\mathrm{i}\Phi(\varepsilon_{\mathrm{CFO}},l)} \\ &= \frac{\sin[\pi(p - k' + \varepsilon_{\mathrm{FFO}})]}{N\sin[\pi(p - k' + \varepsilon_{\mathrm{FFO}})/N]} \mathrm{e}^{\mathrm{i}\frac{\pi(p-k'+\varepsilon_{\mathrm{FFO}})(N-1)}{N}} \mathrm{e}^{\mathrm{i}\Phi(\varepsilon_{\mathrm{CFO}},l)}, \end{aligned} \tag{9.7}$$

$$\mathrm{e}^{\mathrm{i}\Phi(\varepsilon_{\mathrm{CFO}},l)} = \mathrm{e}^{\mathrm{i}\frac{2\pi\varepsilon_{\mathrm{CFO}}l(N+N_{\mathrm{g}})}{N}}, \tag{9.8}$$

where $k' = k - \varepsilon_{\mathrm{IFO}}$ is the true subcarrier index at the receiver, which implies that data symbols transmitted on subcarrier $k' = k - \varepsilon_{\mathrm{IFO}}$ are received on subcarrier k. The function $\Phi(\varepsilon_{\mathrm{CFO}}, l)$ is known as "common phase error," which is identical for all the subcarriers in one OFDM symbol [2]. The second term in Equation (9.5) contains the ICI from the neighboring subcarriers. In summary, CFO has a twofold impact on the received signal:

- A distortion to the desired signal $I(0)$ and ICI.
- An undesirable mismatch of the subcarrier indices between the Transmitter (TX) and the Receiver (RX), denoted by $k' = k - \varepsilon_{\mathrm{IFO}}$.

According to Equation (9.1), the CFO can be compensated by a reverse phase rotation:

$$\bar{r}_{l,n}^{(m)} = r_{l,n}^{(m)} \mathrm{e}^{-\mathrm{i}\frac{2\pi\hat{\varepsilon}_{\mathrm{CFO}}[n+l(N+N_g)]}{N}} \tag{9.9}$$

9.2 Carrier Frequency Offset Estimation in LTE

The frequency synchronization block is located prior to the demodulation of the subcarriers in the receiver, as shown in Figure 9.1. After the signal is received from the N_{R} antennas and down-converted to digital baseband, the time-domain signal is fed into the frequency synchronization block. Following the philosophy outlined in [21], the estimation of the CFO consists of pre-FFT and post-FFT processing. To provide a fast acquisition of the fractional part of the CFO, a pre-FFT algorithm is required. Post-FFT training data in LTE, namely synchronization signals and reference signals, are utilized to estimate the integer and the residual part. The resulting structure with the frequency synchronization block is depicted in Figure 9.1. A brief overview of the three-stage estimation is provided below. The relationship among the three parts can be expressed as

$$\hat{\varepsilon}_{\mathrm{CFO}} = \underbrace{\hat{\varepsilon}_{\mathrm{FFO}} + \hat{\varepsilon}_{\mathrm{RFO}}}_{\approx\varepsilon_{\mathrm{FFO}}} + \underbrace{\hat{\varepsilon}_{\mathrm{IFO}}}_{\approx\varepsilon_{\mathrm{IFO}}} \tag{9.10}$$

Fractional Frequency Offset (FFO) estimation: This step provides an initial estimation of the $\varepsilon_{\mathrm{FFO}}$. It is also referred to as coarse frequency synchronization. Its estimation is conducted in the time domain. A typical estimation range is between −0.5 and +0.5 subcarrier spacing. After this stage, most of the ICI is canceled. Since LTE does not have preambles with repetitive pattern defined, data-aided estimation cannot be applied.

Integer Frequency Offset (IFO) estimation: This step aims to align the subcarrier indices at the receiver to those at the transmitter. We utilize the predefined synchronization signals in the LTE standard to estimate the $\varepsilon_{\mathrm{IFO}}$ in the frequency domain.

Residual Frequency Offset (RFO) estimation: This step is to improve the estimation of the $\varepsilon_{\mathrm{FFO}}$, sometimes referred to as fine frequency synchronization. Owing to the limitation of the estimation performance in the time domain, it is necessary to refine the estimation further in the frequency domain. Regularly distributed in time and frequency, the predefined reference signals in LTE can be employed to improve the estimation performance considerably, especially in frequency-selective fading scenarios.

9.2.1 *Standardized Training Symbols in LTE*

In this section we briefly describe the reserved training symbols in the LTE standard. These training symbols are utilized for frequency synchronization. As depicted in Figure 9.2,

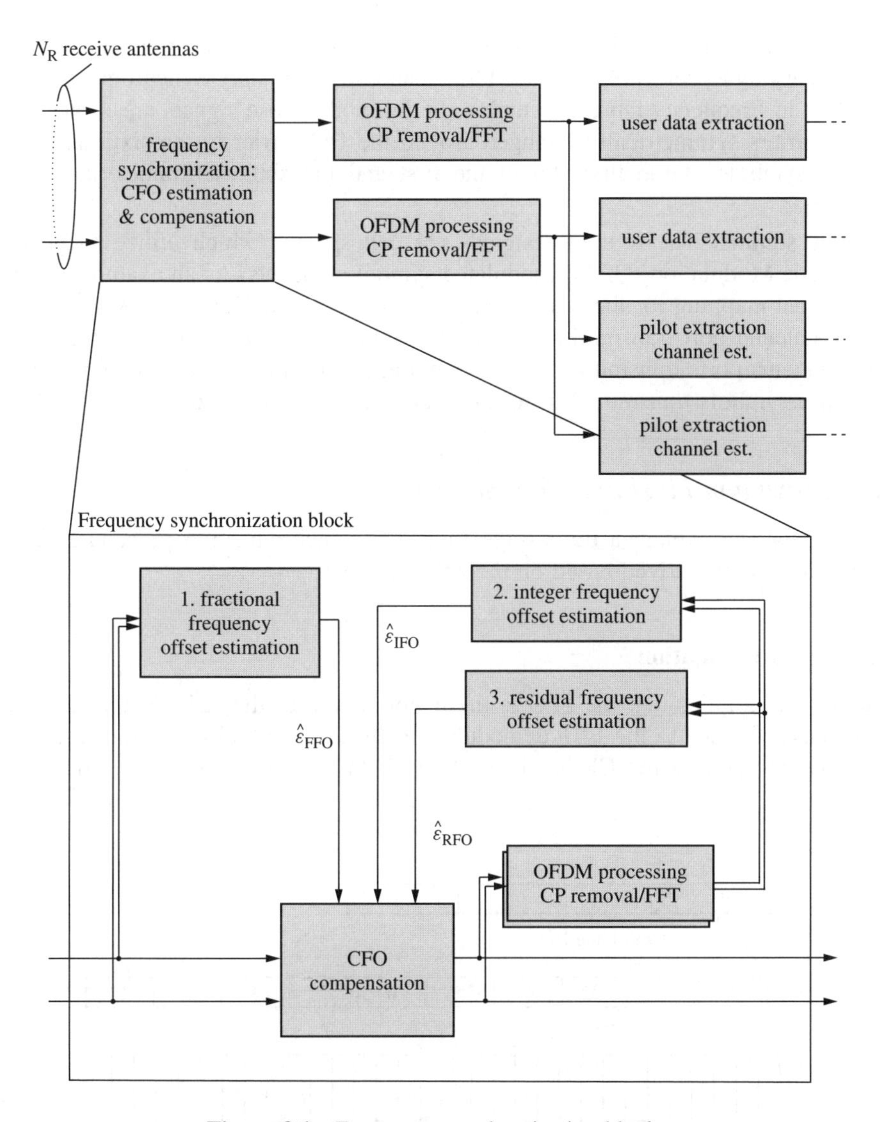

Figure 9.1 Frequency synchronization block.

radio frames in an LTE DL transmission have a duration of $T_{\text{frame}} = 10\,\text{ms}$. Each frame consists of 10 subframes of equal length. Each subframe contains two consecutive slots of time duration $T_{\text{slot}} = 0.5\,\text{ms}$. The smallest scheduling unit is called one Resource Block (RB), which is a time–frequency grid of 12 subcarriers over $N_s = 7$ OFDM symbols for the normal CP length and $N_s = 6$ for the extended CP length. In the LTE standard [1], there are two kinds of training symbols specified (mentioned as physical signals) in the DL: the synchronization signal and the reference signal. These signals do not carry information originating from higher layers and can be utilized for frequency synchronization at the receiver.

Synchronization signals: The LTE standard defines two synchronization signals, namely the Primary Synchronization Signal (PSCH) and the Secondary Synchronization Signal (SSCH). In Frequency Division Duplex (FDD) mode, these signals are located on the 62 subcarriers symmetrically arranged around the DC-carrier in the sixth and seventh OFDM symbols of the first slot in the first and the sixth subframes, as shown in Figure 9.2.

Reference signals: The reference signals are cell specific which utilize 4-Quadrature Amplitude Modulation (QAM) modulated symbol alphabets. As an example, the reference signal mapping for the four-antenna case is shown in Figure 9.3. Whenever there is one antenna port transmitting a reference symbol on one resource element, all the other antenna ports transmit a "zero" symbol at this position. Thus, interference of the reference symbols transmitted from different antennas is avoided.

9.2.2 *Maximum Likelihood Estimators*

In this section we explain a three-stage frequency synchronization procedure [27]. All three estimators are derived based on the ML principle.

9.2.2.1 FFO Estimation

Equation (9.7) implies that the FFO is the major source of the ICI. Therefore, it has to be compensated prior to the FFT demodulation. In [26], a blind estimator based on the duplicative property of the CP has been derived. In the context of slow fading, where

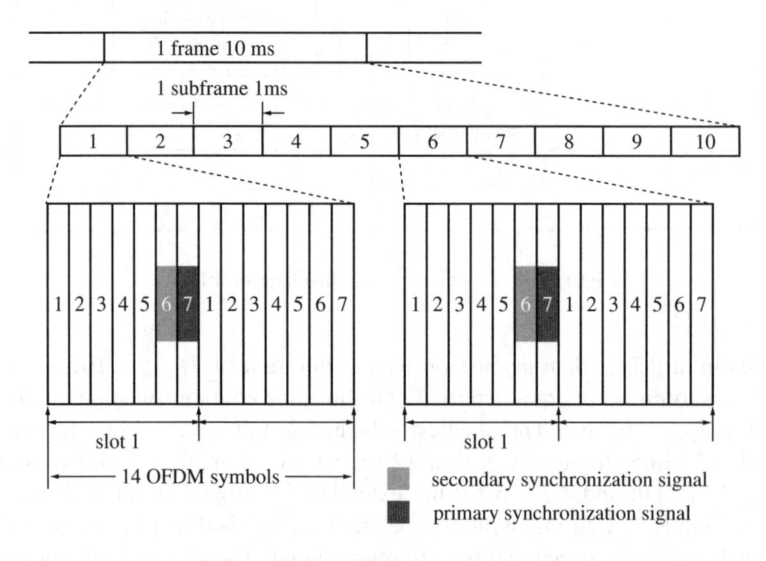

Figure 9.2 LTE frame structure.

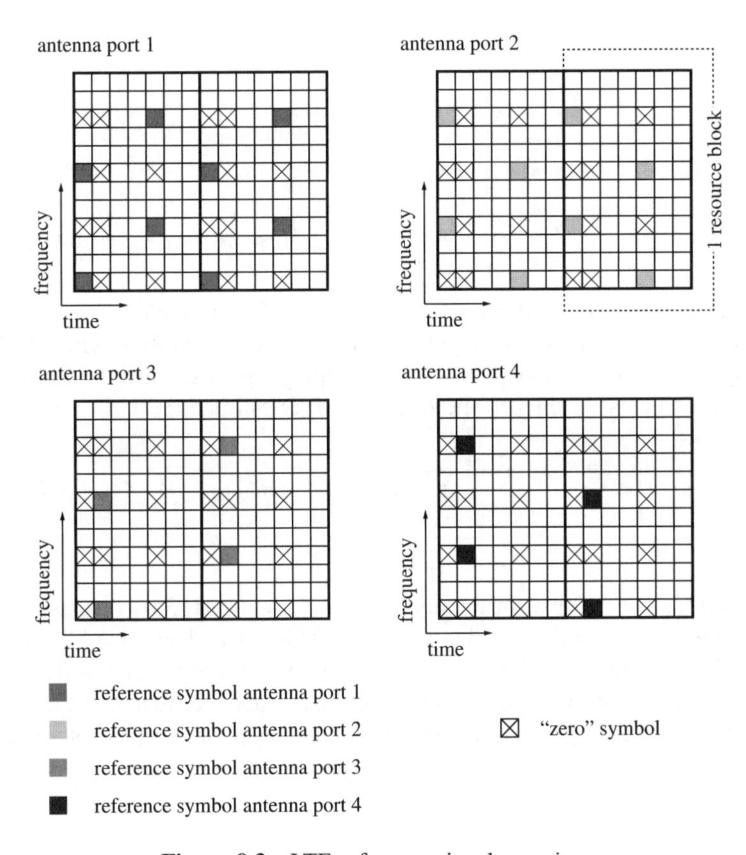

Figure 9.3 LTE reference signal mapping.

the channel coherence time is greater than one subframe, this method can be extended as

$$\hat{\varepsilon}_{\mathrm{FFO}} = -\frac{1}{2\pi} \arg\left\{ \sum_{m=1}^{N_{\mathrm{R}}} \sum_{l=1}^{N_{\mathrm{f}}} \sum_{n=1}^{N_{\mathrm{g}}} r_{l,n}^{(m)} r_{l,n+N}^{(m)*} \right\}, \tag{9.11}$$

where N_{R} denotes the number of receive antennas, N_{f} the number of OFDM symbols in one subframe, and N_{g} the CP length. The estimation range (normalized to the subcarrier spacing) of this stage is $(-0.5, 0.5)$, which is determined by the $\arg\{\cdot\}$ operation.

9.2.2.2 IFO Estimation

After the FFO estimation stage, it can be assumed that the fractional part of the CFO has been mostly compensated. Knowing that identical synchronization signals are transmitted from all transmit antennas, the received secondary and primary synchronization signals become

$$R^{(m)}_{\mathrm{SSCH},k} = \mathrm{e}^{\mathrm{i}\frac{2\pi\varepsilon_{\mathrm{IFO}} l(N+N_{\mathrm{g}})}{N}} X_{\mathrm{SSCH},k-\varepsilon_{\mathrm{IFO}}} \sum_{q=1}^{N_{\mathrm{T}}} H^{(m,q)}_{\mathrm{SSCH},k-\varepsilon_{\mathrm{IFO}}} + \tilde{V}^{(m)}_{\mathrm{SSCH},k}, \tag{9.12}$$

$$R^{(m)}_{\mathrm{PSCH},k} = \mathrm{e}^{\mathrm{i}\frac{2\pi\varepsilon_{\mathrm{IFO}}(l+1)(N+N_{\mathrm{g}})}{N}} X_{\mathrm{PSCH},k-\varepsilon_{\mathrm{IFO}}} \sum_{q=1}^{N_{\mathrm{T}}} H^{(m,q)}_{\mathrm{PSCH},k-\varepsilon_{\mathrm{IFO}}} + \tilde{V}^{(m)}_{\mathrm{PSCH},k} \tag{9.13}$$

In [24], an IFO estimator is derived based on the ML principle. The symbols of the estimation can be either pre-defined pilot symbols or Phase-Shift Keying (PSK) modulated data symbols. In the context of LTE, we apply this algorithm to the standardized synchronization signals, leading to the following IFO estimator:

$$\hat{\varepsilon}_{\mathrm{IFO}} = \arg\max_{s} \left\{ \Re\left[\mathrm{e}^{\frac{\mathrm{i}2\pi s(N+N_{\mathrm{g}})}{N}} \sum_{m=1}^{N_{\mathrm{R}}} \sum_{k\in\mathcal{K}_{\mathrm{SCH}}} \left(R^{(m)*}_{\mathrm{SSCH},k+s} R^{(m)}_{\mathrm{PSCH},k+s} \right) \left(X_{\mathrm{SSCH},k} X^{*}_{\mathrm{PSCH},k} \right) \right] \right\}, \tag{9.14}$$

where $\Re\{\cdot\}$ returns the real part of the argument and $s \in (-31, 31)$ corresponds to the set of potential integer offsets that can be estimated. This set is determined by the length of the synchronization signals in LTE. The set $\mathcal{K}_{\mathrm{SCH}}$ represents the subcarrier indices that contain the synchronization signals. Note that due to the fact that in LTE the synchronization signals only exist in every fifth subframe, the $\varepsilon_{\mathrm{IFO}}$ can only be estimated in these subframes.

9.2.2.3 RFO Estimation

The RFO is the error of the FFO estimation. As a pre-FFT processing, the performance of the FFO estimation degrades when the channel disperses in time. Therefore, it is necessary to improve the estimation by using the cell-specific reference signals in the frequency domain after the FFT. As shown in Figure 9.3, LTE has the advantage that the reference signals from different transmit antennas do not overlap with each other. If one Resource Element (RE) is reserved for the reference signals, there is only one antenna transmitting the reference signal while all others keep silence by sending zero. This easily reduces the Multiple-Input Multiple-Output (MIMO) processing to an Single-Input Single-Output (SISO) case. We assume the interference term is small enough to be neglected and the subcarriers correctly aligned. The received signal on the reference signal positions can be expressed as

$$R^{(m)}_{l,k} = \mathrm{e}^{\mathrm{i}\frac{2\pi\varepsilon_{\mathrm{RFO}} l(N+N_{\mathrm{g}})}{N}} X^{(q)}_{l,k} H^{(m,q)}_{l,k} + \tilde{V}^{(m)}_{l,k} \tag{9.15}$$

If we observe the two reference signals on subcarrier k in the OFDM symbols l and $l + N_{\mathrm{s}}$, we obtain

$$W^{(m)}_{l,k} = \mathbb{E}\left\{ R^{(m)}_{l,k} R^{(m)*}_{l+N_{\mathrm{s}},k} \left(X^{(q)}_{l,k} X^{(q)*}_{l+N_{\mathrm{s}},k} \right)^{*} \right\}$$

$$= \mathrm{e}^{\mathrm{i}\frac{2\pi\varepsilon_{\mathrm{RFO}}l(N+N_{\mathrm{g}})}{N}} |X_{l,k}^{(q)}|^2 |X_{l+N_{\mathrm{s}},k}^{(q)}|^2 |H_{l,k}^{(m,q)}|^2 \tag{9.16}$$

Here, the block-fading assumption holds; that is, the channel coherence time is greater than one subframe duration. This implies that $H_{l,k}^{(m,q)} = H_{l+N_{\mathrm{s}},k}^{(m,q)}$. Given the common phase error, which linearly increases with the OFDM symbol index l, the RFO can be estimated by

$$\hat{\varepsilon}_{\mathrm{RFO}} = -\frac{1}{2\pi}\frac{N}{N_{\mathrm{s}}(N+N_{\mathrm{g}})}\arg\left\{\sum_{m}\sum_{(k,l)\in\mathcal{K}_{\mathrm{P}}} W_{l,k}^{(m)}\right\} \tag{9.17}$$

The estimation is performed on a subframe basis with the set $\mathcal{K}_{\mathrm{P}}$ corresponding to the joint set of subcarrier indices and OFDM symbol indices on which the reference symbols are located. It is noticed from Figure 9.3 that the pair-wise reference signals $X_{l,k}^{(q)}$ and $X_{l+N_{\mathrm{s}},k}^{(q)}$ are only available for antenna ports 1 and 2. Therefore, the reference signals of antenna ports 3 and 4 are not utilized in this case. Similar to the FFO estimation, the estimation range for $\varepsilon_{\mathrm{RFO}}$ is $(-N/[2N_{\mathrm{s}}(N+N_{\mathrm{g}})], N/[2N_{\mathrm{s}}(N+N_{\mathrm{g}})])$. Specifically, for the 1.4 MHz mode with normal CP length, $\hat{\varepsilon}_{\mathrm{RFO}} \in (-0.066, 0.066)$.

9.3 Performance Evaluation

As mentioned at the beginning of this chapter, the quality of the frequency offset estimators is mostly measured by MSE. In the past decade, efforts have been made in evaluating the performance degradation caused by the CFO in terms of BER or Symbol Error Ratio (SER) analytically [5, 7, 8, 17–19]. Some consider the effect of CFO only and give analytical expressions in terms of uncoded BER for the Additive White Gaussian Noise (AWGN) channel [5, 17, 19]. This work is extended to frequency-selective fading channels in [18]. The authors of [7, 8] further develop the calculation for uncoded BER and link capacity under the aggregate effect of time-variant impairments, namely CFO, imperfect channel knowledge, and I/Q imbalance. In this section, as illustrated in Figure 9.4, we provide a typical analysis of the theoretical MSE for the given FFO and RFO estimator. Then, the Signal to Interference and Noise Ratio (SINR) after FFT is calculated and simulated. Furthermore, we investigate the impact of the synchronization error on the SINR after the equalizer. Eventually, we choose to simulate the coded throughput in order to measure the link performance degradation due to the CFO.

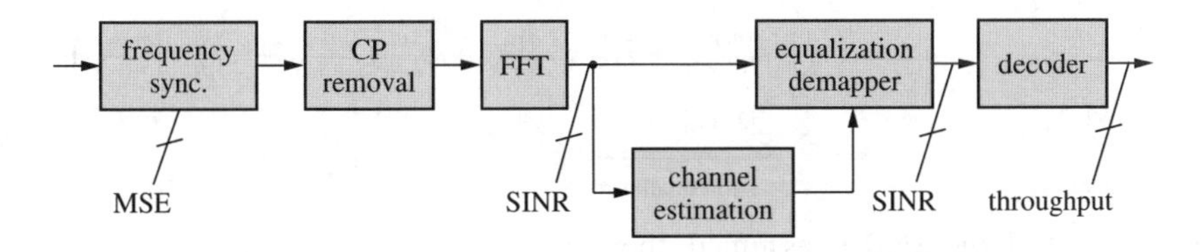

Figure 9.4 An evaluation chain of the frequency synchronization.

9.3.1 Estimation Performance

9.3.1.1 Theoretical MSE

The estimation performance of the FFO is expressed in terms of the MSE, which is given by

$$\mathrm{MSE}_{\mathrm{FFO}} = \mathbb{E}\left\{|\varepsilon_{\mathrm{FFO}} - \hat{\varepsilon}_{\mathrm{FFO}}|^2\right\} \tag{9.18}$$

and for the RFO

$$\mathrm{MSE}_{\mathrm{RFO}} = \mathbb{E}\left\{|\varepsilon_{\mathrm{FFO}} - \hat{\varepsilon}_{\mathrm{FFO}} - \hat{\varepsilon}_{\mathrm{RFO}}|^2\right\} \tag{9.19}$$

In order to derive the theoretical MSE of the FFO estimator, we rewrite Equation (9.11) as

$$\hat{\varepsilon}_{\mathrm{FFO}} = -\frac{1}{2\pi} \arctan \frac{\Im\left\{\sum_{m=1}^{N_{\mathrm{R}}} \sum_{l=1}^{N_{\mathrm{f}}} \sum_{n=1}^{N_{\mathrm{g}}} r_{l,n}^{(m)} r_{l,n+N}^{(m)*}\right\}}{\Re\left\{\sum_{m=1}^{N_{\mathrm{R}}} \sum_{l=1}^{N_{\mathrm{f}}} \sum_{n=1}^{N_{\mathrm{g}}} r_{l,n}^{(m)} r_{l,n+N}^{(m)*}\right\}}, \tag{9.20}$$

where $\Re\{\cdot\}$ and $\Im\{\cdot\}$ are operators to extract the real and imaginary parts of the argument respectively. The estimation error can be written as

$$\varepsilon_{\mathrm{FFO}} - \hat{\varepsilon}_{\mathrm{FFO}} = \frac{1}{2\pi} \arctan \frac{\Im\left\{\sum_{m=1}^{N_{\mathrm{R}}} \sum_{l=1}^{N_{\mathrm{f}}} \sum_{n=1}^{N_{\mathrm{g}}} r_{l,n}^{(m)} r_{l,n+N}^{(m)*} \mathrm{e}^{\mathrm{i}2\pi\varepsilon_{\mathrm{FFO}}}\right\}}{\Re\left\{\sum_{m=1}^{N_{\mathrm{R}}} \sum_{l=1}^{N_{\mathrm{f}}} \sum_{n=1}^{N_{\mathrm{g}}} r_{l,n}^{(m)} r_{l,n+N}^{(m)*} \mathrm{e}^{\mathrm{i}2\pi\varepsilon_{\mathrm{FFO}}}\right\}} \tag{9.21}$$

When the error is small, Equation (9.21) can be approximated to

$$\varepsilon_{\mathrm{FFO}} - \hat{\varepsilon}_{\mathrm{FFO}} \cong \frac{1}{2\pi} \frac{\Im\left\{\sum_{m=1}^{N_{\mathrm{R}}} \sum_{l=1}^{N_{\mathrm{f}}} \sum_{n=1}^{N_{\mathrm{g}}} r_{l,n}^{(m)} r_{l,n+N}^{(m)*} \mathrm{e}^{\mathrm{i}2\pi\varepsilon_{\mathrm{FFO}}}\right\}}{\Re\left\{\sum_{m=1}^{N_{\mathrm{R}}} \sum_{l=1}^{N_{\mathrm{f}}} \sum_{n=1}^{N_{\mathrm{g}}} r_{l,n}^{(m)} r_{l,n+N}^{(m)*} \mathrm{e}^{\mathrm{i}2\pi\varepsilon_{\mathrm{FFO}}}\right\}} \tag{9.22}$$

In order to identify the real and imaginary terms, we rewrite Equation (9.1) for simplicity as

$$r_{l,n}^{(m)} = \left(\bar{r}_{l,n}^{(m)} + v_{l,n}^{(m)}\right) \mathrm{e}^{\mathrm{i}2\pi\hat{\varepsilon}_{\mathrm{FFO}}(n+l(N+N_{\mathrm{g}}))/N}, \tag{9.23}$$

with $\bar{r}_{l,n}^{(m)} = \sum_{q=1}^{N_{\mathrm{T}}} x_{l,n}^{(q)} * h_{l,n}^{(m,q)}$ denoting the CFO-free received signal. Thus, we obtain

$$\begin{aligned} r_{l,n}^{(m)} r_{l,n+N}^{(m)*} \mathrm{e}^{\mathrm{i}2\pi\varepsilon_{\mathrm{FFO}}} &= \left(\bar{r}_{l,n}^{(m)} + v_{l,n}^{(m)}\right)\left(\bar{r}_{l,n+N}^{(m)} + v_{l,n+N}^{(m)}\right)^* \mathrm{e}^{\mathrm{i}2\pi(\varepsilon_{\mathrm{FFO}} - \hat{\varepsilon}_{\mathrm{FFO}})} \\ &= \left(\bar{r}_{l,n}^{(m)} \bar{r}_{l,n+N}^{(m)*} + \bar{r}_{l,n}^{(m)} v_{l,n+N}^{(m)*} + v_{l,n}^{(m)} \bar{r}_{l,n+N}^{(m)*} + v_{l,n}^{(m)} v_{l,n+N}^{(m)*}\right) \\ &\quad \times \underbrace{\mathrm{e}^{\mathrm{i}2\pi(\varepsilon_{\mathrm{FFO}} - \hat{\varepsilon}_{\mathrm{FFO}})}}_{\approx 1 \text{ for small error}} \end{aligned} \tag{9.24}$$

When relatively large SNR is assumed, there is

$$\Re\left\{\bar{r}_{l,n}^{(m)} \bar{r}_{l,n+N}^{(m)*}\right\} \gg \Re\left\{\bar{r}_{l,n}^{(m)} v_{l,n+N}^{(m)*} + v_{l,n}^{(m)} \bar{r}_{l,n+N}^{(m)*} + v_{l,n}^{(m)} v_{l,n+N}^{(m)*}\right\} \tag{9.25}$$

Thus, the denominator in Equation (9.22) becomes

$$\Re\left\{\sum_{m=1}^{N_R}\sum_{l=1}^{N_f}\sum_{n=1}^{N_g} r_{l,n}^{(m)} r_{l,n+N}^{(m)*} e^{i2\pi\varepsilon_{FFO}}\right\} = \sum_{m=1}^{N_R}\sum_{l=1}^{N_f}\sum_{n=1}^{N_g} \bar{r}_{l,n}^{(m)} \bar{r}_{l,n+N}^{(m)*} = N_R N_f N_g \sigma_r^2, \tag{9.26}$$

where σ_r^2 is the average signal power of the received signal in the time domain. Knowing that $\mathbb{E}\{v_{l,n}^{(m)}\} = 0$, $\mathbb{E}\{|v_{l,n}^{(m)}|^2\} = \sigma_n^2$, the variance of the nominator in Equation (9.22) can be found by straightforward manipulation:

$$\mathbb{E}\{|\Im\{\cdot\}|^2\} = N_R N_f N_g \left(\sigma_n^2 \sigma_r^2 + \frac{1}{2}\sigma_n^4\right) \tag{9.27}$$

Therefore, the MSE of the FFO is given by

$$\begin{aligned} \text{MSE}_{\text{FFO}} = \mathbb{E}\{|\varepsilon_{FFO} - \hat{\varepsilon}_{FFO}|^2\} &= \frac{1}{4\pi^2}\frac{\mathbb{E}\{|\Im\{\cdot\}|^2\}}{(N_R N_f N_g \sigma_r^2)^2} = \frac{2\sigma_n^2\sigma_r^2 + \sigma_n^4}{8\pi^2 N_R N_f N_g \sigma_r^4} \\ &= \frac{2\gamma_t + 1}{8\pi^2 N_R N_f N_g \gamma_t^2} \approx \frac{1}{4\pi^2 N_R N_f N_g \gamma_t}, \end{aligned} \tag{9.28}$$

with $\sigma_r^2 = \mathbb{E}\{|r_{l,n}^{(m)}|^2\}$ representing the average received signal power in the time domain and $\gamma_t = \sigma_r^2/\sigma_n^2 \gg 1$ the averaged SNR of the received sequence. As shown above, the MSE of the FFO estimation is determined by the number of RX antennas, the CP length, the number of OFDM symbols in one (sub-)frame, and the received SNR in the time domain. In addition, for the transmission modes with higher bandwidths which have higher sampling rates, better estimation performance can be achieved. Following a similar procedure, the MSE of the RFO estimation can be derived as

$$\text{MSE}_{\text{RFO}} = \mathbb{E}\{|\varepsilon_{FFO} - \hat{\varepsilon}_{FFO} - \hat{\varepsilon}_{RFO}|^2\} \approx \frac{N^2}{4\pi^2 N_s^2 (N + N_g)^2 N_R K \gamma_f}, \tag{9.29}$$

with $\sigma_R^2 = \mathbb{E}\{|R_{l,k}^{(m)}|^2\}$ representing the average received signal power in the frequency domain and $\gamma_f = \sigma_R^2/\sigma_n^2$ the SNR of the received signal at each subcarrier. In LTE, the factor K is the total number of reference symbols in one slot. Specifically, for the 1.4 MHz mode we have $K = 24$ for one transmit antenna port and $K = 48$ for two transmit antenna ports. Since the RFO estimator relies on the pair-wise reference symbols located on the same subcarrier, the reference signals on antenna ports 3 and 4 do not contribute.

9.3.1.2 Simulation Results

Figure 9.5 plots the calculated and simulated MSE curves of an LTE system at 1.4 MHz. For the 1×1 SISO case, a simple AWGN channel model is considered. For the 2×2 transmission, we define an equivalent AWGN channel matrix of rank 2 for subcarrier k as

$$\mathbf{H}_k = \begin{bmatrix} 1 & 1 \\ 1 & -1 \end{bmatrix} \tag{9.30}$$

In most of the SNR range, the results from calculation and simulation are in excellent agreement. Confidence intervals of 95% are expressed by vertical lines on the curves.

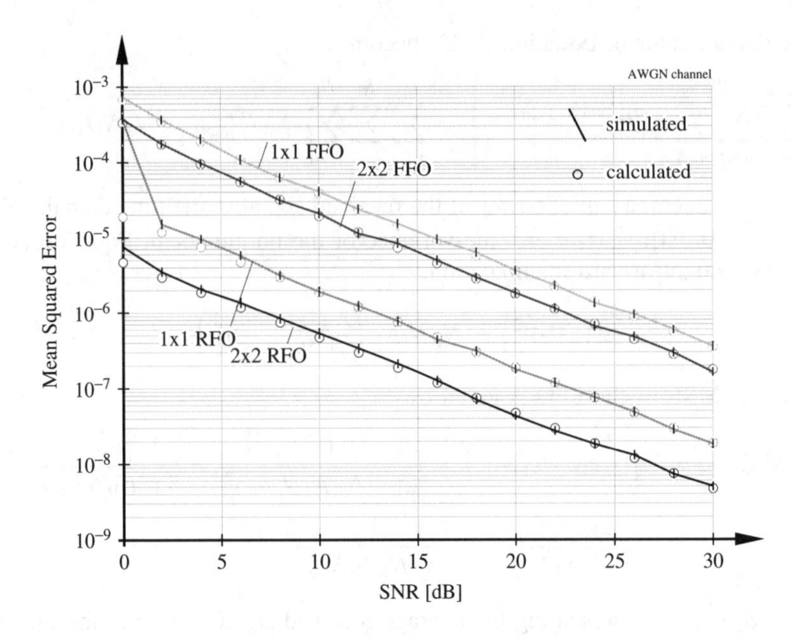

Figure 9.5 Calculated and simulated MSE.

The curves of the FFO and RFO estimators follow a similar trend. For the FFO estimation in the SISO case, as the estimation error (namely the residual CFO,) approaches the estimation range of the RFO estimator, an outlier can be observed at 0 dB SNR. When two antennas are deployed at the transmitter side, twice the number of the reference signals are utilized in total, as shown in Figure 9.3. This brings an extra gain of 3 dB to the estimation performance of the RFO part. However, since pair-wise reference signals that are placed on the same subcarrier are merely available for the antenna ports 1 and 2, higher order antenna configurations do not bring additional gain to the estimation performance.

9.3.2 Post-FFT SINR

As a metric to measure the ICI, post-FFT SINR has been selected to demonstrate the effect of imperfect frequency synchronization to OFDM systems [9, 15, 23]. According to Equations (9.5)–(9.8), we calculate the power of the signal, the interference, and the noise term as

$$E_{\mathrm{S}} = \sigma_{\mathrm{s}}^2 |I(0)|^2 \sum_{q=1}^{N_{\mathrm{T}}} \mathbb{E}\left\{|H_{l,k'}^{(m,q)}|^2\right\} = \sigma_{\mathrm{s}}^2 \left[\frac{\sin(\pi\varepsilon_{\mathrm{FFO}})}{N\sin(\pi\varepsilon_{\mathrm{FFO}}/N)}\right]^2 \sum_{q=1}^{N_{\mathrm{T}}} \mathbb{E}\left\{|H_{l,k'}^{(m,q)}|^2\right\}, \tag{9.31}$$

$$E_{\mathrm{I}} = \sigma_{\mathrm{s}}^2 \left\{1 - \left[\frac{\sin(\pi\varepsilon_{\mathrm{FFO}})}{N\sin(\pi\varepsilon_{\mathrm{FFO}}/N)}\right]^2\right\} \sum_{q=1}^{N_{\mathrm{T}}} \mathbb{E}\left\{|H_{l,k'}^{(m,q)}|^2\right\}, \tag{9.32}$$

$$E_{\mathrm{N}} = \mathbb{E}\left\{|\tilde{V}_{l,k}^{(m)}|^2\right\} = \sigma_{\mathrm{n}}^2, \tag{9.33}$$

where $\sigma_s^2 = \mathbb{E}\{|X_{l,k}^{(m)}|^2\}$ denotes the average transmit power on each subcarrier. This is related to the total transmit power $P_{Tx} = N_T K \sigma_s^2$ when K subcarriers are used for data transmission. Thus, the post-FFT SINR per subcarrier can be written as

$$\begin{aligned}
\mathrm{SINR}_{l,k}^{(m)} &= \frac{E_S}{E_I + E_N} \\
&= \frac{\sigma_s^2 \left[\dfrac{\sin(\pi\varepsilon_{FFO})}{N\sin(\pi\varepsilon_{FFO}/N)}\right]^2 \sum_{q=1}^{N_T} \mathbb{E}\left\{|H_{l,k'}^{(m,q)}|^2\right\}}{\sigma_s^2\left[1 - \left(\dfrac{\sin(\pi\varepsilon_{FFO})}{N\sin(\pi\varepsilon_{FFO}/N)}\right)^2\right] \sum_{q=1}^{N_T} \mathbb{E}\left\{|H_{l,k'}^{(m,q)}|^2\right\} + \sigma_n^2} \\
&\cong \frac{\left[\dfrac{\sin(\pi\varepsilon_{FFO})}{N\sin(\pi\varepsilon_{FFO}/N)}\right]^2}{1 - \left[\dfrac{\sin(\pi\varepsilon_{FFO})}{N\sin(\pi\varepsilon_{FFO}/N)}\right]^2} \quad \text{for high SNR}
\end{aligned} \tag{9.34}$$

Figure 9.6 plots the post-FFT SINR against varying levels of CFO in an 1.4 MHz LTE transmission over an AWGN channel. The simulation result obtained from the "Vienna LTE Link Level Simulator" agrees excellently with the calculation. Note that here and in the following figures, 95% confidence intervals are expressed by the vertical lines on the curve.

9.3.3 Post-equalization SINR and Throughput

Most literature on frequency synchronization evaluates the transmission performance in terms of the estimation error; in other words, the MSE. The MSE indicates how precise

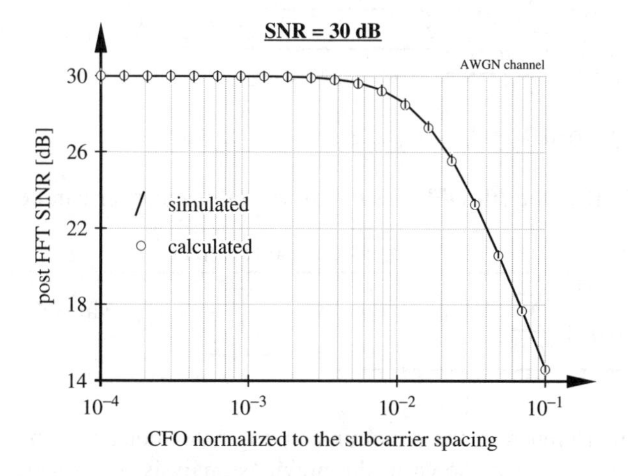

Figure 9.6 Post-FFT SINR loss.

the estimation is. Although the analysis in terms of post-FFT SINR has the merit of being mathematically simple, the performance of the system needs to be evaluated eventually in terms of the coded throughput. In this section we continue to evaluate the link performance of a practical OFDM system, namely LTE, which is impaired by the CFO. First, an analytical derivation of the post-equalization SINR is provided. Afterwards, an evaluation in terms of coded throughput is conducted by a standard-compliant LTE simulator (see Chapter 11).

9.3.3.1 Post-Equalization SINR

In this section the impact of frequency synchronization error is evaluated in terms of post-equalization SINR. A simplified SISO model is employed to demonstrate this degradation analytically. As an example, we consider a Zero Forcing (ZF) equalizer $G_{l,k}$ for the OFDM symbol l and subcarrier k, expressed as

$$G_{l,k} = \left(H_{l,k}^H H_{l,k}\right)^{-1} H_{l,k}^H \tag{9.35}$$

Given the expression of the received signal in Equation (9.5) for the SISO case with the IFO corrected,

$$R_{l,k} = I(0) X_{l,k} H_{l,k} + \sum_{p \neq k} X_{l,p} H_{l,p} \cdot I(p-k) + \tilde{V}_{l,k}^{(m)}, \tag{9.36}$$

the outcoming signal from the equalizer becomes

$$\begin{aligned} Y_{l,k} &= G_{l,k} R_{l,k} = I(0) X_{l,k} \\ &\quad + \underbrace{\left(H_{l,k}^H H_{l,k}\right)^{-1} H_{l,k}^H \sum_{p \neq k} X_{l,p} H_{l,p} \cdot I(p-k)}_{I_{l,k}} + \underbrace{\left(H_{l,k}^H H_{l,k}\right)^{-1} H_{l,k}^H \tilde{V}_{l,k}}_{V'_{l,k}} \\ &= \underbrace{\frac{\sin(\pi \varepsilon_{\text{FFO}})}{N \sin(\pi \varepsilon_{\text{FFO}}/N)}}_{\gamma} \underbrace{\mathrm{e}^{\mathrm{i}\frac{\pi \varepsilon_{\text{FFO}}(N-1)}{N}} \mathrm{e}^{\mathrm{i}\Phi(\varepsilon_{\text{FFO}}, l)}}_{\mathrm{e}^{\mathrm{i}\Phi'(\varepsilon_{\text{FFO}}, l)}} X_{l,k} + I_{l,k} + V'_{l,k} \\ &= \mathrm{e}^{\mathrm{i}\Phi'(\varepsilon_{\text{FFO}}, l)} \gamma X_{l,k} + I_{l,k} + V'_{l,k} \end{aligned} \tag{9.37}$$

The resulting SINR at the lth OFDM symbol and kth subcarrier can be expressed as

$$\text{SINR}_{l,k} = \frac{\sigma_{\text{s}}^2 \gamma^2 \cos^2 \Phi'(\varepsilon_{\text{FFO}}, l)}{\underbrace{\frac{\sigma_{\text{s}}^2}{|H_{l,k}|^2} \sum_{p \neq k} |H_{l,p}|^2 \cdot I^2(p-k)}_{E_{\text{ICI}}} + \underbrace{\frac{\sigma_{\text{n}}^2}{|H_{l,k}|^2}}_{E_N} + \sigma_{\text{s}}^2 \gamma^2 \sin^2 \Phi'(\varepsilon_{\text{FFO}}, l)} \tag{9.38}$$

Therefore, the interference in the denominator originates not only from the ICI and the Gaussian noise, as has been shown in the previous analysis of post-FFT SINR, but also

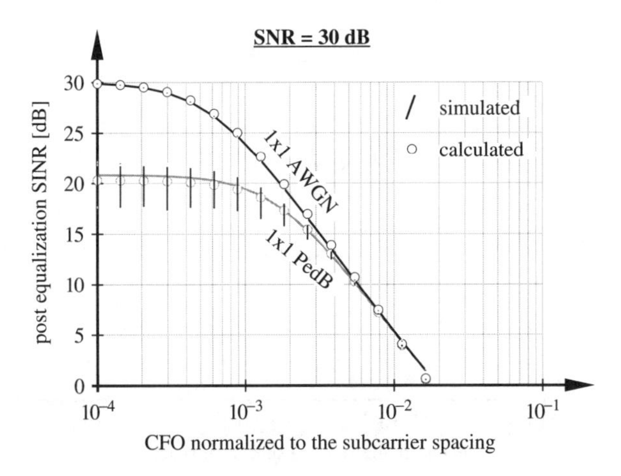

Figure 9.7 Post-equalization SINR.

from the inappropriately equalized signal term. In most referred analysis, the so-called common phase error $\Phi'(\varepsilon_{\text{FFO}}, l)$ is incorporated as an additional rotation to the channel frequency response. Unfortunately, this holds only if the channel estimation can be carried out instantly at each OFDM symbol. For practical systems in reality, this is not the case. Similar to that for post-FFT SINR, the simulated and calculated post-equalization SINR curves are plotted in Figure 9.7. Unlike the previous situation, the SINR after the equalizer drops dramatically as the CFO increases.

9.3.3.2 Coded Throughput

In this section we evaluate the performance degradation in terms of the coded throughput under the CFO by simulation. In the first experiment, we fix the SNR at 30 dB and introduce a logarithmically spaced CFO between $\varepsilon_{\text{CFO}} = 10^{-4}$ and 10^{-1}. A SISO transmission is simulated under AWGN and frequency-selective fading channels without CFO estimation or compensation involved. As shown in Figure 9.8, corresponding to the post-equalization SINR, the coded throughput decreases to zero rapidly as the CFO increases. In a second experiment, a fixed CFO is introduced to an LTE system and estimated/compensated by the method described in Section 9.2.2. Results are obtained using the "Vienna LTE Link Level Simulator" described in Chapter 11. Detailed parameters are presented in Table 9.1. Figure 9.9 shows the performance degradation in terms of coded throughput. The percentage losses are plotted in the lower figure. It can be seen that the performance degradation in coded throughput decreases with increasing number of RX antennas. According to the standard [1], when two TXs are employed, the total number of reference symbols that contribute to the RFO estimation is also doubled. In the upper figure, for the 2×2 case, the degradation compared with the perfect synchronization case is hardly visible. The lower figure shows that, in most of the SNR levels, the loss is lower than 5 %. For the higher bandwidth modes, the estimation scheme can be applied

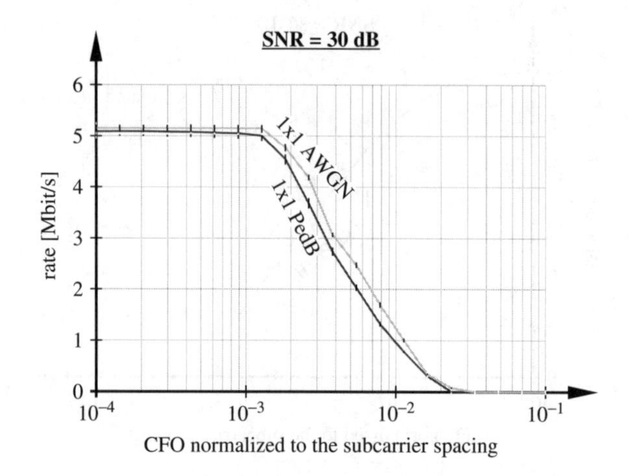

Figure 9.8 Coded throughput without CFO compensation.

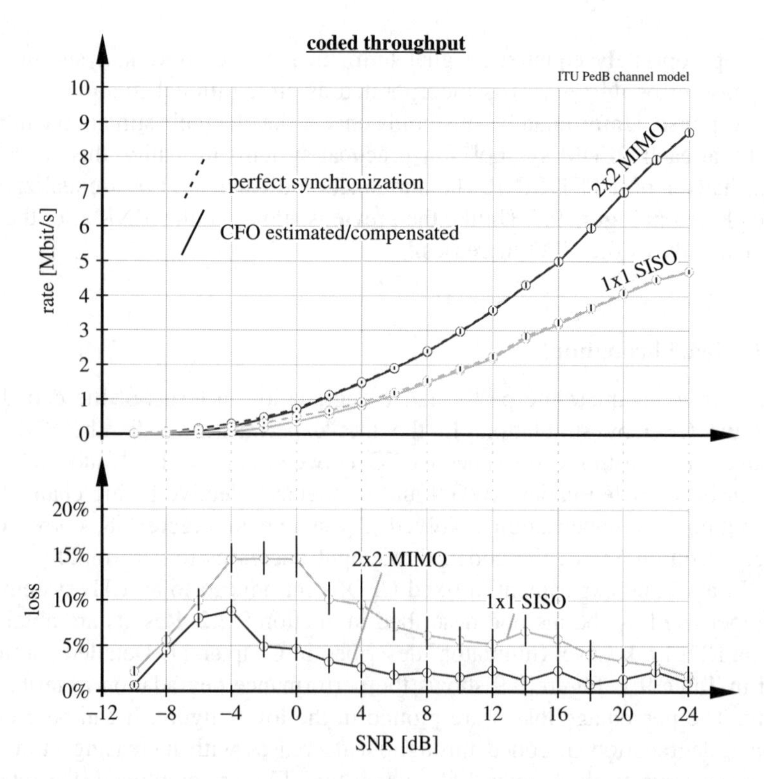

Figure 9.9 Coded throughput of LTE 1.4 MHz mode, perfect frequency synchronization versus estimated/compensated CFO.

Table 9.1 Simulation parameters

	SISO	OLSM
Antennas $N_R \times N_T$	1×1	2×2
Channel	PedB [6]	
Bandwidth	1.4 MHz	
Receiver	soft sphere decoder	
CFO (ε_{CFO})	0 / 3.1415 ...	
Symbol timing	perfect	
Channel knowledge	perfect	
Modulation and coding scheme	optimally adaptive	

directly. Since the number of reference symbols distributed in the frequency domain increases correspondingly with the number of RBs, the estimation performance will show proportional improvements.

Since the cell-specific reference signals are transmitted in all the subframes in the LTE DL, all users receive an equal number of reference signals. The same is true for the synchronization signals. Therefore, the degradations due to the CFO for different users are at the same level. When Coordinated Multi-Point (CoMP) transmission is considered in LTE-Advanced (LTE-A), it is usually assumed that the carrier frequencies at different base stations are ideally synchronized. This eases the problem of frequency synchronization in more sophisticated scenarios.

References

[1] 3GPP (2009) Technical Specification TS 36.211 Version 8.7.0 'Evolved universal terrestrial radio access (E-UTRA); physical channels and modulation; (resease 8),' http://www.3gpp.org/ftp/Specs/html-info/36211.htm.

[2] Armada, A. G. (2001) 'Understanding the effects of phase noise in orthogonal frequency division multiplexing (OFDM),' *IEEE Transactions on Broadcasting*, **47** (2), 153–159.

[3] Bölcskei, H. (2001) 'Blind estimation of symbol timing and carrier frequency offset in wireless OFDM systems,' *IEEE Transactions on Communications*, **49** (6), 988–999. Available from http://www.nari.ee.ethz.ch/commth/pubs/p/synch.

[4] Chen, B. and Wang, H. (2002) 'Maximum likelihood estimation of OFDM carrier frequency offset,' in *IEEE International Conference on Communications, (ICC 2002)*, volume 1, pp. 49–53.

[5] Dharmawansa, P., Rajatheva, N., and Minn, H. (2009) 'An exact error probability analysis of OFDM systems with frequency offset,' *IEEE Transactions on Communications*, **57** (1), 26–31.

[6] ITU (1997) 'Recommendation ITU-R M.1225: Guidelines for evaluation of radio transmission technologies for IMT-2000,' Technical report.

[7] Krondorf, M. and Fettweis, G. (2008) 'OFDM link performance analysis under various receiver impairments,' *EURASIP Journal on Wireless Communications and Networking*, **2008**, article ID 145279, doi: 10.1155/2008/145279.

[8] Krondorf, M., Liang, T., and Fettweis, G. (2007) 'Symbol error rate of OFDM systems with carrier frequency offset and channel estimation error in frequency selective fading channels,' in *IEEE International Conference on Communications, (ICC 2007)*, pp. 5132–5136.

[9] Lee, J., Lou, H.-L., Toumpakaris, D., and Cioffi, J. (2006) 'SNR analysis of OFDM systems in the presence of carrier frequency offset for fading channels,' *IEEE Transactions on Wireless Communications*, **5** (12), 3360–3364.

[10] Li, J., Liu, G., and Giannakis, G. B. (2001) 'Carrier frequency offset estimation for OFDM-based WLANs,' *IEEE Signal Processing Letters*, **8** (3), 80–82.

[11] Liu, H. and Tureli, U. (1998) 'A high-efficiency carrier estimator for OFDM communications,' *IEEE Communications Letters*, **2** (4), 104–106.

[12] Ma, X., Tepedelenlioglu, C., Giannakis, G. B., and Barbarossa, S. (2001) 'Non-data-aided carrier offset estimators for OFDM with null subcarriers: identifiability, algorithms and performance,' *IEEE Journals on Selected Areas*, **19** (12), 2504–2515.

[13] Minn, H., Bhargava, V. K., and Letaief, K. B. (2003) 'A robust timing and frequency synchronization for OFDM systems,' *IEEE Transactions on Wireless Communications*, **2** (4), 822–839.

[14] Mo, R., Chew, Y. H., Tjhung, T. T., and Ko, C. C. (2008) 'A new blind joint timing and frequency offset estimator for OFDM systems over multipath fading channels,' *IEEE Transactions on Vehicular Technology*, **57** (5), 2947–2957.

[15] Moose, P. H. (1994) 'A technique for orthogonal frequency division multiplexing frequency offset correction,' *IEEE Transactions on Communications*, **42** (10), 2908–2914.

[16] Morelli, M. and Mengali, U. (1999) 'An improved frequency offset estimator for OFDM applications,' in *Communication Theory Mini-Conference*, Canada, pp. 106–109.

[17] Pollet, T., Van Bladel, M., and Moeneclaey, M. (1995) 'BER sensitivity of OFDM systems to carrier frequency offset and wiener phase noise,' *IEEE Transactions on Communications*, **43** (234), 191–193.

[18] Rugini, L. and Banelli, P. (2005) 'BER of OFDM systems impaired by carrier frequency offset in multipath fading channels,' *IEEE Transactions on Wireless Communications*, **4** (5), 2279–2288.

[19] Sathananthan, K. and Tellambura, C. (2001) 'Probability of error calculation of OFDM systems with frequency offset,' *IEEE Transactions on Communications*, **49** (11), 1884–1888.

[20] Schmidl, T. and Cox, D. (1997) 'Robust frequency and timing synchronization for OFDM,' *IEEE Transactions on Communications*, **45** (12), 1613–1621.

[21] Speth, M., Fechtel, S., Fock, G., and Meyr, H. (1999) 'Optimum receiver design for wireless broadband systems using OFDM. Part I,' *IEEE Transactions on Communications*, **47** (11), 1668–1677.

[22] Speth, M., Fechtel, S., Fock, G., and Meyr, H. (2001) 'Optimum receiver design for OFDM-based broadband transmission. II. A case study,' *IEEE Transactions on Communications*, **49** (4), 571–578.

[23] Stantchev, B. and Fettweis, G. (2000) 'Time-variant distortion in OFDM,' *IEEE Communications Letters*, **4**, 312–314.

[24] Toumpakaris, D., Lee, J., and Lou, H. (2009) 'Estimation of integer carrier frequency offset in OFDM systems based on the maximum likelihood principle,' *IEEE Transactions on Broadcasting*, **55**, 95–108.

[25] Tureli, U., Liu, H., and Zoltowski, M. (2000) 'OFDM blind carrier offset estimation: ESPRIT,' *IEEE Transactions on Communications*, **48** (9), 1459–1461.

[26] Van de Beek, J. J., Sandell, M., and Borjesson, P. O. (1997) 'ML estimation of time and frequency offset in OFDM system,' *IEEE Transactions on Signal Processing*, **45** (7), 1800–1805.

[27] Wang, Q., Mehlführer, C., and Rupp, M. (2010) 'Carrier frequency synchronization in the downlink of 3GPP LTE,' in *Proceedings of the 21st Annual IEEE International Symposium on Personal, Indoor and Mobile Radio Communications (PIMRC'10)*, Istanbul, Turkey. Available from http://publik.tuwien.ac.at/files/PubDat_187636.pdf

[28] Yao, Y. and Giannakis, G. (2005) 'Blind carrier frequency offset estimation in SISO, MIMO, and multiuser OFDM systems,' *IEEE Transactions on Communications*, **53** (1), 173–183.

10

LTE Performance Evaluation

Contributed by Stefan Schwarz
Vienna University of Technology (TU Wien), Austria

This chapter presents link-level simulation results for Long-Term Evolution (LTE) and explains the signal processing algorithms involved. In particular, we focus on the link between a base station, which is called an Evolved base station (eNodeB) in LTE, and a single User Equipment (UE). Such a simulation scenario enables the comparison and performance investigation of different single-user transmitter and receiver algorithms (for example, channel estimation [34], equalizers [44], time and frequency synchronization [43], UE feedback calculation [26]). Specifically, we consider the computation of the UE feedback indicators defined in the LTE specification in Section 10.4. Furthermore, a mathematical model of the physical layer of the LTE Downlink (DL) is proposed in Section 10.1, the performance of which is later evaluated by simulations in Section 10.6. The simulations are performed with the "Vienna LTE Link Level Simulator" which is described in Chapter 11. In Section 10.2, a generic receiver structure for an LTE UE is described and an overview about related signal processing algorithms is provided. The post-equalization Signal to Interference and Noise Ratio (SINR) of a linear receiver is derived in Section 10.3 and, furthermore, the important concept of SINR averaging is introduced.

Additionally, several performance bounds for the physical layer data throughput are introduced in Section 10.5. Starting with channel capacity, as the ultimate upper bound on the amount of information that can be transmitted reliably through a channel, step by step we develop tighter bounds on the achievable data throughput by taking into account practical system design constraints. This enables one to systematically identify the major sources of performance losses incurred in the LTE system. The investigation reveals flaws of the LTE design and points to directions to efficiently improve performance in the design of future wireless communication systems.

Evaluation of HSDPA and LTE: From Testbed Measurements to System Level Performance, First Edition.
Sebastian Caban, Christian Mehlführer, Markus Rupp and Martin Wrulich.

10.1 Mathematical Model of the Physical Layer

To derive a suitable mathematical model of the physical layer of the LTE DL, we resort to the DL physical channel processing chain introduced in Chapter 2. The schematic of the signal processing chain is depicted in Figure 2.7. Our mathematical model comprises the precoding at the transmitter, the Orthogonal Frequency-Division Multiplexing (OFDM) signal processing at the transmitter and receiver, the effect of channel impairments, and the equalization at the receiver. It does not incorporate the LTE-specified scrambling of the coded bits, mapping of the scrambled bits to modulation symbols (modulation mappers in Figure 2.7), and the mapping between codewords and spatial transmission layers (layer mapper). As these processing steps are one-to-one mappings, they can simply be inverted at the receiver.

The coded user data bits define the input of the signal processing chain, depicted in Figure 2.7. As explained in Chapter 2, all user data bits of a subframe are collected in a transport block and jointly coded to deliver a single codeword. This codeword then is scrambled, which amounts to modulo 2 addition to a cell-specific Pseudo Noise (PN) sequence in order to whiten the transmitted bits. Because different PN sequences are employed in different cells, the interference between cells is randomized after descrambling at the receiver, which improves the performance of the channel code [12].

The support of multiple antennas at the transmitter and receiver side (Multiple-Input Multiple-Output (MIMO)) enables the transmission of multiple data streams, so-called transmission layers, in parallel, provided that the wireless channel is well conditioned. Compared with a Single-Input Single-Output (SISO) system, this enables an increased spectral efficiency. LTE restricts the number of transmit antennas and consequently also the number of layers to at most four [2]. As depicted in Figure 2.7, the number of codewords per user is limited to one or two. It follows that a single codeword can be mapped onto multiple spatial transmission layers. It is important to take that into account for the calculation of the UE feedback values, because different transmission layers in general experience different channel quality (see Section 10.4). The standard defines a one-to-one mapping between the transmit symbols of the individual codewords and layers [2], implemented in the layer mapper. Before transmission over the antennas, the data symbols are precoded, utilizing a standard defined codebook of unitary precoding matrices [2].

The DL modulation and multiple access of LTE is realized through Orthogonal Frequency-Division Multiple Access (OFDMA). An OFDM system converts a broadband frequency-selective channel into K narrowband, orthogonal, frequency-flat channels, called subcarriers, by means of a Fast Fourier Transform (FFT) and application of a Cyclic Prefix (CP) of appropriate length. In addition to dividing the system bandwidth into a set of subcarriers, the time domain is also partitioned into orthogonal bins, namely OFDM symbols. In that way, a time–frequency grid of OFDM samples, so-called Resource Elements (REs), is spanned onto which the user data is mapped by means of the RE mappers (see Chapter 2). Multiple access is handled by assigning orthogonal resources to different users. The major part of the signal processing at the transmitter and the receiver (precoding, equalization, channel estimation, feedback calculation) occurs on an individual RE basis. Hence, we derive a mathematical model, describing the signal processing part and the effect of the channel distortions, for each RE individually.

We denote the transmit symbol vector on RE r as $\mathbf{x}_r \in a^{L\times 1}$. Here, L denotes the number of spatial transmission layers. Furthermore, $a \in \mathcal{A}$ refers to one of the supported modulation alphabets $\mathcal{A} = \{\mathcal{A}_4, \mathcal{A}_{16}, \mathcal{A}_{64}\}$, namely 4-, 16-, or 64-Quadrature Amplitude Modulation (QAM) [2]. This symbol vector is mapped onto the transmit antennas by means of unitary precoders. Depending on the eNodeB setup, different precoders can be employed on different REs. Alternatively, the same precoder can also be utilized for a larger part of the system bandwidth, called a subband. The size of a subband is configured by the eNodeB. The unitary precoder on RE r is denoted $\mathbf{W}_r \in \mathcal{W}^{(L)}$, with $\mathcal{W}^{(L)}$ referring to the transmission-mode-dependent codebook [2] (see Chapter 2 for details on the defined transmission modes). The employed $\mathcal{W}^{(L)}$ codebook also depends on the number of transmission layers L (also called the transmission rank). The unitary precoders define an energy-preserving mapping for the transmit symbols from spatial layers to transmit antennas; therefore, $\mathcal{W}^{(L)} \subset \mathbb{C}^{N_\mathrm{T}\times L}$. Here, N_T denotes the number of active transmit antennas. The obtained signal vector $\mathbf{t}_r$ on RE r is given by

$$\mathbf{t}_r = \mathbf{W}_r\mathbf{x}_r, \quad r \in \{1, \ldots, R\} \tag{10.1}$$

The set $\{1, \ldots, R\}$ denotes the set of REs spanning one subframe. After the baseband time-domain signal is generated by means of an Inverse Fast Fourier Transform (IFFT) and CP insertion, the passband signal is obtained by modulation onto a carrier and transmitted over the wireless multipath channel. At the receiver, the CP is discarded and the signals obtained, received on the N_R receive antennas, are transformed back to the frequency domain via an FFT. As long as the CP is long enough to cover the full delay spread of the wireless channel, inter-symbol interference between OFDM symbols is avoided and each RE experiences a frequency-flat channel response denoted $\mathbf{H}_r \in \mathbb{C}^{N_\mathrm{R}\times N_\mathrm{T}}$. With this channel matrix, the input–output relation of the transmission system, incorporating the OFDM signal processing, can be written as

$$\mathbf{y}_r = \mathbf{H}_r\mathbf{t}_r + \mathbf{n}_r = \mathbf{H}_r\mathbf{W}_r\mathbf{x}_r + \mathbf{n}_r \tag{10.2}$$

The vector $\mathbf{n}_r \sim \mathcal{CN}(0, \sigma_\mathrm{n}^2 \cdot \mathbf{I}_{N_\mathrm{R}})$ denotes the additive white Gaussian receiver noise with zero mean and variance σ_n^2. The noise at different receive antennas is assumed to be uncorrelated. The received vector $\mathbf{y}_r \in \mathbb{C}^{N_\mathrm{R}\times 1}$ is processed at the receiver in order to separate the spatial transmission layers, which are mixed up by the channel matrix. The corresponding receiver processing is explained next.

10.2 Receiver

The goal of the receiver is to estimate the transmitted symbol vectors $\mathbf{x}_r$ of all REs, in order to be able to decode the transmitted data. For that purpose, the distortions, introduced by the channel, must be equalized and the signal processing of the transmitter inverted. We focus on the baseband signal processing part only. We briefly discuss means for channel estimation and data detection, but for detailed descriptions the reader is referred to the vast

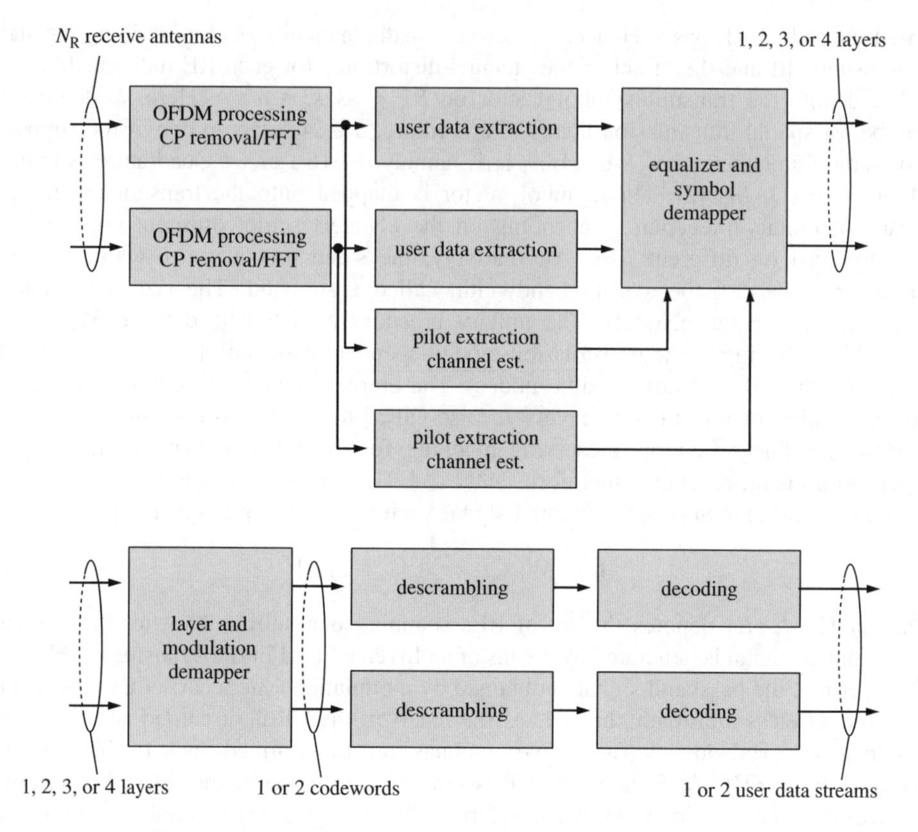

Figure 10.1 Block diagram of a generic LTE receiver.

amount of literature available on that topic (several citations are stated in the corresponding sections). Furthermore, the receiver has to provide Channel State Information (CSI) to the transmitter via a feedback channel, in order to enable link adaptation at the transmitter. Link adaptation allows the adaptation of important transmission parameters (code rate, modulation alphabet, transmission rank, precoding) to the current channel quality, such as to achieve a target Block Error Ratio (BLER). LTE defines three feedback indicators for that purpose that are discussed in detail in Section 10.4; algorithms to estimate the feedback indicators are derived in Section 10.4 as well. A generic receiver architecture for LTE is shown in Figure 10.1. The different processing blocks shown in Figure 10.1 invert the signal processing applied at the transmitter side (see Section 2.3 for details on the processing involved). Throughout this section we will refer to this architecture and describe its components.

10.2.1 Channel Estimation

After the signal is received at the N_R antennas and down-converted to baseband, the time-domain signal is transformed to the frequency domain via an FFT. The OFDM signal processing deals with the frequency distortions, introduced by the channel, and provides a frequency-flat channel response on each subcarrier, as long as the CP is long enough to cover the full delay spread of the wireless channel. For that purpose, two different CP lengths are defined in the LTE specification: a normal CP with a length of approximately 5 μs and an extended CP of approximately 17 μs duration [2]. In a MIMO system, in addition to frequency equalization, spatial equalization is also necessary in order to achieve a diversity, beamforming, or multiplexing gain, depending on the MIMO mode of operation. This requires coherent detection of the data symbols and necessitates knowledge of the MIMO channel matrix $\mathbf{H}_r \in \mathbb{C}^{N_\mathrm{R} \times N_\mathrm{T}}$ experienced on each RE. This knowledge can be obtained by means of channel estimation, exploiting the known pilot symbols inserted into the RE grid by the transmitter on each antenna port; for example, see Chapter 2. An important channel property that strongly influences the complexity of a channel estimator is the channel coherence time. If the coherence time of the channel is longer than the Transmission Time Interval (TTI), the channel appears as block fading and it is sufficient to estimate it only once per TTI. On the other hand, if the coherence time is smaller than a TTI, the channel changes significantly over one block of data (fast fading) and a single channel estimate is not sufficient to obtain good throughput performance [34].

Channel estimation for OFDM systems in the case of block fading is a well-studied topic [10, 18, 41]. The Least Squares (LS) and Linear Minimum Mean Square Error (LMMSE) channel estimators are typical examples of channel estimator classes defined in that context. LMMSE estimators also exist [13, 16, 36] for fast fading environments. These estimators suffer from high computational complexity. An Approximate Linear Minimum Mean Square Error (ALMMSE) channel estimator with scalable complexity, that allows the tradeoff of estimation accuracy versus computational complexity, is proposed in [34]. Furthermore, for very high speed scenarios the orthogonality between the individual OFDM subcarriers is lost, due to the Doppler spread incurred, leading to significant Inter-Carrier Interference (ICI). In that case, ICI estimation is required to reduce the corresponding throughput performance degradation [20, 21, 35].

10.2.2 Data Detection

In a MIMO system the data symbols transmitted from each transmit antenna interfere at the receive antennas. In order to separate the spatial streams from each other, post-processing of the received data symbols is required. Furthermore, the received symbols are corrupted by additive, complex Gaussian-distributed receiver noise. The task of data detection is to combat these impairments and provide an estimate of the transmitted symbol vector. There are several means to accomplish this job. Ideally, the transmit symbol vector $\mathbf{x}_r$ is estimated by utilizing a Maximum Likelihood (ML) detector that solves the optimization problem [22]:

$$\hat{\mathbf{x}}_r = \arg \min_{\mathbf{x} \in a^{L \times 1}} ||\mathbf{y}_r - \mathbf{H}_r \mathbf{W}_r \mathbf{x}_r||_2^2 \tag{10.3}$$

Here, $|| \cdot ||_2$ denotes the Euclidean norm. Employing a brute-force search over all possible transmit symbols vectors, to find the solution to that problem, implies an exponential computational complexity in the number of transmission layers L. Methods exist that reduce the complexity of finding the ML solution to be polynomial in the number of layers by decreasing the number of computed hypotheses in a sophisticated way; for example, the Soft Sphere Decoder (SSD) algorithm [5, 14, 15]. In that way, a hard-decision estimate of the transmitted symbol vector is obtained, meaning that all information about the reliability of a specific symbol is lost. This approach is suitable if a hard-input decoder is employed that cannot deal with reliability information, but better decoding performance can be achieved by utilizing a soft-input decoder [19]. In that case, the received symbols must not be hard demapped to the defined symbol alphabet a, but reliability information about each transmit bit has to be provided by means of Log-Likelihood Ratios (LLRs). This can be achieved by employing an SSD [37].

In many practical cases the ML receiver, even utilizing the SSD algorithm, has too high a computational complexity. Then, simple linear receive equalizer filters are employed instead to separate the transmission layers. Typical examples are the Zero Forcing (ZF) equalizer, which nulls all interference between different spatial layers, or the Minimum Mean Square Error (MMSE) equalizer, which trades off inter-layer interference versus noise enhancement, caused by the equalizer filter itself [22]. Denoting with $\mathbf{F}_r$ the equalizer on RE r, the received symbol vector after equalization is given by

$$\mathbf{r}_r = \mathbf{F}_r \mathbf{y}_r = \underbrace{\mathbf{F}_r \mathbf{H}_r \mathbf{W}_r}_{\mathbf{K}_r} \mathbf{x}_r + \mathbf{F}_r \mathbf{n}_r \tag{10.4}$$

From this vector a hard decision estimate of the transmit symbol vector $\mathbf{x}_r$ is obtained by mapping $\mathbf{r}_r$ to the nearest neighbor symbol constellation point from the symbol alphabet $a^{L\times 1}$. Alternatively, LLRs for each bit decision can also be computed employing a soft-output symbol demapper; for example [39]. If the performance of a simple linear equalizer filter is not good enough, then a system utilizing Successive Interference Cancelation (SIC) is an alternative. In that case, simple linear filters are combined with feedback loops for interference cancelation in the equalizer, rendering the whole system nonlinear. It can be shown that an MMSE SIC receiver is capable of achieving the information theoretic optimum sum rate of the single-user MIMO channel [40].

10.2.3 Further Receiver Processing

The equalizer and symbol demapper provide an (either hard- or soft-decision) estimate of the transmitted symbol vector. In order to decode the data, all symbols or LLRs corresponding to one codeword must be gathered. In Figure 10.1, this is achieved by the layer demapper. Additionally, the modulation demapper inverts the mapping between bits and symbols, which is necessary if LLRs for each bit are not already provided by the data detection step. Afterwards, the codeword is descrambled and decoded, by means of a hard- or soft-input decoder, providing the final bit decisions. The UE additionally has to compute estimates for the feedback indicators defined in the LTE standard (see Section 10.4 for details).

10.3 Physical Layer Modeling

In this section we define the post-equalization SINR for an LTE system under linear receive filters. Furthermore, we introduce the method of Effective Signal to Interference and Noise Ratio Mapping (ESM) to obtain an average SINR value for multiple REs. This technique is essential for the computation of the UE feedback values, as explained in Section 10.4.

10.3.1 Post-equalization SINR

One important measure for the performance of a transmission system employing linear spatial equalizers at the receivers is the post-equalization SINR obtained on each transmission layer, because it is directly related to the theoretically possible throughput I_l on layer l via Shannon's formula:

$$I_l = \log_2 \det\left(1 + \mathrm{SINR}_{r,l}\right) \tag{10.5}$$

Here, $\mathrm{SINR}_{r,l}$ denotes the SINR on layer l of RE r. This value can easily be computed for a linear equalizer and equals

$$\mathrm{SINR}_{r,l} = \frac{P_l \cdot |\mathbf{K}_r[l,l]|^2}{\sum_{i \neq l} P_i \cdot |\mathbf{K}_r[l,i]|^2 + \sigma_\mathrm{n}^2 \sum_i |\mathbf{F}_r[l,i]|^2} \tag{10.6}$$

In this equation, P_l denotes the transmit power on layer l and $\mathbf{K}_r[l,i]$ refers to the element in the lth row and ith column of matrix $\mathbf{K}_r$ defined in (10.4). Correspondingly, $|\mathbf{K}_r[l,l]|^2$ is the channel power gain experienced on layer l. The term in the numerator of Equation (10.6) gives the received power of layer l. The first term in the denominator is the interference caused by the other transmission layers after equalization. For ZF equalization this term vanishes. The right term in the denominator of Equation (10.6) provides the noise power on layer l after enhancement by the equalizer filter $\mathbf{F}_r$. Expressions exist for the post-equalization SINR achieved by an MMSE SIC receiver as well [6].

10.3.2 SINR Averaging

In many cases it is important to obtain a single representative SINR value for several REs. An important example, which we will encounter in more detail in Section 10.4, is link adaptation based on UE CSI feedback. The goal in the course of this is to feed back a single quantized Channel Quality Indicator (CQI), which is employed to choose an appropriate Modulation and Coding Scheme (MCS) for transmission, in order to achieve a given target BLER over a broadband channel spanning several subcarriers (see also Section 5.4). Thereby, Additive White Gaussian Noise (AWGN) channel BLER performance curves are utilized to estimate the BLER obtained for each MCS. Simply linearly averaging the subcarrier SINRs to obtain an average Signal to Noise Ratio (SNR) does not make sense in that respect, because the BLER is not linearly related to the SINRs.

Two methods currently dominate SINR averaging for link performance modeling and link quality prediction: Exponential Effective Signal to Interference and Noise Ratio Mapping (EESM) [24] and Mutual Information Effective Signal to Interference and Noise Ratio Mapping (MIESM) [42]. In both cases an average, effective SNR, also referred to as AWGN-equivalent SNR, is obtained by mapping the SINRs experienced on the

R REs of interest into some other domain via a bijective function $f(\cdot)$, computing a linear average in this domain and mapping that value back to the SNR domain:

$$\mathrm{SNR}_{\mathrm{eff}} = \beta f^{-1}\left[\frac{1}{R}\sum_{r=1}^{R} f\left(\frac{\mathrm{SINR}_r}{\beta}\right)\right] \tag{10.7}$$

The former method, EESM, is based on the Chernoff upper bound for the error probability of binary phase shift keying transmission over an AWGN channel [24]. In that case, the averaging function f turns out to be the exponential function: $f(\cdot) = \mathrm{e}^{(\cdot)}$. In case of MIESM, the Bit-Interleaved Coded Modulation (BICM) capacity [7] is employed as averaging function. The idea behind this approach is to identify the frequency-selective channel, experienced on the R REs, with an AWGN channel that achieves the same average BICM capacity.

The factor β, appearing in Equation (10.7), is a calibration factor that is necessary to adapt the method to the different MCSs. It is obtained from extensive link-level simulations, employing different channel models [9]. For MIESM this calibration factor is very close to one, and a good SNR approximation is obtained even without calibration simply by setting $\beta = 1$. EESM, however, requires careful calibration of β to obtain an accurate SNR estimate.

10.4 User Equipment Feedback Calculation

LTE supports Adaptive Modulation and Coding (AMC) in order to adapt the transmission parameters (code rate and modulation order) to the current channel conditions. Additionally, the spatial preprocessing steps employed (transmission rank, precoding) are also adaptive, so as to maximize the possible MIMO gains. For that purpose, the transmitter requires CSI, which is obtained by means of UE feedback, utilizing three distinct feedback indicators. Not all of these indicators are employed for all transmission modes, as defined in [1]. In this section, we first give an overview of the different transmission modes and the corresponding UE feedback indicators utilized. Furthermore, we derive algorithms to calculate the feedback indicators at the UEs, with the aim of maximizing the estimated user throughput. These algorithms require a post-equalization SINR and, therefore, are limited to receivers for which this value can be estimated (see Section 10.3.1). Evaluation of the performance of the feedback estimators, by means of link-level throughput simulations, is postponed until Section 10.6 in order to be able to compare with the throughput bounds derived in Section 10.5.

10.4.1 User Equipment Feedback Indicators

Obtaining accurate CSI at the transmitter is crucial in a modern wireless communication system so that a high system throughput can be achieved. In LTE, quantized CSI is provided by means of the CQI, Precoding Matrix Indicator (PMI), and Rank Indicator (RI). The purpose of these feedback values is as follows:

- The CQI provides information about the quality of the current channel realization in terms of a quantized SINR or equivalently as estimated achievable throughput. This information is utilized at the transmitter to choose the appropriate MCS to achieve a

given target BLER at the first Hybrid Automatic Repeat reQuest (HARQ) transmission, typically BLER ≤ 0.1. Furthermore, the CQI information of multiple users can be employed to decide about the allocation of resources to users, so as to maximize the overall cell throughput [27, 28].
- The PMI is utilized for MIMO preprocessing. It informs the eNodeB about the preferred precoder, stemming from the standard defined unitary codebook [2], in order to maximize the user throughput. LTE and especially LTE-Advanced (LTE-A) also support rank-one Multi User MIMO (MU-MIMO). In that case, the PMI is utilized as a Channel Direction Indicator (CDI), providing quantized information about the favored beamforming direction (see Chapter 15 for more details).
- The purpose of the RI is to signal the preferred transmission rank given the current channel conditions. At low SNR, it is in general better to utilize the multiple transmit antennas to achieve a beamforming SNR gain, while a larger throughput can be achieved at high SNR by spatially multiplexing several parallel data streams [40].

Whether these feedback indicators are required or not depends on the eNodeB - configured transmission mode [1]:

- In an SISO system, spatial preprocessing is unnecessary and, therefore, only the CQI is required.
- If the transmission system is configured in transmit diversity mode, the precoding is fixed by the standard to the Alamouti scheme [3]. Furthermore, the number of transmission layers only depends on the number of transmit antennas [2]. Therefore, the only necessary feedback indicator is the CQI.
- When Open-Loop Spatial Multiplexing (OLSM) is activated, the precoding is also fixed by the standard to a specified pattern, so as to achieve a multiplexing and/or diversity gain [2]. The transmission rank can be adapted in that case (depending on the eNodeB setup), which requires RI feedback in addition to the CQI. This spatial multiplexing mode is employed in cases in which only outdated, and therefore useless, spatial preprocessing feedback is available (high feedback delay/high-mobility scenarios).
- The Closed-Loop Spatial Multiplexing (CLSM) mode adapts the transmission rank, precoders, and MCS to the current channel conditions and, therefore, requires all three feedback indicators.

The wireless channel, in general, is frequency and time selective. This implies that the preferred feedback values vary over time and frequency. The structure of the LTE signal processing chain, shown in Figure 2.7, implies that the same MCS is utilized for all resources belonging to one codeword. Therefore, a single CQI for each codeword would be sufficient per TTI and for the total system bandwidth. Nevertheless, it can be beneficial to employ frequency-selective CQI feedback in order to achieve a multi-user diversity gain at UE scheduling [28]. Also, the same transmission rank is utilized on all resources allocated to a user. Therefore, a single RI per TTI is sufficient as well. The precoders, on the other hand, can vary from one RE to the next, requiring frequency-selective PMI feedback. The eNodeB configures the granularity of the feedback indicators in the time domain in multiples of TTIs and in the frequency domain through the subband size.

10.4.2 Calculation of the CQI, PMI, and RI

For the calculation of the UE feedback indicators we consider the most complicated case of the CLSM transmission mode, which requires CQI, PMI, and RI feedback. If not all of these values are utilized in the system considered, optimization with respect to the missing values can simply be skipped and the appropriate standard defined values substituted. The following feedback calculation algorithms are based on [26, 29, 31]. The goal for the choice of the feedback indicators at a UE is to maximize the estimated spectral efficiency, assuming transmission on all REs $r \in \{1, \ldots, R\}$. Furthermore, as a side constraint, the chosen CQI has to satisfy an upper bound on the BLER. We next derive the corresponding optimization problem that needs to be solved in order to find the appropriate feedback indicators. As mentioned in the previous section, CQI and PMI feedback can be subband or wideband specific, meaning that single values are computed for each subband or just for the total system bandwidth. We explicitly distinguish the cases of wideband and subband specific CQI feedback. Furthermore, we apply a simple trick to jointly capture the cases of wideband- and subband-specific PMI feedback, by defining the precoder for subband s as follows:

$$\mathbf{W}_s = \mathbf{W}^{(1)} \cdot \mathbf{W}_s^{(2)} \tag{10.8}$$

We denote the standard defined, rank L-dependent precoder codebook, defined in [2], as $\mathcal{W}^{(L)}$. In Equation (10.8), $\mathbf{W}^{(1)}$ denotes the wideband precoder and $\mathbf{W}_s^{(2)}$ the subband precoder for subband s. If we consider the case of wideband feedback, then $\mathbf{W}^{(1)} \in \mathcal{W}_1^{(L)}$ stems from the standard defined codebook $\mathcal{W}_1^{(L)} = \mathcal{W}^{(L)}$ and the codebook for the subband precoder $\mathbf{W}_s^{(2)} \in \mathcal{W}_2^{(L)}$ is set as $\mathcal{W}_2^{(L)} = \{1\}$. Otherwise, in the case of subband-specific PMI feedback, the wideband precoder codebook is set equal to the trivial codebook $\mathcal{W}_1^{(L)} = \{1\}$ and the subband precoder codebook is identified with the standard defined codebook $\mathcal{W}_2^{(L)} = \mathcal{W}^{(L)}$.[1] The eNodeB is configured in such a way that the total system bandwidth, consisting of R REs, is divided into S subbands. The set of REs belonging to subband s is denoted as $\mathcal{R}_s$. Furthermore, we define a surjective mapping $\rho : \{1, \ldots, R\} \to \{1, \ldots, S\}$ which assigns an RE r to the corresponding subband s. Because the precoder now only depends on the subband index, the input–output relation of the LTE system from Equation (10.2) is slightly modified to

$$\mathbf{y}_r = \mathbf{H}_r \mathbf{W}_s \mathbf{x}_r + \mathbf{n}_r, \quad r \in \{1, \ldots, R\}, s = \rho(r) \tag{10.9}$$

The post-equalization SINR according to Equation (10.6) depends on the precoder:

$$\mathrm{SINR}_{r,l}\left(\mathbf{W}_s\right) = \frac{P_l \cdot |\mathbf{K}_r[l,l]|^2}{\sum_{i \neq l} P_i \cdot |\mathbf{K}_r[l,i]|^2 + \sigma_{\mathrm{n}}^2 \sum_i |\mathbf{F}_r[l,i]|^2}, \tag{10.10}$$

$$\mathbf{K}_r = \mathbf{F}_r \mathbf{H}_r \mathbf{W}_s, r \in \{1, \ldots, R\}, l \in \{1, \ldots, L\}, s = \rho(r) \tag{10.11}$$

Furthermore, the SINR is rank L dependent as well, due to the rank dependency of the precoder codebook $\mathcal{W}^{(L)}$. As mentioned in Section 10.1, it can happen that a single

[1] As shown in Chapter 15, this notation is especially suitable for the eight-transmit-antenna codebook of LTE-A. In that case the standard defines both a wideband and a subband precoder, each with a nontrivial codebook.

codeword is mapped onto multiple transmission layers, because the number of codewords is limited to at most two, while the number of transmission layers can be up to four. The set of layers belonging to a codeword c is denoted $\mathcal{L}_c$. Furthermore, the set of standard defined MCSs is denoted $\mathcal{M}$ (see Chapter 2 for details on the defined MCSs).

In order to choose the appropriate feedback values, the spectral efficiency and BLER for each MCS $m \in \mathcal{M}$ need to be estimated. This is achieved by employing ESM techniques (see Section 10.3.2) to obtain an effective, AWGN channel equivalent, SNR from the rank- and precoder-dependent post-equalization SINRs given by Equation (10.10). The effective SNR is then mapped to the corresponding BLER and spectral efficiency via precomputed look-up tables. The exact mathematical formulation of this process is provided next for the two cases of wideband- and subband-specific CQI feedback.

10.4.2.1 Subband-specific CQI Feedback

For the case of subband-specific CQI feedback, the BLER constraint must be satisfied for each subband individually. We denote the target BLER as $P_{\mathrm{b}}^{(t)}$. The effective SNR for subband s and codeword c is computed by means of ESM:

$$\mathrm{SNR}_s^c(\mathbf{W}_s, m) = f_m^{-1}\left[\frac{1}{|\mathcal{R}_s||\mathcal{L}_c|}\sum_{r\in\mathcal{R}_s, l\in\mathcal{L}_c} f_m\left(\mathrm{SINR}_{r,l}(\mathbf{W}_s)\right)\right] \tag{10.12}$$

In this equation, $f_m(\cdot)$ denotes the ESM averaging function for MCS m. This function incorporates the calibration factor β given in Equation (10.7). Furthermore, for MIESM the averaging function also depends on the modulation alphabet corresponding to MCS m, because the BICM capacity is modulation alphabet dependent. The ESM averaging in Equation (10.12) is performed over all resources $r \in \mathcal{R}_s$, corresponding to subband s, and over all layers $l \in \mathcal{L}_c$ that belong to codeword c. The cardinality of a set is denoted by $|\cdot|$. The effective SNR obtained depends on the MCS, the precoder $\mathbf{W}_s$, and, via the precoder, on the transmission rank L. Via a precomputed look-up table, the AWGN channel BLER corresponding to this SNR can be obtained. An example of such a mapping is shown in Figure 10.2. In the figure, the leftmost curve corresponds to MCS 1, which utilizes 4-QAM modulation and a code rate of 0.08, while the rightmost curve gives the BLER performance of MCS 15, utilizing 64-QAM and a code rate of 0.93. Furthermore, the target BLER of $P_{\mathrm{b}}^{(t)} = 0.1$ is shown as a gray line. We denote the MCS-dependent mapping between SNR and BLER as $g_m(\mathrm{SNR})$. With this mapping, the estimated BLER corresponding to the effective SNR equals

$$P_s^c(\mathbf{W}_s, m) = g_m\left(\mathrm{SNR}_s^c(\mathbf{W}_s, m)\right) \tag{10.13}$$

Next, we define a function $h_m(\mathrm{BLER})$ that delivers the spectral efficiency e_m of MCS m, if the BLER constraint is satisfied and zero otherwise:

$$h_m(\mathrm{BLER}) = \begin{cases} e_m; & \mathrm{BLER} \le P_{\mathrm{b}}^{(t)} \\ 0; & \mathrm{BLER} > P_{\mathrm{b}}^{(t)} \end{cases} \tag{10.14}$$

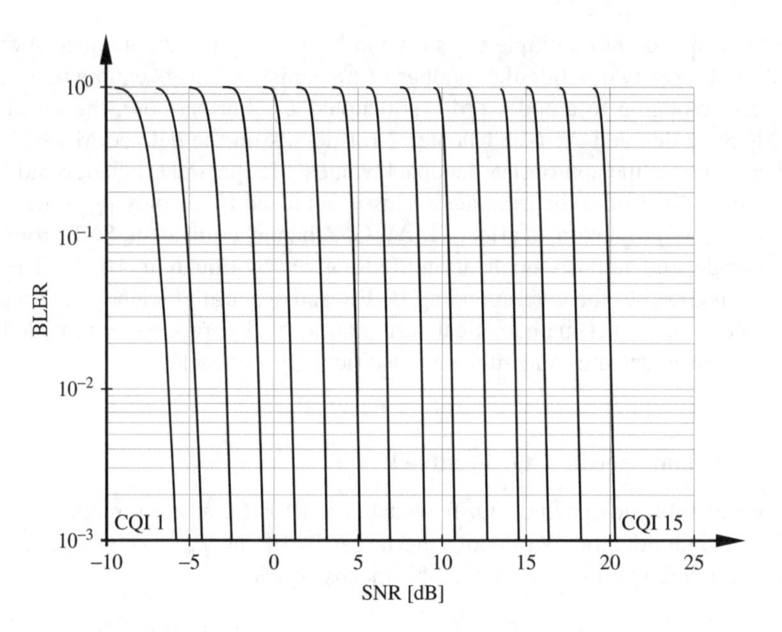

Figure 10.2 BLERs obtained from link-level simulations for the 15 LTE standard defined MCSs over an AWGN channel.

The spectral efficiencies of the 15 MCSs defined in the LTE specifications can be found in Chapter 2. This allows an estimate of the spectral efficiency for subband s and codeword c:

$$E_s^c(\mathbf{W}_s, m) = h_m\left(P_s^c(\mathbf{W}_s, m)\right) \cdot \left(1 - P_s^c(\mathbf{W}_s, m)\right) \tag{10.15}$$

If the BLER target is not satisfied, the estimated spectral efficiency is zero, indicating that the corresponding MCS must not be utilized. Otherwise, the estimated spectral efficiency is given by the product of the MCS spectral efficiency e_m and the probability that the transmission is error free. The estimated spectral efficiency also depends on the MCS, the precoder, and the transmission rank. The optimal feedback values are obtained by maximizing the sum spectral efficiency over all subbands $s \in \{1, \ldots, S\}$ and codewords $c \in \mathcal{C}_L$. Here, $\mathcal{C}_L$ denotes the number of codewords corresponding to transmission rank L. The constrained "subband optimization problem" for the feedback indicators then reads

$$\begin{aligned}
[\hat{L}, \{\hat{\mathbf{W}}_s\}_S, \{\hat{\mathbf{m}}_s\}_S] = \quad & \underset{L, \mathbf{W}^{(1)}, \mathbf{W}_s^{(2)}, \mathbf{m}_s}{\operatorname{argmax}} \sum_{s=1}^{S} \sum_{c=1}^{C_L} E_s^c(\mathbf{W}_s, \mathbf{m}_s[c]) \\
\text{subject to: } & L \leq L_{\max} \\
& \mathbf{W}^{(1)} \in \mathcal{W}_1^{(L)} \\
& \mathbf{W}_s^{(2)} \in \mathcal{W}_2^{(L)} \\
& \mathbf{m}_s \in \mathcal{M}^{C_L \times 1}.
\end{aligned} \tag{10.16}$$

The value $\hat{\mathbf{m}}_s[c]$ denotes the optimal MCS for subband s and codeword c. These values are stacked into the vectors $\hat{\mathbf{m}}_s$, whose lengths is equal to the number of codewords.

Furthermore, $\{\hat{\mathbf{m}}_s\}_S$ refers to the set of optimal MCS vectors for all S subbands. This set corresponds to the preferred set of CQIs. The size S set of optimal precoders is given by $\{\hat{\mathbf{W}}_s\}_S$ and the PMIs correspond to their codebook indices. Finally, the preferred RI equals $\hat{L}$. The constraint on the number of transmission layers $L \leq L_{\max}$ follows from the rank of the channel matrices:

$$L_{\max} = \min_{r \in \{1,\ldots,R\}} \operatorname{rank}(\mathbf{H}_r). \tag{10.17}$$

In the case of subband-specific PMI feedback, the optimization problem decouples into individual optimization problems for each subband s. For wideband PMI feedback, on the other hand, all terms in sum (10.16) are coupled via the wideband precoder.

10.4.2.2 Wideband-specific CQI Feedback

If just a single wideband CQI is computed for each codeword, the CQIs have to be chosen such, that the BLER constraint is satisfied for all resources jointly. Therefore, ESM averaging is applied to the total system bandwidth to deliver an effective SNR for each codeword:

$$\mathrm{SNR}^c(\{\mathbf{W}_s\}_S, m) = f_m^{-1}\left[\frac{1}{R|\mathcal{L}_c|}\sum_{r=1}^{R}\sum_{l\in\mathcal{L}_c} f_m\left(\mathrm{SINR}_{r,l}(\mathbf{W}_s)\right)\right], \quad s = \rho(r) \tag{10.18}$$

The difference from Equation (10.12) is that the sum here comprises all R REs. Therefore, the effective SNR obtained also depends on the complete set of S subband precoders. Again, the AWGN channel equivalent SNR is utilized to estimate the BLER via the mapping $g_m(\mathrm{SNR})$:

$$P^c(\{\mathbf{W}_s\}_S, m) = g_m(\mathrm{SNR}^c(\{\mathbf{W}_s\}_S, m)) \tag{10.19}$$

Then the codeword-specific spectral efficiency is estimated by the function $h_m(\mathrm{BLER})$ via

$$E^c(\{\mathbf{W}_s\}_S, m) = h_m(P^c(\{\mathbf{W}_s\}_S, m)) \cdot (1 - P^c(\{\mathbf{W}_s\}_S, m)) \tag{10.20}$$

Finally, the "wideband optimization problem" is formulated by maximizing the sum spectral efficiency over all codewords:

$$\begin{aligned}
[\hat{L}, \{\hat{\mathbf{W}}_s\}_S, \hat{\mathbf{m}}] = \quad & \underset{L,\mathbf{W}^{(1)},\mathbf{W}_s^{(2)},\mathbf{m}}{\operatorname{argmax}} \sum_{c=1}^{C_L} E_s^c(\{\mathbf{W}_s\}_S, \mathbf{m}[c]) \\
\text{subject to: } & L \leq L_{\max} \\
& \mathbf{W}^{(1)} \in \mathcal{W}_1^{(L)} \\
& \mathbf{W}_s^{(2)} \in \mathcal{W}_2^{(L)} \\
& \mathbf{m} \in \mathcal{M}^{C_L \times 1}
\end{aligned} \tag{10.21}$$

As we are considering wideband CQI feedback, the MCS vector $\hat{\mathbf{m}}$ has no subband dependency. In this case, it is not possible to split the optimization problem into individual subband problems, because all resources are interconnected via the ESM averaging employed in Equation (10.18). Solving either of the two optimization problems (10.16) or (10.21) implies very high computational complexity, because they are combinatorial problems. In general, the solution requires an exhaustive search over all possible combinations of precoders, MCSs, and transmission ranks. Considering the short time frame of one TTI (1 ms) available for feedback computation, as well as the limited signal processing hardware, in many practical cases it will not be possible to solve these optimization problems. Therefore, we resort to the simplified, suboptimal algorithm presented next.

10.4.2.3 Approximate Sequential Solution of the Optimization Problems

To reduce the complexity of the optimization problems described in the previous section, we resort to the sequential optimization suggested in [1]. In this approach the precoders and transmission rank are chosen in advance, and afterwards, with knowledge of these values, the optimal CQIs are selected by substituting the solution for the precoders into (10.16) or (10.21) and solving the optimization problems. In the first step of the sequential solution we try to find the preferred subband precoders. This choice is based on maximizing the spectral efficiency of a subband, estimated by means of the BICM capacity [7], with respect to the subband precoder. For that purpose we first need to compute the spectral efficiency of RE r, denoted I_r, for all combinations of precoders and transmission ranks. Because the BICM capacity is modulation alphabet dependent, we utilize a function $f(\mathrm{SNR})$ that delivers the maximum efficiency over all modulation alphabets:

$$f(\mathrm{SNR}) = \max_{a \in \mathcal{A}} f_a(\mathrm{SNR}) \tag{10.22}$$

In this equation, $f_a(\mathrm{SNR})$ denotes the BICM capacity of modulation alphabet a and maximization is performed with respect to all defined modulation alphabets $a \in \mathcal{A} = \{\mathcal{A}_4, \mathcal{A}_{16}, \mathcal{A}_{64}\}$, as already defined in Section 10.1. Figure 10.3 shows the BICM capacities of 4/16/64-QAM modulation and the resulting function $f(\cdot)$ as the envelope of the capacities. This step is necessary in order to make the choice of the subband precoders independent of the MCSs. The estimated spectral efficiency of RE r then equals

$$I_r\left(\mathbf{W}_s, L\right) = \sum_{l=1}^{L} f\left(\mathrm{SINR}_{r,l}\left(\mathbf{W}_s\right)\right) \tag{10.23}$$

The most expensive step to obtain this value is the computation of the post-equalization SINR according to Equation (10.6), because it involves matrix inversions in order to calculate the receive equalizer filters. As soon as the SINR is computed, the spectral efficiency is easily obtained by means of a simple look-up table. In order to choose the subband precoders $\mathbf{W}_s^{(2)}$, the sum spectral efficiency for each subband is maximized:

$$I_s(\mathbf{W}_s, L) = \sum_{r \in \mathcal{R}_s} I_r(\mathbf{W}_s, L), \tag{10.24}$$

$$\hat{\mathbf{W}}_s^{(2)}(\mathbf{W}^{(1)}, L) = \underset{\mathbf{W}_s^{(2)} \in \mathcal{W}_2^{(L)}}{\operatorname{argmax}} \; I_s(\mathbf{W}_s, L) \tag{10.25}$$

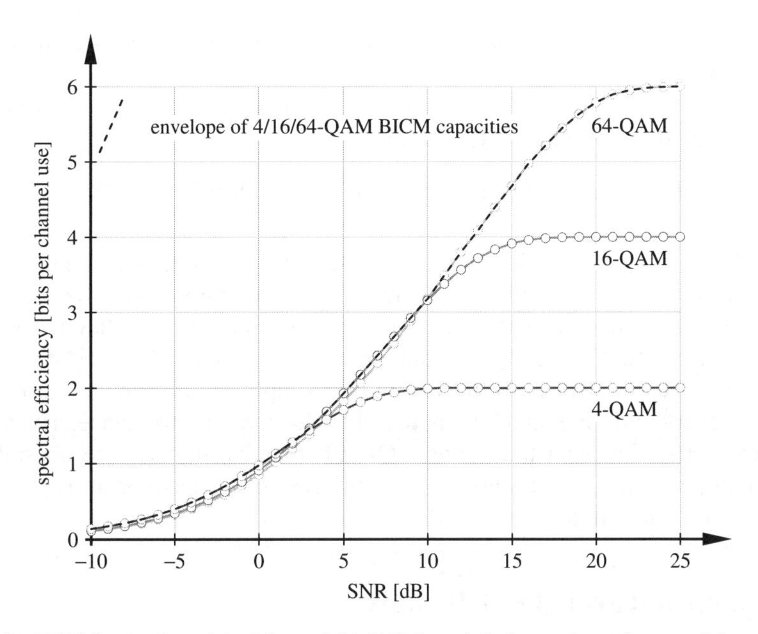

Figure 10.3 BICM capacity of 4-, 16-, and 64-QAM modulation and maximum of these capacities according to Equation (10.22).

In Equation (10.24), I_s denotes the sum spectral efficiency of subband s. This efficiency is maximized with respect to the subband precoder, and is dependent on the transmission rank and wideband precoder choice. Because a closed-form solution of this maximization cannot in general be found, the optimum precoder for each possible choice of transmission rank and wideband precoder is stored. Utilizing this pre-knowledge of the subband precoders, it is possible to obtain the corresponding optimal wideband precoder, rank, and CQIs from (10.16) or (10.21). Alternatively, the optimization problem can be further simplified by choosing the wideband precoder in advance as well. The wideband precoder $\hat{\mathbf{W}}_1$ follows from the sum spectral efficiency over all subbands:

$$I(\mathbf{W}_s, L) = \sum_{s=1}^{S} I_s(\mathbf{W}_s, L), \tag{10.26}$$

$$\hat{\mathbf{W}}^{(1)}(L) = \underset{\mathbf{W}^{(1)} \in \mathcal{W}_1^{(L)}}{\operatorname{argmax}} \; I(\mathbf{W}_s, L) \tag{10.27}$$

The pre-knowledge of the subband precoder is utilized to solve the optimization problem (10.27). The solution for the wideband precoder depends on the choice of the transmission rank. Again, because no closed-form solution can be found, the optimum wideband precoder is computed and stored for all rank possibilities. By plugging this precoder back into the solutions for the subband precoders (10.25) we obtain the corresponding rank-dependent optimum subband precoders $\hat{\mathbf{W}}_s^{(2)}(L) = \hat{\mathbf{W}}_s^{(2)}(\hat{\mathbf{W}}^{(1)}(L), L)$. Again, this pre-knowledge of the wideband and subband precoders can be utilized to simplify the solution of the optimization problems (10.16) and (10.21) for the corresponding optimal rank and CQIs.

If the complexity of solving (10.16) or (10.21) is still too high, one can even go a step further to choose the rank in advance as well:

$$\hat{L} = \underset{L \leq L_{\max}}{\operatorname{argmax}}\, I(\hat{\mathbf{W}}_s(L), L), \quad \hat{\mathbf{W}}_s(L) = \hat{\mathbf{W}}^{(1)}(L)\hat{\mathbf{W}}_s^{(2)}(L) \tag{10.28}$$

Similar to the wideband precoder, the rank is chosen to maximize the sum spectral efficiency of all R resources. With knowledge of the optimum rank, the corresponding wideband and subband precoders can be found from the stored solutions of the optimization problems (10.25) and (10.27). The only variables left for optimization in (10.16) and (10.21) are then the CQIs. We decided not to simplify the computation of the CQIs, because we observed during simulations that they are more performance critical than the RIs and the PMIs. Choosing the wrong CQI either underestimates the possible performance (leading to a waste of resources) or, in the case of overestimated performance, the BLER bound will no longer be satisfied.

10.5 Practical Throughput Bounds

The ultimate goal for any communication system is to utilize the available resources as efficiently as possible. Equivalently, this goal can be formulated as the strive to operate as close to channel capacity as possible. The channel capacity for the single-user Gaussian MIMO broadcast channel has been known for many years [38], but still modern wireless communication systems are far from achieving throughputs that come close to capacity. This fact is exposed with the measurement results of Chapter 5 for HSDPA and by means of link-level simulation results for LTE in Section 10.6.

The goal of this section is to reveal explanations for the severe throughput loss of practical systems compared with channel capacity. This is achieved by developing theoretical throughput bounds that take into account practical system design constraints [30], such as the finite code block length, the limited set of supported MCSs, and the restricted codebook of precoders. Starting from channel capacity, step by step, tighter, but also more restrictive, practical throughput bounds for LTE are constructed. Each practical constraint can then be equated with a corresponding throughput loss. In that way, we systematically identify the major sources of performance losses. The bounds can more generally be applied to any BICM-based MIMO OFDM communication system. In Section 10.6 we compare the single-user throughput of LTE with the derived performance bounds by means of link-level simulations.

10.5.1 *Channel Capacity*

The amount of information that any communication system can transmit reliably over a given channel is upper bounded by the well-known channel capacity. For completeness, we state the channel capacity of the single-user MIMO channel [38]. We assume an OFDM system with N_T transmit and N_R receive antennas that splits the total system bandwidth into K orthogonal subcarriers. The MIMO channel matrix on subcarrier k is denoted $\mathbf{H}_k \in \mathbb{C}^{N_\mathrm{R} \times N_\mathrm{T}}$. We do not consider REs as in the previous sections, because we are not interested in the temporal frame structure now. Consider the Singular Value Decomposition (SVD) of the MIMO channel matrix $\mathbf{H}_k$ on subcarrier k:

$$\mathbf{H}_k = \mathbf{U}_k \Sigma_k \mathbf{V}_k^H, \quad \Sigma_k = \mathrm{diag}\left\{\sqrt{\lambda_{k,l}}\right\}, \tag{10.29}$$

$$l = 1 \ldots L_k, \quad L_k = \mathrm{rank}(\mathbf{H}_k) \tag{10.30}$$

Here, $\mathbf{U}_k \in \mathbb{C}^{N_\mathrm{R} \times L_k}$ and $\mathbf{V}_k \in \mathbb{C}^{N_\mathrm{T} \times L_k}$ are unitary matrices, composed of the left and right singular vectors of $\mathbf{H}_k$, and the values $\lambda_{k,l}$ are the singular values of the channel matrix. By applying the matrix $\mathbf{V}_k$ as precoder at the transmitter and the matrix $\mathbf{U}_k^H$ as receive filter at the receiver, the MIMO channel splits into L_k parallel SISO channels with power gain $\lambda_{k,l}$. In this way the MIMO OFDM system can equivalently be regarded as $\sum_{k=1}^{K} L_k$ parallel SISO channels. The "channel capacity" C can be achieved by optimally distributing the available transmit power P_Tx over these parallel SISO subchannels:

$$C = \max_{\{P_{k,l}\}} \sum_{k=1}^{K} \sum_{l=1}^{L_k} \log_2 \left(1 + P_{k,l} \frac{\lambda_{k,l}}{\sigma_\mathrm{n}^2}\right) \tag{10.31}$$

$$\text{subject to:} \sum_{k=1}^{K} \sum_{l=1}^{L_k} P_{k,l} = P_\mathrm{Tx} \tag{10.32}$$

Here, σ_n^2 denotes the variance of the white Gaussian receiver noise and $P_{k,l}$ the power on layer l of subcarrier k. This optimization problem can be solved by utilizing the famous water-filling power allocation algorithm [33, 40].

10.5.2 Open-Loop Mutual Information

Many wireless communication systems are based on frequency-division duplexing, in which case the transmitter does not have full channel knowledge. If the transmitter does not have any channel knowledge at all, it was shown that the Mutual Information (MI) is the appropriate tight upper bound on the throughput [22]. The MI is given by

$$I^{(\mathrm{MI})} = \sum_{k=1}^{K} \log_2 \det \left(\mathbf{I}_{N_\mathrm{R}} + \frac{P_\mathrm{Tx}}{N_\mathrm{T} K} \frac{1}{\sigma_\mathrm{n}^2} \mathbf{H}_k \mathbf{H}_k^H\right) \tag{10.33}$$

In this case, the transmit power P_Tx is equally distributed over all transmit antennas and subcarriers and no precoding is applied.

This bound is not applicable for LTE because, even in the OLSM mode, without PMI and RI feedback, MIMO precoding is applied in order to achieve a diversity gain. This diversity gain is obtained by applying Cyclic Delay Diversity (CDD) precoding [32]. With this technique the frequency diversity of the channel is artificially increased by transmitting delayed versions of the same signal from different transmit antennas, leading to a frequency-dependent interference pattern of the signal received at the UE. Taking this precoding of the Open Loop (OL) system into account leads to the "Open-Loop Mutual

Information (OLMI) bound":

$$I^{(\mathrm{OL})} = \sum_{k=1}^{K} \log_2 \det \left(\mathbf{I}_{N_\mathrm{R}} + \frac{P_\mathrm{Tx}}{N_\mathrm{T} K} \frac{1}{\sigma_\mathrm{n}^2} \mathbf{H}_k \mathbf{W}_k \mathbf{W}_k^H \mathbf{H}_k^H \right) \tag{10.34}$$

In this equation, the transmit power is still uniformly distributed over all resources, but precoding according to the standard defined pattern [2] is included. Theoretically, power loading over the different subcarriers would be possible utilizing the CQI feedback, provided it is frequency selective. LTE does not employ that kind of channel adaptation, but rather adapts the MCS.

10.5.3 Closed-Loop Mutual Information

By means of receiver feedback, an LTE system obtains CSI at the transmitter. This information is utilized in the CLSM transmission mode to adapt the spatial preprocessing (transmission rank and precoding) in order to increase the throughput. For such a Closed-Loop (CL) system, the previously described OLMI bound is not appropriate. Instead, one must consider the possibility of choosing the appropriate precoder for the current channel realization from the finite precoder codebook. In contrast to channel capacity, an optimum power allocation over the transmission layers is not possible, because the channel matrix will in general not diagonalize by applying one of the unitary precoders. Anyway, power loading is not considered in the LTE specification. Therefore, we define the Closed-Loop Mutual Information (CLMI) via the following optimization problem:

$$I^{(\mathrm{CL})} = \max_{L \leq L_\mathrm{max}} \sum_{k=1}^{K} \max_{\mathbf{W}_k \in \mathcal{W}(L)} \log_2 \det \left(\mathbf{I}_{N_\mathrm{R}} + \frac{P_\mathrm{Tx}}{N_\mathrm{T} K} \frac{1}{\sigma_\mathrm{n}^2} \mathbf{H}_k \mathbf{W}_k \mathbf{W}_k^H \mathbf{H}_k^H \right) \tag{10.35}$$

In this bound, different precoders on individual subcarriers are allowed, but the same transmission rank is required for all resources. The optimization is simply carried out with an exhaustive search over all rank and precoder combinations.

As described in Section 10.4, LTE does not allow one to change the precoder on a subcarrier basis, only on a subband basis. This can simply be taken into account by fixing the precoder in the bound (10.35) for subbands of appropriate size as well. Considering, for example, wideband (WB) precoding leads to the Wideband Closed-Loop Mutual Information (WB-CLMI):

$$I^{(\mathrm{CL,WB})} = \max_{L \leq L_\mathrm{max}} \max_{\mathbf{W} \in \mathcal{W}(L)} \sum_{k=1}^{K} \log_2 \det \left(\mathbf{I}_{N_\mathrm{R}} + \frac{P_\mathrm{Tx}}{N_\mathrm{T} K} \frac{1}{\sigma_\mathrm{n}^2} \mathbf{H}_k \mathbf{W} \mathbf{W}^H \mathbf{H}_k^H \right) \tag{10.36}$$

The bounds derived until now all assumed an ideal ML detector at the receiver (see Section 10.2). Owing to their high computational complexity, in most cases such detectors will not be utilized in practice. Instead, simple linear receive equalizers are mainly employed. In that case, the impact of the receive filter on the theoretically achievable throughput has to be taken into account. The theoretical throughput is then determined by the SINR achieved after equalization, as stated in Equation (10.5). The post-equalization SINR achieved on each layer and subcarrier (10.6) determines the "WB-CLMI-Wideband Closed-Loop Mutual Information with Linear Receiver (WB-CLMI-LR) bound":

$$I^{(\text{CL,WB,LR})} = \max_{L \leq L_{\max}} \max_{\mathbf{W} \in \mathcal{W}^{(L)}} \sum_{k=1}^{K} \sum_{l=1}^{L} \log_2(1 + \text{SINR}_{k,l}) \tag{10.37}$$

The bound can be trivially generalized to subband- or subcarrier-specific precoding by appropriately modifying the sums and/or shifting the maximization with respect to the precoder into the sum, as in (10.35).

10.5.4 BICM Bounds

The throughput bounds derived in the previous sections focused on the impact of the structure of the MIMO precoding, as prescribed by the standard, on channel capacity. To achieve these mutual information and capacity bounds, it is necessary to employ signaling based on Gaussian codebooks [11]. On the other hand, the practical communication systems we consider employ a finite set of MCSs for signaling. In particular, the architecture of LTE is based on BICM [8], employing 4-, 16-, or 64-QAM and a rate-1/3 turbo code that is appropriately rate matched to achieve the desired code rates as defined in [1] (see Chapter 2 for details). The following bounds investigate how the BICM architecture influences the performance of the system. Furthermore, we consider the throughput degradation caused by the finite block length of the code and the finite number of MCSs defined by the standard.

10.5.4.1 BICM Bound

A BICM system divides channel coding and modulation mapping into two independent entities that are connected via an ideal bit interleaver. This architecture is very popular in practice, because it allows the combination of any channel code with any arbitrary modulation alphabet. The capacity of BICM systems is well known, albeit not in closed form [7]. For an SISO AWGN channel it can easily be evaluated by means of Monte Carlo simulations. The result of such simulations is shown in Figure 10.3. In Equation (10.22) of Section 10.4, we introduced a function $f(\text{SNR})$ that delivers the maximum of the BICM capacities over the implemented modulation alphabets. This function is now utilized to replace the Gaussian signaling capacity in Equation (10.37) in order to compute the BICM capacity of a closed-loop communication system, employing the finite set $\mathcal{A}$ of modulation alphabets, wideband precoding, and a linear receive equalizer. This defines the "BICM" bound:

$$\text{BI}^{(\text{CL,WB,LR})} = \max_{L \leq L_{\max}} \max_{\mathbf{W} \in \mathcal{W}^{(L)}} \sum_{k=1}^{K} \sum_{l=1}^{L} f(\text{SINR}_{k,l}(\mathbf{W})) \tag{10.38}$$

Similar bounds can be straightforwardly derived, for example, for closed-loop systems with subband-specific precoding by appropriately modifying the sums and maximizations, or for open-loop systems by omitting the maximizations and employing the standard defined precoders. This specific bound is stated, as it will be utilized for comparisons in Section 10.6.

10.5.4.2 Shifted BICM Bound

Achieving the bounds proposed in the previous sections requires, in general, a code with an infinite block length. Of course, in a practical system the block length of the code is finite, and determined by the number of resources assigned to a user and the length of a TTI. Typical block lengths for an LTE system are in the order of a few hundred to several thousands of bits. Optimal Shannon codes achieve the Shannon capacity limit for infinite code block length and there exist practical turbo and LDPC codes that perform close to these ideal codes if the block length is sufficiently large [25]. Nevertheless, a finite code block length prevents even an ideal Shannon code from achieving capacity.

Here, we include the effect of a finite code block length on the throughput performance of a communication system, employing an ideal Shannon code. Consider for that purpose a code of rate R_S, in information bits per channel symbol. If the code block length is infinite, then the Block Error Probability (BLEP) of the code is given by a step function, dropping from one to zero at exactly the SNR that corresponds to a channel capacity of $C = R_S$. If the channel capacity is less than R_S, the BLEP equals one, otherwise a BLEP of zero is achieved. A finite block length N (in channel symbols) causes this step in the BLEP to be replaced with a continuous decrease over SNR. For a Shannon code there exist upper and lower bounds on the BLEP P_B. One upper bound is given by the Gallager bound [25]:

$$P_B < 2^{-N \cdot E(R_S)}, \tag{10.39}$$

$$E(R_S) = \max_q \max_{0 \le \rho \le 1} [E_0(\rho, q) - \rho R_S], \tag{10.40}$$

$$E_0(\rho, q) = -\log_2 \int_y \left[\sum_x q(x) p(y|x)^{1/(1+\rho)} \right]^{1+\rho} \mathrm{d}y \tag{10.41}$$

In Equation (10.41) x denotes the channel input signal, $q(x)$ is the channel input distribution, y is the channel output signal, and $p(y|x)$ refers to the channel transition probability. Furthermore, ρ is employed as an auxiliary variable and $E(R_S)$ is called the Gallager exponent. A very similar lower bound on the BLEP of a Shannon code is given by the sphere packing bound [4]:

$$P_B > 2^{-N \cdot E(R_S)} \tag{10.42}$$

$$E(R_S) = \max_q \max_{0 \le \rho} [E_0(\rho, q) - \rho R_S] \tag{10.43}$$

$$E_0(\rho, q) = -\log_2 \int_y \left[\sum_x q(x) p(y|x)^{1/(1+\rho)} \right]^{1+\rho} \mathrm{d}y \tag{10.44}$$

The only difference to the Gallager bound is that ρ in this case is only lower bounded by zero and not upper bounded at all. This means that, if the optimizing ρ lies between zero and one, the two bounds are equal and exactly define the BLEP of an ideal code.

Employing BICM at the transmitter renders the channel input signal uniformly distributed over the QAM alphabet. Therefore, the maximization with respect to $q(x)$ in the two bounds can be skipped. Furthermore, assuming a linear receiver, the MIMO OFDM channel decomposes into parallel SISO AWGN channels with SNR given by Equation (10.6). To apply the proposed bounds for such a system, therefore, one needs to evaluate them for an AWGN channel with uniform input distribution over the QAM constellation of interest and for a given code rate R_{S}. This task can be performed numerically and delivers rate-dependent bounds on the BLEP. We computed the bounds for modulation alphabets defined in the LTE specification (4-, 16-, and 64-QAM) and several code rates. In all cases the optimal ρ was below one and, therefore, the exact BLEP of an ideal Shannon code was obtained.

We next utilize the computed BLEP of a Shannon code to obtain a mapping between SNR and spectral efficiency, similar to the BICM capacity shown in Figure 10.3, but valid for a code of finite block length. Because codes of finite block length are considered, obtaining a BLEP of zero is no longer possible. Therefore, the BLEP is fixed to the target value $P_{\mathrm{b}} = 10^{-1}$, frequently employed in wireless communication systems. For a given SNR we can then utilize the computed Shannon code BLEP to find the code rate R_{S} that achieves the target BLEP. This defines the desired mapping from SNR to spectral efficiency.

Figure 10.4 shows an example of the spectral efficiency obtained for a system employing 64-QAM and a code block length of $N = 1000$ channel symbols. The figure compares the spectral efficiency obtained with the BICM capacity of the same system (which assumes infinite block length). As expected, the system with finite block length performs worse than the BICM capacity for SNR $>$ 10 dB. Unexpectedly, at low SNR, the spectral efficiency obtained is higher than the BICM capacity.

The reason for this behavior is that the Gallager and sphere packing bounds hold for a Coded Modulation (CM) but not a BICM system. In CM, the channel code and modulation mapping are jointly designed, leading to better throughput performance at low SNR compared with a BICM system [8]. Nevertheless, it is known that CM and BICM perform equally well at high SNR [7]. The spectral efficiency obtained from the bounds can, therefore, be utilized to figure out how much the BICM capacity needs to be shifted in order to take into account the finite code block length. This shift is obtained by matching the BICM capacity and the spectral efficiency obtained at high SNR. In Figure 10.4, the shift equals merely 0.35 dB, which shows that for moderate block lengths the corresponding loss is almost negligible. We denote the shifted BICM capacity for modulation alphabet a as $f_{\mathrm{S}}^{(a)}(\mathrm{SNR})$. Similar to Equation (10.22), the maximum shifted spectral efficiency over all modulation alphabets is computed and denoted $f_{\mathrm{S}}(\mathrm{SNR})$. This mapping is now employed instead of the BICM capacity in Equation (10.38), to define the Shifted BICM (SBICM) bound:

$$\mathrm{SBICM}^{(\mathrm{CL,WB,LR})} = \max_{L \le L_{\max}} \max_{\mathbf{W} \in \mathcal{W}^{(L)}} \sum_{k=1}^{K} \sum_{l=1}^{L} f_{\mathrm{S}}(\mathrm{SINR}_{k,l}(\mathbf{W})) \tag{10.45}$$

10.5.4.3 Quantized and Shifted BICM Bound

To achieve the SBICM bound, the communication system must support any possible code rate. In practical systems, typically just a small set of possible MCSs is employed (15 in LTE). To still guarantee the desired BLER constraint in such a system, the supported MCS with largest spectral efficiency less than or equal to the SBICM bound has to be utilized. Mathematically, this can be formulated as "quantizing" the SBICM bound to the nearest MCS with lower or equal spectral efficiency:

$$\mathrm{QSBICM}^{(\mathrm{CL,WB,LR})} = \lfloor \mathrm{SBICM}^{(\mathrm{CL,WB,LR})} \rfloor_{\mathcal{M}} \tag{10.46}$$

The operator $\lfloor\cdot\rfloor_{\mathcal{M}}$ means flooring with respect to the MCSs defined in the codebook $\mathcal{M}$ of the system considered. We denote this bound the Quantized and Shifted BICM (QSBICM) bound. Alternatively, if the finite code block length is not to be incorporated, the BICM bound can be quantized in a similar way to obtain the Quantized BICM (QBICM) bound.

10.5.5 Achievable Throughput Bounds

In general, the MIMO channel is varying over time. Assuming ergodicity, we obtain values for the (ergodic) bounds (for example, ergodic capacity [22]) by means of Monte Carlo simulations for a chosen channel model in Section 10.6. During these Monte Carlo simulations we take into account the most obvious source of effective user throughput loss, namely the insertion of system overhead (for example, for synchronization, channel estimation), by skipping the appropriate subcarriers, as defined by the standard

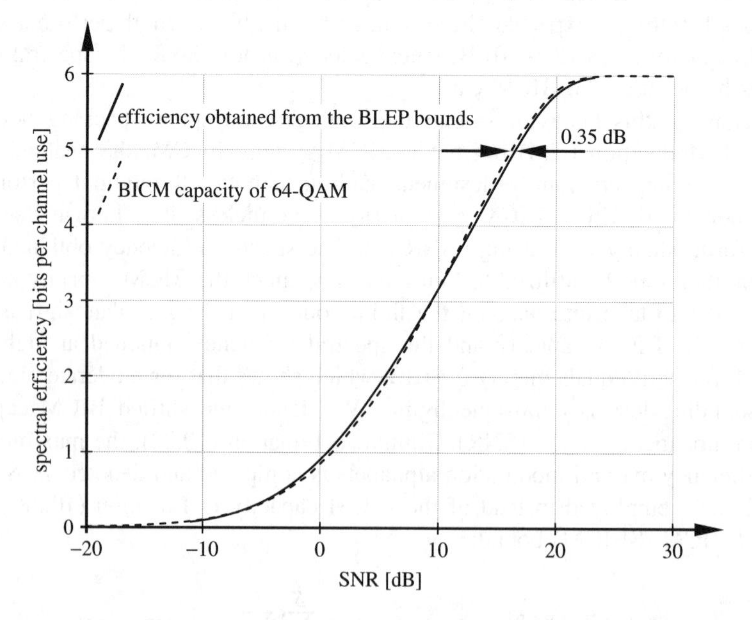

Figure 10.4 Spectral efficiency obtained via the Gallager and sphere packing bounds compared with BICM capacity of a system transmitting coded 64-QAM symbols over an AWGN channel.

considered. We refer to the bounds obtained in that way as "achievable" bounds; for example, achievable mutual information [23].

10.5.6 Prediction of the Optimal Performance

Even the tightest throughput bound, derived in the previous sections, still overestimates the performance of a practical communication system, as we show by means of simulations in Section 10.6. This is mainly caused by the suboptimality of the channel code employed, which cannot be taken into account theoretically. In Section 10.6, we obtain the optimal performance of an LTE system by means of exhaustive search simulations. This means that simulations with all possible MCS, precoder, and rank combinations are performed for a given channel realization and the best of these combinations (in terms of throughput) gives the optimal performance. Such simulations are very time consuming, especially for systems with large antenna configurations.

In this section, therefore, we present a method to predict the optimal performance of a practical system that proved to be accurate for LTE (see Section 10.6). The method is based on the shifted BICM efficiencies $f_{\mathrm{S}}^{(a)}(\mathrm{SNR})$. In the first step, an average efficiency over all subcarriers is computed for each possible combination of precoders, transmission ranks, and modulation alphabets $a \in \mathcal{A}$. Assuming wideband precoding, this average efficiency is computed according to

$$E(\mathbf{W}, a, L) = \frac{1}{KL} \sum_{k=1}^{K} \sum_{l=1}^{L} f_{\mathrm{S}}^{(a)}(\mathrm{SINR}_{k,l}(\mathbf{W})) \tag{10.47}$$

The SINR is computed according to Equation (10.10). Similar to ESM, the average efficiency $E(\mathbf{W}, a, L)$ is employed to obtain an effective SNR value of an AWGN channel that achieves the same spectral efficiency. This SNR value is computed via the inverse function of $f_{\mathrm{S}}^{(a)}(\mathrm{SNR})$:

$$\mathrm{SNR}_{\mathrm{eff}}(\mathbf{W}, a, L) = f_{\mathrm{S}}^{(a)^{-1}}(E(\mathbf{W}, a, L)) \tag{10.48}$$

Afterwards, a modulation alphabet a-dependent look-up table, denoted $e_a(\mathrm{SNR})$, is employed to estimate the spectral efficiency of the practical communication system. This look-up table is obtained from link-level simulations of an AWGN channel, as the maximum spectral efficiency over all MCSs that utilize modulation alphabet a. Figure 10.5 shows an example of such a mapping for an LTE system. By multiplying this spectral efficiency with the number of subcarriers and layers, an estimate of the throughput of each rank, precoder, and alphabet combination is computed:

$$\hat{I}(\mathbf{W}, a, L) = KLe_a(\mathrm{SNR}_{\mathrm{eff}}(\mathbf{W}, a, L)) \tag{10.49}$$

Maximization of this value with respect to the transmission rank, precoder, and alphabet then yields the predicted optimal performance of the practical communication system:

$$\hat{I}_{\mathrm{opt}} = \max_{a \in \mathcal{A}, L \leq L_{\max}, \mathbf{W} \in \mathcal{W}_L} \hat{I}(\mathbf{W}, a, L) \tag{10.50}$$

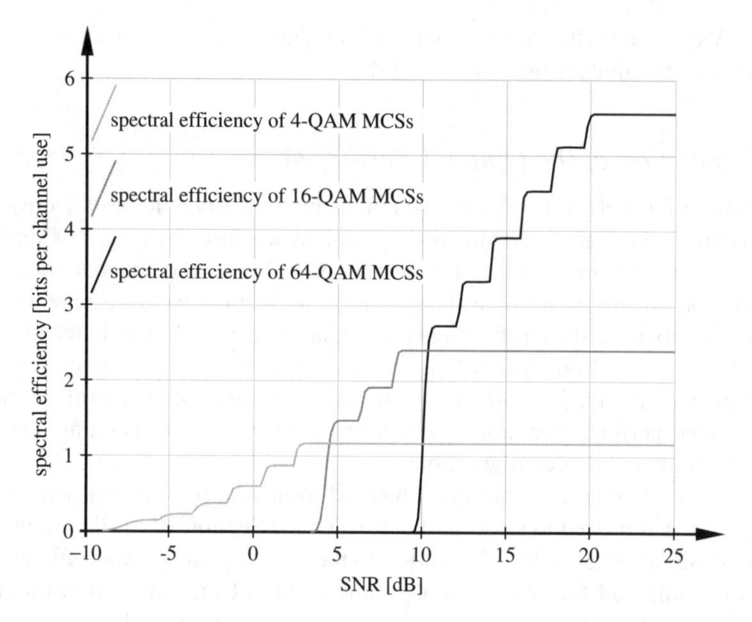

Figure 10.5 Spectral efficiency obtained via link-level simulations of an SISO LTE system over an AWGN channel.

10.6 Simulation Results

This section presents simulation results for an LTE system obtained with the "Vienna LTE Link Level Simulator" (described in Chapter 11). Single-user simulations are considered and results for three different antenna configurations, in combination with different transmission modes, are presented. The user is equipped with a simple ZF equalizer (see Section 10.2) and employs the feedback calculation methods described in Section 10.4. Furthermore, ideal channel knowledge (perfect channel estimation) and perfect timing and frequency synchronization of the transmitter and receiver are assumed.

The throughput results, obtained by simulations, are compared to the throughput bounds derived in Section 10.5. The throughput loss incurred by each of the practical system design constraints, taken into account by the bounds, is analyzed. This investigation can help researchers and system designers deciding in which directions to increase effort in order to improve the system performance.

Important simulation parameters for the three settings presented are summarized in Table 10.1. The 3rd Generation Partnership Project (3GPP) VehA power-delay-profile-based channel model is utilized to generate channel realizations. Furthermore, we employ wideband-specific precoding. The performance of the described feedback method is compared with the optimal system performance, computed by exhaustive search simulations (see Section 10.5.6). All simulation results are averaged over 2 000 independent channel and noise realizations, to obtain statistically accurate results, as can be verified by means of the 95 % confidence intervals shown inside the markers of the figures. The UE

Table 10.1 Simulation parameters for the three simulator setups considered

	SISO	OLSM	CLSM
Channel	VehA [17]	VehA	VehA
Bandwidth (MHz)	1.4	1.4	1.4
Subcarriers	72	72	72
Receiver	ZF	ZF	ZF
Antennas $N_R \times N_T$	1×1	2×2	4×4
CQI feedback	✓	✓	✓
RI feedback	×	✓	✓
PMI feedback	×	×	✓

feedback is provided before transmission, in order to avoid influences of outdated feedback. Furthermore, the feedback indicators are provided with a temporal granularity of one TTI (1 ms) and a frequency granularity of one value per system bandwidth (1.4 MHz).

10.6.1 SISO Transmission

In an SISO system, the UE only has to provide CQI feedback to the eNodeB, for adapting the MCS to the current channel quality. Precoding is not utilized for SISO transmission,

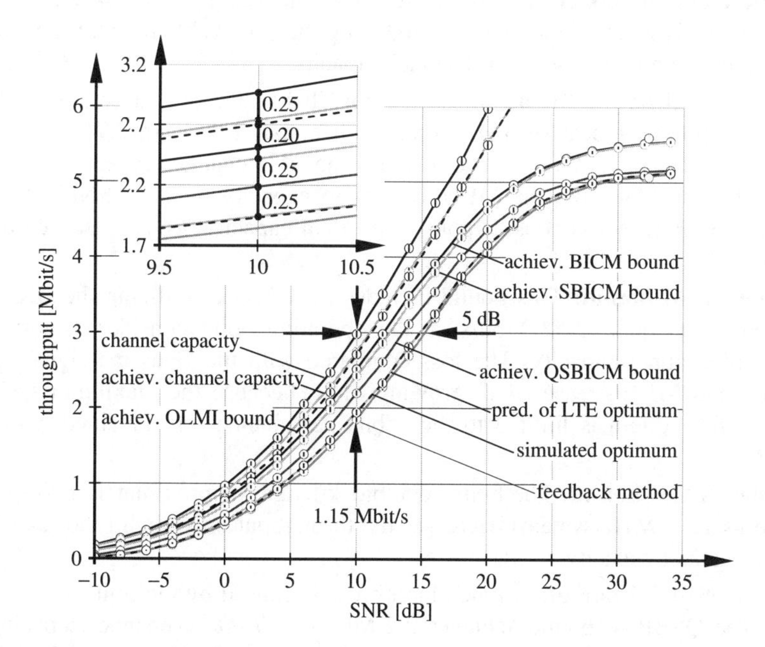

Figure 10.6 Comparison of throughput bounds and simulated system performance for a single-user SISO LTE system.

as it is noneffective. Additionally, the transmission rank is confined to $L = 1$, meaning that PMI and RI feedback are not employed.

Figure 10.6 shows the simulation results obtained for the SISO LTE system. The leftmost curve, achieving the largest throughput, corresponds to channel capacity, while the rightmost curve shows the simulated performance of LTE, employing the feedback methods derived in this chapter. The inset in the figure shows a detail of the results obtained around a typical operating point of 10 dB SNR. At this operating point we observe the following:

- The largest possible theoretical throughput, given by the channel capacity, equals 3 Mbit/s.
- The simulated LTE system, utilizing the presented feedback methods, achieves a throughput of 1.85 Mbit/s, corresponding to 62 % of channel capacity.
- The simulated optimal performance of the LTE system equals 1.95 Mbit/s or 65 % of channel capacity. The feedback method achieves 95 % of this optimal value.
- The predicted optimal performance, obtained by means of the method described in Section 10.5.6, overlaps the simulated optimal performance.

Next, we employ the performance bounds, presented in Section 10.5, to analyze the reasons for the large throughput gap between channel capacity and the simulated system, at the considered operating point of SNR = 10 dB:

Achievable channel capacity: This bound takes into account the system overhead for reference and synchronization signals, reducing the achievable user throughput to 2.75 Mbit/s, corresponding to 92 % of channel capacity.

Achievable OLMI **bound for an** SISO **system:** This bound only accounts for the effect of uniform power allocation and, therefore, it equals MI. With 2.71 Mbit/s or 90 % of channel capacity it is very close to the achievable channel capacity, which can be explained by the fact that the MI converges to capacity for high SNR. On the other hand, at low SNR, MI is considerably worse than capacity, as can also be observed in the figure.

Achievable BICM **bound:** This bound includes the effect of utilizing the BICM system architecture. It equals 2.52 Mbit/s, at the operating point considered, corresponding to 84 % of channel capacity. The loss compared with the Gaussian signaling bounds (capacity and MI) increases with growing SNR, because the maximum spectral efficiency of the system is limited to the 6 bits per channel use obtained with uncoded 64-QAM.

Achievable SBICM **bound:** The finite code block length (approximately 800 symbols for a single user 1.4 MHz system) decreases the throughput only slightly to 2.43 Mbit/s or 81 % of channel capacity.

Achievable QSBICM **bound:** A much larger loss is caused by the finite set of supported MCSs. The QSBICM bound achieves 2.2 Mbit/s or 73 % of channel capacity. This is the tightest throughput bound, presented in Section 10.5. The gap of 0.25 Mbit/s to the optimal LTE performance can be attributed to the performance of the channel code utilized.

As the inset in Figure 10.6 shows, none of the practical constraints dominates the performance difference between channel capacity and the simulated throughput. The throughput gap of 1.15 Mbit/s divides into almost equally sized parts, caused by the system overhead, the BICM architecture, the finite set of supported MCSs, and the performance of the channel code. The other practical constraints (equal power allocation, finite code block length) cause only minor performance degradations.

10.6.2 OLSM Transmission

An OLSM system with two transmit and two receive antennas supports up to two spatial transmission layers. In order to enable dynamic rank adaptation, we employ RI feedback additionally to CQI feedback. Still, the PMI is not utilized because OLSM employs CDD precoding (see Section 10.4 and Chapter 2 for details).

Figure 10.7 shows the simulation results and corresponding throughput bounds for that system setup. We analyze the system performance at an operating point of SNR = 10.4 dB, at which point the channel capacity compared with the SISO system at 10 dB is doubled to 6 Mbit/s. The inset in the figure shows a detail of the results around this operating point. Following observations can be made at the chosen operating point:

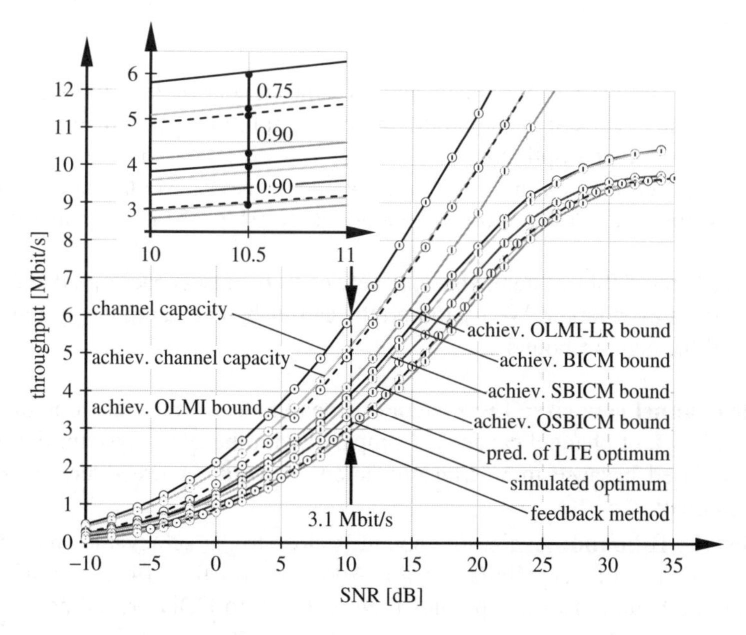

Figure 10.7 Comparison of throughput bounds and simulated system performance for a single-user $N_R \times N_T = 2 \times 2$ OLSM LTE system.

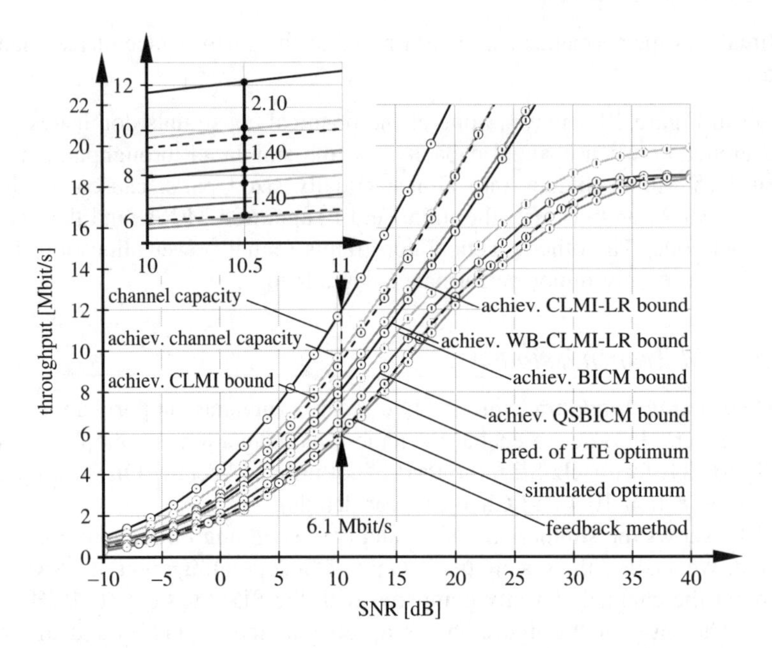

Figure 10.8 Comparison of throughput bounds and simulated system performance for a single-user $N_R \times N_T = 4 \times 4$ CLSM LTE system.

- The simulated optimal performance achieves a throughput of 3.1 Mbit/s corresponding to 52 % of channel capacity. The prediction of the optimum performance overlaps very well with the simulated optimum.
- With the feedback method presented, a throughput of 2.91 Mbit/s (48.5 % of channel capacity) is obtained. This performance amounts to 94 % of the optimal value.

Compared with the SISO system, the gap between channel capacity and the simulated throughput grows even larger. We try to find the causes by analyzing the system by means of the derived throughput bounds:

Achievable channel capacity: The system overhead reduces the user throughput to 5.25 Mbit/s or 87.5 % of channel capacity. Compared with the SISO system, the percentage loss is increased because more transmit antennas require more reference signals for channel estimation.

Achievable OLMI bound: Again, this bound is close to the achievable channel capacity with a throughput of 5.1 Mbit/s (85 %). Additional to the effect of uniform power allocation, the bound also incorporates the restriction to CDD precoding.

Achievable OLMI-LR bound: This bound considers the impact of employing a linear ZF equalizer at the receiver. It obtains a throughput value of merely 4.25 Mbit/s, amounting to 70 % of channel capacity. A large portion (15 %) of the total performance gap of approximately 50 %, between the channel capacity and the simulated performance, therefore, can be attributed to the simple equalizer. In the case of SISO transmission, we did not consider this bound because there it is equal to the OLMI bound, as the ZF receiver obtains ML performance.

Achievable BICM bound: With the BICM transmitter architecture the throughput decreases further to 4 Mbit/s (66.7 %).

Achievable SBICM bound: The finite code block length reduces the throughput by 0.2 Mbit/s to 3.8 Mbit/s (63.3 %), corresponding to a loss of 3.4 % of channel capacity.

Achievable QSBICM **bound:** A larger loss of 5.8 % is caused by restricting the MCSs to the 15 standard defined combinations. The achievable QSBICM bound equals 3.45 Mbit/s (57.5 %).

As this analysis shows, the proposed throughput bounds are able to explain a large part of the performance difference between the simulated system and the channel capacity. From the inset of Figure 10.7 we observe two dominating causes for this throughput loss. One is the system overhead (reference and synchronization signals) and the other one is the linear ZF equalizer. While the former loss cannot be avoided, the latter can, by employing a more sophisticated receiver (MMSE-SIC, SSD, ML detector). Correcting the simulated performance for the loss caused by the ZF receiver (the 15 % between the OLMI and the OLMI-LR bound), a throughput of 4 Mbit/s (67 %) could be achieved with an ML detector, which is very similar to the SISO system (65 %).

Although, at the operating point of 10.4 dB considered, the 2×2 system doubles the channel capacity compared with the SISO system, this is not true for the simulated throughput of the LTE system, which increases merely by a factor of 1.57. This is partly caused by the increased system overhead and confined precoding, but mainly by the ZF equalizer. This can be recognized from the OLMI bound, which assumes ML detection and increases by a factor of 1.88, compared with the SISO system.

10.6.3 CLSM Transmission

As a final example, we analyze the largest antenna configuration defined in the LTE specification, of four transmit and four receive antennas. Additionally, we employ CLSM transmission, which allows one to adapt not only the MCS and transmission rank employed, but also the precoding. For that purpose, it requires CQI, RI, and PMI feedback.

In Figure 10.8, the throughput performance obtained for that system by means of link-level simulations is compared with the throughput bounds of Section 10.5. We consider an operating point of 10.4 dB SNR, in order to achieve a twofold increase of the channel capacity to 12 Mbit/s, compared with the 2×2 OLSM system.

- The simulated optimal performance of the 4×4 CLSM LTE system equals 6 Mbit/s, corresponding to 50 % of channel capacity. Again, the predicted optimal performance is very close to the simulated one.
- Employing the realistic feedback method from Section 10.4, a throughput of 5.9 Mbit/s is obtained (49 % of channel capacity). The loss compared with the optimal performance is only 2 %. Note that, although the absolute gap between the optimal performance and the throughput obtained with the feedback method increases with SNR, the percentage

loss stays almost constant. At SNR = 28 dB the loss amounts to 2.5 %. In comparison, a 4 × 4 OLSM system, utilizing the feedback method, only obtains a throughput of 4.5 Mbit/s, or 37.5 % of channel capacity, at the operating point considered (not shown in the figure).

Analyzing the reasons for the throughput gap between channel capacity and the simulated LTE system, we observe:

Achievable channel capacity: As the 4 × 4 system requires even more pilots for channel estimation, the achievable channel capacity amounts to 10 Mbit/s, or 83 % of channel capacity. This means that 17 % of the user throughput is lost for system overhead, not including any signaling channels that further decrease the user throughput.

Achievable CLMI bound: For the closed-loop system we have to consider the CLMI bound, which optimizes the precoder choice for each subcarrier. This bound equals 9.5 Mbit/s, corresponding to 79 % of channel capacity.

Achievable CLMI-LR bound: Compared with the OLSM system, the loss induced by utilization of a ZF equalizer is not that severe. The bound equals 8.7 Mbit/s (72.5 %), which amounts to a loss of 6.5 % compared with the achievable CLMI bound, in contrast to the 15 % of the OLSM system. This means that the adaptive precoding already reduces the inter-layer interference at the transmitter, reducing the noise enhancement caused by the receive filter.

Achievable WB-CLMI-LR bound: The simulated system utilizes wideband precoding. The effect of that is considered with the WB-CLMI-LR bound. It leads to a throughput reduction to 8.2 Mbit/s (68 %). All following bounds also assume wideband precoding.

Achievable BICM bound: The BICM architecture reduces the possible throughput to 7.6 Mbit/s, or 63 % of channel capacity. The achievable SBICM bound is not shown in the figure, to limit the number of curves. The loss, induced by the finite code block length, is on the order of 3 %, which is similar to the other two simulated systems.

Achievable QSBICM bound: The value of this bound equals 6.75 Mbit/s, corresponding to 56 % of channel capacity.

The simulated 4 × 4 CLSM system is capable of doubling the throughput of the 2 × 2 OLSM system, although the system overhead grows. This comes at the price of an increased feedback overhead for the PMI of 4 bit/TTI. Compared with a 4 × 4 OLSM system, a throughput gain of 1/3 is achieved at the SNR = 10.4 dB considered, assuming a feedback delay of zero. Therefore, CLSM is preferable to OLSM in situations where it can be guaranteed that the feedback is not outdated (low-mobility scenarios, small feedback delay, large channel coherence time).

References

[1] 3GPP (2010) Technical Specification TS 36.213 ‘Evolved universal terrestrial radio access (E-UTRA); physical layer procedures (release 10),’ www.3gpp.org.

[2] 3GPP (2009) Technical Specification TS 36.211 Version 8.7.0 'Evolved universal terrestrial radio access (E-UTRA); physical channels and modulation; (release 8),' http://www.3gpp.org/ftp/Specs/html-info/36211.htm
[3] Alamouti, S. (1998) 'A simple transmit diversity technique for wireless communications,' *IEEE Journal on Selected Areas in Communications*, **16** (8), 1451–1458, doi: 10.1109/49.730453.
[4] Babich, F. (2004) 'On the performance of efficient coding techniques over fading channels,' *IEEE Transactions on Wireless Communications*, **3** (1), 290–299, doi: 10.1109/TWC.2003.821138.
[5] Barbero, L. and Thompson, J. (2006) 'A fixed-complexity MIMO detector based on the complex sphere decoder,' in *Proceedings of the IEEE 7th Workshop on Signal Processing Advances in Wireless Communications (SPAWC 2006)*, doi: 10.1109/SPAWC.2006.346388. Available from http://ieeexplore.ieee.org/stamp/stamp.jsp?tp = &arnumber = 4153876.
[6] Böhnke, R. and Kammeyer, K. (2006) 'SINR analysis for V-BLAST with ordered MMSE-SIC detection,' in *Proceedings of the 2006 International Conference on Wireless Communications and Mobile Computing IWCMC06*, New York.
[7] Caire, G., Taricco, G., and Biglieri, E. (1996) 'Capacity of bit-interleaved channels,' *Electronics Letters*, **32** (12), 1060–1061.
[8] Caire, G., Taricco, G., and Biglieri, E. (1998) 'Bit-interleaved coded modulation,' *IEEE Transactions on Information Theory*, **44** (3), 927–946.
[9] Cipriano, A. M., Visoz, R., and Sälzer, T. (2008) 'Calibration issues of PHY layer abstractions for wireless broadband systems,' in *Proceedings of the IEEE Vehicular Technology Conference (Fall VTC2008)*, Calgary, Canada, pp. 1–5.
[10] Coleri, S., Ergen, M., Puri, A., and Bahai, A. (2002) 'Channel estimation techniques based on pilot arrangement in OFDM systems,' *IEEE Transactions on Broadcasting*, **48** (3), 223–229, doi: 10.1109/TBC.2002.804034.
[11] Cover, T. and Thomas, J. (1991) *Elements of Information Theory*, John Wiley & Sons.
[12] Dahlman, E., Parvall, S., Skold, J., and Beming, P. (2007) *3G Evolution: HSPA and LTE for Mobile Broadband*, Academic Press.
[13] Fertl, P. and Matz, G. (2006) 'Efficient OFDM channel estimation in mobile environments based on irregular sampling,' in *Fortieth Asilomar Conference on Signals, Systems and Computers (ACSSC'06)*, pp. 1777–1781, doi: 10.1109/ACSSC.2006.355067.
[14] Hassibi, B. and Vikalo, H. (2005) 'On the sphere-decoding algorithm I. Expected complexity,' *IEEE Transactions on Signal Processing*, **53** (8), 2806–2818.
[15] Hochwald, B. M. and Brink, S. T. (2003) 'Achieving near-capacity on a multiple-antenna channel,' *IEEE Transactions on Communications*, **51** (3), 389–399. Available from http://ieeexplore.ieee.org/iel5/26/26869/01194444.pdf?tp = &arnumber = 1194444&isnumber = 26869.
[16] Hoeher, P., Kaiser, S., and Robertson, P. (1997) 'Pilot-symbol-aided channel estimation in time and frequency,' in *Proceedings of the IEEE Global Telecommunications Conference (GLOBECOM'97), Communication Theory Mini-Conference*, pp. 90–96.
[17] ITU (1997) 'Recommendation ITU-R M.1225: Guidelines for evaluation of radio transmission technologies for IMT-2000,' Technical report, ITU.
[18] Li, Y. (2000) 'Pilot-symbol-aided channel estimation for OFDM in wireless systems,' *IEEE Transactions on Vehicular Technology*, **49** (4), 1207–1215, doi: 10.1109/25.875230.
[19] Lin, S. and Costello, D. (2004) *Error Control Coding: Fundamentals and Applications*, Pearson–Prentice Hall.
[20] Mostofi, Y. and Cox, D. (2005) 'ICI mitigation for pilot-aided OFDM mobile systems,' *IEEE Transactions on Wireless Communications*, **4** (2), 765–774, doi: 10.1109/TWC.2004.840235.
[21] Ni, J.-H. and Liu, Z.-M. (2009) 'A joint ICI estimation and mitigation scheme for OFDM systems over fast fading channels,' in *Proceedings of the Global Mobile Congress*, pp. 1–6.
[22] Paulraj, A., Nabar, R., and Gore, D. (2003) *Introduction to Space–Time Wireless Communications*, first edition, Cambridge University Press, Cambridge, UK.
[23] Rupp, M., Mehlführer, C., and Caban, S. (2010) 'On achieving the Shannon bound in cellular systems,' in *Proceedings of the 20th International Conference Radioelektronika*, Brno, Czech Republic, doi: 10.1109/RADIOELEK.2010.5478551. Available from http://publik.tuwien.ac.at/files/PubDat_185403.pdf.

[24] Sandanalakshmi, R., Palanivelu, T., and Manivannan, K. (2007) 'Effective SNR mapping for link error prediction in OFDM based systems,' in *Proceedings of the IET-UK International Conference on Information and Communication Technology in Electrical Sciences ICTES*.

[25] Schlegel, C. and Perez, L. (1999) 'On error bounds on turbo-codes,' *IEEE Communication Letters*, **3** (7), 205–207.

[26] Schwarz, S., Mehlführer, C., and Rupp, M. (2010) 'Calculation of the spatial preprocessing and link adaption feedback for 3GPP UMTS/LTE,' in *Proceedings of the IEEE 6th Conference on Wireless Advanced (WiAD)*, London, UK. Available from http://publik.tuwien.ac.at/files/PubDat_186497.pdf.

[27] Schwarz, S., Mehlführer, C., and Rupp, M. (2010) 'Low complexity approximate maximum throughput scheduling for LTE,' in *44th Annual Asilomar Conference on Signals, Systems, and Computers*, Pacific Grove, CA. Available from http://publik.tuwien.ac.at/files/PubDat_187402.pdf.

[28] Schwarz, S., Mehlführer, C., and Rupp, M. (2011) 'Throughput maximizing multiuser scheduling with adjustable fairness,' in *Proceedings of the International Conference on Communications ICC 2011*, Kyoto, Japan.

[29] Schwarz, S. and Rupp, M. (2011) 'Throughput maximizing feedback for MIMO OFDM based wireless communication systems,' in *Proceedings of the Signal Processing Advances in Wireless Communications (SPAWC'11)*, San Francisco, CA.

[30] Schwarz, S., Simko, M., and Rupp, M. (2011) 'On performance bounds for MIMO OFDM based wireless communication Systems,' in *Proceedings of the Signal Processing Advances in Wireless Communications (SPAWC'11)*, San Francisco, CA.

[31] Schwarz, S., Wrulich, M., and Rupp, M. (2010) 'Mutual information based calculation of the precoding matrix indicator for 3GPP UMTS/LTE,' in *Proceedings of the IEEE Workshop on Smart Antennas*, Bremen, Germany. Available from http://publik.tuwien.ac.at/files/PubDat_184424.pdf.

[32] Sesia, S., Baker, M., and Toufik, I. (2009) *LTE, the UMTS Long Term Evolution: From Theory to Practice*, John Wiley & Sons.

[33] Shannon, C. E. (1949) 'Communication in the presence of noise,' *Proceedings of the IRE*, **37**, 10–21.

[34] Šimko, M., Mehlführer, C., Wrulich, M., and Rupp, M. (2010) 'Doubly dispersive channel estimation with scalable complexity,' in *Proceedings of the International ITG Workshop on Smart Antennas (WSA 2010)*, Bremen, Germany, pp. 251–256, doi: 10.1109/WSA.2010.5456443. Available from http://publik.tuwien.ac.at/files/PubDat_181229.pdf.

[35] Šimko, M., Mehlführer, C., Zemen, T., and Rupp, M. (2011) 'Inter carrier interference estimation in MIMO OFDM systems with arbitrary pilot structure,' in *Proceedings of the 73rd IEEE Vehicular Technology Conference (VTC2011-Spring)*, Budapest, Hungary.

[36] Song, W.-G. and Lim, J.-T. (2003) 'Pilot-symbol aided channel estimation for OFDM with fast fading channels,' *IEEE Transactions on Broadcasting*, **49** (4), 398–402, doi: 10.1109/TBC.2003.819049.

[37] Studer, C., Wenk, M., Burg, A. P., and Bölcskei, H. (2006) 'Soft-output sphere decoding: performance and implementation aspects,' in *Conference Record of the 40th Asilomar Conference on Signals, Systems and Computers*, Pacific Grove, CA. Available from http://ieeexplore.ieee.org/iel5/4176490/4176491/04176942.pdf?tp = &arnumber = 4176942.

[38] Telatar, I. E. (1999) 'Capacity of multi-antenna Gaussian channels,' *European Transactions on Telecommunications*, **10** (6), 585–596.

[39] Tosato, F. and Bisaglia, P. (2002) 'Simplified soft-output demapper for binary interleaved COFDM with application to HIPERLAN/2,' in *IEEE International Conference on Communications*, volume 2, pp. 664–668, doi: 10.1109/ICC.2002.996940.

[40] Tse, D. and Viswanath, P. (2008) *Fundamentals of Wireless Communications*, Cambridge University Press.

[41] Van de Beek, J.-J., Edfors, O., Sandell, M. *et al.* (1995) 'On channel estimation in OFDM systems,' in *IEEE 45th Vehicular Technology Conference*, volume 2, pp. 815–819, doi: 10.1109/VETEC.1995.504981.

[42] Wan, L., Tsai, S., and Almgren, M. (2006) 'A fading-insensitve performance metric for a unified link quality model,' in *Proceedings of the IEEE Wireless Communications & Networking Conference WCNC*.

[43] Wang, Q., Mehlführer, C., and Rupp, M. (2010) 'Carrier frequency synchronization in the downlink of 3GPP LTE,' in *Proceedings of the 21st Annual IEEE International Symposium on Personal, Indoor*

and Mobile Radio Communications (PIMRC 2010), Istanbul, Turkey. Available from http://publik.tuwien.ac.at/files/PubDat_187636.pdf.

[44] Wrulich, M., Mehlführer, C., and Rupp, M. (2008) 'Interference aware MMSE equalization for MIMO TxAA,' in *Proceedings of the 3rd International Symposium on Communications, Control and Signal Processing*, St. Julians, Malta, pp. 1585–1589, doi: 10.1109/ISCCSP.2008.4537480. Available from http://publik.tuwien.ac.at/files/pub-et_13657.pdf.

Part Four

Simulators for Wireless Systems

Introduction

Useful modeling in the context of wireless networks is in general a difficult task, which requires careful investigations of many different parameter settings and a design that tries to avoid leading to potentially wrong conclusions. "System-level modeling" is particularly demanding in this sense, because, in addition to the design requirements, the computational complexity plays a crucial role.

So what is system-level modeling all about? One thing that emerges when investigating the term "system level" is that it is quite extensively[1] used in the literature. As a matter of fact, every comprehensive investigation of a technical system – not only in the context of a cellular system – may be denoted system-level analysis. Examples are electronic design methodology, virtualization for operating systems, hardware – software co-design for Very High Speed Integrated Circuits (VHSIC), or network synthesis. In the scope of this book, however, the term "system-level model" describes:

> A model capable of representing the physical layer of a wireless transmission system in an abstract, yet accurate, way that is computationally less complex to evaluate than computing all the algorithms involved in the physical layer processing in their full detail, and can be described by a low number of parameters.

Given this rather conceptional definition, what are system-level models needed for? Usually, physical-layer simulations are used to identify and evaluate promising transmission techniques. Whereas these investigations are suitable for the deployment of receiver algorithms, feedback strategies, coding design, and so on, they are not directly capable of reflecting network issues such as cell planning, scheduling, and interference situations in the context of massive multi-user operation [23]. Therefore, to understand the network and user performance under typical operating conditions in various deployment scenarios, network simulations are crucial [4, 7, 14]. Such wireless networks define the "system" to which the term "system level" is referring to. Accordingly, "system-level simulations" are built to comprise full networks, thus trying to cover at least

1. network deployment issues, including network performance [11, 12, 24];
2. multi-user and multi-NodeB (inter-cell) interference [3, 6, 16];
3. Radio Link Control (RLC) and admission control algorithms [13, 15]; and
4. scheduling [18, 19].

One of the major difficulties of system-level analyses, however, is the computational complexity involved in evaluating the performance of the radio links between all base

[1] For example, a search for the term "system level" on Google returned approximately 3 160 000 hits.

stations and mobile terminals. Performing such a large number of individual physical-layer – or also often called "link-level" – simulations is clearly prohibitive in terms of the computational complexity. Thus, a system-level model has to simplify the physical link in a sufficiently accurate way to capture the essential behavior [9, 10, 17, 20].

To summarize, an ideal system-level model has to be

1. sufficiently accurate;
2. flexible in terms of the scenarios it is capable of representing, as well as applicable for analytic investigations;
3. able to be evaluated with very low computational complexity; and
4. able to be represented by a reasonable number of parameters.

While link-level performance can be evaluated experimentally by our testbed method, system-level evaluations require simulators, as such experiments would hardly be possible, requiring several base stations and a multitude of users. Part IV thus takes a different approach: the simulative evaluation of cellular systems.

In Chapter 11 we present link- and system-level simulator approaches and show how both simulation worlds can cooperate in order to validate results, as otherwise the system-level simulators may only be based on assumptions and their predicted performance may drift far away from the final system performance. The "Vienna Long-Term Evolution (LTE) Link Level Simulator" supports three modes: (1) single-downlink, (2) single-cell multi-user, and (3) multi-cell multi-user scenarios. The link-level simulator, furthermore, serves as the basis for the development end verification of physical-layer models that can be employed in the "Vienna LTE System Level Simulator." Both the Vienna LTE simulators presented are freely available for research purposes under an educational license from http://www.nt.tuwien.ac.at/ltesimulator/. Prior to the book being released, several thousand downloads had taken place with many valuable feedback of the first users.[2]

Reproducibility is one of the pillars of scientific research. Whereas reproducibility has a long tradition in theoretical sciences, such as mathematics and most nature sciences, it is only recently that reproducible research has become more and more important in the field of signal processing [1, 21]. In contrast to results in fields of purely theoretical sciences, results of signal processing research papers can only be reproduced if a comprehensive description of the algorithms investigated (including the setting of all necessary parameters), as well as eventually required input data, are fully available. Owing to lack of space, a fully comprehensive description of the algorithm is often omitted in research papers. Even if an algorithm is explained in detail, for instance by a pseudo code, initialization values are often not fully defined. Also, sometimes it is simply not possible to include in a paper all necessary resources, such as data that was processed by the algorithms presented. Ideally, all resources, including source code of the algorithms presented, should be made available for download to enable other researchers (and also reviewers of papers) to reproduce the results presented.

In recent years several researchers have started to build up online resource databases in which simulation code and data are provided; for example, see [2, 22]. However, it is still not a common practice in signal processing research, a circumstance that has

[2] By April 2011 we have had more than 7000 downloads for each of the Vienna LTE simulators.

recently been complained about quite openly in [5]. We are furthermore convinced that reproducibility should also play an important role in the review process of a paper. Reproducibility becomes even more important when the simulated systems become more and more complex, as is the case, for example, in the evaluation of wireless communication systems. When algorithms for wireless systems are evaluated, authors often claim to employ a standard-compliant transmission system and simply reference to the corresponding technical specification. Since technical specifications are usually extensive, including a cornucopia of options, it is not always clear which parts of a specification were actually implemented and which parts were omitted for simplicity reasons. Without knowing such details, comparisons of algorithms developed by different researchers are very difficult, if not impossible, to carry out. A way out of this dilemma is to refer to a publicly available simulation environment. In this work we present such an open-source simulation environment that allows link- and system-level simulations of the Universal Mobile Telecommunications System (UMTS) LTE. The development and publishing of this LTE simulation environment is based on our previous, very good experience with a WiMAX physical-layer simulator [8].[3]

In Chapter 12 a link-level simulation of High-Speed Downlink Packet Access (HSDPA) is presented. First, a computationally efficient link-quality model is developed, which analytically describes the Multiple-Input Multiple-Output (MIMO) HSDPA link quality in the network context. The proposed model is very flexible in describing various transmission setups, including even higher order spatial multiplexing schemes than the standardized Double Transmit Antenna Array (D-TxAA) mode. Furthermore, the interference terms in the link-quality model are described by means of so-called equivalent fading parameters, which can be evaluated prior to the system-level simulation. This allows a dramatic reduction of the computational complexity during the run time of the system-level simulation. The link-quality model is then validated against link-level simulation results, showing an almost perfect approximation of the true Signal to Interference and Noise Ratio (SINR). By analytical analysis, a fundamental performance limit of the D-TxAA Spatial Division Multiple Access (SDMA) user separation capabilities is identified, when utilizing the standard 3rd Generation Partnership Project (3GPP) precoding codebook. The virtue of the proposed model is highlighted by elaborating on the modeling deficiencies of a pure statistical approach and the computational complexity reduction in a semi-analytical framework. Second, a concept for the link-performance model which describes the Block Error Ratio (BLER) performance of the system is presented. In principle, the model is composed of a transmission parameter adaptive mapping/compression of the link-quality model parameters and a successive Additive White Gaussian Noise (AWGN)-based BLER performance prediction. Consecutively, the training of the link-performance model, meaning the tuning of the parameters within the compression/mapping description utilized, is conducted. The necessary link-level simulations also serve to validate the link-performance model.[4]

[3] Also available for download at https://www.nt.tuwien.ac.at/downloads/featured-downloads.

[4] This chapter is based on the Ph.D. thesis of Martin Wrulich. The full thesis can be downloaded from EURASIP's open Ph.D. library at http://www.arehna.di.uoa.gr/thesis/ or directly from http://publik.tuwien.ac.at/files/PubDat_181165.pdf.

References

[1] Barni, M. and Perez-Gonzalez, F. (2005) 'Pushing science into signal processing,' *IEEE Signal Processing Magazine*, **22** (4), 120–119, doi: 10.1109/MSP.2005.1458324. Available from http://ieeexplore.ieee.org/stamp/stamp.jsp?tp=&arnumber=1458324.

[2] Buckheit, J. B. and Donoho, D. L. (1995) 'Wavelab and reproducible research,' Technical report, Department of Statistics, Stanford University, Technical Report 474. Available from http://www-stat.stanford.edu/~donoho/Reports/1995/wavelab.pdf.

[3] Castañeda, M., Ivrlac, M., Nossek, J. et al. (2007) 'On downlink intercell interference in a cellular system,' in *Proceedings of the IEEE 18th International Symposium on Personal, Indoor and Mobile Radio Communications (PIMRC)*, pp. 1–5, doi: 10.1109/PIMRC.2007.4394052.

[4] Czylwik, A. and Dekorsy, A. (2004) 'System-level performance of antenna arrays in CDMA-based cellular mobile radio systems,' *EURASIP Journal of Applied Signal Processing*, **2004** (9), 1308–1320.

[5] Dohler, M., Heath Jr., R., Lozano, A. et al. (2011) 'Is the PHY layer dead?' *IEEE Communications Magazine*, **49** (4), 159–165.

[6] Gkonis, P., Kaklamani, D., and Tsoulos, G. (2008) 'Capacity of WCDMA multicellular networks under different radio resource management strategies,' in *Proceedings of the 3rd International Symposium on Wireless Pervasive Computing (ISWPC)*, pp. 60–64, doi: 10.1109/ISWPC.2008.4556166.

[7] Haring, L., Chalise, B., and Czylwik, A. (2003) 'Dynamic system level simulations of downlink beamforming for UMTS FDD,' in *Proceedings of the IEEE Global Telecommunications Conference (GLOBECOM)*, volume 1, pp. 492–496, doi: 10.1109/GLOCOM.2003.1258286.

[8] Mehlführer, C., Caban, S., and Rupp, M. (2008) 'Experimental evaluation of adaptive modulation and coding in MIMO WiMAX with limited feedback,' *EURASIP Journal on Advances in Signal Processing*, **2008**, article ID 837102, doi: 10.1155/2008/837102. Available from http://publik.tuwien.ac.at/files/pub-et_13762.pdf.

[9] Moltchanov, D., Koucheryavy, Y., and Harju, J. (2005) 'Simple, accurate and computationally efficient wireless channel modeling algorithm,' in *Wired/Wireless Internet Communications* (eds. T. Braun, G. Carle, Y. Koucheryavy, and V. Tsaoussidis), volume 3510 of *Lecture Notes in Computer Science*, Springer, Berlin, pp. 234–245.

[10] Moltchanov, D., Koucheryavy, Y., and Harju, J. (2006) 'Cross-layer modeling of wireless channels for data-link and IP layer performance evaluation,' *Computer Communications*, **29** (7), 827–841.

[11] Nihtila, T. and Haikola, V. (2008) 'HSDPA MIMO system performance in macro cell network,' in *Proceedings of the IEEE Sarnoff Symposium*, doi: 10.1109/SARNOF.2008.4520092.

[12] Pedersen, K. I., Lootsma, T. F., Støttrup, M. et al. (2004) 'Network performance of mixed traffic on high speed downlink packet access and dedicated channels in WCDMA,' in *Proceedings of the IEEE 60th Vehicular Technology Conference (VTC)*, volume 6, pp. 4496–4500, doi: 10.1109/VETECF.2004.1404930.

[13] Peppas, K., Al-Gizawi, T., Lazarakis, F. et al. (2004) 'System level evaluation of reconfigurable MIMO techniques enhancements for HSDPA,' in *Proceedings of the IEEE Global Telecommunications Conference (GLOBECOM)*, volume 5, pp. 2869–2873.

[14] Peppas, K., Alexiou, A., Lazarakis, F., and Al-Gizawi, T. (2005) 'Performance evaluation at the system level of reconfigurable space–time coding techniques for HSDPA,' *EURASIP Journal of Applied Signal Processing*, **2005** (11), 1656–1667.

[15] Pollard, A. and Heikkila, M. (2004) 'A system level evaluation of multiple antenna schemes for high speed downlink packet access,' in *Proceedings of the IEEE 15th International Symposium on Personal, Indoor and Mobile Radio Communications (PIMRC)*, volume 3, pp. 1732–1735, doi: 10.1109/PIMRC.2004.1368296.

[16] Sai A, P. and Furse, C. (2008) 'System level analysis of noise and interference analysis for a MIMO system,' in *Proceedings of the IEEE Antennas and Propagation Society International Symposium (AP-S)*, pp. 1–4, doi: 10.1109/APS.2008.4619499.

[17] Seeger, A., Sikora, M., and Klein, A. (2003) 'Variable orthogonality factor: a simple interface between link and system level simulation for high speed downlink packet access,' in *Proceedings of the IEEE 58th Vehicular Technology Conference Fall (VTC)*, pp. 2531–2534.

[18] Shuping, C., Huibinu, L., Dong, Z., and Asimakis, K. (2007) 'Generalized scheduler providing multimedia services over HSDPA,' in *Proceedings of the IEEE International Conference on Multimedia and Expo*, pp. 927–930, doi: 10.1109/ICME.2007.4284803.

[19] Skoutas, D., Komnakos, D., Vouyioukas, D., and Rouskas, A. (2008) 'Enhanced dedicated channel scheduling optimization in WCDMA,' in *Proceedings of the 14th European Wireless Conference*.

[20] Staehle, D. and Mader, A. (2007) 'A model for time-efficient HSDPA simulations,' in *Proceedings of the IEEE 66th Vehicular Technology Conference Fall (VTC)*, pp. 819–823.

[21] Vandewalle, P., Kovačević, J., and Vetterli, M. (2009) 'Reproducible research in signal processing,' *IEEE Signal Processing Magazine*, **26** (3), 37–47. Available from http://ieeexplore.ieee.org/stamp/stamp.jsp?tp=&arnumber=4815541.

[22] Vandewalle, P., Süsstrunk, S., and Vetterli, M. (2006) 'A frequency domain approach to registration of aliased images with application to super-resolution,' *EURASIP Journal on Applied Signal Processing*, **2006**, article ID 71459, doi: 10.1155/ASP/2006/71459. Available from http://downloads.hindawi.com/journals/asp/2006/071459.pdf.

[23] Wrulich, M. and Rupp, M. (2009) 'Computationally efficient MIMO HSDPA system-level modeling,' *EURASIP Journal on Wireless Communications and Networking*, **2009**, article ID 382501, doi: 10.1155/2009/382501.

[24] Wrulich, M., Weiler, W., and M. Rupp (2008) 'HSDPA performance in a mixed traffic network,' in *Proceedings of the IEEE Vehicular Technology Conference Spring (VTC)*, pp. 2056–2060, doi: 10.1109/VETECS.2008.462. Available from http://publik.tuwien.ac.at/files/pub-et_13769.pdf.

11

LTE Link- and System-Level Simulation

Contributed by Josep Colom Ikuno
Vienna University of Technology (TU Wien), Austria

During[1] the development, standardization, and further improvement of modern cellular systems such as Long-Term Evolution (LTE), simulations are necessary to test and optimize algorithms and procedures prior to their implementation process of equipment manufacturers. This chapter elaborates on an open simulator suite developed at the Technical University of Vienna: the "Vienna LTE Simulators", which currently comprises two simulators: the "Vienna LTE Link Level Simulator" and the "Vienna LTE System Level Simulator." Accurate simulations of simple setups, as well as simulations of more complex systems via abstracted models are necessary in order to assess system performance at different levels. To this end, simulations have to be carried out on both the physical layer (link level) and in the network (system level) context. If no computation limitations were to exist, both simulation types would be achievable by a single simulation tool. Realistically, each of the two simulation types has different objectives, which are met by different simulation tools:

Link-level simulations are basically a software implementation of one or multiple links between the Evolved base station (eNodeB) and the User Equipments (UEs), with a channel model to reflect the actual transmission of the waveforms generated. This results in very computationally intensive simulations, as transmitter and receiver procedures, which are normally performed by specialized hardware, as well as the generation of appropriate channel coefficients, are then performed in software. Such simulations allow for the investigation of channel estimation, tracking, and prediction algorithms [73], synchronization algorithms [52, 83, 88], Multiple-Input Multiple-Output (MIMO) gains, Adaptive Modulation and Coding (AMC) and feedback [37, 43, 66], receiver

[1] Reproduced [32], J. Ikuno, M. Wrulich, M. Rupp, "System Level Simulation of LTE networks," in VTC 2010-Spring, by permission of © 2010 IEEE.

Evaluation of HSDPA and LTE: From Testbed Measurements to System Level Performance, First Edition.
Sebastian Caban, Christian Mehlführer, Markus Rupp and Martin Wrulich.

structures [13], modeling of channel encoding and decoding [31], and physical-layer modeling for system-level simulations [91]. They typically neglect inter-cell interference and the impact of scheduling, as this increases simulation complexity and run time dramatically. In short, any physical-layer procedure can be evaluated by means of link-level simulations. Although MIMO broadcast channels have been investigated quite extensively over recent years [12, 21, 87], there are still a lot of open questions that need to be resolved, both in theory and in practical implementations. LTE offers the flexibility to adjust many transmission parameters, but it is not clear up to now how to exploit its available Degrees of Freedom (DoFs) [33] to achieve the optimum performance. Some recent theoretical results point towards interesting directions [27, 53, 87], but practical results for LTE are still missing.

System-level simulations focus on simulating large networks comprising multiple eNodeBs and UEs. For this, and in order to keep computational requirements under feasible limits, a different approach has to be employed. Physical-layer procedures have to be abstracted by accurate but also low-complexity models. While such an approach requires a prior analysis and modeling at link level, it enables the investigation of more network-related issues, such as resource allocation and scheduling [67, 70, 74], Multi-User (MU) handling, mobility management, admission control [59, 60], interference management [17, 23, 64], and network planning optimization [54, 58, 92]. On top of that, in an MU-oriented system such as LTE, it is not directly clear which figures of merit properly describe the performance of the system. Classical measures of (un)coded Bit Error Ratio (BER), Block Error Ratio (BLER), and throughput do not cover MU scenario properties. More comprehensive measures of the LTE performance, such as "fairness" [19] or "multi-user diversity gain" are also used. However, these theoretical concepts have to be mapped to performance values that can be evaluated by means of simulations [55, 56].

The two simulators described in this chapter [32, 44] are freely available, including source code under an academic use license (http://www.nt.tuwien.ac.at/ltesimulator). This enables academic researchers access to an open LTE simulation framework, which facilitates collaboration and the sharing of algorithms from different universities and research facilities. Without the need of having to implement a full LTE transmission chain, developed algorithms can be tested faster and shared when published, making the comparison of algorithms easier, reproducible, and therefore more credible.

To the best of the authors' knowledge, the "Vienna LTE Simulators" are the first to be offered publicly. Thus, the simulators provide opportunities for many institutions to directly apply their ideas and algorithms in the context of LTE. Although some commercial products (Steepest Ascent [75] and MimoOn[2]) and noncommercial platforms do exist [65], as well as vendor-developed proprietary solutions, they do not offer open access of their source code. The availability of the simulators, furthermore, enables researchers to quickly reproduce published results [79]. To this end, our published work after the release of each simulator [29, 30, 35, 66, 68, 72, 73, 85] not only contains the formal description of such work and results, but also a reference to the simulator and any additional data and/or code required to reproduce it. In this manner, any researcher can readily evaluate

[2] mi!Mobile, available from http://www.mimoon.de/pages/Products/miMobile/.

the work, compare it, build on it, or discuss its correctness. Although a complete testing is not possible, it nevertheless makes the work presented more transparent to the review process and, therefore, more credible. The more complex the system is, the more important reproducibility becomes.

11.1 The Vienna LTE Link Level Simulator

This section describes the overall structure of the "Vienna LTE Link Level Simulator" (current release: January 2011, v1.6r917). Through its description, the simulator's capabilities are presented, and some application examples are given.

11.1.1 Structure of the Simulator

The "Vienna LTE Link Level Simulator" is divided into three basic building blocks, namely "transmitter (TX)," "channel model," and "receiver (RX)" (see Figure 11.1). Depending on the type of simulation, one or several instances of these basic building blocks are employed. The transmitter and receiver blocks are linked by the channel model and evaluate the transmitted data, while signaling as well as UE feedback is assumed to be error free, but with a configurable-delay Uplink (UL).

Transmitter: The layout of the transmitter is shown in Figure 11.2, and depicts the implementation of the transmitter description given in the TS 36.x standard series [5–7]. Based on UE feedback values, a scheduling algorithm assigns each UE specific Resource Blocks (RBs) a Modulation and Coding Scheme (MCS), an MIMO transmission mode, and an appropriate precoding matrix/number of spatial layers. A discrete set of coding rates, specified in [7] as Transport Block (TB) sizes with 4-QAM, 16-QAM, or 64-QAM modulation alphabets, can be employed. Supported MIMO modes are Transmit Diversity (TxD), Open-Loop Spatial Multiplexing (OLSM), and Closed-Loop Spatial Multiplexing (CLSM). Such a channel adaptive scheduling allows for the exploitation of frequency diversity, time diversity, spatial diversity, and MU diversity. Given the number of available DoFs, the specific implementation of the scheduler has a large impact on the system performance and is a hot topic in research [24, 39, 68, 86] (see Section 11.4.2 on performance evaluations).

Channel model: The "Vienna LTE Link Level Simulator" supports both block-fading and fast-fading channels, which are used for DL transmissions. In the block-fading case, the channel is constant during the duration of one subframe (1 ms). In the fast-fading case, time-correlated channel impulse responses are generated for each sample of the transmit signal. The following options are possible as channel models:

1. Additive White Gaussian Noise (AWGN);
2. flat Rayleigh fading;
3. power-delay-profile-based channel models such as ITU Pedestrian B, ITU Vehicular A [46], or typical COST 259 [20] realizations such as typical urban or hilly terrain [3];
4. Winner Phase II+ channel model [26].

The most sophisticated model and the choice for simulating MIMO setups is the Winner Phase II+ channel model. It is an evolution of the 3rd Generation Partnership Project (3GPP) Spatial Channel Model (SCM) and SCM Extension (SCME) [10, 2] and, as such, it is a Geometry-based Stochastic Channel Model (GSCM). It extends the SCM/SCME methodology to make it applicable to the frequency ranges and extended system bandwidths in LTE, and introduces additional features such as support for arbitrary three-dimensional (3D) antenna patterns [51].

Receiver: The receiver implementation offers a diversity of options in terms of receiver algorithm, channel estimation, and feedback calculation, among others. Therefore, similar to other wireless technologies, it is not specified in the standard. Nevertheless, an LTE receiver will, regardless of its implementation, follow the common structure shown in Figure 11.3. After Cyclic Prefix (CP) removal and FFT, the RBs assigned to the UE are disassembled and passed on to the receiver, which in parallel receives information from the channel estimator and precoding signaling. The detected soft bits are subsequently decoded to obtain the data bits and figures of merit, such as throughput, coded/uncoded BER, and BLER. The simulator currently supports Zero Forcing (ZF), Linear Minimum Mean Square Error (LMMSE), and Soft Sphere Decoder (SSD) as detection algorithms; and regarding channel estimation, four different types of channel estimator are supported: (i) Least Squares (LS), (ii) Minimum Mean Square Error (MMSE), (iii) approximate LMMSE [41, 71], and (iv) genie-driven (near) perfect channel knowledge based on all transmitted symbols.

With the results from channel estimation, feedback calculation can be performed, which includes the Channel Quality Indicator (CQI) for all modes, the Rank Indicator (RI) for the Spatial Multiplexing (SM) modes and additionally the Precoding Matrix Indicator (PMI) for the CLSM mode. Together with ACK/NACK reports, this information forms the UE feedback, which is sent back to the eNodeB via a configurable-delay error-free channel. Given this receiver structure, the simulator allows the investigation of various aspects, such as frequency synchronization [85], channel estimation [71], or interference awareness.

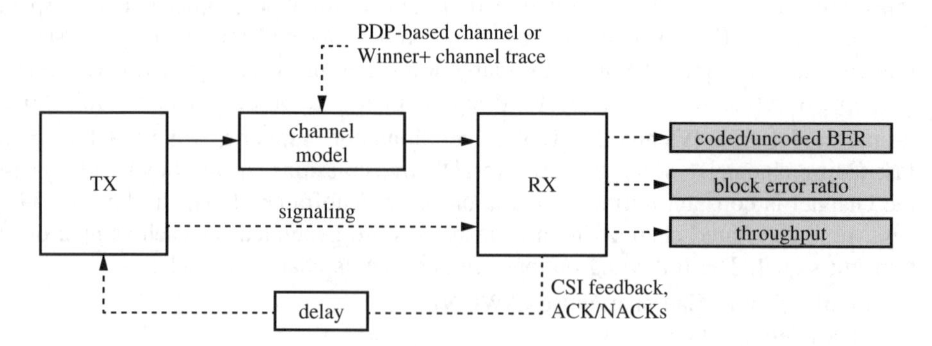

Figure 11.1 Functional block diagram of the structure of the "Vienna LTE Link Level Simulator".

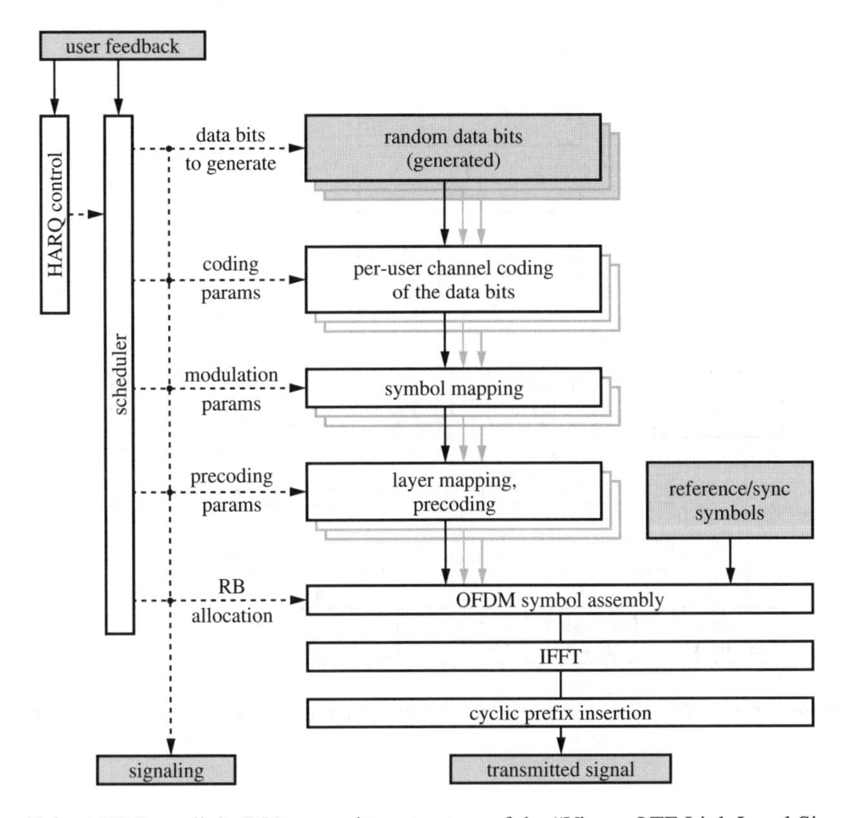

Figure 11.2 LTE Downlink (DL) transmitter structure of the "Vienna LTE Link Level Simulator".

11.1.2 Complexity

Link-level simulators such as the "Vienna LTE Link Level Simulator" are a software implementation of the actual Physical (PHY) layer procedures the transmission chain is comprised of, with the difference of a transmission model being used instead of the actual transmission of the waveforms. The implementation follows the TS 36.x series standards and implements the PHY layer procedures, including segmentation, channel coding, MIMO, transmit signal generation, pilot patterns, and synchronization sequences. Therefore, implementation complexity and run times are elevated.

To obtain a simulator with readable and maintainable code, a high-level language (MATLAB®) has been chosen. While it could be argued that MATLAB® is not the most efficient language, this choice enabled the development in a fraction of the time required for an implementation in other languages, such as C, while maintaining readability and cross-platform compatibility [80]. Although MATLAB® programs certainly run slower than C programs, simulation run time can be greatly reduced by means of code vectorization and straightforward parallelization with the MATLAB® Parallel/Distributed

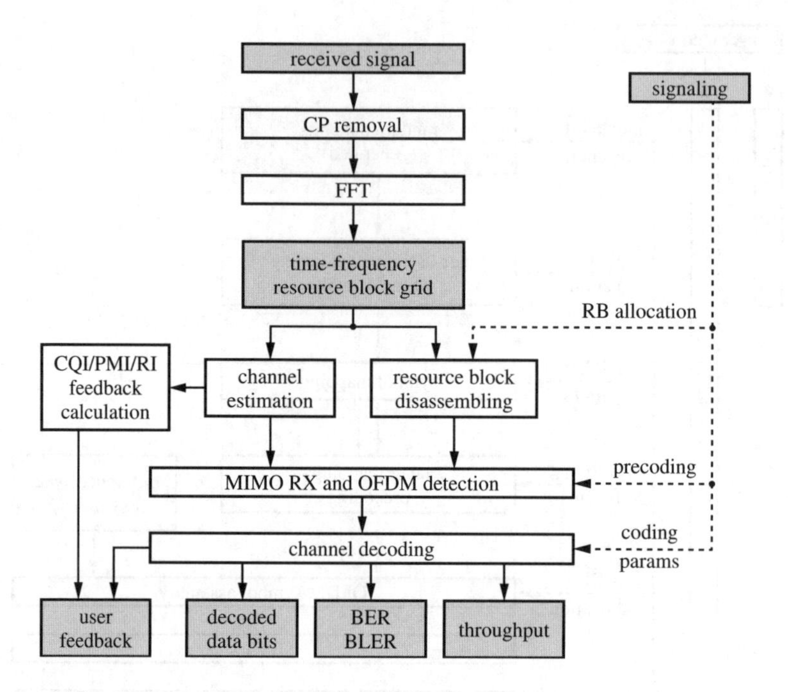

Figure 11.3 LTE DL receiver structure of the "Vienna LTE Link Level Simulator".

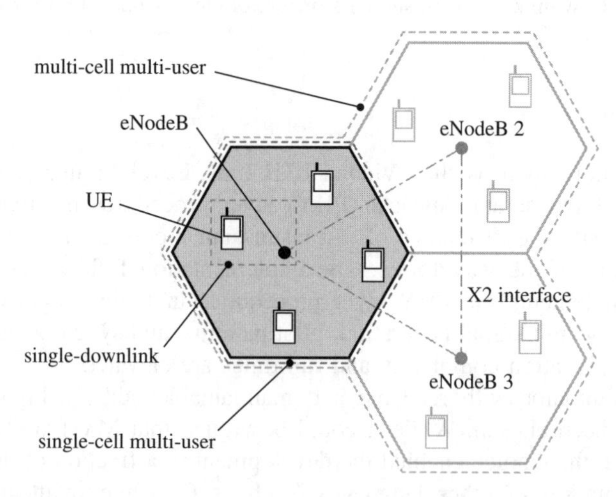

Figure 11.4 The scale of the simulation is supported by three different scenarios in the "Vienna LTE Link Level Simulator".

Computing Toolbox. To further decrease execution time and unnecessary reimplementation, the critical, often-called, difficult to vectorize, or already freely available in optimized form functional parts of the simulator – such as channel decoding,[3] Cyclic Redundancy Check (CRC) computation,[4] and soft sphere decoding – are implemented in C and linked to the MATLAB® code via MEX functions. Depending on the simulation needs, three basic simulation scenarios with increasing computational complexity are contemplated. These are single-DL, single-cell MU, and multi-cell MU, which are shown in Figure 11.4.

Single-downlink: This simulation type only covers the link between one eNodeB and UE. Such a setup allows for the investigation of channel tracking, channel estimation [71], synchronization [83, 84], MIMO gains, AMC and feedback optimization [38], receiver structures [14] (neglecting interference and the impact of the scheduling[5]), modeling of channel encoding and decoding [29, 31], and physical-layer modeling [43, 89], which can support system-level abstraction of the physical layer.

Single-cell multi-user: This setup covers the links between a single eNodeB and multiple UEs, additionally allowing for the investigation of receiver structures that take into account the influence of scheduling, MU precoding [62], resource allocation, and MU gains. Furthermore, it offers researchers the ability to investigate practically achievable MU rate regions. Supplementary to the requirements of the single-downlink scenario, these simulations need a fully functional feedback loop and AMC. The simulator implementation fully evaluates the receivers of all users; thus, complexity relative to the previous case grows de facto linearly with the number of simulated links. A portion of the PHY procedures on the transmit side is not performed on a per-user basis, but their impact complexity-wise is not significant.

Multi-cell multi-user: This simulation is by far the most computationally demanding scenario and covers the links between multiple eNodeBs and UEs. This setup allows for the realistic investigation of interference-aware receiver techniques [45, 90], interference management, which includes Coordinated Multi-Point (CoMP), and cooperative transmissions [28], and interference alignment [16, 40, 77], and network-based algorithms such as joint resource allocation and scheduling. Furthermore, despite the large computational efforts needed, such simulations are crucial to verify system-level simulations, as they represent an overlapping point between feasible computationally but complex link-level simulations, and computationally simple system-level simulations, both simulating an equivalent scenario.

In the simulator package, example simulations for each of the three described scenarios can be run by means of the following MATLAB® scripts: LTE_sim_batch_single_downlink.m, LTE_sim_batch_single_cell_multi_user.m, and LTE_sim_batch_multi_cell_multi_user.m.

[3] Iterative Solutions Coded Modulation Library (ISCML), available from http://www.iterativesolutions.com/.

[4] pycrc CRC calculator and C source code generator, available from http://www.tty1.net/pycrc/.

[5] Note that the scheduler in an MU system will change the statistics of the individual user's channel, thus significantly influencing receiver performance.

11.2 The Vienna LTE System Level Simulator

In this section, we describe the overall structure of the "Vienna LTE System Level Simulator" (current release: November 2010, v.1.3r427), as well as the link-to-system model, which abstracts the PHY layer procedures at low complexity.

11.2.1 Structure of the Simulator

System-level simulations aim at evaluating the performance of much more complex networks than those tested in typical link-level simulations. While still accurately capturing the PHY layer procedures via low-complexity models, system-level simulations try to capture the network structure. In LTE, such a network consists of a number of eNodeBs that cover a specific area, defined by a Region Of Interest (ROI), in which UEs are placed and subsequently move. In the case of the "Vienna LTE System Level Simulator", a DL transmission is then simulated for each UE. Figure 11.5 depicts an exemplary setting for an LTE system-level simulation. A single-link simulation of a length of 5000 subframes, performed with the LTE link-level simulator, takes on the order of a few hours, depending on the number of MCSs. Understandably, a setting such as that at Figure 11.5, although technically possible, requires an extensive computation time to complete. On the other hand, such a setup in system level only requires on the order of minutes to be completed without losing much accuracy. Section 11.3 provides more insight into the accuracy and validation of the link-to-system model.

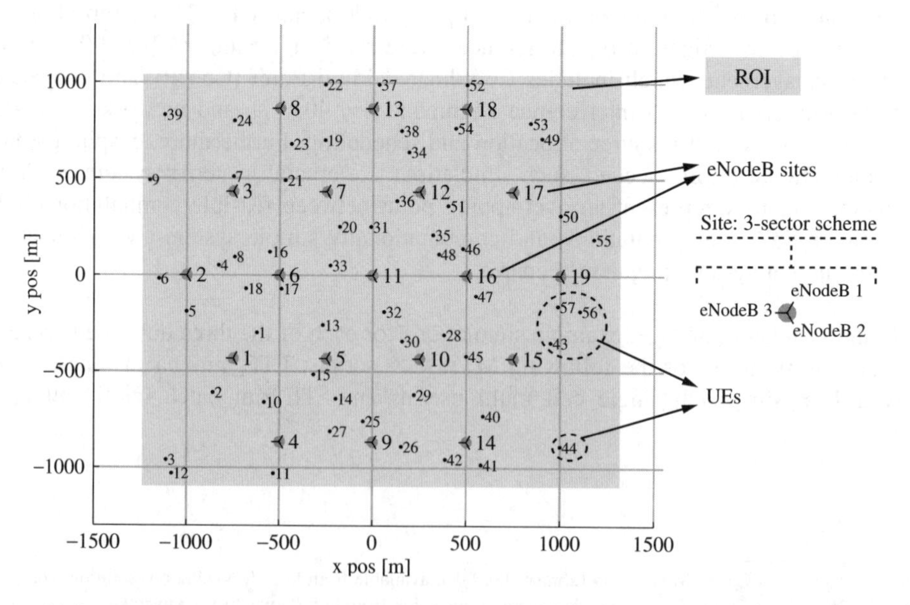

Figure 11.5 Example setting for an LTE system-level simulation. This simulation encompasses two rings of sites, each with three eNodeBs (sectorized site) and one UEs per site.

While simulations of individual physical-layer links allow for the investigation of PHY procedures [71, 85], AMC feedback [66, 69], channel coding and/or retransmission modeling [29, 94], it is not possible to reflect the effects of cell planning, scheduling, or interference on a large scale. As mentioned, the size of the simulated system makes it impossible to simply perform physical layer simulations of the radio links between all terminals and base stations. At the same time, a bigger scale than a single/few link-level simulations is necessary to evaluate new designs such as scheduling.

The design of the LTE system-level simulator adapts the already known link-measurement and link-performance model [11, 15, 25, 89] to the specifics of Orthogonal Frequency-Division Multiplexing (OFDM) and LTE. Thus, the simulator consists of two parts: (i) a link-measurement model (sometimes also referred to as a link-quality model) and (ii) a link-performance model. The link-measurement model, as its name implies, reflects the quality of the link. On the other hand, and based on the quality metric of the link-measurement model, the link-performance model outputs system-dependent performance metrics such as BLER and throughput. Figure 11.6 illustrates the interaction between the two models and the several physical-layer parameters. In our simulator, the link quality measure is post-equalization subcarrier Signal to Interference and Noise Ratio (SINR). Thus, it includes an analytical model of the receiver, which in our case is the ZF receiver. In order to obtain the quality of the channel, not only the channel itself is needed, but also which portion of the channel is assigned to each specific UE.

The link-quality model then maps the set of output SINRs to a single AWGN-equivalent SINR value via an appropriate compressing function such as Exponential Effective Signal to Interference and Noise Ratio Mapping (EESM) or Mutual Information Effective Signal to Interference and Noise Ratio Mapping (MIESM) [36, 48, 78, 82] (for more details on

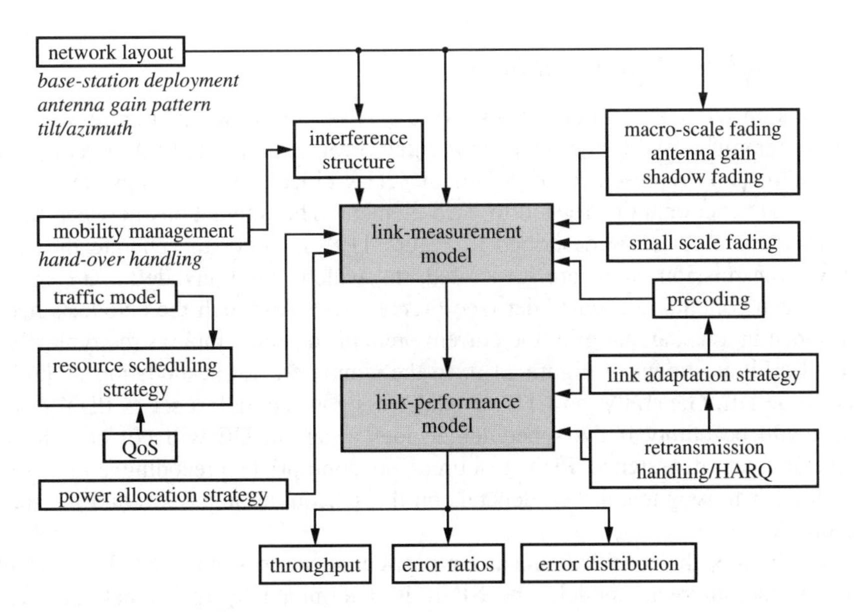

Figure 11.6 Schematic block diagram of the "Vienna LTE System Level Simulator." From [32]. Reproduced by IEEE © 2010.

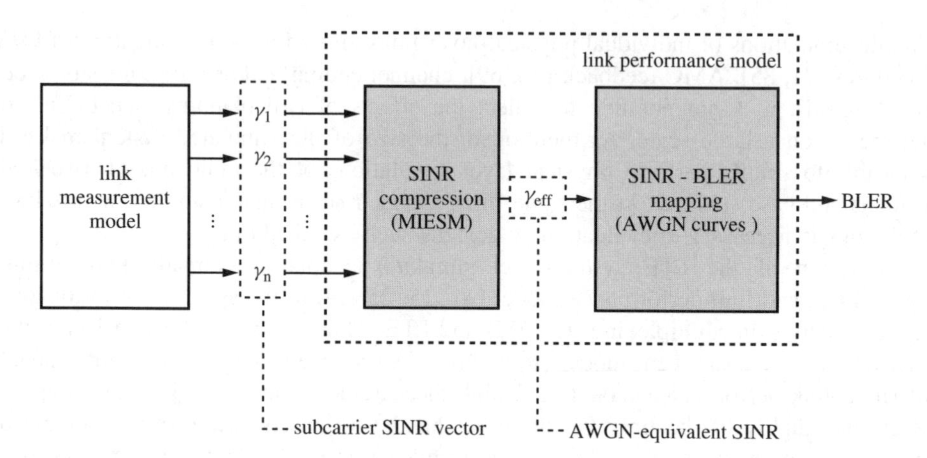

Figure 11.7 Link measurement and link performance model. The SINR vector output from the link measurement model is compressed to an AWGN-equivalent SINR and then mapped to BLER.

MIESM, see also Section 12.3.1). By using an SINR compression algorithm to obtain a fading-insensitive metric, there is no need to generate different performance curves mapping SINR to BLER for every single channel type. In this way, only one AWGN BLER curve for every MCS is needed. Figure 11.7 depicts the two-step process which generates the BLER. As the curves are obtained from single-user AWGN-channel link-level simulations, the BLER performance curves effectively connect the link-level and system-level simulator concepts.

11.2.2 Simulator Implementation

Implementation-wise, the simulator follows the structure shown in Figure 11.8. Being entirely programmed with the Object-Oriented Programming (OOP) paradigm, each network element is represented by a suitable class object which encapsulates the data structures and functionalities related to each element. The interactions between the main simulator object classes are described in Figure 11.8. In order to generate the network topology, transmission sites are generated, to which three eNodeBs are appended (sectors), each containing a scheduler (see Figure 11.8). Although the eNodeBs need not be positioned in a hexagonal grid, the current predefined setup employs the typically used hexagonal grid, as shown in Figure 11.9. In the simulator, traffic modeling assumes full buffers in the DL. Generally, an LTE scheduler assigns each UE a set of PHY resources (the set could be empty if the scheduler decides a certain UE will not be scheduled), and for each set of assigned PHY resources an appropriate precoding and a suitable MCS. The actual assignment then depends on the scheduling algorithm and the received UE feedback.

At the UE side, the received subcarrier post-equalization symbol SINR is calculated by the link-measurement model. The SINR is determined by the signal, interference, and noise power levels, which are dependent on the cell layout (defined by the eNodeB positions, large-scale pathloss, shadow fading [4]) and the time-variant small-scale fading

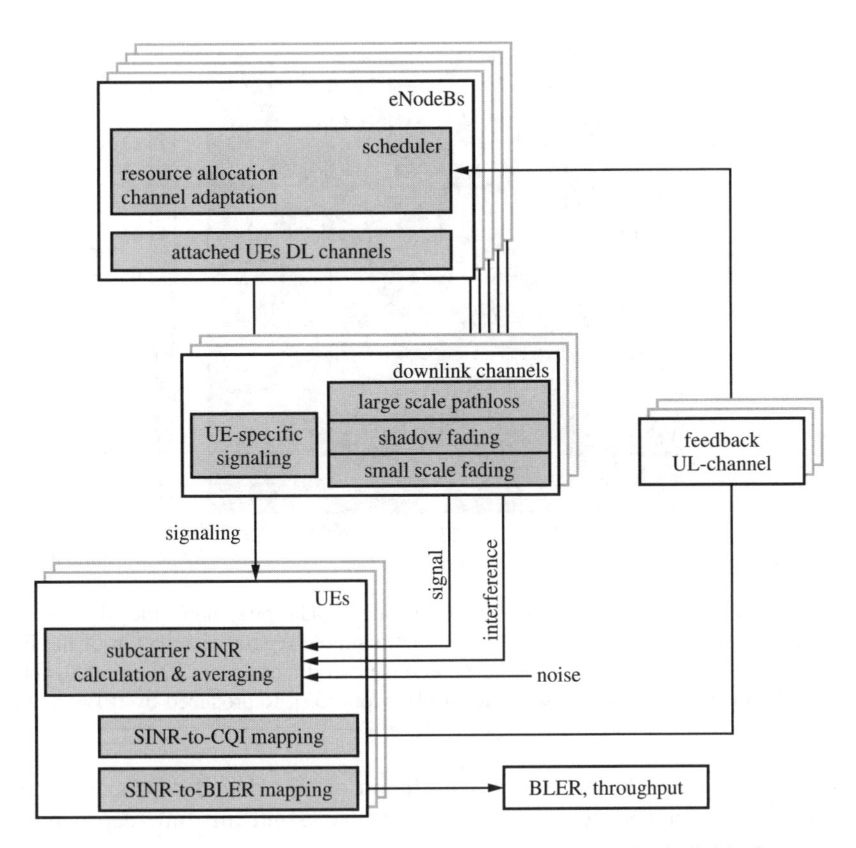

Figure 11.8 Schematic class diagram showing the relation between the individual components of the "Vienna LTE System Level Simulator".

[95, 96]. Further detail on the actual implementation of each is found in Section 11.2.3. The CQI feedback report is calculated based on the subcarrier SINRs and an SINR-to-CQI mapping obtained from the 10 % BLER points of the AWGN BLER curves [32] and made available to the eNodeB via an error-free feedback channel with adjustable delay.

In the link-performance model, the SINR is mapped to BLER [32, 44], the value of which acts as a probability for computing ACK/NACKs. The ACK/NACKs are then recombined with the Transport Block Size (TBS) to compute the link throughput. As results, the simulator outputs traces of the link throughput and the link error ratios for all UEs and eNodeBs. From these traces, cell throughput and error ratio distributions can be computed.

11.2.3 Complexity

A desirable functionality of a system-level simulator is its ability to precalculate as much as possible of the simulation that is needed at run time. This not only reduces the computational load while carrying out a simulation, but also offers repeatability by making it possible to load precalculated simulation scenarios. The precalculations involved in

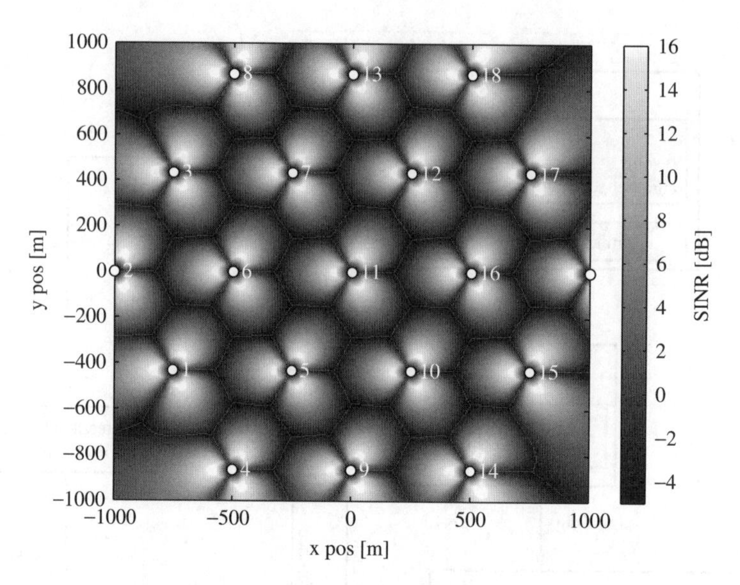

Figure 11.9 Hexagonal grid positioning of the eNodeBs. The plot shows the SINR in decibels over the ROI. The overlayed cell boundaries are not a simple regular tessellation of the ROI, but with the actual boundaries obtained from the SINRs; the two-dimensional (2D) antenna gain pattern and large-scale (macroscopic) pathloss are as in [4]. From [32]. Reproduced by IEEE © 2010.

the LTE system level simulator are the generation of (i) eNodeB-dependent large-scale pathloss maps, (ii) site-dependent shadow fading maps, and (iii) time-dependent small-scale fading traces for each eNodeB-UE pair.

11.2.3.1 Pathloss and Fading Maps

The large-scale pathloss and the shadow fading are modeled as position-dependent maps. The large-scale pathloss is calculated according to well-known models [1, 4], and combined with the antenna gain pattern of the corresponding eNodeB. Both 2D and 3D antenna gain patterns (obtained from horizontal and vertical antenna gain cuts and then interpolated [76]) are possible. Thus, both electrical and mechanical down-tilting and/or measured antenna gain patterns can be applied. The spatially correlated shadow fading is obtained from a log-normal random distribution by means of a low-complexity variant of the Cholesky decomposition [18], with inter-site map correlation being introduced with a similar method. Figure 11.10 shows large-scale pathloss and shadow fading maps.

11.2.3.2 Time-dependent Fading Trace

While large-scale pathloss and shadow fading are modeled as position dependent, the small-scale fading is modeled as a time-dependent trace. The calculation of this trace is based on the ZF receiver and includes the MIMO channel matrix (normalized channel matrix $\mathbf{H}$), the receive filter, and the precoding. Although our system-level simulator only supports the linear ZF receiver, any other linear receiver could be easily integrated given

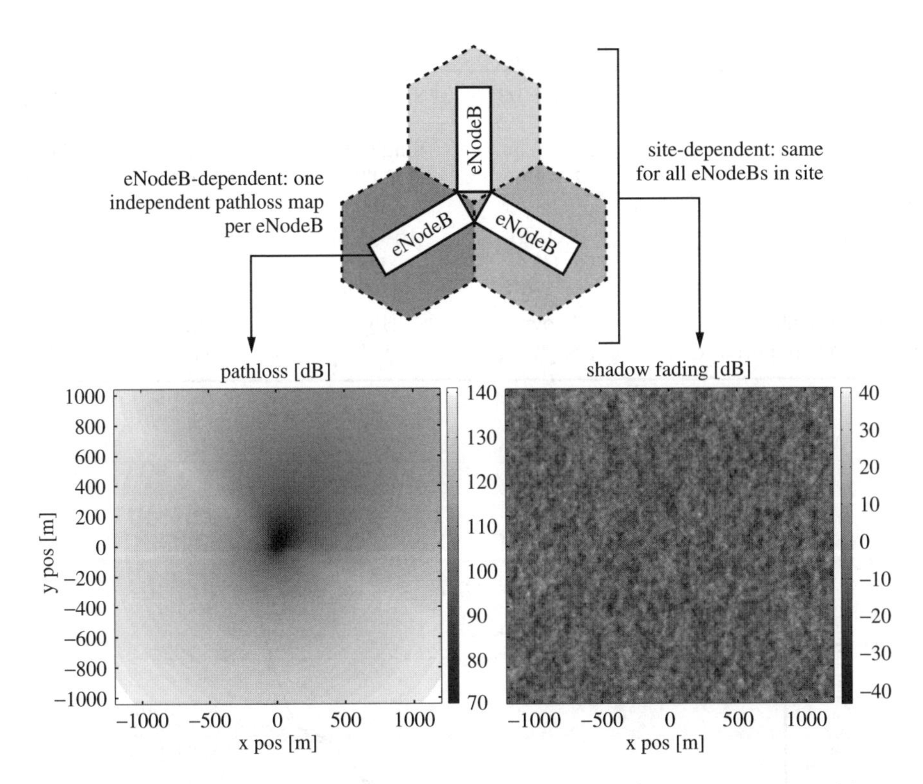

Figure 11.10 Left: Large-scale pathloss and antenna gain map [dB] at one eNodeB. Right: space-correlated shadow fading [dB] at one site. From [32]. Reproduced by IEEE © 2010.

an appropriate SINR modeling. The small-scale fading trace consists of the signal power and the interference power after the receive filter. The breakdown into these two parts significantly reduces the computational effort, since it avoids many complex multiplications required when working directly with MIMO channel matrices on a system level [32, 89, 91].

In order to reduce memory and computational requirements, as the trace has to be stored in memory, block fading is assumed and frequency decimation of the trace is performed [47]. Thus, the channel is assumed constant over the duration of one subframe and instead of storing all data subcarriers, only one out of six is taken (two per RB). From the generated frequency-domain fading parameters trace, independent traces are taken for each link by taking random starting points. Given a sufficiently long fading trace, these different trace realizations can be assumed to be independent.

11.3 Validation of the Simulators

The first validation of the simulators was performed by comparing the link-level throughput with the minimum performance requirements stated by 3GPP in the technical specification TS 36.101 [8] (shown in Table 11.1 and Figure 11.11). Subsequently, results were compared with RAN results in [9]. Finally, simulation results from the link-

Table 11.1 Test scenarios of 3GPP TS 36.101

	8.2.1.1.1/1	8.2.1.1.1/8	8.2.1.2.1/1	8.2.1.3.2/1
TX mode	Single ant.	Single ant.	TxD	OLSM
Channel	EVehA	ETU	EVehA	EVehA
Doppler frequency (Hz)	5	300	5	70
Modulation	4-QAM	16-QAM	16-QAM	16-QAM
Code rate	1/3	1/2	1/2	1/2
Antenna configuration	1 × 2	1 × 2	2 × 2	4 × 2
Antenna correlation	Low	High	Medium	Low
Channel SNR required (dB)	−1	9.4	6.8	14.3

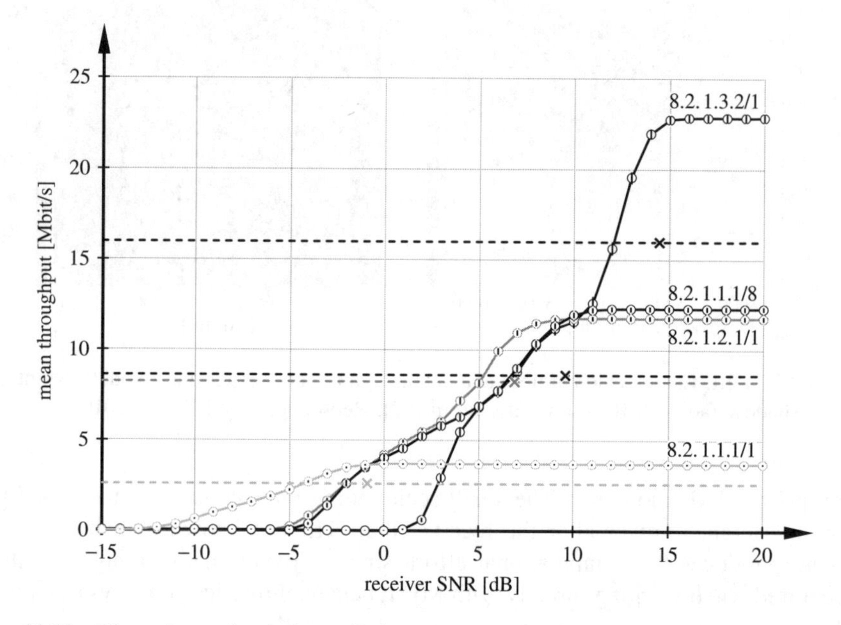

Figure 11.11 Throughput simulations of the test scenarios in 3GPP TS 36.101 and comparison with the minimum performance requirements (marked with crosses). The small vertical bars within the circular markers indicate the 99 % confidence intervals. Reproducible by running Reproducibility_RAN_sims.m.

and system-level simulators were cross-validated against each other in order to test the accuracy of the link-to-system model. Since no other open LTE simulation platforms exist, it was not cross-validated against other simulators.

Although not a formal validation, our link-level and system-level simulators have been openly available since May 2009 and March 2010, respectively. Since then, many small bugs and discrepancies have been resolved and discussed with the help of the research community. Allowing open access to the source code of the simulators has allowed anyone to check the code for correctness, as well as making it possible for many to adapt it to their individual requirements [22, 50, 93]. An online forum is also provided (http://www.nt.tuwien.ac.at/forum) for problem/bugs discussion and exchange of results.

11.3.1 3GPP Minimum Performance Requirements

The technical specification TS 36.101 [8] defines minimum performance requirements for a UE that utilizes a dual-antenna receiver. These requirements have to be met by real devices and, therefore, have to be surpassed by the simulator results, especially if channel estimation, frequency and timing synchronization, and other nonideal effects, such as quantization, are not considered. In particular, TS 36.101 specifies reference measurement channels for the Physical Downlink Shared Channel (PDSCH) (comprising bandwidth, AMC scheme, overhead) and propagation conditions (power delay profiles, Doppler frequencies, antenna correlation). The simulation scenarios considered are completely specified by referring to sections and test numbers in TS 36.101. For example, in TS 36.101 Section 8.2.1.1.1 tests are defined for a single-transmit-antenna scenario. By referring to test number 1 in this section, the AMC mode is defined as 4-QAM[6] with a target coding rate of 1/3. Furthermore, the propagation channel is defined as Extended Vehicular A (EVehA) with a Doppler frequency of 5 Hz and low antenna correlation. For the simulations presented here, we selected four test scenarios with a bandwidth of 10 MHz but different transmit modes (single antenna port transmission, OLSM, and TxD), different AMC schemes, and different channel models. Hybrid Automatic Repeat reQuest (HARQ) is supported with at most three retransmissions. The most important parameters of the test scenarios are listed in Table 11.1. The first scenario (8.2.1.1.1/1) refers to the test scenario described above. The OLSM scenario (8.2.1.3.2/1) utilizes a rank 2 transmission; that is, transmission of two spatial streams.

Simulation results for the scenarios considered are shown in Figure 11.11. The dashed horizontal lines correspond to 70 % of the maximum throughput values for which TS 36.101 defines a channel Signal to Noise Ratio (SNR) requirement (shown as crosses in Figure 11.11). For all test scenarios considered, the link-level simulator outperforms the minimum requirements by approximately 2–3 dB, offering space for implementation losses. The small vertical bars within the markers in Figure 11.11 are the 99 % confidence intervals of the simulated mean throughput. Since the confidence intervals are much smaller than the distances between the individual throughput curves, a repeated simulation with different seeds of the random number generators will lead to similar results. Figure 11.11 can be reproduced by calling the script Reproducibility_RAN_sims.m.

11.3.2 Link- and System-Level Cross-Comparison

In this section, a performance cross-comparison of the link- and system-level simulators is shown. We consider a single-user single-cell scenario with different antenna configurations and transmit modes, as summarized in Table 11.2. Depending on the channel conditions, we adapt the AMC scheme, the transmission rank, and the matrices. For this purpose, we utilize the UE feedback schemes originally presented in [66, 69]. In order to create an equivalent simulation scenario on link and system levels, no shadow fading is employed. Whereas on the link level the SNR is usually directly specified, on the system level the SNR is a function of the user location in the cell. Without shadow fading, the

[6] Note that this is not a 4-QAM or Quadrature Phase Shift Keying (QPSK) modulation; only the signal points are referred to. In this case, both schemes are identical.

Table 11.2 Test scenarios to compare link- and system-level simulations

	SISO	TxD	OLSM	CLSM
Channel[a]	TU	TU	TU	TU
Bandwidth (MHz)	1.4	1.4	1.4	1.4
Antenna configuration	1×1	2×2	2×2	4×2
CQI feedback	✓	✓	✓	✓
RI feedback	×	×	✓	✓
PMI feedback	×	×	×	✓

[a]TU: Typical Urban channel model [3].

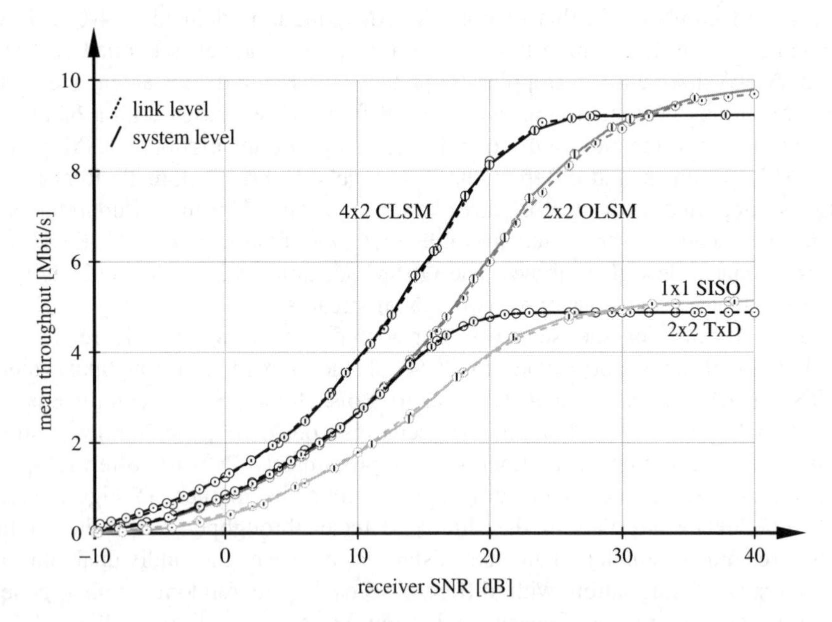

Figure 11.12 Cross-comparison of throughput results obtained with the link-level and the system-level simulators. CLSM: closed-loop spatial multiplexing; OLSM: open-loop spatial multiplexing; SISO: single antenna system, 2×2 TxD: Alamouti coding on two transmit antennas and two receive antennas. The small vertical bars within the circular markers indicate the 99 % confidence intervals. Reproducible by running Reproducibility_LLvsSL_batch.m.

UE-perceived SNR on the system level becomes a function of the distance between the base station and the user. This can be utilized to indirectly select appropriate SNR values in the system-level simulator. The results of the link- and system-level comparison are shown in Figure 11.12. For all simulation scenarios considered, an excellent match between the results of the two simulators is obtained, confirming the validity of our PHY modeling. Figure 11.12 can be reproduced by running the script Reproducibility_LLvsSL_batch.m provided in the system-level simulator package. Further comparisons between link- and system-level simulator results are shown in Section 11.4.2.

11.4 Exemplary Results

In this section, we show two exemplary simulation results obtained with the "Vienna LTE simulators." First, we present a comparison between link-level throughput simulations of different MIMO schemes against theoretical bounds. Based on this simulation setup, researchers can investigate algorithms such as channel estimation, detection, or synchronization and evaluate the results against theoretical upper bounds. Second, we compare the performance of different state-of-the-art schedulers in a single-cell MU environment. These schedulers serve as reference for researchers investigating advanced scheduling techniques.

11.4.1 Link-Level Throughput

Before presenting the link-level throughput results of the different LTE MIMO schemes, we first have to introduce theoretical bounds. We identify three bounds; namely, the channel capacity, the Mutual Information (MI), and the so-called achievable MI. Depending on the type of channel state information available at the transmitter (full, quantized, or only receive SNR), an ideal transmission system is expected to attain one of these bounds.

11.4.1.1 Channel Capacity

For calculating the channel capacity of a frequency selective MIMO channel [57, 61], consider the singular value decomposition [49] of the channel matrix $\mathbf{H}_k$ scaled by the standard deviation σ_n of the additive white Gaussian noise impairment:

$$\frac{1}{\sigma_\mathrm{n}}\mathbf{H}_k = \mathbf{U}_k \Sigma_k \mathbf{V}_k^\mathrm{H}; \quad \text{with} \tag{11.1}$$

$$\Sigma_k = \mathrm{diag}\left\{\sqrt{\lambda_{k,m}}\right\}; \quad m = 1 \ldots \min\left(N_\mathrm{R}, N_\mathrm{T}\right)$$

The capacity-achieving, frequency-dependent optimum at the transmitter is given by the unitary matrix $\mathbf{V}_k$. If this precoding matrix is applied at the transmitter and also the optimum receive filter $\mathbf{U}_k^\mathrm{H}$ is employed, the MIMO channel is separated into $\min\left(N_\mathrm{R}, N_\mathrm{T}\right)$ independent Single-Input Single-Output (SISO) channels, each with a gain of $\sqrt{\lambda_{k,m}}$, $m = 1 \ldots \min\left(N_\mathrm{R}, N_\mathrm{T}\right)$, $k = 1 \ldots N_\mathrm{tot}$. Here, N_R denotes the number of receive antennas and N_T the number of transmit antennas. The channel capacity is obtained by optimally distributing the available transmit power over these parallel SISO subchannels. The optimum power distribution $P_{k,m}$ is the solution of the optimization problem

$$C = \max_{P_{k,m}} \frac{1}{N_\mathrm{tot}} \sum_{m=1}^{\min(N_\mathrm{R},N_\mathrm{T})} \sum_{k=1}^{N_\mathrm{tot}} \log_2\left(1 + P_{k,m}\lambda_{k,m}\right) \tag{11.2}$$

$$\text{subject to} \quad \sum_{m=1}^{\min(N_\mathrm{R},N_\mathrm{T})} \sum_{k=1}^{N_\mathrm{tot}} P_{k,m} = P_\mathrm{t}$$

Here, the second equation is a transmit power constraint that ensures an average transmit power proportional to the number of data subcarriers: $P_\mathrm{t} = N_\mathrm{tot}$. Note that, owing to the definition of $\sqrt{\lambda_{k,m}}$ in Equation (11.1), the power distribution $P_{k,m}$ and, thus, P_t remain dimensionless. The power coefficients maximizing Equation (11.2) are calculated by means of the waterfilling algorithm described in [57, 61]. In order to achieve a throughput equal to the channel capacity, the transmitter needs full channel state information and has to apply the optimum precoder. Furthermore, the receiver needs to apply the optimum receive filter in order to separate the parallel SISO subchannels.

11.4.1.2 Mutual Information

The MI is the theoretic bound for the data throughput if only the receive SNR but no further channel state information is available at the transmitter side [42]:

$$I = \sum_{k=1}^{N_\mathrm{tot}} B_\mathrm{sub} \log_2 \det \left(\mathbf{I}_{N_\mathrm{R}} + \frac{1}{\sigma_\mathrm{n}^2} \mathbf{H}_k \mathbf{H}_k^\mathrm{H} \right) \tag{11.3}$$

Here, B_sub denotes the bandwidth occupied by a single data subcarrier, $\mathbf{H}_k$ is the $N_\mathrm{R} \times N_\mathrm{T}$-dimensional MIMO channel matrix of the kth subcarrier, σ_n^2 is the energy of noise and interference at the receiver, N_tot is the total number of usable subcarriers, and $\mathbf{I}_{N_\mathrm{R}}$ is an identity matrix of size equal to the number of receive antennas N_R. In Equation (11.3), we normalized the transmit power to one, $P_\mathrm{TX} = 1$, and the channel matrix according to $E\{\|\mathbf{H}_k\|_2^2\} = 1$. Therefore, Equation (11.3) does not show a dependence on the transmit power and the number of antennas (compare Equation (1.1)). The bandwidth B_sub of a subcarrier is calculated as

$$B_\mathrm{sub} = \frac{N_\mathrm{sc} N_\mathrm{s} N_\mathrm{rb}}{T_\mathrm{sub} N_\mathrm{tot}}, \tag{11.4}$$

where $N_\mathrm{sc} = 12$ is the number of subcarriers in one RB, N_s is the number of OFDM symbols in one subframe (usually equal to 14 when the normal CP length is selected), N_rb is the number of resource blocks that fit into the selected system bandwidth, and $T_\mathrm{sub} = 1$ ms is the subframe duration. Note that we are calculating the mutual information for all "usable" subcarriers of the OFDM system, thereby taking into account the loss in spectral efficiency caused by the guard band carriers. If different transmission systems that apply different modulation formats are to be compared, however, a fair comparison would require calculating the mutual information over the entire system bandwidth instead of calculating only over the usable bandwidth. In order to achieve a throughput equal to the MI, in theory the transmitter only has to adjust the transmission data rate to the receive SNR.

11.4.1.3 Achievable Mutual Information

Both MI and channel capacity do not consider system design losses caused, for example, by the transmission of CP or reference symbols, or the quantization of the transmitter. In order to obtain a tighter bound for the link-level throughput, therefore, we consider these effects in the definition of the so-called achievable MI. In the case of open-loop transmission, in which space–time coding is employed at the transmitter, we obtain for the achievable MI

$$I_{\mathrm{a}}^{(\mathrm{OL})} = \sum_{k=1}^{N_{\mathrm{tot}}} F B_{\mathrm{sub}} \frac{1}{N_{\mathrm{L}}} \log_2 \det\left(\mathbf{I}_{N_{\mathrm{R}}N_{\mathrm{L}}} + \frac{1}{\sigma_{\mathrm{n}}^2} \tilde{\mathbf{H}}_k \tilde{\mathbf{H}}_k^{\mathrm{H}}\right), \tag{11.5}$$

with N_{L} denoting the number of spatial transmission layers. The $N_{\mathrm{R}}N_{\mathrm{L}} \times N_{\mathrm{T}}$ dimensional matrix $\tilde{\mathbf{H}}_k$ is the effective (virtual) channel matrix including the space–time coding. The factor F is an efficiency factor accounting for the inherent system losses due to the transmission of the CP and the reference symbols. In detail, the factor F is calculated as

$$F = \underbrace{\frac{T_{\mathrm{sub}} - T_{\mathrm{cp}}}{T_{\mathrm{sub}}}}_{\text{CP loss}} \cdot \underbrace{\frac{N_{\mathrm{sc}} \cdot N_{\mathrm{s}}/2 - N_{\mathrm{ref}}}{N_{\mathrm{sc}} \cdot N_{\mathrm{s}}/2}}_{\text{reference symbols loss}}, \tag{11.6}$$

where T_{cp} is the time required for the transmission of the CPs within one subframe and N_{ref} is the number of reference symbols per resource block. In LTE, the number of reference symbols depends on the number of transmit antennas. The efficiency factor F, therefore, decreases with increasing number of transmit antennas (see Table 11.3). In the case of closed-loop transmission, a channel-adapted precoding matrix $\mathbf{W}$ is chosen from a set $\mathcal{W}$ (defined in the standard) and applied to the transmit signal. We calculate the achievable mutual information for closed-loop transmission as

$$I_{\mathrm{a}}^{(\mathrm{CL})} = \max_{\mathbf{W}\in\mathcal{W}} \sum_{k=1}^{N_{\mathrm{tot}}} F B_{\mathrm{sub}} \log_2 \det\left(\mathbf{I}_{N_{\mathrm{R}}} + \frac{1}{\sigma_{\mathrm{n}}^2} \mathbf{H}_k \mathbf{W}\mathbf{W}^{\mathrm{H}} \mathbf{H}_k^{\mathrm{H}}\right). \tag{11.7}$$

Figure 11.13 shows the throughput of a 2 × 2 LTE system with 5 MHz bandwidth, perfect channel knowledge, and an SSD receiver, together with the corresponding theoretic bounds. The difference between channel capacity and mutual information is only small; therefore, even knowledge of the full channel state information at the transmitter does not considerably increase the potential performance. In contrast, the difference between the mutual information and the achievable mutual information is quite large, resulting in a loss of 56 % at an SNR of 15 dB. Most (41 %) of this loss is due to the restrictions implied by the standard, as indicated by the achievable MI curves in Figure 11.13. At a rate of 16 Mbit/s, the difference between achievable mutual information and simulated throughput

Table 11.3 Pilot symbols and efficiency factor F in LTE

Transmit antennas N_{T}	Reference symbols[a] N_{ref}	Efficiency factor F (%)
1	4	88.88
2	8	84.44
4	12	80
8	12	80

[a]Here, only cell-specific reference symbols are considered.

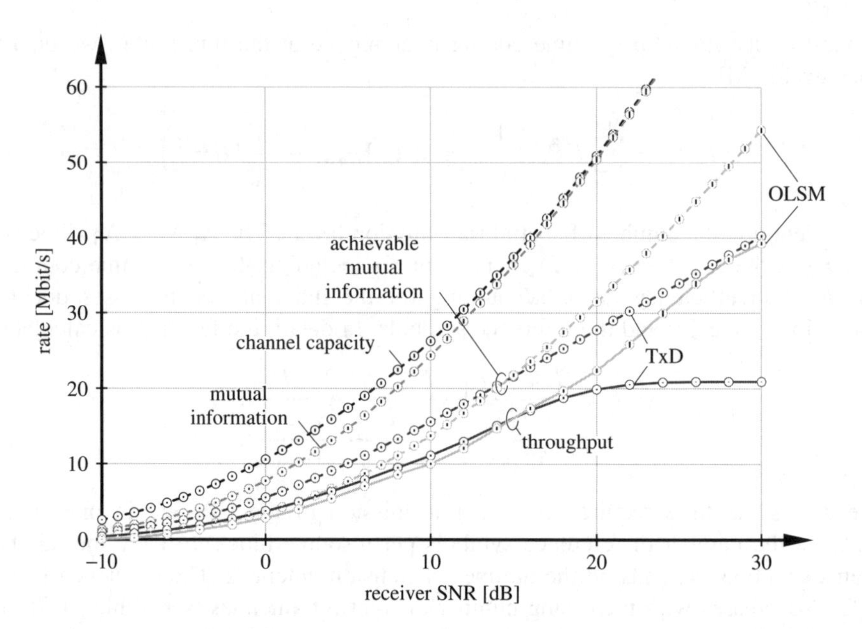

Figure 11.13 Throughput of a 2 × 2 system with 5 MHz bandwidth compared with the channel capacity, the mutual information, and the achievable mutual information. The small vertical bars within the circular markers indicate the 99 % confidence intervals. Reproducible by running Physical_Layer_batch.m.

is approximately 4 dB. These findings are similar to the results [63] we obtained when analyzing the performance of Worldwide Inter-operability for Microwave Access (WiMAX) and High-Speed Downlink Packet Access (HSDPA) (compare with Chapter 8 and Chapter 10). Furthermore, Figure 11.13 shows that, for SNRs lower than 14 dB, the TxD mode outperforms OLSM. Only at larger SNRs above 20 dB, where the throughput of the TxD mode saturates, does OLSM benefit from the second spatial stream and outperform TxD. Figure 11.13 can be reproduced by executing the script Physical_Layer_batch.m provided in the "Vienna LTE Link Level Simulator" package.

11.4.2 LTE Scheduling

In this section, the performance of various MU LTE scheduling techniques is compared by means of link-level and system-level simulations. By appropriately selecting the simulation parameters at link level and system level, equivalent simulations can be obtained from both simulators, which serve the purpose of further validating the link-to-system model. Once simpler setups are validated by both simulations, more complex settings are then possible with the system-level simulator. In particular, we consider in the "Vienna LTE System Level Simulator" one sector of a single-cell SISO system with 20 randomly positioned users. The user positions yield the large-scale pathloss and shadow fading coefficients of all users and, as a consequence, the average receive SNRs, distributed in the range 2.7–36 dB. The average receive SNRs of the 20 users are set in

Table 11.4 Link and system level parameters for the scheduling simulations

Parameter	Value
System bandwidth (MHz)	5
Number of subcarriers	300
Number of resource blocks	50
Number of users	20
Channel model	3GPP TU [3]
Channel realizations	2 500
Antenna configuration	1 transmit, 1 receive (1 $\times$ 1)
Receiver	Zero Forcing (ZF)
Schedulers	Best CQI (BCQI)
	Maxmin
	Proportional fair
	Resource fair
	Round robin

the "Vienna LTE Link Level Simulator" to ensure the same propagation environment as on the system level simulation. Further simulation parameters of both simulators are summarized in Table 11.4. The simulation results are averaged over 2500 small-scale fading and noise realizations. In order to guarantee exactly the same channel realizations for all scheduler simulations on the system level, the user positions, and large- and small-scale fading realizations are loaded from pregenerated files. On the link level, the seeds of the random number generators for fading and noise generation are set at the beginning of each simulation. A performance comparison of different scheduling strategies is shown in Figures 11.14 and 11.15 in terms of total sector throughput and fairness [34]. The figures show that the results produced by the link- and system-level simulators are very similar for both throughput and fairness. The largest difference between the results of the two simulators is less than 2 %, while the 99 % confidence intervals (too small to be identified in the figures) of the simulated throughput are much smaller. The schedulers considered pursue different goals for resource allocations. The best CQI scheduler tries to maximize total throughput and completely ignores fairness by assigning resources to the users with the best channel conditions. This is reflected in Figures 11.14 and 11.15 showing the highest system throughput and the lowest fairness for the best CQI scheduler. In contrast, the maxmin-scheduler assigns the resources so that equal throughput for all users is guaranteed, thereby maximizing Jain's fairness index [34]. Round-robin scheduling cyclically assigns the same amount of PHY resources to each user regardless of feedback information. Ignoring the user equipment feedback results in the worst throughput performance of all schedulers. The proportional fair scheduler emphasizes MU diversity by scheduling the user who has the best current channel realization relative to its own average [81]. The resource fair scheduling strategy guarantees an equal amount of resources for all users while trying to maximize the total throughput. In the simulations, the proportional fair strategy outperforms resource fair in terms of throughput and fairness, thereby resulting in a good tradeoff between throughput and fairness. Further details about the schedulers implemented, as well as more simulation results, can be found in [67]. Comparing the

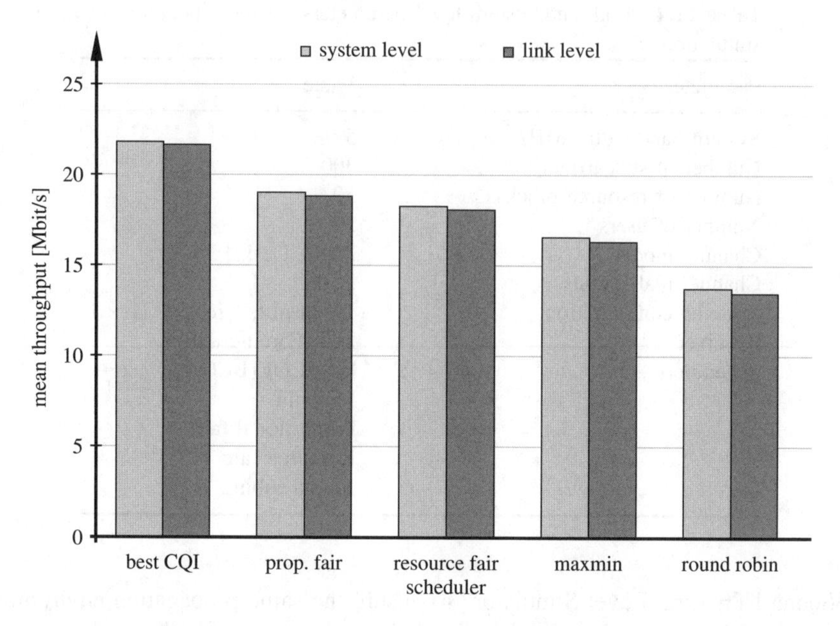

Figure 11.14 Comparison of system throughput obtained with different scheduling strategies with link and system level simulations. Reproducible by running Reproducibility_Schedulers_batch.m.

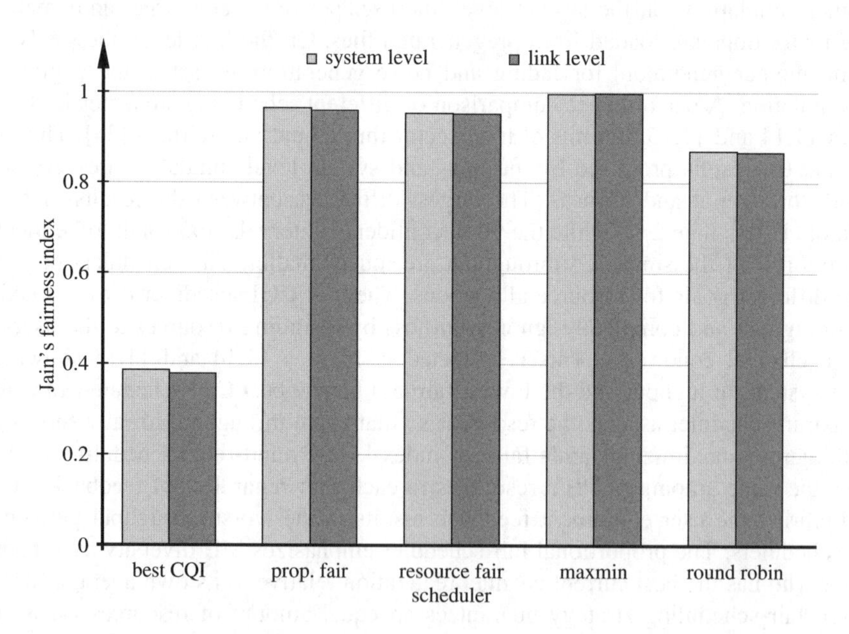

Figure 11.15 Comparison of fairness obtained with different scheduling strategies with link and system level simulations. Reproducible by running Reproducibility_Schedulers_batch.m.

simulation times of both simulators renders the system-level simulator about 15 times faster than the link-level simulator when running both simulators on a single CPU core. The link-level simulator supports multiple CPU cores, which boosts the simulation speed by a factor approximately equal to the number of cores when simulating different SNR values. The simulation results presented can be reproduced by executing the script Reproducibility_Schedulers_batch.m that can be found in the directory "paper scripts" of the link-level and the system-level simulators.

References

[1] 3GPP (2006) Technical Specification TS 25.814 'Physical layer aspects for E-UTRA,' www.3gpp.org.
[2] 3GPP (2007) Technical Specification TS 25.996 Version 7.0.0 'Spatial channel model for multiple-input multiple-output (MIMO) simulations,' www.3gpp.org.
[3] 3GPP (2008) Technical Specification TS 25.943 Version 8.0.0 'Deployment aspects (release 8),' www.3gpp.org.
[4] 3GPP (2008–2009) Technical Specification TS 36.942 'E-UTRA; LTE RF system scenarios,' www.3gpp.org.
[5] 3GPP (2009) Technical Specification TS 36.212 'Evolved universal terrestrial radio access (E-UTRA); multiplexing and channel coding,' www.3gpp.org.
[6] 3GPP (2009) Technical Specification TS 36.211 Version 8.7.0 'Evolved universal terrestrial radio access (E-UTRA); physical channels and modulation; (release 8),' http://www.3gpp.org/ftp/Specs/html-info/36211.htm.
[7] 3GPP (2009) Technical Specification TS 36.213 'Evolved universal terrestrial radio access (E-UTRA); physical layer procedures,' www.3gpp.org.
[8] 3GPP (2009) Technical Specification TS 36.101 Version 8.5.1 'Evolved universal terrestrial radio access (E-UTRA); user equipment (UE) radio transmission and reception,' www.3gpp.org.
[9] Alcatel-Lucent (2007) 'DL E-UTRA performance checkpoint,' Technical Report R1-071967.
[10] Baum, D. S., Salo, J., Del Galdo, G. *et al*. (2005) 'An interim channel model for beyond-3G systems,' in *Proceedings of the IEEE Vehicular Technology Conference (VTC05)*, Stockholm, Sweden.
[11] Blankenship, Y., Sartori, P., Classon, B. *et al*. (2004) 'Link error prediction methods for multicarrier systems,' in *Proceedings of the IEEE 60th Vehicular Technology Conference (VTC2004-Fall)*, volume 6, pp. 4175–4179.
[12] Blum, R. S. (2003) 'MIMO capacity with interference,' *IEEE Journal on Selected Areas in Communications*, **21** (5), 793–801, doi: 10.1109/JSAC.2003.810345.
[13] Boher, L., Legouable, R., and Rabineau, R. (2008) 'Performance analysis of iterative receiver in 3GPP/LTE DL MIMO OFDMA system,' in *Proceedings of the IEEE 10th International Symposium on Spread Spectrum Techniques and Applications (ISSSTA)*, pp. 103–108, doi: 10.1109/ISSSTA.2008.26.
[14] Boher, L., Legouable, R., and Rabineau, R. (2008) 'Performance analysis of iterative receiver in 3GPP/LTE DL MIMO OFDMA system,' in *Proceedings of the IEEE 10th International Symposium on Spread Spectrum Techniques and Applications 2008 (ISSSTA 2008)*, pp. 103–108, doi: 10.1109/ISSSTA.2008.26. Available from http://ieeexplore.ieee.org/stamp/stamp.jsp?arnumber=4621385.
[15] Brueninghaus, K., Astely, D., Salzer, T. *et al*. (2005) 'Link performance models for system level simulations of broadband radio access systems,' in *Proceedings of the IEEE 16th International Symposium on Personal, Indoor and Mobile Radio Communications (PIMRC 2005)*, volume 4, pp. 2306–2311, doi: 10.1109/PIMRC.2005.1651855.
[16] Cadambe, V. and Jafar, S. (2008) 'Interference alignment and degrees of freedom of the *k*-user interference channel,' *IEEE Transactions on Information Theory*, **54** (8), 3425–3441, doi: 10.1109/TIT.2008.926344. Available from http://ieeexplore.ieee.org/stamp/stamp.jsp?tp=&arnumber=4567443.
[17] Castañeda, M., Ivrlac, M., Nossek, J. *et al*. (2007) 'On downlink intercell interference in a cellular system,' in *Proceedings of the IEEE 18th International Symposium on Personal, Indoor and Mobile Radio Communications (PIMRC)*, pp. 1–5, doi: 10.1109/PIMRC.2007.4394052.

[18] Claussen, H. (2005) 'Efficient modelling of channel maps with correlated shadow fading in mobile radio systems,' in *International Symposium on Personal, Indoor and Mobile Radio Communications*.

[19] Corporation, D. E. (1984) 'A quantitative measure of fairness and discrimination for resource allocation in shared computer systems,' Technical Report TR-301, Digital Equipment Corporation.

[20] Correia, L. M. (2001) *Wireless Flexible Personalized Communications COST 259: European Co-operation in Mobile Radio Research*, John Wiley & Sons.

[21] Cover, T. M. (1998) 'Comments on broadcast channels,' *IEEE Transactions on Information Theory*, **44** (6), 2524–2530, doi: 10.1109/18.720547.

[22] Feng, Y., Krewski, A., and Schroeder, W. L. (2010) 'Simulation based comparison of metrics and measurement methodologies for OTA test of MIMO terminals,' in *Proceedings of the Fourth European Conference on Antennas and Propagation (EuCAP 2010)*. Available from http://ieeexplore.ieee.org/stamp/stamp.jsp?tp=&arnumber=5505730.

[23] Gkonis, P., Kaklamani, D., and Tsoulos, G. (2008) 'Capacity of WCDMA multicellular networks under different radio resource management strategies,' in *Proceedings of the 3rd International Symposium on Wireless Pervasive Computing (ISWPC)*, pp. 60–64, doi: 10.1109/ISWPC.2008.4556166.

[24] Gotsis, A., Komnakos, D., and Constantinou, P. (2009) 'Linear modeling and performance evaluation of resource allocation and user scheduling for LTE-like OFDMA networks,' in *Proceedings of the IEEE 6th International Symposium on Wireless Communication Systems*, pp. 196–200.

[25] Haring, L., Chalise, B., and Czylwik, A. (2003) 'Dynamic system level simulations of downlink beamforming for UMTS FDD,' in *Proceedings of the IEEE Global Telecommunications Conference (GLOBECOM)*, volume 1, pp. 492–496, doi: 10.1109/GLOCOM.2003.1258286.

[26] Hentilä, L., Kyösti, P., Käske, M. *et al*. (2007) 'MATLAB implementation of the WINNER phase II channel model ver1.1,' Available from http://www.ist-winner.org/phase_2_model.html.

[27] Holliday, T., Goldsmith, A. J., and Poor, H. V. (2008) 'Joint source and channel coding for MIMO systems: Is it better to be robust or quick?' *IEEE Transactions on Information Theory*, **54** (4), 1393–1405, doi: 10.1109/TIT.2008.917725.

[28] Ibing, A. and Jungnickel, V. (2008) 'Joint transmission and detection in hexagonal grid for 3GPP LTE,' in *Proceedings of the International Conference on Information Networking*. Available from http://ieeexplore.ieee.org/stamp/stamp.jsp?arnumber=4472825.

[29] Ikuno, J. C., Mehlführer, C., and Rupp, M. (2011) 'A novel link error prediction model for OFDM systems with HARQ,' in *Proceedings of the IEEE International Conference on Communications (ICC 2011)*, Japan.

[30] Ikuno, J. C., Schwarz, S., and Šimko, M. (2011) 'LTE rate matching performance with code block balancing,' in *Proceedings of the European Wireless Conference*.

[31] Ikuno, J. C., Wrulich, M., and Rupp, M. (2009) 'Performance and modeling of LTE H-ARQ,' in *Proceedings of the International ITG Workshop on Smart Antennas (WSA 2009)*, Berlin, Germany. Available from http://publik.tuwien.ac.at/files/PubDat_173876.pdf.

[32] Ikuno, J. C., Wrulich, M., and Rupp, M. (2010) 'System level simulation of LTE networks,' in *Proceedings of the 2010 IEEE 71st Vehicular Technology Conference (VTC2010-Spring)*, Taipei, Taiwan, doi: 10.1109/VETECS.2010.5494007. Available from http://publik.tuwien.ac.at/files/PubDat_184908.pdf.

[33] Jafar, S. A. and Fakhereddin, M. J. (2007) 'Degrees of freedom for the MIMO interference channel,' *IEEE Transactions on Information Theory*, **53** (7), 2637–2642.

[34] Jain, R., Chiu, D., and Hawe, W. (1984) 'A Quantitative Measure of Fairness and Discrimination for Resource Allocation in Shared Computer Systems,' Technical Report TR-301.

[35] Khairy, M. S., Mehlführer, C., and Rupp, M. (2010) 'Boosting sphere decoding speed through graphic processing units,' in *Proceedings of the 16th European Wireless Conference (EW 2010)*, Lucca, Italy, pp. 99–104, doi: 10.1109/EW.2010.5483399. Available from http://publik.tuwien.ac.at/files/PubDat_184823.pdf.

[36] Kim, J., Ashikhmin, A., van Wijngaarden, A. *et al*. (2004) 'On efficient link error prediction based on convex metrics,' in *Proceedings of the IEEE 60th Vehicular Technology Conference (VTC2004-Fall)*, volume 6, pp. 4190–4194, doi: 10.1109/VETECF.2004.1404869.

[37] Kolehmainen, N., Puttonen, J., Kela, P. *et al*. (2008) 'Channel quality indication reporting schemes for UTRAN long term evolution downlink,' in *Proceedings of the IEEE Vehicular Technology Conference Spring (VTC)*, pp. 2522–2526.

[38] Kolehmainen, N., Puttonen, J., Kela, P. *et al.* (2008) 'Channel quality indication reporting schemes for UTRAN long term evolution downlink,' in *Proceedings of the 67th IEEE Vehicular Technology Conference (VTC'08)*, pp. 2522–2526. Available from http://ieeexplore.ieee.org/stamp/stamp.jsp?arnumber=4526111.

[39] Kwan, R., Leung, C., and Zhang, J. (2008) 'Multiuser scheduling on the downlink of an LTE cellular system,' *Research Letters in Communications*, **2008**, article ID 323048, doi: 10.1155/2008/323048. Available from http://downloads.hindawi.com/journals/jece/2008/323048.pdf.

[40] Maddah-Ali, M., Motahari, A., and Khandani, A. (2006) 'Signaling over MIMO multi-base systems: Combination of multi-access and broadcast schemes,' in *IEEE International Symposium on Information Theory 2006*, pp. 2104–2108, doi: 10.1109/ISIT.2006.261922. Available from http://ieeexplore.ieee.org/stamp/stamp.jsp?tp=&arnumber=4036340.

[41] Mehlführer, C., Caban, S., and Rupp, M. (2008) 'An accurate and low complex channel estimator for OFDM WiMAX,' in *Proceedings of the 3rd International Symposium on Communications, Control and Signal Processing*, Malta, pp. 922–926, doi: 10.1109/ISCCSP.2008.4537355. Available from http://publik.tuwien.ac.at/files/pub-et_13650.pdf.

[42] Mehlführer, C., Caban, S., and Rupp, M. (2010) 'Measurement-based performance evaluation of MIMO HSDPA,' *IEEE Transactions on Vehicular Technology*, **59** (9), 4354–4367, doi: 10.1109/TVT.2010.2066996. Available from http://publik.tuwien.ac.at/files/PubDat_187112.pdf.

[43] Mehlführer, C., Caban, S., Wrulich, M., and Rupp, M. (2008) 'Joint throughput optimized CQI and precoding weight calculation for MIMO HSDPA,' in *Conference Record of the 42nd Asilomar Conference on Signals, Systems and Computers*, Pacific Grove, CA, pp. 1320–1325, doi: 10.1109/ACSSC.2008.5074632. Available from http://publik.tuwien.ac.at/files/PubDat_167015.pdf.

[44] Mehlführer, C., Wrulich, M., Ikuno, J. C. *et al.* (2009) 'Simulating the long term evolution physical layer,' in *Proceedings of the 17th European Signal Processing Conference (EUSIPCO 2009)*, Glasgow, Scotland, UK, pp. 1471–1478. Available from http://publik.tuwien.ac.at/files/PubDat_175708.pdf.

[45] Mehlführer, C., Wrulich, M., and Rupp, M. (2008) 'Intra-cell interference aware equalization for TxAA HSDPA,' in *Proceedings of the 3rd IEEE International Symposium on Wireless Pervasive Computing (ISWPC 2008)*, Santorini, Greece, pp. 406–409, doi: 10.1109/ISWPC.2008.4556239. Available from http://publik.tuwien.ac.at/files/pub-et_13749.pdf.

[46] Members of ITU (1997) 'Recommendation ITU-R M.1225: guidelines for evaluation of radio transmission technologies for IMT-2000,' Technical report, International Telecommunication Union (ITU).

[47] Members of WINNER (2003) 'Assessment of advanced beamforming and MIMO technologies,' Technical Report IST-2003-507581 D2.7 v1.0, WINNER.

[48] Moisio, M. and Oborina, A. (2006) 'Next generation teletraffic and wired/wireless advanced networking,' chapter Comparison of Effective SINR Mapping with Traditional AVI Approach for Modeling Packet Error Rate in Multistate Channel, LNCS 4712, Springer, Berlin.

[49] Moon, T. K. and Stirling, W. C. (2000) *Mathematical Methods and Algorithms for Signal Processing*, first edition, Prentice Hall, Upper Saddle River, NJ.

[50] Muntean, V. H., Otesteanu, M., and Muntean, G.-M. (2010) 'QoS parameters mapping for the e-learning traffic mix in LTE networks,' in *Proceedings of the International Joint Conference on Computational Cybernetics and Technical Informatics (ICCC-CONTI 2010)*, pp. 299–304. Available from http://ieeexplore.ieee.org/stamp/stamp.jsp?tp=&arnumber=5491266.

[51] Narandzic, M., Schneider, C., Thomä, R. S. *et al.* (2007) 'Comparison of SCM, SCME, and WINNER channel models,' in *Proceedings of the IEEE Vehicular Technology Conference (VTC07)*.

[52] Nguyen-Le, H., Le-Ngoc, T., and Ko, C. C. (2009) 'Joint channel estimation and synchronization for MIMO-OFDM in the presence of carrier and sampling frequency offsets,' *IEEE Transactions on Vehicular Technology*, **58** (6), 3075–3081.

[53] Niesen, U., Gupta, P., and Shah, D. (2010) 'The balanced unicast and multicast capacity regions of large wireless networks,' *IEEE Transactions on Information Theory*, **56** (5), 2249–2271, doi: 10.1109/TIT.2010.2043979.

[54] Nihtila, T. and Haikola, V. (2008) 'HSDPA MIMO system performance in macro cell network,' in *Proceedings of the IEEE Sarnoff Symposium*, doi: 10.1109/SARNOF.2008.4520092.

[55] Nokia & Nokia-Siemens-Networks (2007) 'LTE performance benchmarking,' Technical Report R1-071960, 3rd Generation Partnership Project (3GPP).

[56] Nortel (2008) 'Performance evaluation of CL MIMO under different UE speed,' Technical Report R1-080384, 3rd Generation Partnership Project (3GPP).

[57] Paulraj, A., Nabar, R., and Gore, D. (2003) *Introduction to Space–Time Wireless Communications*, first edition, Cambridge University Press, Cambridge, UK.

[58] Pedersen, K. I., Lootsma, T. F., Støttrup, M. *et al*. (2004) 'Network performance of mixed traffic on high speed downlink packet access and dedicated channels in WCDMA,' in *Proceedings of the IEEE 60th Vehicular Technology Conference (VTC)*, volume 6, pp. 4496–4500, doi: 10.1109/VETECF.2004.1404930.

[59] Peppas, K., Al-Gizawi, T., Lazarakis, F. *et al*. (2004) 'System level evaluation of reconfigurable MIMO techniques enhancements for HSDPA,' in *Proceedings of the IEEE Global Telecommunications Conference (GLOBECOM)*, volume 5, pp. 2869–2873.

[60] Pollard, A. and Heikkila, M. (2004) 'A system level evaluation of multiple antenna schemes for high speed downlink packet access,' in *Proceedings of the IEEE 15th International Symposium on Personal, Indoor and Mobile Radio Communications (PIMRC)*, volume 3, pp. 1732–1735, doi: 10.1109/PIMRC.2004.1368296.

[61] Raleigh, G. G. and Cioffi, J. M. (1998) 'Spatio-temporal coding for wireless communication,' *IEEE Transactions on Communications*, **46** (3), 357–366. Available from http://www-isl.stanford.edu/cioffi/dsm/wlpap/mimocioffi98.pdf.

[62] Ribeiro, C., Hugl, K., Lampinen, M., and Kuusela, M. (2008) 'Performance of linear multi-user MIMO precoding in LTE system,' in *Proc. 3rd International Symposium on Wireless Pervasive Computing 2008 (ISWPC 2008)*, pp. 410–414. Available from http://ieeexplore.ieee.org/stamp/stamp.jsp?arnumber=4556240.

[63] Rupp, M., Mehlführer, C., and Caban, S. (2010) 'On achieving the Shannon bound in cellular systems,' in *Proceedings of the 20th International Conference Radioelektronika*, Brno, Czech Republic, doi: 10.1109/RADIOELEK.2010.5478551. Available from http://publik.tuwien.ac.at/files/PubDat_185403.pdf.

[64] Sai A, P. and Furse, C. (2008) 'System level analysis of noise and interference analysis for a MIMO system,' in *Proceedings of the IEEE Antennas and Propagation Society International Symposium (AP-S)*, pp. 1–4, doi: 10.1109/APS.2008.4619499.

[65] Sánchez, J. J., Gómez, G., Morales-Jiménez, D., and Entrambasaguas, J. T. (2006) 'Performance evaluation of OFDMA wireless systems using WM-SIM platform,' in *Proceedings of the ACM 4th International Workshop on Mobility Management and Wireless Access (MobiWac)*, pp. 131–134, doi: 10.1145/1164783.1164808.

[66] Schwarz, S., Mehlführer, C., and Rupp, M. (2010) 'Calculation of the spatial preprocessing and link adaption feedback for 3GPP UMTS/LTE,' in *Proceedings of the IEEE 6th Conference on Wireless Advanced (WiAD)*, London, UK, doi: 10.1109/WIAD.2010.5544947. Available from http://publik.tuwien.ac.at/files/PubDat_186497.pdf.

[67] Schwarz, S., Mehlführer, C., and Rupp, M. (2010) 'Low complexity approximate maximum throughput scheduling for LTE,' in *Conference Record of the 44th Asilomar Conference on Signals, Systems and Computers*, Pacific Grove, CA, USA. Available from http://publik.tuwien.ac.at/files/PubDat_187402.pdf.

[68] Schwarz, S., Mehlführer, C., and Rupp, M. (2011) 'Throughput maximizing multiuser scheduling with adjustable fairness,' in *Proceedings of the IEEE International Conference on Communications (ICC 2011)*, Japan.

[69] Schwarz, S., Wrulich, M., and Rupp, M. (2010) 'Mutual information based calculation of the precoding matrix indicator for 3GPP UMTS/LTE,' in *Proceedings of the IEEE Workshop on Smart Antennas (WSA 2010)*, Bremen, pp. 52–58, doi: 10.1109/WSA.2010.5456388. Available from http://publik.tuwien.ac.at/files/PubDat_184424.pdf.

[70] Shuping, C., Huibinu, L., Dong, Z., and Asimakis, K. (2007) 'Generalized scheduler providing multimedia services over HSDPA,' in *Proceedings of the IEEE International Conference on Multimedia and Expo*, pp. 927–930, doi: 10.1109/ICME.2007.4284803.

[71] Šimko, M., Mehlführer, C., Wrulich, M., and Rupp, M. (2010) 'Doubly dispersive channel estimation with scalable complexity,' in *Proceedings of the International ITG Workshop on Smart Antennas (WSA 2010)*, Bremen, Germany, pp. 251–256, doi: 10.1109/WSA.2010.5456443. Available from http://publik.tuwien.ac.at/files/PubDat_181229.pdf.

[72] Šimko, M., Mehlführer, C., Zemen, T., and Rupp, M. (2011) 'Inter carrier interference estimation in MIMO OFDM systems with arbitrary pilot structure,' in *Proceedings of the 73rd IEEE Vehicular Technology Conference (VTC2011-Spring)*, Budapest, Hungary.

[73] Šimko, M., Wu, D., Mehlführer, C. *et al*. (2011) 'Implementation aspects of channel estimation for 3GPP LTE terminals,' in *Proceedings of the 17th European Wireless Conference*, Vienna, Austria.

[74] Skoutas, D., Komnakos, D., Vouyioukas, D., and Rouskas, A. (2008) 'Enhanced dedicated channel scheduling optimization in WCDMA,' in *Proceedings of the 14th European Wireless Conference*.

[75] Steepest Ascent Ltd (2009) '3GPP LTE toolbox and blockset.' Available from www.steepestascent.com/content/default.asp?page=s2_10.

[76] Thiele, L., Wirth, T., Boerner, K. *et al.* (2009) 'Modeling of 3D field patterns of downtilted antennas and their impact on cellular systems,' in *ITG International Workshop on Smart Antennas (WSA)*, Berlin, Germany.

[77] Tresch, R. and Guillaud, M. (2009) 'Cellular interference alignment with imperfect channel knowledge,' in *IEEE International Conference on Communications Workshops*. Available from http://ieeexplore.ieee.org/stamp/stamp.jsp?tp=&arnumber=5208018.

[78] Tsai, S. and Soong, A. (2003) 'Effective-SNR mapping for modeling frame error rates in multiple-state channels,' Technical Report 3GPP2-C30-20030429-010, 3GPP2.

[79] Vandewalle, P., Kovačević, J., and Vetterli, M. (2009) 'Reproducible research in signal processing,' *IEEE Signal Processing Magazine*, **26** (3), 37–47. Available from http://ieeexplore.ieee.org/stamp/stamp.jsp?tp=&arnumber=4815541.

[80] Venkataraman, P. (2009) *Applied Optimization with MATLAB Programming*, John Wiley & Sons, Inc., Hoboken, NJ.

[81] Viswanath, P., Tse, D., and Laroia, R. (2002) 'Opportunistic beamforming using dumb antennas,' *IEEE Transactions on Information Theory*, **48** (6), 1277–1294.

[82] Wan, L., Tsai, S., and Almgren, M. (2006) 'A fading-insensitive performance metric for a unified link quality model,' in *Proceedings of the IEEE Wireless Communications and Networking Conference (WCNC2006)*, volume 4, pp. 2110–2114, doi: 10.1109/WCNC.2006.1696622.

[83] Wang, Q., Caban, S., Mehlführer, C., and Rupp, M. (2009) 'Measurement based throughput evaluation of residual frequency offset compensation in WiMAX,' in *Proceedings of the 51st International Symposium ELMAR-2009*, Zadar, Croatia, pp. 233–236. Available from http://publik.tuwien.ac.at/files/PubDat_176679.pdf.

[84] Wang, Q., Mehlführer, C., and Rupp, M. (2009) 'SNR optimized residual frequency offset compensation for WiMAX with throughput evaluation,' in *Proceedings of the 17th European Signal Processing Conference (EUSIPCO 2009)*, Glasgow, Scotland, UK. Available from http://publik.tuwien.ac.at/files/PubDat_176678.pdf.

[85] Wang, Q., Mehlführer, C., and Rupp, M. (2010) 'Carrier frequency synchronization in the downlink of 3GPP LTE,' in *Proceedings of the 21st Annual IEEE International Symposium on Personal, Indoor and Mobile Radio Communications (PIMRC 2010)*, Istanbul, Turkey. Available from http://publik.tuwien.ac.at/files/PubDat_187636.pdf.

[86] Wang, X., Giannakis, G., and Marques, A. (2007) 'A unified approach to QoS-guaranteed scheduling for channel-adaptive wireless networks,' *Proceedings of the IEEE*, **95** (12), 2410–2431, doi: 10.1109/JPROC.2007.907120. Available from http://ieeexplore.ieee.org/stamp/stamp.jsp?tp=&arnumber=4383376.

[87] Weingarten, H., Steinberg, Y., and Shamai, S. (2006) 'The capacity region of the Gaussian multiple-input multiple-output broadcast channel,' *IEEE Transactions on Information Theory*, **52** (9), 3936–3964, doi: 10.1109/TIT.2006.880064.

[88] Williams, C., McLaughlin, S., and Beach, M. A. (2009) 'Exploiting multiple antennas for synchronization,' *IEEE Transactions on Vehicular Technology*, **58** (2), 773–787.

[89] Wrulich, M., Eder, S., Viering, I., and Rupp, M. (2008) 'Efficient link-to-system level model for MIMO HSDPA,' in *Proceedings of the 4th IEEE Broadband Wireless Access Workshop*, New Orleans, LA, doi: 10.1109/GLOCOMW.2008.ECP.83. Available from http://publik.tuwien.ac.at/files/PubDat_170334.pdf.

[90] Wrulich, M., Mehlführer, C., and Rupp, M. (2008) 'Interference aware MMSE equalization for MIMO TxAA,' in *Proceedings of the 3rd International Symposium on Communications, Control and Signal Processing*, St. Julians, Malta, pp. 1585–1589, doi: 10.1109/ISCCSP.2008.4537480. Available from http://publik.tuwien.ac.at/files/pub-et_13657.pdf.

[91] Wrulich, M. and Rupp, M. (2009) 'Computationally efficient MIMO HSDPA system-level modeling,' *EURASIP Journal on Wireless Communications and Networking*, **2009**, article ID 382501, doi: 10.1155/2009/382501.

[92] Wrulich, M., Weiler, W., and M. Rupp (2008) 'HSDPA performance in a mixed traffic network,' in *Proceedings of the IEEE Vehicular Technology Conference Spring (VTC)*, pp. 2056–2060, doi: 10.1109/VETECS.2008.462. Available from http://publik.tuwien.ac.at/files/pub-et_13769.pdf.

[93] Wu, D., Eilert, J., Asghar, R., and Liu, D. (2010) 'VLSI implementation of a fixed-complexity soft-output MIMO detector for high-speed wireless,' *EURASIP Journal on Wireless Communications and Networking*, **2010**, article ID 893184. Available from http://downloads.hindawi.com/journals/wcn/2010/893184.pdf.
[94] Wu, P. and Jindal, N. (2010) 'Performance of hybrid-ARQ in block-fading channels: A fixed outage probability analysis,' *IEEE Transactions on Communications*, **58** (4), 1129–1141, doi: 10.1109/TCOMM.2010.04.080622.
[95] Zemen, T. and Mecklenbräuker, C. (2005) 'Time-variant channel estimation using discrete prolate spheroidal sequences,' *IEEE Transactions on Signal Processing*, **53** (9), 3597–3607.
[96] Zheng, Y. R. and Xiao, C. (2003) 'Simulation models with correct statistical properties for Rayleigh fading channels,' *IEEE Transactions on Communications*, **51** (6), 920–928.

12

System-Level Modeling for MIMO-Enhanced HSDPA

12.1 Concept of System-Level Modeling

The problem in system-level modeling is to find a suitable mathematical abstraction of the physical layer. Generally speaking, such a model consists of two concatenated parts. Figure 12.1 shows a very abstract view of the transmission in a wireless communication system. Here, the whole transmission chain is divided into two parts. One part handles the physical layer processing, which includes in High-Speed Downlink Packet Access (HSDPA)

- spreading/despreading, scrambling/descrambling;
- modulation;
- precoding, space–time coding;
- channel estimation, synchronization; and
- receiver processing.

The second part handles the channel encoding and decoding. This part includes in HSDPA

- turbo coding/decoding;
- interleaving, rate-matching, Hybrid Automatic Repeat reQuest (HARQ); and
- demapping.

Note that both parts incorporate the transmitter and the receiver. When introducing such a split, the physical-layer processing can be modeled by a "link-measurement model," which describes the physical-layer quality, and a "link-performance model," describing the transmission performance with respect to the channel coding and decoding. For system-level purposes, the link-measurement model has to incorporate:

- the interference structure; that is, intra-cell and inter-cell interference, incorporating possible multi-user interference (again, see Figure 12.1);
- the influence of channels other than the one evaluated; for example, the synchronization channel or other Dedicated Channel (DCH) channels;

Evaluation of HSDPA and LTE: From Testbed Measurements to System Level Performance, First Edition.
Sebastian Caban, Christian Mehlführer, Markus Rupp and Martin Wrulich.

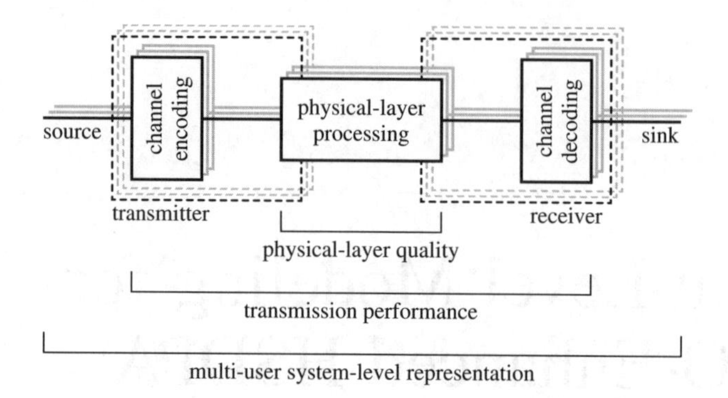

Figure 12.1 Abstract illustration of the transmission chain of a wireless communication system.

- the power allocation in the cell, as well as in the whole network; and
- the channel effects, in particular the large-scale pathloss, the antenna gain, the shadow-fading, and the small-scale fading.

At this point it is important to note that the small-scale[1] fading has to be covered by the system-level model to allow for the accurate evaluation of NodeB-based schedulers, which operate on a 2 ms Transmission Time Interval (TTI) basis in HSDPA. If those effects are covered, such a system-level model is often referred to as an "actual value interface" [9, 10].

The link-performance model then – based on the evaluations of the link-measurement model – has to represent the coding/decoding performance in terms of the Block Error Ratio (BLER), given a link-adaptation strategy [6]. Considering the flexibility of the HSDPA physical layer, this model has to cover a wide range of conditions. Together, both models provide means to assess the figures of merit in a system-level context, and they are often commonly referred to as "system-level interface."

Figure 11.6 depicts the relations of the two models in a system-level approach. On top of the already mentioned influencing factors, the figure also shows the connection to the algorithms and deployment issues that influence the system performance. In particular:

- the network layout and the mobility management in the network affect the channel and interference structure;
- the power allocation strategy can adaptively enhance the physical-layer conditions in the cell; and
- the resource scheduling strategy – which depends on the traffic and the Quality of Service (QoS) settings – controls the performance in a direct way.

[1] Sometimes in the system-level literature, for example [8, 9], the small-scale fading is denoted "fast fading" in order to point out the difference from slowly changing channel effects, such as shadow-fading. However, the notation "fast" depends strongly on the relative speed of the transmitter, the receiver, and the channel, thus rendering this term rather inaccurate. Furthermore, the classical literature [21] defines fast fading to be fading which changes the channel condition within the duration of one symbol, which is usually not modeled in the system-level context. Accordingly, in this book the term "small-scale fading" for Rayleigh fading effects is used.

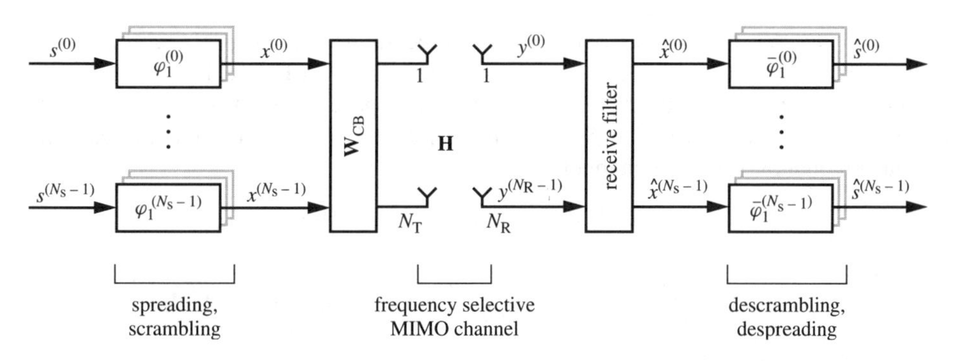

Figure 12.2 System model for a WCDMA system, reflecting the parallel transmission of N_S streams over N_T transmit antennas and reception with N_R receive antennas utilizing a receive filter.

12.2 Computationally Efficient Link-Measurement Model

To be able to derive an accurate and computationally efficient system-level model, an analytical model of the Multiple-Input Multiple-Output (MIMO) HSDPA link quality is needed. The Double Transmit Antenna Array (D-TxAA) mode [1] is shown in Figure 1.9. To represent this transmission setting, the framework of [22] is adapted in order to reflect one individual link between a Base station (NodeB) and a User Equipment (UE). Figure 12.2 depicts the model where the transmitter and the receiver are equipped with N_T and N_R antennas respectively. Note that this description allows for the modeling of more than two spatial multiplexed data streams, which would be at maximum supported by Rel'7 MIMO HSDPA. Accordingly, as illustrated in Figure 12.2, N_S parallel data streams $s^{(0)}, \ldots, s^{(N_S-1)}$ enter the system. Each data stream $s^{(n)}$, $n = 1, \ldots, N_S$ is individually encoded and modulated, as shown in Figure 1.9, and is then being spread by a number of spreading sequences $\varphi_k^{(n)}$, with k specifying the individual spreading code. This represents multi-code usage on each stream. The set of individual spreading codes utilized for stream n is denoted Φ_n; thus, $\varphi_k^{(n)} \in \Phi_n$. The streams are, of course, scrambled as well, but for the sake of clarity the scrambling/descrambling is not depicted. The scrambling sequence utilized for this particular link can still be denoted by ψ.

Assumption 12.1 *There is only one scrambling code active for each link.*

Accordingly, both streams are scrambled with the same scrambling sequence;[2] however, the NodeB could still apply different scrambling codes for different users. Note that by this convention, the model does not allow individual streams to be separately scrambled. The inverse operations of the spreading and scrambling at the receiver side are consequently denoted $\bar{\varphi}_k^{(n)}$ and $\bar{\psi}$. The spread and scrambled sequences $x^{(n)}, n = 1, \ldots, N_S$, are then

[2] The scrambling code, as opposed to the spreading code, is not targeted for a spreading of the utilized bandwidth, but rather to separate cells in the Downlink (DL) and users in the Uplink (UL). The scrambling codes are orthogonal, such as the spreading codes, but also keep their orthogonality properties when shifted by an arbitrary delay (Gold codes).

mapped to the N_T transmit antennas by the precoding matrix $\mathbf{W}_{CB} \in \mathbb{C}^{N_T \times N_S}$, which contains the precoding weights $w_1, \ldots, w_{N_T N_S}$, similar to in Figure 1.9. At the receiver, the signals are gathered with N_R antennas and chip space sampled before they enter the discrete-time receive filter. The MIMO channel $\mathbf{H} \in \mathbb{C}^{N_R \times N_T L_h}$ is modeled as a time-discrete, frequency-selective channel:

$$\mathbf{H} = \begin{bmatrix} h_0^{(1,1)} \cdots h_0^{(1,N_T)} & h_1^{(1,1)} \cdots h_1^{(1,N_T)} & \cdots & h_{L_h-1}^{(1,N_T)} \cdots h_{L_h-1}^{(1,N_T)} \\ \vdots \quad\quad \vdots & \vdots \quad\quad \vdots & \ddots & \vdots \quad\quad \vdots \\ h_0^{(N_R,1)} \cdots h_0^{(N_R,N_T)} & h_1^{(N_R,1)} \cdots h_1^{(N_R,N_T)} & \cdots & h_{L_h-1}^{(N_R,N_T)} \cdots h_{L_h-1}^{(N_R,N_T)} \end{bmatrix} \quad (12.1)$$

The entry $h_l^{(n_r,n_t)}$, $l = 1, \ldots, L_h$, $n_r = 1, \ldots, N_R$, $n_t = 1, \ldots, N_T$, denotes the lth sampled chip of the channel impulse response from transmit antenna n_t to receive antenna n_r, with a total channel length of L_h chip intervals. Note that pulse shaping, transmit and receive band-filtering, and the sampling operation can be incorporated in the MIMO channel matrix. To account for the precoding, an equivalent time-discrete channel $\mathbf{H}_w \in \mathbb{C}^{N_R \times N_S L_h}$ is defined that includes the precoding matrix $\mathbf{W}_{CB}$ and the MIMO channel,

$$\mathbf{H}_w \triangleq \mathbf{H}\left(\mathbf{I}_{L_h} \otimes \mathbf{W}_{CB}\right), \quad (12.2)$$

with $\mathbf{I}_{L_h}$ denoting the identity matrix of size $L_h \times L_h$ and $\otimes$ being the Kronecker product. Furthermore, the transmit and receive vectors can be defined as

$$\mathbf{y}_i \triangleq \left[y_i^{(1)} \cdots y_i^{(N_R)}\right]^T, \quad (12.3)$$

$$\mathbf{x}_i \triangleq \left[x_i^{(0)} \cdots x_i^{(N_S-1)} \; x_{i-1}^{(0)} \cdots x_{i-1}^{(N_S-1)} \cdots x_{i-L_h+1}^{(0)} \cdots x_{i-L_h+1}^{(N_S-1)}\right]^T, \quad (12.4)$$

and the noise vector as $\mathbf{n}_i \triangleq [\mathrm{n}_i^{(1)} \; \cdots \; \mathrm{n}_i^{(N_R)}]^T$, with $i \in \mathbb{N}$ denoting the discrete-time index, as in Chapter 1. Then the input–output relation for one link, formulated by means of the equivalent MIMO channel matrix, is given by

$$\mathbf{y}_i = \mathbf{H}_w \mathbf{x}_i + \mathbf{n}_i \quad (12.5)$$

The description in Equation (12.5) certainly allows for the description of the D-TxAA scheme which implies that $N_T = N_R = 2$ and $N_S = 1, 2$. However, the input–output relation can be used for more than two independent data streams, thus supporting possible future extensions of the scheme, or other transmission schemes [24].

12.2.1 Receive Filter

To derive a suitable link-measurement model the receive filter also has to be taken into account.

Assumption 12.2 *The model presented is restricted to linear receive filters.*

This accordingly excludes Successive Interference Cancelation (SIC) receivers as well as Maximum Likelihood (ML)-based structures. Note that, currently, Rel'5 HSDPA devices utilize a rake-receiver [20] and Rel'7 HSDPA devices are recommended to employ Minimum Mean Square Error (MMSE) equalizers [19]. Thus, current networks can be well represented by the model presented. For the derivation of the receive filter, the input–output relation in Equation (12.5) has to be extended for L_f received chips at the N_R receive antennas, specifying the filter length. By defining the "stacked" versions of the parameters introduced so far,

$$\tilde{\mathbf{y}}_i \triangleq \left[\mathbf{y}_i^\mathrm{T} \; \cdots \; \mathbf{y}_{i-L_\mathrm{f}+1}^\mathrm{T}\right]^\mathrm{T}, \tag{12.6}$$

$$\tilde{\mathbf{x}}_i \triangleq \left[x_i^{(0)} \cdots x_i^{(N_\mathrm{S}-1)} \; x_{i-1}^{(0)} \cdots x_{i-1}^{(N_\mathrm{S}-1)} \cdots x_{i-L_\mathrm{h}-L_\mathrm{f}+2}^{(0)} \cdots x_{i-L_\mathrm{h}-L_\mathrm{f}+2}^{(N_\mathrm{S}-1)}\right]^\mathrm{T}, \tag{12.7}$$

and $\tilde{\mathbf{n}}_i \triangleq [\mathbf{n}_i^\mathrm{T} \; \cdots \; \mathbf{n}_{i-L_\mathrm{f}+1}^\mathrm{T}]^\mathrm{T}$, as well as the stacked equivalent MIMO channel matrix $\tilde{\mathbf{H}}_\mathrm{w} \in \mathbb{C}^{N_\mathrm{R}L_\mathrm{f} \times N_\mathrm{S}(L_\mathrm{h}+L_\mathrm{f}-1)}$,

$$\tilde{\mathbf{H}}_\mathrm{w} \triangleq \begin{bmatrix} \mathbf{H}_\mathrm{w} & \mathbf{0}_{N_\mathrm{R},N_\mathrm{S}} & \cdots & \mathbf{0}_{N_\mathrm{R},N_\mathrm{S}} \\ \mathbf{0}_{N_\mathrm{R},N_\mathrm{S}} & \mathbf{H}_\mathrm{w} & & \mathbf{0}_{N_\mathrm{R},N_\mathrm{S}} \\ \vdots & & \ddots & \\ \mathbf{0}_{N_\mathrm{R},N_\mathrm{S}} & \mathbf{0}_{N_\mathrm{R},N_\mathrm{S}} & \cdots & \mathbf{H}_\mathrm{w} \end{bmatrix}, \tag{12.8}$$

where $\mathbf{0}_{N_\mathrm{R},N_\mathrm{S}}$ denotes the all-zero matrix of dimension $N_\mathrm{R} \times N_\mathrm{S}$, the input–output relation becomes

$$\tilde{\mathbf{y}}_i = \tilde{\mathbf{H}}_\mathrm{w}\tilde{\mathbf{x}}_i + \tilde{\mathbf{n}}_i \tag{12.9}$$

Note, however, that the stacked equivalent channel matrix $\tilde{\mathbf{H}}_\mathrm{w}$ cannot be represented by a Kronecker product because the matrix does not have a block diagonal structure, as indicated by the size of the matrices $\mathbf{0}_{N_\mathrm{R},N_\mathrm{S}}$. One particular example of a receive filter that – as already mentioned is recommended by the 3rd Generation Partnership Project (3GPP) for Rel'7 HSDPA – is the MMSE equalizer. For the input–output relation of Equation (12.9), the cost function of the MMSE receiver can be formulated as

$$J(\mathbf{F}) = \mathbb{E}\left\{\left\|\mathbf{F}\tilde{\mathbf{y}}_i - \mathbf{x}_{i-\tau}^\mathrm{S}\right\|^2\right\}, \tag{12.10}$$

where τ defines the delay of the MMSE filter, which has to fulfill $\tau \geq L_\mathrm{h}$ due to causality arguments. The vector of transmit chips to be estimated is defined as

$$\mathbf{x}_{i-\tau}^\mathrm{S} \triangleq \left[x_{i-\tau}^{(0)} \; \cdots \; x_{i-\tau}^{(N_\mathrm{S}-1)}\right]^\mathrm{T}, \tag{12.11}$$

thus representing the chips of all transmitted streams at time index $i - \tau$. A filter satisfying Equation (12.10) tries to restore the orthogonality of the individual spreading codes caused by the multipath propagation. This problem has already been investigated, for example in [13, 22], and its solution turns out to be the suitably formulated Wiener–Hopf equation

$$\mathbf{F} = \mathbf{R}_{\mathbf{x}_{i-\tau}^\mathrm{S}\tilde{\mathbf{x}}_i}\tilde{\mathbf{H}}_\mathrm{w}^\mathrm{H}(\tilde{\mathbf{H}}_\mathrm{w}\mathbf{R}_{\tilde{\mathbf{x}}_i}\tilde{\mathbf{H}}_\mathrm{w}^\mathrm{H} + \mathbf{R}_{\tilde{\mathbf{n}}_i})^{-1} \tag{12.12}$$

Assuming white, i.i.d., uncorrelated transmit sequences, the covariance matrices can be evaluated to have the following structure:

$$\mathbf{R}_{\mathbf{x}^{\mathrm{S}}_{i-\tau}\tilde{\mathbf{x}}_i} = \left[\mathbf{0}_{N_\mathrm{S}\times N_\mathrm{S}\tau} \;\; \mathbf{P} \;\; \mathbf{0}_{N_\mathrm{S}\times N_\mathrm{S}(L_\mathrm{f}+L_\mathrm{h}-\tau+2)}\right], \tag{12.13}$$

$$\mathbf{R}_{\tilde{\mathbf{x}}_i} = \mathbf{I}_{L_\mathrm{f}+L_\mathrm{h}-1} \otimes \mathbf{P}, \tag{12.14}$$

$$\mathbf{P} \triangleq \mathrm{diag}\{\left[P_1 \;\cdots\; P_n \;\cdots\; P_{N_\mathrm{S}}\right]\}; \qquad P_n = \mathbb{E}\left\{\left|x_i^{(n)}\right|^2\right\} = \sigma^2_{x_i^{(n)}} \tag{12.15}$$

specifying the powers utilized on each stream n. The equalizer length L_f and the detection delay τ are important parameters, influencing the performance of the system, but their optimization is not within the scope of this book.

Assumption 12.3 *The equalizer delay is assumed to be $\tau = L_\mathrm{f}/2$; for example, see [19].*

Note that this choice does not affect the system-level modeling itself, but does the simulation results presented in Section 12.2.4.

12.2.2 WCDMA MIMO in the Network Context

In a multi-cell, multi-user scenario, Equation (12.9) has to be extended to cover the signals from all sectors and all users, as well as all spreading codes that are active at that moment. To do so, the transmit (and, accordingly, the receive sequence) in Equation (12.9) are defined to correspond to a set of utilized spreading and scrambling sequences

$$\tilde{\mathbf{x}}_i = \sum_{\varphi_k\in\Phi} \tilde{\mathbf{x}}_i^{(\varphi_k,\psi)}, \tag{12.16}$$

where $\varphi_k \in \Phi$, $\Phi \triangleq \Phi_0 \cup \cdots \cup \Phi_{N_\mathrm{S}-1}$. The entries of $\tilde{\mathbf{x}}_i^{(\varphi_k,\psi)}$ additionally have to fulfill

$$x_i^{(n),(\varphi_k,\psi)} \triangleq \begin{cases} x_i^{(n)}; & \varphi_k^{(n)} = \varphi_k, \\ 0; & \varphi_k^{(n)} \neq \varphi_k, \end{cases} \tag{12.17}$$

which allows for representing cases in which different streams utilize only partly overlapping, or even nonoverlapping, spreading code sets $\Phi_n, n = 0, \ldots, N_\mathrm{S} - 1$. Furthermore, the users of a sector are indexed by the pair (u, b) with $u = 1, \ldots, U(b)$ because the number of active users can be different for each sector b, $b = 1, \ldots, B$. Here, the term "sector" is used because potentially every sector of a NodeB can be independently controlled in terms of its radio resources; that is, the power and spreading code allocation. In principle, a link in a sectored network is specified by the triple $(u, z, b) \in$ {user, sector, base station}, where $z = 1, \ldots, Z(b)$ is the sector and b the base station index. For the sake of notational simplicity, however, the sectors are indexed individually, denoted by b. In most practical networks, the number of sectors per base station is constant, thus not imposing the need to identify the links by three individual parameters. For a single-sector network, the pair (u, b) is obviously sufficient as well. Thus, the transmit signal of one particular link utilizing a given spreading and scrambling code pair is denoted $\tilde{\mathbf{x}}_i^{(\varphi_k,\psi),(u,b)}$. With the individual link being specified by (u, b), the set of utilized spreading codes on that link is denoted by $\Phi^{(u,b)}$. This, however, would require the pair of spreading and scrambling sequences to be indexed by (u, b); that is, $\varphi_k \rightarrow \varphi_k^{(u,b)}$ and $\psi \rightarrow \psi^{(u,b)}$ to be mathematically precise. Nevertheless, these indices are omitted, again for the sake of notational simplicity. By these definitions, the power $P^{(u,b)}_{n,\varphi_k^{(n)}}$ of stream n and spreading code $\varphi_k^{(n)}$ on link (u, b) is given by

$$
\begin{aligned}
P_n^{(u,b)} &= \mathbb{E}\left\{\left|\sum_{\varphi_k^{(n)}\in\Phi_n^{(u,b)}} x_i^{(n),(\varphi_k^{(n)},\psi),(u,b)}\right|^2\right\} \\
&= \sum_{\varphi_k^{(n)}\in\Phi_n^{(u,b)}} \sigma^2_{x_i^{(n),(\varphi_k^{(n)},\psi),(u,b)}} = \sum_{\varphi_k^{(n)}\in\Phi_n^{(u,b)}} P^{(u,b)}_{n,\varphi_k^{(n)}}
\end{aligned}
\tag{12.18}
$$

Assumption 12.4 *For Equation (12.18) to hold, uncorrelated data sequences on the individual spreading codes are assumed.*

The interference on a particular user u_0, which is called the "desired user," is received over B frequency-selective MIMO channels. This is the user for whom the physical-layer quality has to be evaluated. The base station/sector with index b_0 to which this user is attached will be called the "target sector" in the following. However, owing to the user-specific precoding, the equivalent channel matrix $\tilde{\mathbf{H}}_{\mathrm{w}}$ will be different for different individual users. Accordingly, the channel matrices involved in the network context are denoted $\tilde{\mathbf{H}}_{\mathrm{w}}^{(u,b)}$. With these definitions, the total received signal at the desired user u_0 can be evaluated by summation over all active sectors and users, as well as spreading codes:

$$
\tilde{\mathbf{y}}_i^{(u_0)} = \sum_{b=1}^{B}\sum_{u=1}^{U(b)}\sum_{\varphi_k\in\Phi^{(u,b)}} \tilde{\mathbf{H}}_{\mathrm{w}}^{(u,b)}\tilde{\mathbf{x}}_i^{(\varphi_k,\psi),(u,b)} + \tilde{\mathbf{n}}_i
\tag{12.19}
$$

The received signal is then passed through the receive filter, which leads to the useful post-equalization signal given by

$$
\hat{\mathbf{x}}_i^{(u_0)} = \sum_{b=1}^{B}\sum_{u=1}^{U(b)}\sum_{\varphi_k\in\Phi^{(u,b)}} \mathbf{F}\tilde{\mathbf{H}}_{\mathrm{w}}^{(u,b)}\tilde{\mathbf{x}}_i^{(\varphi_k,\psi),(u,b)} + \mathbf{F}\tilde{\mathbf{n}}_i
\tag{12.20}
$$

To analyze this relation and identify the different interference terms, it is necessary to decompose the receive filter into the according filters responsible for each stream,

$$
\mathbf{F} = \left[\mathbf{f}^{(0)} \ \cdots \ \mathbf{f}^{(n)} \ \cdots \ \mathbf{f}^{(N_{\mathrm{S}}-1)}\right]^{\mathrm{T}},
\tag{12.21}
$$

as well as the equivalent channel matrix into its columns,

$$
\tilde{\mathbf{H}}_{\mathrm{w}}^{(u,b)} = \left[\mathbf{h}_0^{(u,b)} \ \cdots \ \mathbf{h}_m^{(u,b)} \ \cdots \ \mathbf{h}^{(u,b)}_{N_{\mathrm{S}}(L_{\mathrm{f}}+L_{\mathrm{h}}-1)-1}\right],
\tag{12.22}
$$

where m is the index of the transmit chips for all streams entering the receive filter. Accordingly, the post-equalization signal of (12.20) for stream n can be written as

$$\hat{x}_i^{(n),(u_0)} = \sum_{b=1}^{B}\sum_{u=1}^{U(b)}\sum_{\varphi_k\in\Phi^{(u,b)}}\sum_{m=0}^{N_S(L_f+L_h-1)-1}(\mathbf{f}^{(n)})^T\mathbf{h}_m^{(u,b)}x_{i-\lfloor m/N_S\rfloor}^{(\lfloor m/N_S\rceil),(\varphi_k,\psi),(u,b)}+(\mathbf{f}^{(n)})^T\tilde{\mathbf{n}}_i, \tag{12.23}$$

where $\lfloor m/N_S\rceil$ denotes the remainder of the integer division which represents the index of the substream and $\lfloor m/N_S\rfloor$ denotes the largest integer smaller than m/N_S, which corresponds to the delay of the transmit chip. After, the receive filtering $\hat{x}_i^{(n),(u_0)}$ is multiplied with the complex conjugated scrambling and spreading codes and integrated over the period of a symbol[3] to obtain the estimated transmit symbols $\hat{s}^{(n)}$ for stream n.

12.2.3 Equivalent Fading Parameters Description

Nearly all relevant performance metrics in wireless communications are evaluated with respect to Signal to Interference and Noise Ratio (SINR). Accordingly, a suitable description of the SINR as observed in the network context to represent the physical-layer quality has to be derived. The applicability of the SINR in order to predict the BLER performance of the overall transmission scheme also seems intuitive.

Assumption 12.5 *In the following, it is assumed that the NodeB utilizes only one scrambling sequence for all of its links.*

Note that this reflects a typical WCDMA scenario as currently implemented for Rel'5 HSDPA and simplifies the notation, so the index ψ will be omitted in the following. For the derivation of the SINR, Equation (12.23) has to be decomposed into the different relevant interference and desired signal power terms. Note that all of the following power terms apply to the desired user u_0 and target sector b_0.

12.2.3.1 Desired Signal

The power of the desired signal on stream n and spreading code φ_k is given by the resulting power after the receive filter at delay τ,

$$P_{n,\varphi_k}^{S} \triangleq P_{n,\varphi_k}^{S,(u_0,b_0)} = \left|(\mathbf{f}^{(n)})^T\mathbf{h}_{\tau N_S+n}^{(u_0,b_0)}\right|^2 P_{n,\varphi_k^{(n)}}^{(u_0,b_0)}, \tag{12.24}$$

where the indication of the desired user and sector in the resulting power term is omitted for sake of notational simplicity. The idea now is to define the desired signal power as the multiplication of the power on stream n and spreading code $\varphi_k^{(n)}$ – which is known in a system-level simulation – with a so-called "equivalent fading parameter." Here, the

[3] Which corresponds to SF chip periods T_c, with $T_c = 1/3\,840\,000$ s in WCDMA systems.

equivalent fading parameter of the desired signal on stream n, G_n^{s}, also called "desired signal gain," is defined as

$$G_n^{\mathrm{s}} \triangleq \left|(\mathbf{f}^{(n)})^{\mathrm{T}}\mathbf{h}_{\tau N_{\mathrm{S}}+n}^{(u_0,b_0)}\right|^2 \Rightarrow P_{n,\varphi_k}^{\mathrm{S}} = G_n^{\mathrm{s}} P_{n,\varphi_k^{(n)}}^{(u_0,b_0)} \tag{12.25}$$

This fading parameter contains the full information about the consequences of the physical layer onto the desired signal.

12.2.3.2 Intra-cell Interference

The intra-cell interference is composed of a number of terms; in particular:

- the remaining Inter-Symbol Interference (ISI) after equalization,

$$P_{n,\varphi_k}^{\mathrm{ISI}} \triangleq P_{n,\varphi_k}^{\mathrm{ISI},(u_0,b_0)} = \sum_{\substack{m=0 \\ m\neq[\tau N_{\mathrm{S}},\tau N_{\mathrm{S}}+N_{\mathrm{S}}-1]}}^{N_{\mathrm{S}}(L_{\mathrm{f}}+L_{\mathrm{h}}-1)-1} \left|(\mathbf{f}^{(n)})^{\mathrm{T}}\mathbf{h}_m^{(u_0,b_0)}\right|^2 P_{\lfloor m/N_{\mathrm{S}}\rceil,\varphi_k^{(\lfloor m/N_{\mathrm{S}}\rceil)}}^{(u_0,b_0)}; \tag{12.26}$$

- the inter-code interference when the same scrambling but a different spreading code is used,

$$\begin{aligned} P_{n,\varphi_k}^{\mathrm{IC}} &\triangleq P_{n,\varphi_k}^{\mathrm{IC},(u_0,b_0)} \\ &= \sum_{\substack{m=0 \\ m\neq[\tau N_{\mathrm{S}},\tau N_{\mathrm{S}}+N_{\mathrm{S}}-1]}}^{N_{\mathrm{S}}(L_{\mathrm{f}}+L_{\mathrm{h}}-1)-1} \left|(\mathbf{f}^{(n)})^{\mathrm{T}}\mathbf{h}_m^{(u_0,b_0)}\right|^2 \sum_{\substack{\tilde{\varphi}_k\in\Phi^{(u,b)} \\ \tilde{\varphi}_k\neq\varphi_k}} P_{\lfloor m/N_{\mathrm{S}}\rceil,\tilde{\varphi}_k^{(\lfloor m/N_{\mathrm{S}}\rceil)}}^{(u_0,b_0)}; \end{aligned} \tag{12.27}$$

- the time-aligned intra-cell interference from users with the same spreading code as the desired user,

$$P_{n,\varphi_k}^{\mathrm{intra}_1} \triangleq P_{n,\varphi_k}^{\mathrm{intra}_1,(u_0,b_0)} = \sum_{\substack{u=1 \\ u\neq u_0}}^{U(b_0)} \sum_{m=\tau N_{\mathrm{S}}}^{\tau N_{\mathrm{S}}+N_{\mathrm{S}}-1} \left|(\mathbf{f}^{(n)})^{\mathrm{T}}\mathbf{h}_m^{(u,b_0)}\right|^2 P_{\lfloor m/N_{\mathrm{S}}\rceil,\varphi_k^{(\lfloor m/N_{\mathrm{S}}\rceil)}}^{(u,b_0)}; \tag{12.28}$$

- the not time-aligned intra-cell interference from users with the same spreading code caused by multi-path propagation,

$$\begin{aligned} P_{n,\varphi_k}^{\mathrm{intra}_2} &\triangleq P_{n,\varphi_k}^{\mathrm{intra}_2,(u_0,b_0)} \\ &= \sum_{\substack{u=1 \\ u\neq u_0}}^{U(b_0)} \sum_{\substack{m=0 \\ m\neq[\tau N_{\mathrm{S}},\tau N_{\mathrm{S}}+N_{\mathrm{S}}-1]}}^{N_{\mathrm{S}}(L_{\mathrm{f}}+L_{\mathrm{h}}-1)-1} \left|(\mathbf{f}^{(n)})^{\mathrm{T}}\mathbf{h}_m^{(u,b_0)}\right|^2 P_{\lfloor m/N_{\mathrm{S}}\rceil,\varphi_k^{(\lfloor m/N_{\mathrm{S}}\rceil)}}^{(u,b_0)}; \end{aligned} \tag{12.29}$$

- and the intra-cell interference from users with the same scrambling code, but different spreading code

$$P^{\text{intra}_3}_{n,\varphi_k} \triangleq P^{\text{intra}_3,(u_0,b_0)}_{n,\varphi_k} = \sum_{\substack{u=1\\u\neq u_0}}^{U(b_0)} \sum_{\substack{m=0\\m\neq[\tau N_S,\tau N_S+N_S-1]}}^{N_S(L_f+L_h-1)-1} \left|(\mathbf{f}^{(n)})^{\mathrm{T}}\mathbf{h}_m^{(u,b_0)}\right|^2 \sum_{\substack{\tilde{\varphi}_k\in\Phi^{(u,b)}\\\tilde{\varphi}_k\neq\varphi_k}} P^{(u,b_0)}_{\lfloor m/N_S\rceil,\tilde{\varphi}_k^{(\lfloor m/N_S\rceil)}} \tag{12.30}$$

As already mentioned, owing to the assumption that there is only one scrambling code per link, as well as the more restrictive assumption that the NodeB uses only one scrambling code overall, no interference term for inter-scrambling code interference appears in the context of the intra-cell interference. Furthermore, in a practical system implementation of MIMO HSDPA, it is not very likely that other users will be scheduled in parallel to the desired user utilizing the same or even overlapping spreading code sets. In such a case, the users would only be separated by means of Spatial Division Multiple Access (SDMA), and the resulting interference would see a spreading gain.

Assumption 12.6 *No pure SDMA user separation is performed, which implies that*

$$P^{\text{intra}_1}_{n,\varphi_k} = 0 \tag{12.31}$$

From the remaining terms, $P^{\text{ISI}}_{n,\varphi_k}$ and $P^{\text{IC}}_{n,\varphi_k}$ represent the intra-cell interference generated by the desired user themselves, which can be called "self-interference,"

$$P^{\text{self}}_{n,\varphi_k} \triangleq P^{\text{ISI}}_{n,\varphi_k} + P^{\text{IC}}_{n,\varphi_k} = \sum_{\substack{m=0\\m\neq[\tau N_S,\tau N_S+N_S-1]}}^{N_S(L_f+L_h-1)-1} \left|(\mathbf{f}^{(n)})^{\mathrm{T}}\mathbf{h}_m^{(u_0,b_0)}\right|^2 P^{(u_0,b_0)}_{\lfloor m/N_S\rceil} \tag{12.32}$$

Similarly, $P^{\text{intra}_2}_{n,\varphi_k}$ and $P^{\text{intra}_3}_{n,\varphi_k}$ specify the intra-cell interference generated by all other users in the sector, which may be called "other-user interference,"

$$P^{\text{other}}_{n,\varphi_k} \triangleq P^{\text{intra}_2}_{n,\varphi_k} + P^{\text{intra}_3}_{n,\varphi_k} = \sum_{\substack{u=1\\u\neq u_0}}^{U(b_0)} \sum_{\substack{m=0\\m\neq[\tau N_S,\tau N_S+N_S-1]}}^{N_S(L_f+L_h-1)-1} \left|(\mathbf{f}^{(n)})^{\mathrm{T}}\mathbf{h}_m^{(u,b_0)}\right|^2 P^{(u,b_0)}_{\lfloor m/N_S\rceil} \tag{12.33}$$

Assumption 12.7 *For the overall intra-cell interference, all streams $n = 0, \ldots, N_S - 1$ transmitted to a user u have the same power,*

$$P_n^{(u,b)} \equiv \frac{1}{N_S} P^{(u,b)}, \tag{12.34}$$

with $P^{(u,b)}$ denoting the total power spent for user u by sector b.

Note that this limits the applicability of the model to cases where no dynamic power loading for the individual streams takes place. More advanced power resource allocation (for example, waterfilling-based algorithms) are not covered. Consequently, the total intra-cell interference, as seen by the desired user u_0 on stream n and spreading code φ_k, becomes

$$P^{\text{intra}}_{n,\varphi_k} = P^{\text{self}}_{n,\varphi_k} + P^{\text{other}}_{n,\varphi_k} = \underbrace{\frac{1}{N_\text{S}} P^{(u_0,b_0)} \sum_{\substack{m=0 \\ m \neq [\tau N_\text{S}, \tau N_\text{S}+N_\text{S}-1]}}^{N_\text{S}(L_\text{f}+L_\text{h}-1)-1} \left|(\mathbf{f}^{(n)})^\text{T} \mathbf{h}_m^{(u_0,b_0)}\right|^2}_{\text{self-interference}} + \underbrace{\frac{1}{N_\text{S}} \sum_{\substack{u=1 \\ u \neq u_0}}^{U(b_0)} P^{(u,b_0)} \sum_{\substack{m=0 \\ m \neq [\tau N_\text{S}, \tau N_\text{S}+N_\text{S}-1]}}^{N_\text{S}(L_\text{f}+L_\text{h}-1)-1} \left|(\mathbf{f}^{(n)})^\text{T} \mathbf{h}_m^{(u,b_0)}\right|^2}_{\text{other-user interference}} \quad (12.35)$$

So far, Equation (12.35) still does not allow for a decoupling into equivalent fading parameters and power terms because $\mathbf{h}_m^{(u,b_0)}$ depends on the user index, and in particular on the choice of the users regarding their precoding. To obtain a model with low computational complexity, this mechanism, however, should be contained in the link-measurement model itself, so that in a system-level simulation no calculations of the precoding and the user-specific equivalent MIMO channels has to be performed.

Assumption 12.8 *For an equivalent fading parameter decomposition of the intra-cell interference, the actual intra-cell interference is approximated by its expected value with respect to the precoding choices of the other users,*

$$\bar{P}^{\text{intra}}_{n,\varphi_k} = \mathbb{E}_w\left\{P^{\text{intra}}_{n,\varphi_k}\right\} \quad (12.36)$$

Accordingly, by defining the "intra-cell orthogonality" of stream n, o_n^{intra}, as

$$o_n^{\text{intra}} \triangleq \frac{1}{N_\text{S}} \frac{1}{G_n^\text{s}} \sum_{\substack{m=0 \\ m \neq [\tau N_\text{S}, \tau N_\text{S}+N_\text{S}-1]}}^{N_\text{S}(L_\text{f}+L_\text{h}-1)-1} \left|(\mathbf{f}^{(n)})^\text{T} \mathbf{h}_m^{(u_0,b_0)}\right|^2 = \frac{1}{N_\text{S}} \frac{1}{G_n^\text{s}} \gamma_\text{s}, \quad (12.37)$$

and the "beamforming orthogonality gain" of stream n, o_n^{BF}, as

$$o_n^{\text{BF}} \triangleq \gamma_\text{s}^{-1} \mathbb{E}_w \left\{ \sum_{\substack{m=0 \\ m \neq [\tau N_\text{S}, \tau N_\text{S}+N_\text{S}-1]}}^{N_\text{S}(L_\text{f}+L_\text{h}-1)-1} \left|(\mathbf{f}^{(n)})^\text{T} \mathbf{h}_m^{(u,b_0)}\right|^2 \right\}, \quad (12.38)$$

the intra-cell interference power in Equation (12.35) does not depend on the user's individual choices of the precoding weights. By these definitions, it follows that

$$\bar{P}^{\text{intra}}_{n,\varphi_k} = \left[P^{(u_0,b_0)} + o_n^{\text{BF}} \sum_{\substack{u=1 \\ u \neq u_0}}^{U(b_0)} P^{(u,b_0)} \right] o_n^{\text{intra}} G_n^\text{s} \quad (12.39)$$

The equivalent fading parameters again contain the full information about the physical-layer implications on the power distribution in the target sector. In the practical implementation of the 3GPP Rel'7 MIMO HSDPA standard, the precoding is strongly quantized to limit the amount of feedback needed. The current definition of the precoding codebook for Rel'7 HSDPA is specified in [1]. If the codebook of the precoding utilized for a user u, $\mathbf{W}_{\mathrm{CB}}^{(u)}$, is denoted as Ω, with its cardinality given by $|\Omega|$, the equivalent fading parameter o_n^{BF} can be evaluated as

$$o_n^{\mathrm{BF}} \triangleq \gamma_{\mathrm{s}}^{-1} \frac{1}{|\Omega|} \sum_{\mathbf{W}_{\mathrm{CB}}^{(u)} \in \Omega} \sum_{\substack{m=0 \\ m \neq [\tau N_{\mathrm{S}}, \tau N_{\mathrm{S}}+N_{\mathrm{S}}-1]}}^{N_{\mathrm{S}}(L_{\mathrm{f}}+L_{\mathrm{h}}-1)-1} \left| (\mathbf{f}^{(n)})^{\mathrm{T}} \mathbf{h}_m^{(u,b_0)} \right|^2 \tag{12.40}$$

12.2.3.3 Inter-stream Interference

The interference generated by the parallel transmission of other spatially multiplexed streams is given by the term

$$P_{n,\varphi_k}^{\mathrm{INT}} \triangleq P_{n,\varphi_k}^{\mathrm{INT},(u_0,b_0)} = \sum_{\substack{m=0 \\ m \neq n}}^{N_{\mathrm{S}}-1} \left| (\mathbf{f}^{(n)})^{\mathrm{T}} \mathbf{h}_{\tau N_{\mathrm{S}}+m}^{(u_0,b_0)} \right|^2 P_{m,\varphi_k^{(m)}}^{(u_0,b_0)}, \tag{12.41}$$

from which the equivalent fading parameter for the "inter-stream interference orthogonality" can immediately be defined:

$$o_n^{\mathrm{INT}} \triangleq \sum_{\substack{m=0 \\ m \neq n}}^{N_{\mathrm{S}}-1} \left| (\mathbf{f}^{(n)})^{\mathrm{T}} \mathbf{h}_{\tau N_{\mathrm{S}}+m}^{(u_0,b_0)} \right|^2 \tag{12.42}$$

With this, the inter-stream interference can be calculated by

$$P_{n,\varphi_k}^{\mathrm{INT}} = o_n^{\mathrm{INT}} P_{m,\varphi_k^{(m)}}^{(u_0,b_0)}, \tag{12.43}$$

where the equivalent fading parameter o_n^{INT} now contains all the information about the physical layer regarding the interference suppression between the individual streams being transmitted. Note that the inter-stream interference only occurs for those spreading codes that are used on both streams, and that this interference will thus see a spreading gain at the receiver.

12.2.3.4 Inter-cell Interference

The inter-cell interference is caused by all neighboring sectors in the network. Owing to the scrambling code, the corresponding interference terms will not see a spreading factor gain. In general, the inter-cell interference power derived from Equation (12.23) is given by

$$P_{n,\varphi_k}^{\text{inter}} \triangleq P_{n,\varphi_k}^{\text{INT},(u_0,b)}$$
$$= \sum_{\substack{b=1 \\ b\neq b_0}}^{B} \sum_{u=1}^{U(b)} \sum_{\varphi_k \in \Phi^{(u,b)}} \sum_{m=0}^{N_S(L_f+L_h-1)-1} \left|(\mathbf{f}^{(n)})^{\mathrm{T}}\mathbf{h}_m^{(u,b)}\right|^2 P_{\lfloor m/N_S \rceil, \varphi_k^{(\lfloor m/N_S \rceil)}}^{(u,b)} \tag{12.44}$$

This description does not allow the decomposition into power terms and equivalent fading parameters.

Assumption 12.9 *The precoding applied by the neighboring cells is independent of the user channel; thus:*

$$\mathbf{h}_m^{(u,b)} \to \mathbf{h}_m^{(b)} \tag{12.45}$$

If this assumption is taken into account, the inter-cell interference power can be rewritten as

$$P_{n,\varphi_k}^{\text{inter}} = \sum_{\substack{b=1 \\ b\neq b_0}}^{B} \sum_{m=0}^{N_S(L_f+L_h-1)-1} \left|(\mathbf{f}^{(n)})^{\mathrm{T}}\mathbf{h}_m^{(b)}\right|^2 \underbrace{\sum_{u=1}^{U(b)} \sum_{\varphi_k \in \Phi^{(u,b)}} P_{\lfloor m/N_S \rceil, \varphi_k^{(\lfloor m/N_S \rceil)}}^{(u,b)}}_{\text{total transmit power per stream}}, \tag{12.46}$$

where the total transmit power per stream – this is the power for the transmission of the data channel High-Speed Downlink Shared CHannel (HS-DSCH) on one particular stream – of the sector can be expressed as

$$P_n^{(b)} \triangleq \sum_{u=1}^{U} P_n^{(u,b)} \tag{12.47}$$

If, furthermore, Assumption 12.7 is employed, the inter-cell interference power becomes

$$P_{n,\varphi_k}^{\text{inter}} = \sum_{\substack{b=1 \\ b\neq b_0}}^{B} \sum_{m=0}^{N_S(L_f+L_h-1)-1} \left|(\mathbf{f}^{(n)})^{\mathrm{T}}\mathbf{h}_m^{(b)}\right|^2 \frac{1}{N_S} P^{(b)}, \tag{12.48}$$

with $P^{(b)}$ being the total power spend by sector b for data transmission. The equivalent fading parameter specifying the "inter-cell interference gain" is defined as

$$G_{n,b}^{\text{inter}} \triangleq \frac{1}{N_S} \sum_{m=0}^{N_S(L_f+L_h-1)-1} \left|(\mathbf{f}^{(n)})^{\mathrm{T}}\mathbf{h}_m^{(b)}\right|^2, \tag{12.49}$$

thus rendering the inter-cell interference to be evaluated by

$$P_{n,\varphi_k}^{\text{inter}} = \sum_{\substack{b=1 \\ b\neq b_0}}^{B} G_{n,b}^{\text{inter}} P^{(b)} \tag{12.50}$$

The inter-cell interference gain equivalent fading parameter represents for every neighboring sector the influence of the according transmissions onto stream n of the desired user. Some details about the generation of the equivalent fading parameter $G_{n,b}^{\text{inter}}$ and the implications of Assumption 12.9 can be found in [23].

12.2.3.5 Thermal Noise

The thermal noise is modeled as a white and Gaussian random process, statistically independent and with identical average power on all receive antennas. Accordingly, the power of the noise can be calculated as

$$P^{\text{noise}} = \mathbb{E}\left\{\left\|(\mathbf{f}^{(n)})^{\mathrm{T}}\tilde{\mathbf{n}}_i\right\|^2\right\} = \sigma_{\mathrm{n}}^2\mathbb{E}\left\{\left\|\mathbf{f}^{(n)}\right\|^2\right\} \tag{12.51}$$

12.2.4 Generation of the Equivalent Fading Parameters

The major benefit of the equivalent fading parameter description is that these factors can be evaluated prior to the system-level simulation and stored for multiple applications. To do so, a fading simulation has to be conducted that calculates the receive filter as well as the precoding of the desired user for the intra-cell channel $\tilde{\mathbf{H}}_{\mathrm{w}}^{(u_0,b_0)}$, and evaluates the equivalent fading parameters G_n^{s}, o_n^{intra}, o_n^{BF}, o_n^{INT}, as well as the equivalent fading parameter for the inter-cell interference $G_{n,b}^{\text{inter}}$.

To assess the characteristics of the so-obtained equivalent fading parameter representation, a set of simulations for a number of channels and system setups is conducted. The MIMO channel coefficients are generated according to the improved Zheng model, see [25, 26], and the MMSE equalizer weights with the precoding coefficients are determined, assuming perfect channel knowledge at the receiver. Furthermore, no signalization errors on the signaling channel for the precoding index are assumed. A realistic precoding delay is implemented such that the actual MMSE weights are slightly mismatched to the channel conditions, with the mismatch depending on the coherence time of the channel. Note that in the Jakes fading model utilized, the coherence time is a function of the Doppler spread, directly related to the speed of the UE.

Assumption 12.10 *For the equivalent fading parameter generation:*

- *channel delay profiles of the individual Single-Input Single-Output (SISO) channels of all users are assumed to be equal;*
- *all streams utilize the same set of spreading codes, $\Phi_n \equiv \Phi$, which is required for the 3GPP D-TxAA HSDPA operation.*

Figure 12.3 shows the empirical cumulative density functions (cdfs) of the equivalent fading parameters for the 2×2 MIMO channel when utilizing a double-stream transmission. In Figures 12.3(a) and 12.3(b), it can be observed that the inter-stream interference orthogonality range is shrunk and shifted slightly to higher values in the case of the longer delay spread Pedestrian B (PedB) channel. This indicates that the receive filter cannot resolve the spatial separation of the streams equally well as in the Pedestrian A (PedA) scenario. Furthermore, owing to the open-loop operation, the equivalent fading parameters

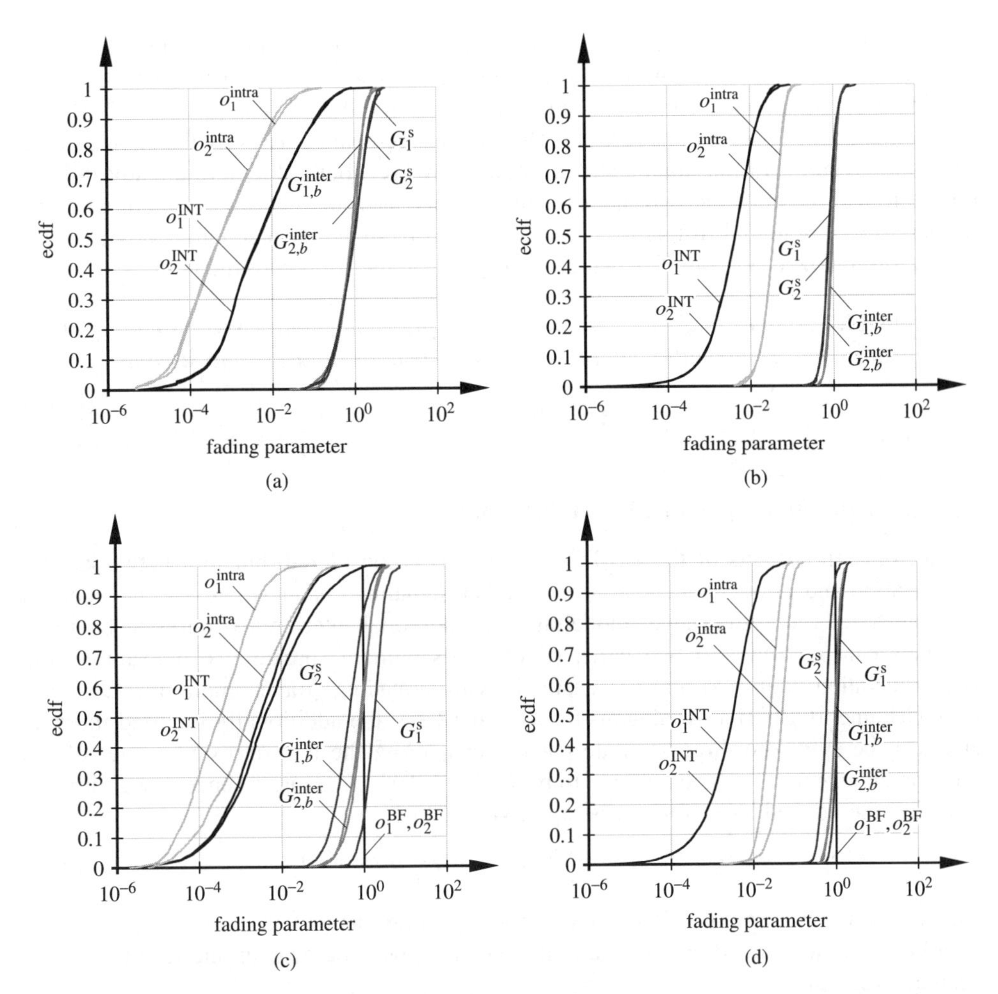

Figure 12.3 Empirical cdfs of the equivalent fading parameters in case of a double-stream transmission over an $N_T \times N_R = 2 \times 2$ MIMO channel: (a) PedA open-loop 2×2 double stream; (b) PedB open-loop 2×2 double stream; (c) PedA closed-loop 2×2 double stream; (d) PedB closed-loop 2×2 double stream.

of both streams are equal, due to fact that no channel state information is exploited to adapt to by means of precoding.

The situation changes, however, when inspecting the empirical cdfs of the equivalent fading parameters for the closed-loop scenario. Figures 12.3(c) and 12.3(d) clearly show a favoring of stream 1 compared with stream 2. For higher delay spread channels, the performance gap between the two streams becomes smaller, rendering the individual performance more equally. The reason for the performance gap between stream 1 and stream 2 lies in the precoding choice as defined by 3GPP [1]. Owing to the strong quantization of the utilizable precoding vectors, as well as the coupling between the precoding for stream 1 and stream 2, stream 1 will always be favored. Details on this issue can be found in Section 1.3.2. Another interesting observation in Figures 12.3(c) and 12.3(d) is that the beamforming orthogonality gain o_n^{BF} is constant and equal to one in these simulation setups. A simple argument for this is that, in a 2×2 MIMO channel, only two Degrees of Freedom (DoFs) exist to separate users by the choice of the precoding vectors. In the double-stream mode, however, both DoFs are already utilized for the transmission of the two independent streams, such that the users cannot gain any further from a potential different choice of their corresponding precoding weights. This can be summarized, the proof of which is in [23], by the following theorem.

Theorem 12.1 *Assume that $N_{\text{S}} = N_{\text{T}} = 2$ and the precoding matrix $\mathbf{W}_{CB}$ is unitary. Then the beamforming orthogonality gain is identical to one for all users and streams:*

$$\o_n^{\text{BF}} \equiv 1 \tag{12.52}$$

12.2.5 Influence of Non-Data Channels

So far, only the effects of the HS-DSCH have been considered, but in a network, synchronization and pilot channels, as well as other signaling channels, are also needed.[4] In the context of the system-level modeling, the additional interference imposed by these channels can be split into two parts: (i) channels without spreading (for example, the synchronization channels) and (ii) channels with spreading (for example, the Common Pilot CHannel (CPICH)). Analogous to Section 12.2.3, the additionally imposed interference can be split into three separate parts. Considering the total transmit power of the non-spread channels in the target cell to be $P^{\text{non-spread},(b_0)}$, the interference power directly affecting the desired stream n can be approximated by

$$P_{n,\varphi_k}^{\text{signaling}_1} \triangleq P_{n,\varphi_k}^{\text{signaling}_1,(b_0)} \approx G_n^{\text{s}} P^{\text{non-spread},(b_0)}, \tag{12.53}$$

which is only an approximation, because the signaling channels are not pre-coded. The interference power added on top of all other streams, which are again interfering with the desired stream, is

$$P_{n,\varphi_k}^{\text{signaling}_2} \triangleq P_{n,\varphi_k}^{\text{signaling}_2,(b_0)} \approx \sum_{\substack{m=0 \\ m \neq n}}^{N_{\text{S}}} o_n^{\text{INT}} P^{\text{non-spread},(b_0)}, \tag{12.54}$$

and the remaining inter-symbol and inter-code interference can be evaluated to be

$$P_{n,\varphi_k}^{\text{signaling}_3} \triangleq P_{n,\varphi_k}^{\text{signaling}_3,(b_0)} \approx o_n^{\text{intra}} G_n^{\text{s}} P^{\text{non-spread},(b_0)} \tag{12.55}$$

Adding these individual parts, the total interference caused by the non-spread signaling channel is given by

$$\sum_{i=1}^{3} P_{n,\varphi_k}^{\text{signaling}_i} \approx P^{\text{non-spread},(b_0)} \left[G_n^{\text{s}} \left(1 + o_n^{\text{intra}}\right) + \sum_{\substack{m=0 \\ m \neq n}}^{N_{\text{S}}} o_n^{\text{INT}} \right] \tag{12.56}$$

[4] These are referred to as "non-data channels"; cf. Section 5.5.

The interference caused by other spread channels can be treated similarly to the intra-cell interference, caused by loss of spreading code orthogonality. Assuming the total power of other spread channels in the cell to be $P^{\text{spread},(b_0)}$, the interference can be evaluated to be

$$P_{n,\varphi_k}^{\text{signaling}_4} \triangleq P_{n,\varphi_k}^{\text{signaling}_4,(b_0)} \approx o_n^{\text{intra}} G_n^{\text{s}} P^{\text{spread},(b_0)}; \tag{12.57}$$

accordingly, the total signaling interference power becomes

$$\begin{aligned} P_{n,\varphi_k}^{\text{signaling}} &= \sum_{i=1}^{4} P_{n,\varphi_k}^{\text{signaling}_i} \\ &= P^{\text{non-spread},(b_0)} \left[G_n^{\text{s}}(1 + o_n^{\text{intra}}) + \sum_{\substack{m=0 \\ m \neq n}}^{N_{\text{S}}} o_n^{\text{INT}} \right] + o_n^{\text{intra}} G_n^{\text{s}} P^{\text{spread},(b_0)} \end{aligned} \tag{12.58}$$

Enhanced interference cancelation receivers are not covered by this modeling; see also Assumption 12.2.

12.2.6 *Resulting SINR Description*

With these findings, the SINR on stream n and spreading code φ_k, as observed after equalization and despreading (also often called symbol-level SINR), can be expressed as

$$\text{SINR}_{n,\varphi_k} = \frac{\text{SF} \cdot P_{n,\varphi_k}^{s} \frac{1}{L^{(b_0)}}}{\begin{array}{c}(\text{SF} \cdot P_{n,\varphi_k}^{\text{INT}} + \bar{P}_{n,\varphi_k}^{\text{intra}} + P_{n,\varphi_k}^{\text{signaling}}) \frac{1}{L^{(b_0)}} \\ + \sum_{\substack{b=1 \\ b \neq b_0}}^{B} G_{n,b}^{\text{inter}} P^{(b)} \frac{1}{L^{(b)}} + P^{\text{noise}}\end{array}}, \tag{12.59}$$

where SF denotes the spreading factor of the HSDPA data channels. The term $1/L^{(b)}$ represents the large-scale and shadow-fading pathloss components of the MIMO channel in the network context. For D-TxAA HSDPA in particular, the inter-stream interference power $P_{n,\varphi_k}^{\text{INT}}$ in Equation (12.59) simplifies to

$$P_{n,\varphi_k}^{\text{INT}} = o_{\neg n}^{\text{INT}} P_{\neg n,\varphi_k^{(\neg n)}}^{(u_0,b_0)}, \tag{12.60}$$

where $\neg n$ denotes the respective second stream of the D-TxAA transmission. The description contains only equivalent fading parameters and power terms, which do not depend on each other. Accordingly, traces for the fading parameters may be generated prior to the system-level simulation; thus, only real-valued scalar multiplications will occur in Equation (12.59) when evaluated on the system level. It is also worth noting that, in principle, only one trace for each equivalent fading parameter has to be generated; statistical independence between different realizations for different users can always be achieved by choosing independent starting indices – for example, drawn uniformly over the length of the respective trace. However, if such an implementation is desired, the equivalent fading parameter traces have to be interpreted as a loop trace, which means that when the current index within the trace reaches its end, it will start from the beginning of the trace again. The index pointing to the current position in the trace thus has to be taken modulo the length of the trace. To avoid "edge"-effects when the trace index jumps from the end to the beginning, some sort of "smoothing" has to be performed; for example, by linear interpolation.

Please note that the proposed structure can be utilized for arbitrary channel models, such as the Spatial Channel Model (SCM) or the SCM Extension (SCME) [3], and that a detailed validation of the link-measurement model showing the applicability for various operation modes of the physical layer can be found in [23].

The largest benefit of a system-level model instead of a full physical-layer simulation is the savings in terms of computational complexity. In [23], the computational complexity is assessed both analytically and by means of simulations, showing that the relative complexity, approximated by the relative difference between the simulation run times, follows a polynomial of third order in L_f and proves huge savings in computation.

12.3 Link-Performance Model

So far, the system-level modeling in this chapter has introduced a way of accurately evaluating the link quality in terms of the post-equalization and despreading SINR. As mentioned in Section 12.2, the link-measurement model depends on numerous factors determined by the network, as well as the transmission scheme selected.

A "link-performance" model, on the other hand, should provide an estimate of the link performance in terms of the BLER or Packet Error Ratio (PER) when the decisions on the Radio Resource Management (RRM), scheduling, and link adaptation are already known [15]. Link-performance models generate bit- or block-error probabilities for performance estimation at the run time of the system-level simulation. This, in turn, can generate, for example, retransmissions and affect the slow link adaptation, as for example the slow outer-loop power control in Rel'4 Universal Mobile Telecommunications Systems (UMTS) systems. In principle, a link-performance model can be viewed as the conditional probability that the transmitted codeword is decoded erroneously given the link measurement during the interleaving period. Since the RRM decisions are already known when the link-performance model is needed in the system-level context, these models are generally less complicated than the link-measurement models, as mismatches due to possibly unknown scheduling decisions are not an issue.

Also note that a suitable link-performance model can be used for a large set of different system realizations. Since the link-measurement model developed represents the complete physical-layer processing in between the channel-coding-related codeword processing – as can be observed, for example, in Figure 12.1 – no changes in the link-performance model are needed when altering details within the link-measurement model. Accordingly, for such investigations the equivalent fading parameters can easily be generated for some variations of the physical-layer processing, allowing for flexible investigations of the system modeled.

In Global System for Mobile communications (GSM), link-performance modeling at the system level was based on the average PER for all transmissions on one link, thus evaluating an average BLER as a function of the linear average SINR, during the time a user was served. This may be adequate as long as every packet encounters similar channel statistics, which would imply very large packet or coding-block lengths with respect to the channel coherence time. However, it is known that specific channel realizations, especially for small transport-block lengths, may result in greatly different performances from those predicted by the average link-quality [14]. This has a significant impact on network-based mechanisms, such as scheduling. Thus, as already mentioned in Section 12.2, useful modeling of the performance of fast scheduling, fast link-adaptation, and HARQ is impossible with average or large-scale parameters and, accordingly, small-scale effects have to be included.

12.3.1 Link-Performance Model Concept

The basic structure of the proposed link-performance model and its connection to the proposed link-measurement model are depicted in Figure 12.4. As illustrated, the link-quality model provides small-scale fading-dependent link-quality estimates, $\mathrm{SINR}_{n,\varphi_k}$, given by Equation (12.59). Note that, as explained in Chapter 1, the packet unit in HSDPA is given by a subframe or transport block with the length of one TTI. A straightforward approach would thus be to evaluate the link quality once every TTI and utilize the resulting measurements as input parameters to the link-performance model. If, however, the link quality changes significantly during one subframe, a link-performance model utilizing TTI-based SINR sampling – corresponding to an "average" SINR evaluation for every subframe – would not reflect these variations and, thus, probably lead to too optimistic performance predictions.

To overcome this problem, an "oversampling" in terms of the link-measurement model estimates can be conducted, thus evaluating $\mathrm{SINR}_{n,\varphi_k}$ more than once per subframe.

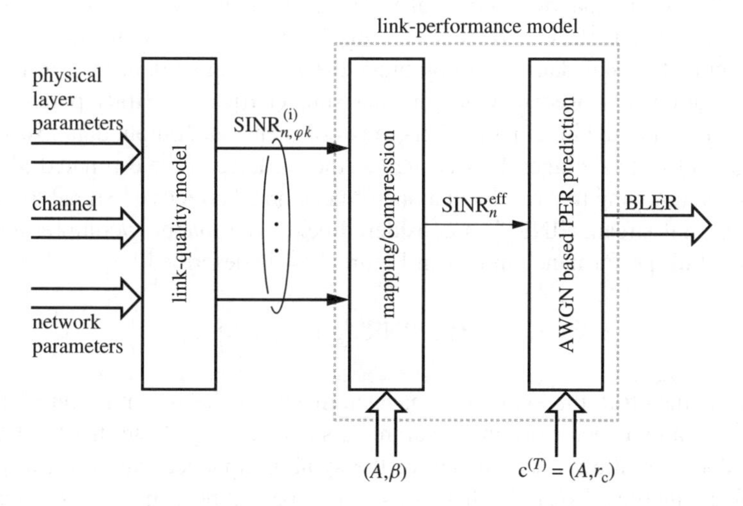

Figure 12.4 Basic concept of the proposed link-performance model, including the input parameter generation by the link-measurement model.

In particular, the SINR shall be evaluated on a slot basis; that is, three times per TTI:

$$\mathrm{SINR}_{n,\varphi_k}^{(i)}, i = 1, 2, 3, \tag{12.61}$$

with i denoting the slot index within one TTI. This obviously allows for an increased accuracy in movement scenarios. To estimate the maximum speed, covered accurately by this link-measurement model sampling rate, the Jakes model [11] can be applied. It is well known [18] that the channel coherence T_C time can be approximated as a function of the Doppler spread D_S; that is, $T_C \approx 1/D_S$. Assuming a Doppler power spectral density as derived by Jakes, the Doppler spread can be approximated by $D_S \approx 2f_{s,max}$, with $f_{s,max}$ denoting the maximum Doppler frequency shift relative to the carrier frequency f_c. This is not always fulfilled in practice, but is a reasonable approximation [4]. The maximum Doppler frequency shift can be calculated according to $f_{s,max} = f_c v/c_0$, with v denoting the relative speed of the transmitter/receiver and the scattering objects, and c_0 being the light speed. Accordingly, the maximum v to result in a channel coherence time, less than one slot duration of 2/3 ms is approximately 31.25 km/h. Considering that the target scenario of HSDPA is static, this seems to be sufficient for most investigation scenarios.

Given the slot-sampled link-measurement model estimations, the link-performance model has to estimate the link performance in terms of the ACKnowledged (ACK)/ Non-ACKnowledged (NACK) ratio report for the individual transport block. In principle, the best (in terms of accuracy) link-performance model approach would be to utilize a multidimensional mapping that directly converts the SINRs to the anticipated BLER,

$$\mathrm{SINR}_{n,\varphi_k}^{(i)} \mapsto \mathrm{BLER}, \tag{12.62}$$

but this would require analytic models of very high complexity, or unfeasibly large look-up tables. A more suitable way of representing the information contained in the sampled link-measurement model data is to compress the input parameters to a few "effective" link-quality parameters. Recent work on that subject tries to extract potential gains in accuracy by mapping the input parameters to two-dimensional output data [5], such as, for example, given by the average SINR as well as the variance of the estimated SINR values. A compression mapping that results in a one-dimensional "effective" Signal to Noise Ratio (SNR) per spatial stream $\mathrm{SINR}_n^{\mathrm{eff}}$ is considered next. The mapping/compression stage of the proposed link-performance model in Figure 12.4 is described by

$$f_M(\mathcal{A}, \boldsymbol{\beta}) : \mathrm{SINR}_{n,\varphi_k}^{(i)} \mapsto \mathrm{SINR}_n^{\mathrm{eff}}, \tag{12.63}$$

with $\mathcal{A}$ and $\boldsymbol{\beta}$ denoting the symbol alphabet utilized for the transmission of the current transport block and a vector of tuning parameters respectively. Note that the HSDPA link adaptation does not allow for a change of the symbol alphabet within a transport block. The mapping function $f_M(\mathcal{A}, \boldsymbol{\beta})$ that turns out to be the best in terms of accuracy and flexibility is the so-called Mutual Information Effective Signal to Interference and Noise Ratio Mapping (MIESM). This mapping was initially proposed in the Radio Access Network (RAN) working meetings of 3GPP [16], and builds upon the Bit-Interleaved

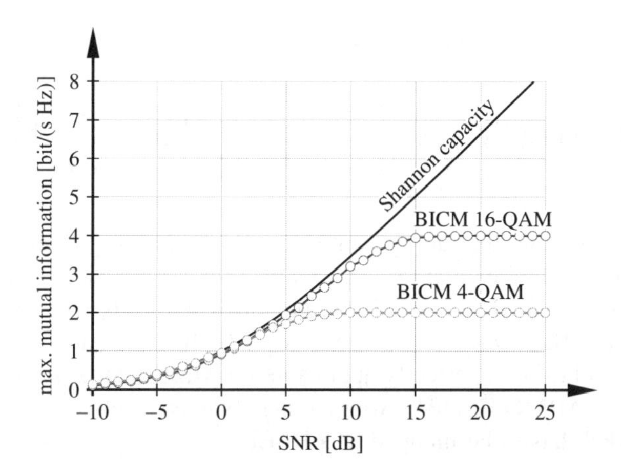

Figure 12.5 Maximum mutual information of BICM systems according to Equation (12.64) for different symbol alphabets $\mathcal{A}$.

Coded Modulation (BICM) capacity as identified in [7]:[5]

$$I(x) = \log_2 |\mathcal{A}| - \mathbb{E}_y \left\{ \frac{1}{|\mathcal{A}|} \sum_{d=1}^{\log_2|\mathcal{A}|} \sum_{b=0}^{1} \sum_{z\in\mathcal{A}_b^{(d)}} \log \frac{\sum_{\hat{x}\in\mathcal{A}} \exp\left[-\left|y - \sqrt{x}\,(\hat{x} - z)\right|^2\right]}{\sum_{\hat{x}\in\mathcal{A}_b^{(d)}} \exp\left[-\left|y - \sqrt{x}\,(\hat{x} - z)\right|^2\right]} \right\}, \tag{12.64}$$

with $|\mathcal{A}|$ denoting the cardinality of the symbol alphabet $\mathcal{A}$, y being a zero-mean unit-variance complex Gaussian variable, and $\mathcal{A}_b^{(d)}$ denoting the set of symbols for which bit d equals b. The resulting Mutual Information (MI) curves in comparison with the Shannon capacity are depicted in Figure 12.5, where it can be observed that the maximum mutual information curves adapt to the Shannon capacity curve for low SNR values and saturate at $\log_2 |\mathcal{A}|$ bit/(sHz), which is the effective information rate limitation imposed by the fixed symbol alphabet. The curves presented are denoted as "maximum mutual information" because they represent an upper bound on the system performance given the limitations of the symbol alphabet and the interleaver structure, but they still rely on infinite-length channel coding arguments.

The expression for the mutual information in Equation (12.64) depends on the demodulator that is utilized. Equation (12.64) holds for the optimum demodulator; other approximate demodulators would need a refinement of the expression. In addition, the interleaver is assumed to be random to obtain the expression for the mutual information, which is valid for large code block lengths. Both of these assumptions are not in conflict with BICM mutual information as the compression function for the link-performance modeling of MIMO HSDPA. With the BICM mutual information being defined in

[5] A list of other mapping functions can be found in [15], including, for example, the well-known Exponential Effective Signal to Interference and Noise Ratio Mapping (EESM). For GSM, some of the proposed mapping functions are based on the so-called "raw" bit-error probability [17], but this is less accurate than other methods [15].

Equation (12.64), the MIESM mapping/compression can be evaluated as

$$f_{\mathrm{M}}(\mathcal{A}, \boldsymbol{\beta}) : \mathrm{SINR}_n^{\mathrm{eff}} \triangleq \beta_1 I^{-1}\left\{\frac{1}{3\,|\Phi_n|}\sum_{k=1}^{|\Phi_n|}\sum_{i=1}^{3} I\left\{\mathrm{SINR}_{n,\varphi_k}^{(i)}\frac{1}{\beta_2}\right\}\right\}, \tag{12.65}$$

with $\boldsymbol{\beta} = \begin{bmatrix}\beta_1 & \beta_2\end{bmatrix}$ defining the tuning parameters. Note that the set of utilized spreading codes is denoted φ_k of stream n by Φ_n again; thus, $\varphi_k \in \Phi_n$, $k = 1, \ldots, |\Phi_n|$. This mapping is well suited for linear receivers, but faces its limits when nonlinear receiver structures, such as ML receivers, are modeled [27]. In accordance with Section 12.2, however, Equation (12.65) is perfectly applicable for MMSE equalizer receiver structures as recommended for MIMO HSDPA. After having obtained an effective SNR by $f_{\mathrm{M}}(\mathcal{A}, \boldsymbol{\beta})$, the resulting $\mathrm{SINR}_n^{\mathrm{eff}}$ has to be mapped to a BLER:

$$f_{\mathrm{AWGN}}(\mathcal{A}, r_{\mathrm{c}}) : \mathrm{SINR}_n^{\mathrm{eff}} \mapsto \mathrm{BLER}_n, \tag{12.66}$$

with r_{c} denoting the code rate of the current subframe. Note that the combined information $(\mathcal{A}, r_{\mathrm{c}})$ uniquely determines the transmission format for the current transport block. If this transmission format corresponds to a configuration as defined in the Channel Quality Indicator (CQI) table, it will be referred to the combined information as "transmission" CQI, $c^{(\mathrm{T})} \sim (\mathcal{A}, r_{\mathrm{c}})$, as already illustrated in Figure 12.4. Details about the CQI tables can be found in Chapter 1.

In this book, Additive White Gaussian Noise (AWGN) performance curves as the mapping from the effective SNR to BLER are utilized. This approach is well known in the literature [6, 15, 27], and has proven to be accurate and robust in terms of applicability for many different transmission scenarios. The fact that the HSDPA link adaptation works on a subframe basis eases the modeling again at this point, because no AWGN performance curves with mixed transmission formats are needed. The AWGN-based mapping $f_{\mathrm{AWGN}}(\mathcal{A}, r_{\mathrm{c}})$ can be acquired by training, meaning a set of AWGN link-level simulations for the corresponding channel coding and rate matching, or by analytic approximations. In this book, a training-based mapping was used. Thus, a set of link-level simulations – for every possible transmission format $c^{(\mathrm{T})}$–forms a look-up table that is utilized as mapping function $f_{\mathrm{AWGN}}(\mathcal{A}, r_{\mathrm{c}})$. Since the link-measurement model and the mapping/compression $f_{\mathrm{M}}(\mathcal{A}, \boldsymbol{\beta})$ evaluate separate performance parameters for every spatial data stream (due to the independent channel coding of the streams), the defined BLER mapping $f_{\mathrm{AWGN}}(\mathcal{A}, r_{\mathrm{c}})$ can be trained by single-stream scenarios and used independently for the transmitted codewords, $n = 0, \ldots, N_{\mathrm{S}} - 1$. To finally obtain a subframe-based ACK/NACK ratio decision that consequently can trigger retransmissions and influences the scheduler, the output BLER of $f_{\mathrm{M}}(\mathcal{A}, \boldsymbol{\beta})$ is mapped to a packet-error event by conducting a binary random experiment with appropriate probability distribution:

$$\xi = \begin{cases} 0 \rightarrow \mathrm{ACK}: & \Pr\{\xi = 0\} = 1 - \mathrm{BLER}, \\ 1 \rightarrow \mathrm{NACK}: & \Pr\{\xi = 1\} = \mathrm{BLER} \end{cases} \tag{12.67}$$

12.3.2 Training and Validation of the Model

With the mapping function being specified by Equation (12.65), it remains to "train" the model by finding the optimum parameter vector $\boldsymbol{\beta}^{\text{opt}}$ and to "validate" the link-performance model in terms of its BLER prediction error. A set of link-level simulations by the HSDPA physical-layer simulator of [12] for different transmission formats is used for training purposes. The simulator is a WCDMA, 3GPP Rel'5 standard-compliant HSDPA simulator that models the complete physical layer. For the training of the MIESM, the simulator was configured to vary the $I_{\text{or}}/I_{\text{oc}}$ ratio, which can be translated to a variation in the post-equalization and despreading SINR per High-Speed Physical Downlink Shared CHannel (HS-PDSCH):

$$\text{SINR}^{\text{HS-PDSCH}} \approx \text{SF}\frac{I_{\text{or}}}{I_{\text{oc}}}\frac{E_{\text{c}}}{I_{\text{or}}}\frac{1}{|\Phi|} \tag{12.68}$$

This implicitly assumes an average DL wireless channel gain of one. As explained in the previous paragraph, only single-stream SISO simulations are necessary to be able to derive separate link-performance models for every spatial stream – which can be interpreted as individual point-to-point rank one transmission; thus, the number of spreading codes for the current transmission is given by the cardinality $|\Phi|$. The instantaneous HS-PDSCH SINR per TTI, needed for the training of the MIESM, however, is extracted from the receive symbols for every received subframe. Varying over $I_{\text{or}}/I_{\text{oc}}$, the physical-layer simulator was utilized to generate a large number of channel realizations and assess the decoding performance for all of those. The fraction of erroneous received packets within a set of channel realizations then approximates the BLER for a given channel condition,

$$\text{BLER}_c^{\text{M}} \approx \frac{\text{no. of NACKs}}{\text{total no. of channel realizations}}, \tag{12.69}$$

referred to as 'measured' BLER, with c denoting the index of the set of channel realizations within the total set of simulated channel characteristics. The channel condition is defined by the average interference and noise density, approximately given by $I_{\text{or}}/I_{\text{oc}}$. Furthermore, the BLER estimated by utilizing the MIESM and the AWGN performance curve is denoted by $\text{BLER}_c^{\text{E}}(\boldsymbol{\beta})$. The individual channel condition and approximated BLER performance pairs are used to train the MIESM to compress/map the measured results to the according AWGN performance curve. Given these definitions, the error of the fitted MIESM in terms of the estimated BLER accuracy can be defined as

$$e_c(\boldsymbol{\beta}) = \text{BLER}_c^{\text{E}}(\boldsymbol{\beta}) - \text{BLER}_c^{\text{M}}, \tag{12.70}$$

or equivalently expressed in the SINR domain as

$$\tilde{e}_c(\boldsymbol{\beta}) = \text{SINR}_c^{\text{eff}}(\boldsymbol{\beta}) - \text{SINR}^{\text{AWGN}} \tag{12.71}$$

This transformation into the SINR domain is possible because the estimated $\text{BLER}_c^{\text{E}}(\boldsymbol{\beta})$ is a one-to-one mapping of the effective SNR, $\text{SINR}_c^{\text{eff}}(\boldsymbol{\beta})$. Thus, given the effective SNR for a specific channel realization c, the model quality depends on the quality of the fit compared with the AWGN reference curve that determines the estimated $\text{BLER}_c^{\text{E}}(\boldsymbol{\beta})$.

Accordingly, the vector of optimum tuning parameters is found by a Least Squares (LS) fit in the SINR domain,

$$\boldsymbol{\beta}^{\text{opt}} = \arg\min_{\boldsymbol{\beta}} \sum_{c=1}^{N_c} |\tilde{e}_c(\boldsymbol{\beta})|^2 , \tag{12.72}$$

with N_c defining the number of channel realization sets within the total set of simulated channel characteristics. This is only one possibility of finding the optimum tuning parameters $\boldsymbol{\beta}$, focusing on a good fit in the overall BLER performance. Other possibilities to find $\boldsymbol{\beta}^{\text{opt}}$ are, for example, treated in [6]. The main advantage of defining the training metric for $\boldsymbol{\beta}$ in the SINR domain is that the whole relevant range of BLER values is equally weighted. In the case that Equation (12.70) is to be utilized, higher BLER values would clearly dominate the optimization procedure to find $\boldsymbol{\beta}^{\text{opt}}$. The simulation parameters utilized for the explained MIESM training were chosen as listed in Table 12.1. The channel realizations set size was chosen to be 200 subframes, thus limiting the applicability of the MIESM training to a BLER range of [1, 0.01]. The LS fit of Equation (12.72) was evaluated utilizing a simplex method as provided by the standard MATLAB command `fminsearch` . Exemplary results of the MIESM training are depicted in Figure 12.6; further training results can be found in [23]. Figures 12.6(a) and 12.6(b) illustrate the MIESM training results in an International Telecommunication Union (ITU) PedA channel profile scenario for the UE capability Table D [2], when the transmission CQIs chosen are $c^{(\mathrm{T})} = 7$ and 14, respectively. According to the 3GPP CQI Table D, these two transmission modes correspond to 4-QAM transmissions, utilizing two and four HS-PDSCH codes respectively. Furthermore, the resulting effective code rates can be evaluated to be 0.34 and 0.67. The two figures show the average BLER performance as a function of the SINR calculated based on the $I_{\text{or}}/I_{\text{oc}}$ approximation in Equation (12.68), the BLER performance as a function of the linear mean post-equalization and despreading SINR evaluated for every channel realizations set, the resulting trained MIESM BLER performance, and the corresponding AWGN performance curve (utilizing the same transmission format) for the sake of showing the MIESM training fit.

Table 12.1 Simulation parameters for the MIESM training

Parameter	Value
Carrier frequency (GHz)	2
Channel type	SISO 1 × 1ITU PedA
UE speed (km/h)	3
Transmission format definitions	3GPP UE CQI Tables D and I [2]
Simulated sub-frames per $I_{\text{or}}/I_{\text{oc}}$ value	2505
HS-PDSCH $E_{\text{c}}/I_{\text{or}}$ (dB)	−3
HARQ retransmissions	Not active
Signaling channels	Perfect, non-interfering
Channel estimation, synchronization	Perfect
Receiver	MMSE, $L_{\text{f}} = 40$ chips
Equalizer channel update	Once per subframe

In principle, the requirement of the MIESM is to map the cloud of individual BLER performances as a function of the post-equalization and despreading SINR to a corresponding effective SNR that matches the AWGN performance curve as closely as possible. When a close fit is achieved, this denotes that the MIESM modeling is capable of mapping/compressing the post-equalization and despreading SINR as provided by the link-measurement model in Equation (12.59) to an effective SNR that can be used for obtaining the predicted BLER from the corresponding AWGN performance curve. Furthermore, please note that Figures 12.6(a) and 12.6(b) also show that a detailed link-measurement model is necessary due to the fact that the I_{or}/I_{oc}-based BLER performance

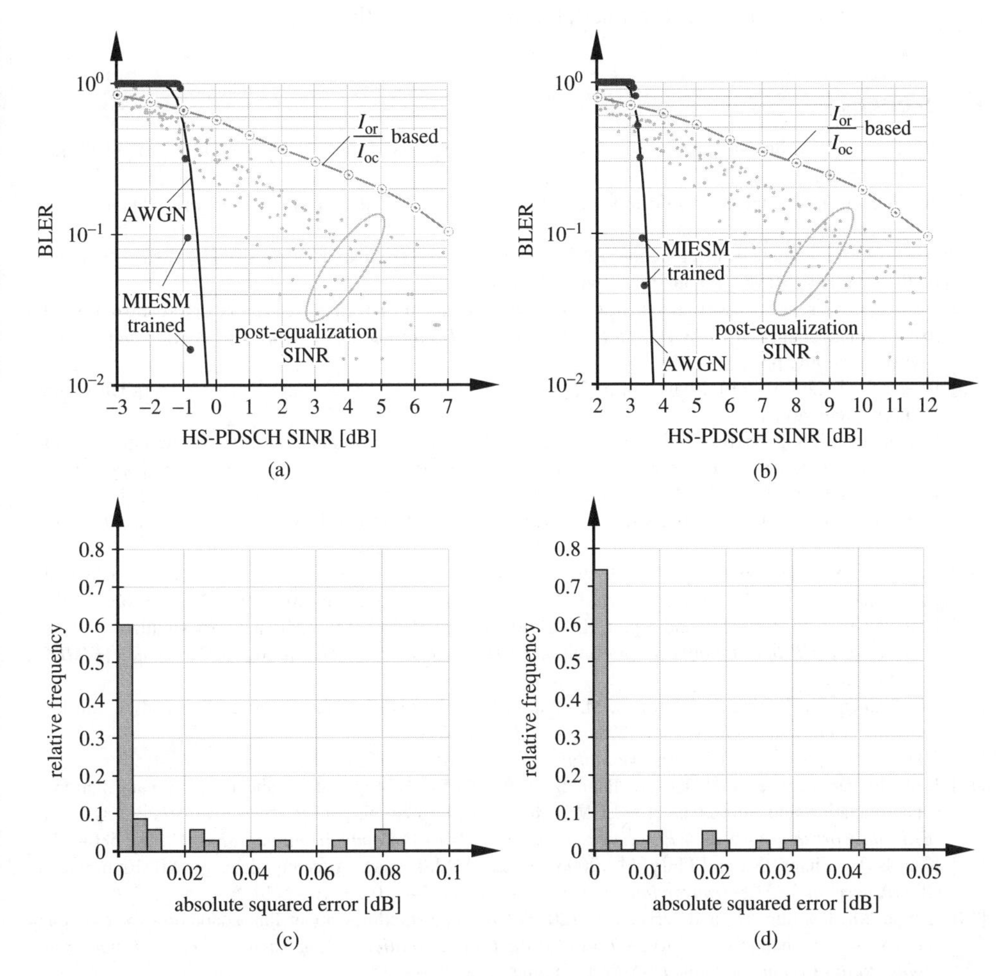

Figure 12.6 MIESM training results and resulting modeling error for the UE capability Table D in a PedA channel environment. The transmit CQIs have been chosen to be $c^{(T)} = 7, 14$, which corresponds to a 4-QAM symbol alphabet with effective code rates of 0.34 and 0.67 respectively. (a) MIESM fitting results for $c^{(T)} = 7$. (b) MIESM fitting results for $c^{(T)} = 14$. (c) Resulting MIESM modeling error $|\tilde{e}_c(\boldsymbol{\beta}^{\text{opt}})|^2$ distribution for $c^{(T)} = 7$. (d) Resulting MIESM modeling error $|\tilde{e}_c(\boldsymbol{\beta}^{\text{opt}})|^2$ distribution for $c^{(T)} = 14$.

curve does not represent the mean of the post-equalization and despreading SINR BLER performance cloud.

In order to assess the modeling error, the estimated BLER performance, $\mathrm{BLER}_c^{\mathrm{E}}(\boldsymbol{\beta}^{\mathrm{opt}})$, is evaluated and compared with the BLER performance as measured from the link-level simulations, $\mathrm{BLER}_c^{\mathrm{M}}$, utilizing the same channel coefficients. The resulting estimation error distribution, again expressed in the SINR domain, $|\tilde{e}_c(\boldsymbol{\beta}^{\mathrm{opt}})|^2$, is depicted in subplots (c) and (d) of Figure 12.6. It can be observed that the error is distributed very close to zero, with a only a few outliers. Please note that the performance of the trained model improves significantly for transmission modes with larger coding block lengths, since the assumption of the random interleaving – utilized in the BICM capacity in Equation (12.64) – is more valid than for smaller code-block lengths [23].

References

[1] 3GPP (2007) Technical Specification TS 25.876 Version 7.0.0 'Multiple-input multiple-output UTRA,' www.3gpp.org.

[2] 3GPP (2007) Technical Specification TS 25.214 Version 7.4.0 'Physical layer procedures (FDD),' www.3gpp.org.

[3] 3GPP (2007) Technical Specification TS 25.996 Version 7.0.0 'Spatial channel model for multiple-input multiple-output (MIMO) simulations,' www.3gpp.org.

[4] Baum, D., Gore, D., Nabar, R. et al. (2000) 'Measurement and characterization of broadband MIMO fixed wireless channels at 2.5GHz,' in *Proceedings of IEEE International Conference on Personal Wireless Communications*, pp. 203–206.

[5] Berger, L. T. (2005) 'Performance of multi-antenna enhanced HSDPA,' Ph.D. thesis, Aalborg Universitet, Aalborg, Denmark.

[6] Brueninghaus, K., Astely, D., Salzer, T. et al. (2005) 'Link performance models for system level simulations of broadband radio access systems,' in *Proceedings of IEEE 16th International Symposium on Personal, Indoor and Mobile Radio Communications (PIMRC 2005)*, volume 4, pp. 2306–2311.

[7] Caire, G., Taricco, G., and Biglieri, E. (1996) 'Capacity of bit-interleaved channels,' *IEE Electronics Letters*, **32** (12), 1060–1061.

[8] Czylwik, A. and Dekorsy, A. (2004) 'System-level performance of antenna arrays in CDMA-based cellular mobile radio systems,' *EURASIP Journal of Applied Signal Processing*, **2004** (9), 1308–1320.

[9] Hämäläinen, S., Holma, H., and Sipilä, K. (1999) 'Advanced WCDMA radio network simulator,' in *Proceedings of IEEE International Symposium on Personal, Indoor and Mobile Radio Conference (PIMRC)*, pp. 951–955.

[10] Hämäläinen, S., Slanina, P., Hartman, M. et al. (1997) 'A novel interface between link and system level simulations,' in *Proceedings of ACTS Summit*, pp. 509–604.

[11] Jakes, W. C. (ed.) (1974) *Microwave Mobile Communications*, IEEE Press.

[12] Kaltenberger, F., Freudenthaler, K., Paul, S. et al. (2005) 'Throughput enhancement by cancellation of synchronization and pilot channel for UMTS high speed downlink packet access,' in *Proceedings of the IEEE 6th Workshop on Signal Processing Advances in Wirless Communications (SPAWC)*, pp. 580–584.

[13] Kim, B.-H., Zhang, X., and Flury, M. (2006) 'Linear MMSE space–time equalizer for MIMO multicode CDMA systems,' *IEEE Transactions on Communications*, **54** (10), 1710–1714.

[14] Lampe, M., Rohling, H., and Zirwas, W. (2002) 'Misunderstandings about link adaptation for frequency selective fading channels,' in *Proceedings of IEEE 13th International Symposium on Personal, Indoor and Mobile Radio Communications (PIMRC)*, volume 2, pp. 710–714.

[15] Members of WINNER (2003) 'Assessment of advanced beamforming and MIMO technologies,' Technical Report IST-2003-507581 D2.7 v1.0, WINNER.

[16] Nortel Networks (2003) 'Effective SIR computation for OFDM system-level simulations,' TSG-RAN WG1 meeting #35 R1-031370, 3rd Generation Partnership Project (3GPP), Lisbon, Portugal.

[17] Olofsson, H., Almgren, M., Johansson, C. et al. (1997) 'Improved interface between link level and system level simulations applied to GSM,' in *Proceedings of IEEE 6th International Conference on Universal Personal Communications*, volume 1, pp. 79–83.

[18] Paulraj, A., Nabar, R., and Gore, D. (2003) *Introduction to Space–Time Wireless Communications*, Cambridge University Press.

[19] Proakis, J. (2000) *Digital Communications*, 4th edition, McGraw-Hill Science Engineering.

[20] Ramos, J., Zoltowski, M. D., and Liu, H. (1997) 'A low-complexity space–time rake receiver for DS-CDMA communications,' *IEEE Signal Processing Letters*, **4** (9), 262–265.

[21] Sklar, B. (1997) 'Rayleigh fading channels in mobile digital communication systems. I. Characterization,' *IEEE Communications Magazine*, **35** (9), 136–146.

[22] Szabo, A., Geng, N., Seegert, A., and Utschick, W. (2003) 'Investigations on link system level interface for MIMO systems,' in *Proceedings of IEEE 3rd International Symposium on Image and Signal Processing and Analysis (ISPA)*, volume 1, pp. 365–369.

[23] Wrulich, M. (2009) 'System-level modeling and optimization for MIMO HSDPA networks,' Ph.D. thesis, Vienna University of Technology. Available from http://publik.tuwien.ac.at/files/PubDat_181165.pdf.

[24] Wrulich, M. and Rupp, M. (2008) 'Efficient link measurement model for system level simulations of Alamouti encoded MIMO HSDPA transmissions,' in *Proceedings of ITG International Workshop on Smart Antennas (WSA)*, Darmstadt, Germany. Available from http://publik.tuwien.ac.at/files/pub-et_13641.pdf

[25] Zemen, T. and Mecklenbräuker, C. (2005) 'Time-variant channel estimation using discrete prolate spheroidal sequences,' *IEEE Transactions on Signal Processing*, **53** (9), pp. 3597–3607.

[26] Zheng, Y. and Xiao, C. (2003) 'Simulation models with correct statistical properties for Rayleigh fading channels,' *IEEE Transactions on Communications*, **51** (6), pp. 920–928.

[27] Zhuang, J., Jalloul, L., Novak, R., and Park, J. (2008) 'IEEE 802.16m evaluation methodology document (EMD),' Technical Report IEEE 802.16m-08/004r1, IEEE.

Part Five

Simulation-Based Evaluation for Wireless Systems

Introduction

This last part of the book is devoted to system-level results based on simulators as explained in more detail in Part Four. In Chapter 13 we show in the example of High-Speed Downlink Packet Access (HSDPA) how difficult and at the same time how complex such simulations are. The proposed system-level model as defined in Part Four serves as the core for the development of a system-level simulator. First, the concept of the system-level simulator, including its advantages and limitations, is explained. Then, network performance prediction results from a set of Double Transmit Antenna Array (D-TxAA) system-level simulations are presented and discussed. The statistics obtained of the average user throughput to fairness measures are related to each other and different network setups by means of their average, and maximum sector throughputs are compared. Also, briefly, the consequences of the double-stream operation on the fairness in the network are discussed. Consequently, a novel Radio Link Control (RLC)-based stream-number decision algorithm with enhanced robustness against signaling errors from the User Equipment (UE) side is proposed. The system-level simulator is also utilized to conduct performance evaluations of an optimized cross-layer scheduler, and the achievable Quality of Experience (QoE) gains in an HSDPA network are investigated. Finally, by extending the proposed system-level model, Common Pilot CHannel (CPICH) power optimizations for various antenna configurations are performed. The results presented provide the optimum average CPICH power values maximizing the HSDPA link quality.

UMTS Rel'6 requires Minimum Mean Square Error (MMSE) equalizers rather than simple rake receiver types. While previous chapters were based on classical MMSE receivers, in Chapter 14 we investigate the structure of the intra-cell interference when multiple users are simultaneously active. In fact, the special form of the interference imposed by the multi-user Transmit Antenna Array (TxAA) transmission requires an enhanced system model to derive an interference-aware MMSE equalizer. The proposed equalizer takes the precoding state of the cell into account. By means of system-level analysis, we investigate the interference suppression capabilities and the theoretical performance limits of the developed receiver when perfect knowledge of the pre-coding state is available. In a practical setup, however, only the pre-coding information of the desired user is known. The basic principle applied to derive a suitable blind estimator is the Gaussian Maximum Likelihood (ML) approach. Some semi-analytical results on the complexity order of the interference-aware equalizer and the precoding state estimator also show that the overall complexity increase is only moderate. Finally, physical-layer and system-level simulations are performed that show significant throughput performance gains of the interference-aware MMSE equalizer compared with the classical single user MMSE equalizer.[1]

The final chapter, Chapter 15, discusses Long-Term Evolution (LTE) advanced aspects as they are being finalized just before this book was written. We start with a discussion of what the standardization bodies International Mobile Telecommunications (IMT) and 3rd Generation Partnership Project (3GPP) have envisioned for the next-generation wireless and how LTE-Advanced (LTE-A) in the form of its Rel'10 satisfies such requirements. In particular, we discuss so-called Coordinated Multi-Point (CoMP) and relaying techniques that promise strong gains. We then focus on Multiple-Input Multiple-Output (MIMO) techniques, as LTE-A will allow up to eight antennas at the transmitter side and receiver side. This, in turn, requires sophisticated codebooks. Based on physical-layer simulations, we demonstrate expected throughput improvements by utilizing eight antennas at the transmit side and compare LTE with LTE-A results. Finally, we also provide results on single-user MIMO in comparison with multiple-user MIMO transmissions.

[1] Various chapters of this book part are based on the Ph.D. theses of Martin Wrulich and Christian Mehlführer. The complete theses can be downloaded from EURASIP's open Ph.D. library at http://www.arehna.di.uoa.gr/thesis/ or directly from http://publik.tuwien.ac.at/files/PubDat_181165.pdf and http://publik.tuwien.ac.at/files/PubDat_181154.pdf repectively.

13

Optimization of MIMO-Enhanced HSDPA

13.1 Network Performance Prediction

Network performance prediction is important to test potential gains of new transmission standards from a system-level perspective. This is of particular interest for network operators, who have to assess whether to invest in new technologies and in which parts of their network. The term "network performance" can include many different figures of merit. Usually, among the metrics investigated are the average throughput performance, the Multiple-Input Multiple-Output (MIMO) utilization, and the fairness of the system. Furthermore, throughput density plots, coverage analyses, or Block Error Ratio (BLER) performance curves can serve to aid the network planning process. The basic functionality of a simulator to evaluate these necessary figures of merit is also crucial to conduct investigations of more advanced algorithms or cross-layer optimizations. Depending on the design of the simulator, the effects of the operation of different technologies can also be investigated.

13.1.1 Simulation Setup

A comprehensive treatment of all simulation possibilities and the resulting figures of merit would exceed the scope of this book. Thus, a subset of interesting performance figures will be commented on and described in greater detail. Most of the simulation results presented will also be of interest for the research in the following sections. First, some comments with respect to the core simulation assumptions are appropriate. All simulation results are based on a Round Robin (RR) scheduler. Furthermore, no sophisticated traffic modeling is applied; rather, a full buffer scenario is simulated. Such investigations result in performance figures that represent the maximum physical-layer capabilities because no buffer limitations are taken into account. However, if traffic statistics are nonuniform, then scheduler and resource allocation algorithms observe more degrees of freedom, which can be beneficial for the handling of fixed Quality of Service (QoS) services; for example, in terms of delay or throughput jitter.

Evaluation of HSDPA and LTE: From Testbed Measurements to System Level Performance, First Edition.
Sebastian Caban, Christian Mehlführer, Markus Rupp and Martin Wrulich.

Table 13.1 Simulation settings for the system-level investigations

Parameter	Value
Interference scenario	Homogeneous network
Number of cells	19, layout Type I
NodeB distance (m)	1000
Carrier frequency (GHz)	1.9
Total transmit power available at NodeB (W)	20
Total Common Pilot CHannel (CPICH) power (W)	0.8
Power of other channels (W)	1.2
Available High-Speed Physical Downlink Shared CHannels (HS-PDSCHs)	15
Large-scale path-loss model	Urban micro
Antenna utilization	$N_T \times N_R = 2 \times 2$
Channel type	ITU PedA
Scheduler	RR
Active users in target sector	25
User mobility	3 km/h, random direction
UE capability class	20
UE receiver type	MMSE, $L_f = 30$ chips, $\tau = 15$ chips
Feedback delay	4 Transmission Time Intervals (TTIs)
Simulation time	50 000 slots, each 2/3 ms

The most important simulation settings can be found in Table 13.1. As illustrated in Figure 13.1, the simulations utilize a 19-site Type I hexagonal network layout in an urban micro-environment, following [2]. The NodeB distance was chosen to be 1000 m and the interfering NodeBs' transmission powers are chosen to represent a homogeneous network scenario. The antenna deployment represents an $N_T \times N_T = 2 \times 2$ MIMO scenario. Accordingly, the network is configured for Double Transmit Antenna Array (D-TxAA) transmission and the User Equipment (UE) capability class is defined to be Double-Stream (DS) compatible. All UEs utilize a Minimum Mean Square Error (MMSE) equalizer. The frequency-selective channel profile was modeled according to an International Telecommunication Union (ITU) Pedestrian A (PedA) channel profile.

The feedback delay for reporting the ACKnowledge (ACK)/Non-ACKnowledged (NACK) and the Channel Quality Indicator (CQI) report has been set according to the timing constraints [1] to four TTIs. Finally, the stream power loading is conducted in a uniform way; thus, no power loading of the individual streams is performed. For the High-Speed Downlink Shared CHannel (HS-DSCH) transmission, it is assumed that all 15 possible HS-PDSCHs are available, which leaves no room for Universal Mobile Telecommunications System (UMTS) traffic in the cell; thus, these results represent a dedicated MIMO High-Speed Downlink Packed Access (HSDPA) network.[1]

13.1.2 Single Network Scenario Investigation

The first results that can be obtained from a system-level simulation are based on a single "network realization." In general, the term "network realization" corresponds to a user

[1] This is due to the fact that one code of spreading factor 16 is reserved for signaling channels.

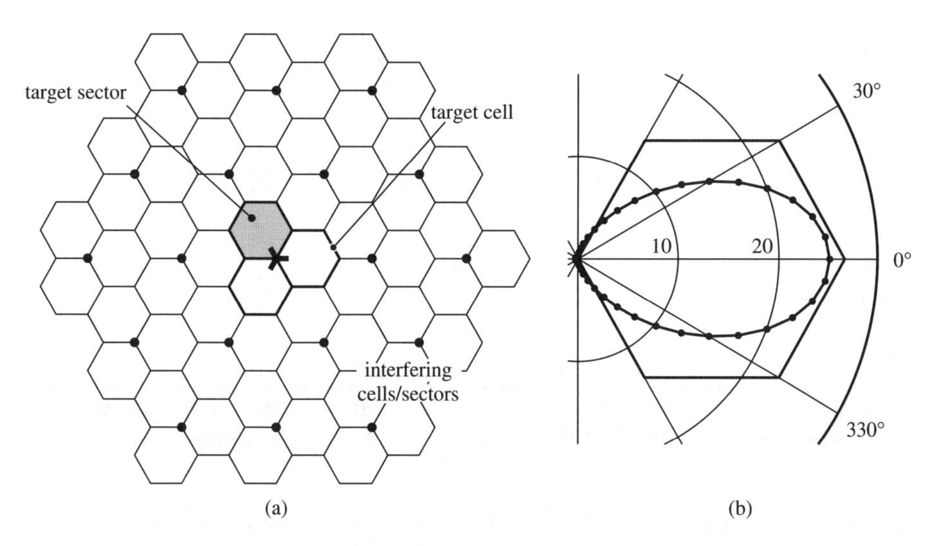

Figure 13.1 Network deployment details in case of a standardized network scenario. The layout corresponds to the definitions in [2]. (a) Three-sector network layout Type I [2]. (b) Antenna gain pattern for a three-sector network according to [2].

positioning and movement realization, given the network layout and algorithm constraints. Although limited in significance for assessing the overall network performance, these results can serve to identify whether and how well the basic adaptation mechanisms of HSDPA work, and to obtain some knowledge about the performance limits of the system.

Figure 13.2 shows the resulting empirical cdfs of the corresponding throughput values and the underlying initial user positioning. The worst user and the empirical cdf of the best user throughput represent the performance bounds for the scheduler, given the network realization. Depending on the scheduler strategy, in particular its fairness constraints, the sector throughput statistics can be closer to the best user performance.

According to the system design [1], and corresponding to the SINR-to-CQI mapping, the average ACK/NACK ratio of the cell should be close to its target value of 0.1.[2] To confirm this behavior in the system-level simulator, Figure 13.3 depicts the ACK/NACK ratio when averaged with a sliding window filter of length 125 TTIs. It is observed that the ACK/NACK ratio fluctuates around its average of 0.11, which is sufficiently close to its desired target value.

When speaking of the ACK/NACK ratio performance, it has to be noted that the link adaptation responsible for stream 2 is much more difficult to calibrate than for stream 1. The reason is that, because of the precoding, stream 2 observes a much higher variance of its SINR and, thus, the link adaptation has to be configured more conservatively. For optimum results, the precoding and the stream decision should be performed in a joint way.

Figure 13.4 lists the distribution of the stream utilization and the distribution of the CQI reports for the Single-Stream (SS) and the DS case respectively. It is observed that the cell utilizes an SS transmission approximately two and a half times as often as the DS transmission. Furthermore, the CQI distributions show that the dynamic range covered

[2] Note that the term ACK/NACK ratio is defined as [no. of NACK/(no. of ACK + no. of NACK)] in this context.

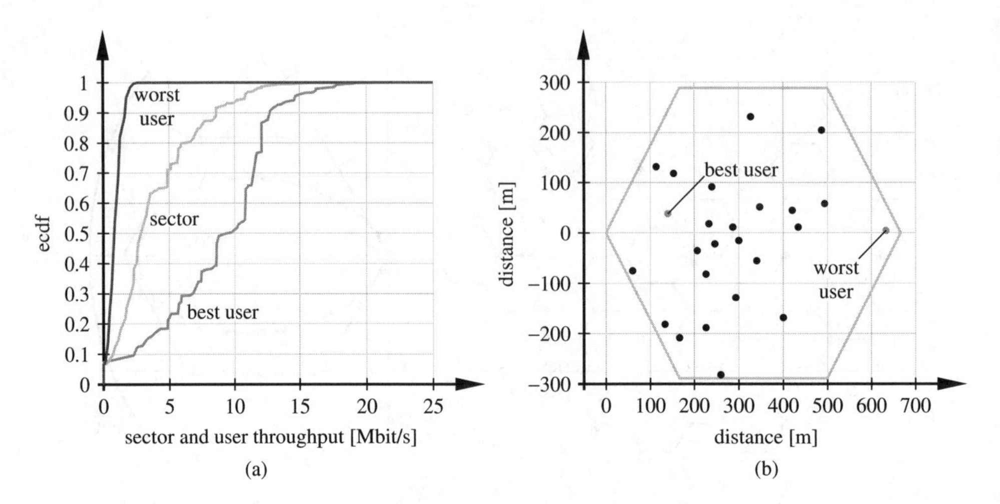

Figure 13.2 Single network realization SINR and throughput results. (a) Empirical cdf of the worst user, the best user, and the sector throughput. (b) UE positioning in the target sector.

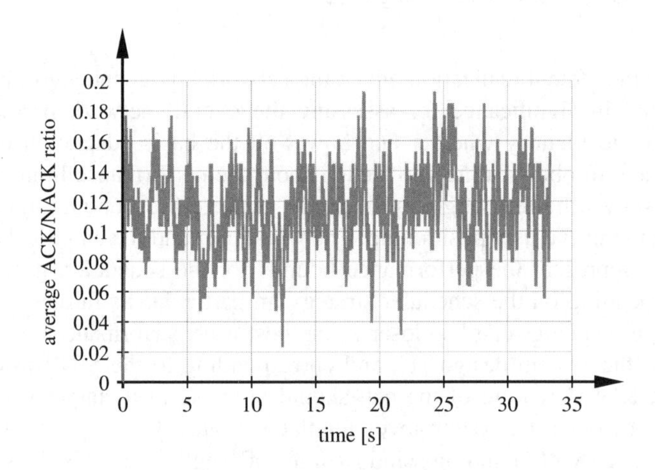

Figure 13.3 BLER results for a single network realization.

by the link adaptation fits the observed SINR statistics in the cell very well; thus, nearly all CQI values are utilized and the distribution is not squeezed to one end. Only the CQI report distribution of stream 2 is skewed to the lower values, which is again a result of the lower average SINR and its higher variance because of the precoding.

13.1.3 Average Network Performance

To assess the network performance, a set of network scenarios has to be simulated to average over different user positions and shadow fading trace realizations. Two exemplary results of such a simulation campaign are shown in Figure 13.5. Figure 13.5(a) illustrates

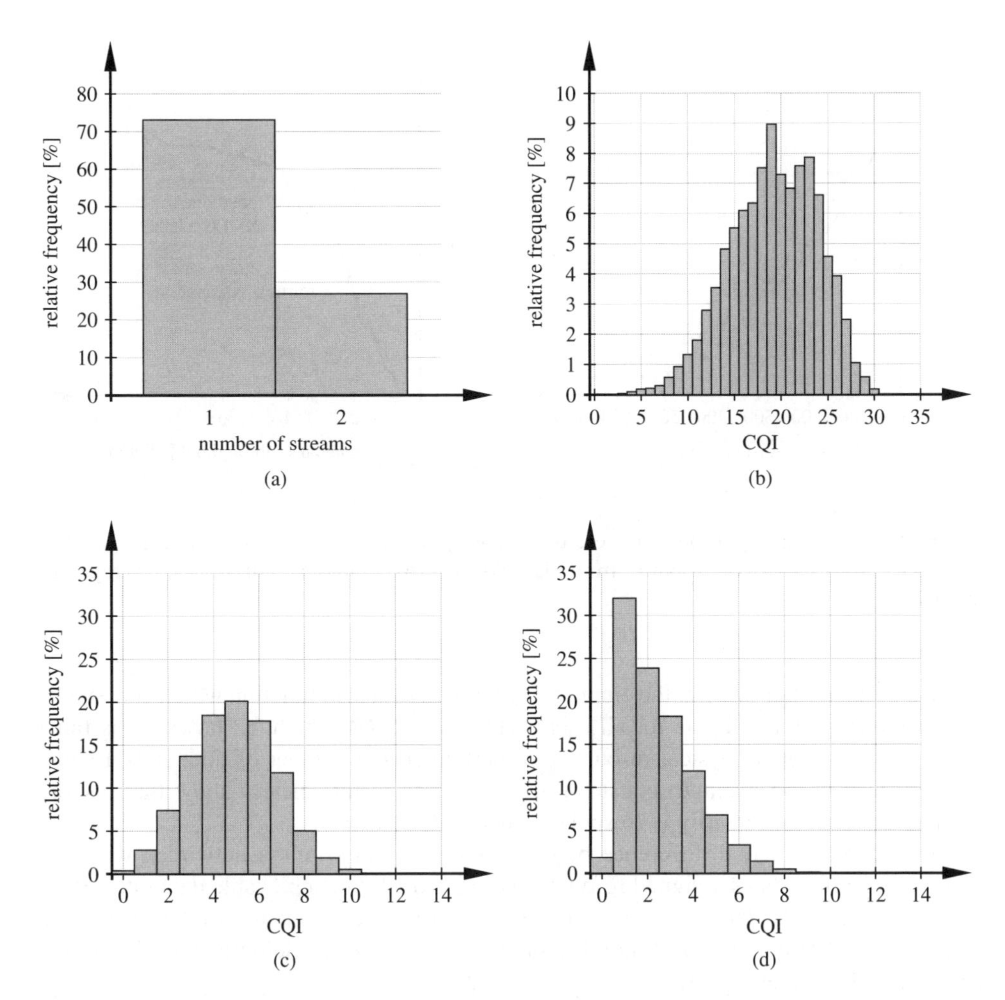

Figure 13.4 Stream utilization and CQI report distributions. (a) Stream utilization in the given network scenario. (b) Distribution of the CQI reports from UEs in SS transmission mode. (c) Distribution of the CQI reports for stream 1 from UEs in DS transmission mode. (d) Distribution of the CQI reports for stream 2 from UEs in DS transmission mode.

the obtained throughput performance over distance. To clarify the underlying trend, a second-order polynomial regression utilizing a Least Squares (LS) fit was performed. This result can serve as a general guideline for NodeB positioning in a wireless network; for example, when the cell edge performance should meet a certain target. If the cell edge performance is to be 2 Mbit/s, the maximum cell radius is given by 465 m.

Figure 13.5(b) then depicts the overall empirical cdf of the average user throughput for the whole simulation campaign. Two important observations can be made in such a figure:

- the steepness of the curve represents a measurement of the fairness of the system and
- Figure 13.5(b) shows a distinctive bend of the empirical cdf around 0.75 in probability.

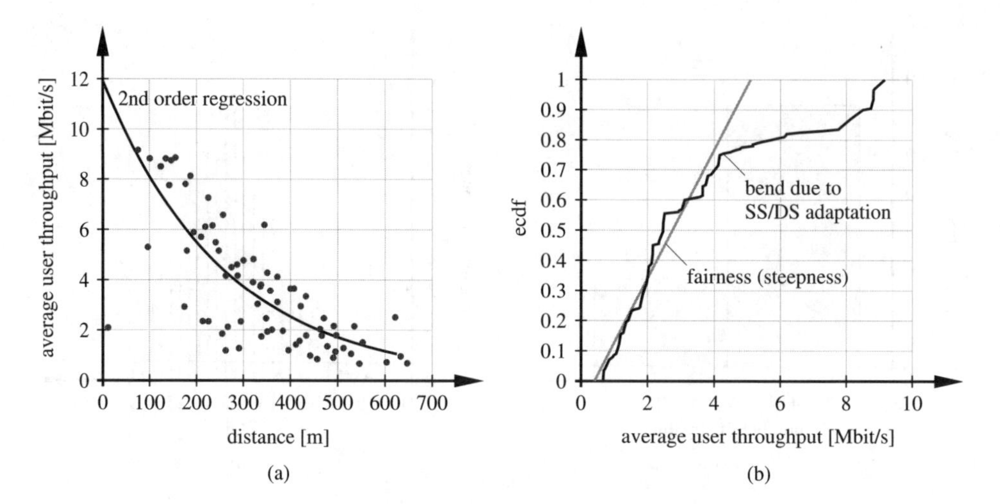

Figure 13.5 Exemplary throughput results for the system-level simulation campaign. (a) Average user throughput over distance from the NodeB. (b) Empirical cdf of the average user throughput.

For a perfectly fair system, the empirical cdf would be a step function which corresponds to equal average throughputs for all users. The utilized RR scheduler, however, is fair in terms of the assigned physical resources. In particular, the number of time slots assigned to every user is (on average) equal for all users. The results clearly show that resource fairness does not necessarily correspond to data rate fairness.

This is a fundamental difference between wireline and wireless networks, where the physical layer translates assigned resources very unequally to achievable data rates. Based on this observation, the literature also defines fairness coefficients for "data-rate fairness" and for "resource allocation fairness" [10]. The standard definition for the fairness coefficient ϑ is given by [5]

$$\vartheta \triangleq \frac{1}{1+\sigma_{\bar{T}}^2}, \tag{13.1}$$

with the throughput variance $\sigma_{\bar{T}}^2$ being

$$\sigma_{\bar{T}}^2 = \mathbb{E}\{(\bar{T}^{(u)})^2\} - \mathbb{E}\{T^{(u)}\}^2 = \frac{1}{U}\sum_{u=1}^{U}\left(\frac{1}{\kappa}\sum_{k=1}^{\kappa}T_k^{(u)}\right)^2 - \left(\frac{1}{U\kappa}\sum_{u=1}^{U}\sum_{k=1}^{\kappa}T_k^{(u)}\right)^2, \tag{13.2}$$

where U denotes the total number of simulated users in the whole ensemble of individual simulated network scenarios, κ represents the total number of simulated TTIs, and $T_k^{(u)}$ is the throughput of user u in TTI k. This definition thus utilizes second-order moments of the throughput results, whereas the steepness of the empirical cdf of the average user throughput compares the average throughput values directly. The RR scheduler on which these results are based is a resource fair scheduler, explaining the low steepness and the accordingly lower data-rate fairness, as illustrated in Figure 13.5(b).

The bend in the empirical cdf in Figure 13.5(b) is a result of the SS/DS adaptation. Not all users are in such favorable channel conditions to benefit from a DS spatial multiplexing transmission that increases their average throughput values significantly. Accordingly, users that obtain two streams can exploit the assigned resources much more efficiently, pushing the throughput to dramatically higher values. This behavior, however, also influences the data-rate fairness, especially when a resource fair RR scheduler is utilized. As a result, the empirical cdf shows a distinct bend in the upper region. Note also that the position of the bend corresponds to the average DS utilization in the simulation campaign, which was 27 %.

13.1.3.1 Performance Comparison for Different Network Configurations

Figure 13.6 illustrates the network performance in terms of the average and maximum sector throughputs for the following four different network deployments and two different channel characteristics (PedA and Pedestrian B (PedB)):

1. $N_T \times N_R = 2 \times 2$ Double-Stream (DS) network with ITU PedA channel profile;
2. 2×2 Single-Stream (SS) network with ITU PedA channel profile;
3. 2×1 SS network with ITU PedA channel profile;
4. 1×1 SS network with ITU PedA channel profile;
5. 2×2 DS network with ITU PedB channel profile.

The 2×2 SS network shows the maximum average sector throughput, but with significantly lower maximum sector throughput. If the 2×2 and 2×1 ITU PedA networks in single-stream operation are compared, it can be observed that the average sector throughput benefits from the second receive antenna. In the case that the channels follow an ITU

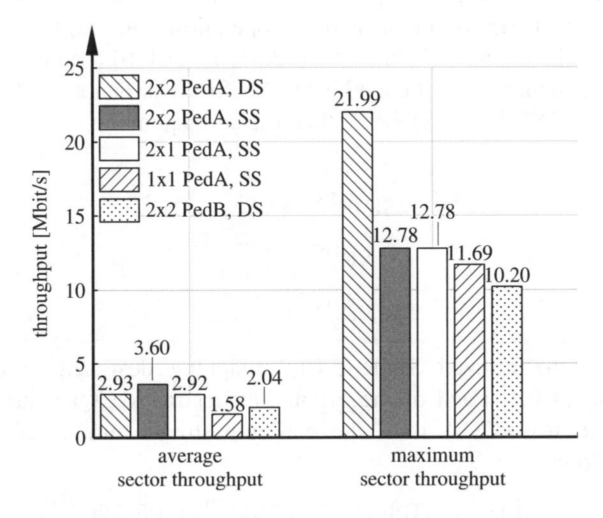

Figure 13.6 Average and maximum sector throughput comparison of five different network deployments.

PedB profile, the network performance both in terms of the average and maximum sector throughput drops significantly. This is because of larger losses due to the less-optimal equalization in these large-delay channels. Still, the network benefits from the utilization of the second stream (which was 19 % in this case). The 1×1 network deployment shows the lowest average sector throughput but a higher maximum sector throughput compared with the 2×2 ITU PedB case, which is again due to deficiencies in terms of diversity.

13.2 RLC-Based Stream Number Decision

The results presented so far are based on a stream utilization as requested by the UE. This means that the UE evaluates the number of streams that maximizes the expected throughput in the next upcoming transmission [16]. However, such an estimation depends on a number of factors and can be very challenging. In addition, the interference structure of the cell (which depends on the number of active streams) is then determined by the UE. This is, in general, undesired by the network operators [14]. Individual enhancements of the user channel quality by means of interference cancelation or interference awareness or similar techniques do not impose such problems. For an efficient and robust network operation, network entities should assign the resources according to overall goals and cost functions – such as, for example, the average cell throughput [6, 21, 30] – and, therefore, actively manage the interference situation in the cells.

13.2.1 UE Decision

Before going into detail on the network-based algorithm presented, a simple UE-based stream-decision algorithm is presented. Therefore, the UE assumes that the NodeB will apply the transmission settings corresponding to the fed back CQI values, $c^{(\mathrm{T})} = c^{(\mathrm{F})}$. With this assumption, the data-rate optimum decision from the viewpoint of the UE is to evaluate the individual SINRs for SS and DS operation – for example, by estimating it with a slightly modified form of Equation (12.59) as in [16] – and compare the resulting expected throughputs. In particular, if the SS SINR is denoted $\mathrm{SINR}^{\mathrm{SS}}$ and the DS SINRs are denoted $\mathrm{SINR}_1^{\mathrm{DS}}$ and $\mathrm{SINR}_2^{\mathrm{DS}}$, then the corresponding feedback CQI values are given by

$$\mathrm{SINR}^{\mathrm{SS}} \rightarrow c_{\mathrm{SS}}^{(\mathrm{F})}, \tag{13.3}$$

$$\mathrm{SINR}_1^{\mathrm{DS}} \rightarrow c_{\mathrm{DS},1}^{(\mathrm{F})}, \tag{13.4}$$

$$\mathrm{SINR}_2^{\mathrm{DS}} \rightarrow c_{\mathrm{DS},2}^{(\mathrm{F})}, \tag{13.5}$$

as determined by the utilized SINR-to-CQI mapping designed to achieve a target ACK/NACK ratio of 0.1. Each of these potential feedback CQI values corresponds to a Transport Block Size (TBS); thus, the optimum stream number $N_{\mathrm{S}}^* = \{1, 2\}$ can be formulated as follows:

$$N_{\mathrm{S}}^* = \begin{cases} 1; & \text{if } \mathrm{TBS}(c_{\mathrm{SS}}^{(\mathrm{F})}) > \mathrm{TBS}(c_{\mathrm{DS},1}^{(\mathrm{F})}) + \mathrm{TBS}(c_{\mathrm{DS},2}^{(\mathrm{F})}), \\ 2; & \text{else.} \end{cases} \tag{13.6}$$

Note that the CQI tables in the case of a double-stream transmission specify a minimum TBS of two times 4581 bits. Thus, in the case the channel becomes very poor, the SS corresponding TBS will strictly be lower and, thus, the algorithm presented will decide on a DS transmission. However, given the low channel quality, a successful DS transmission is very unlikely. Accordingly, there is need for a second constraint, this being

$$\mathrm{SINR}_1^{\mathrm{DS}} < \mathrm{SINR}^{\mathrm{thres}} \Rightarrow N_{\mathrm{S}}^{*} = 1, \tag{13.7}$$

to ensure that an SS transmission is requested when the channel quality is low. The threshold SINR, $\mathrm{SINR}^{\mathrm{thres}}$, has been found by means of simulations to obtain the best algorithm performance, which resulted in a value of 5 dB HS-DSCH SINR.

13.2.2 RLC Decision

The available information of the NodeB for the SS/DS decision is (i) the CQI feedback values and (ii) the ACK/NACK reports of the UEs. This information allows for the design of many different cost functions, of which the "effective TBS," as introduced in [13], is to be introduced. The effective TBS of a particular user is given by

$$\mathrm{TBS}^{(\mathrm{eff})} \triangleq \mathrm{TBS}(c^{(\mathrm{F})}) \cdot (1 - \Upsilon), \tag{13.8}$$

with Υ being a sliding window average of the ACK/NACK ratio, as shown by the example in Figure 13.3. This parameter thus represents the short-term average expected TBS that can successfully be transmitted.[3] With the effective TBS being defined, the optimum number of streams decided by the NodeB is given by

$$N_{\mathrm{S}}^{*} = \begin{cases} 1; & ;\ \text{if } \mathrm{TBS}^{(\mathrm{eff})} < \mathrm{TBS}(c^{(\mathrm{D})}), \\ 2; & ;\ \text{if } \mathrm{TBS}^{(\mathrm{eff})} > \mathrm{TBS}(c^{(\mathrm{U})}), \\ \text{not changed}; & \text{else.} \end{cases} \tag{13.9}$$

The CQI values $c^{(\mathrm{D})}$, $c^{(\mathrm{U})}$ and the corresponding TBS values represent threshold values for the down- and upgrade from DS to SS and vice versa. These two distinct values allow for the realization of a hysteresis which can limit potential oscillation problems of the optimum stream number.

13.2.3 System-Level Simulation Results

To assess the performance of the algorithm presented in comparison with the UE-based stream decision, system-level simulations were conducted. The same simulation settings as specified in Table 13.1 also hold for these simulation campaigns of different network realizations.

Figure 13.7 depicts the average sector throughput for varying up- and down-grade CQI values. Note that the average sector throughput in the case of the UE-based algorithm should be a straight line. Owing to independent network realizations for all simulated

[3] If the TBS were mapped to a throughput by dividing by the duration of the TTI of 2 ms, this parameter could also be interpreted as the short-term average expected throughput.

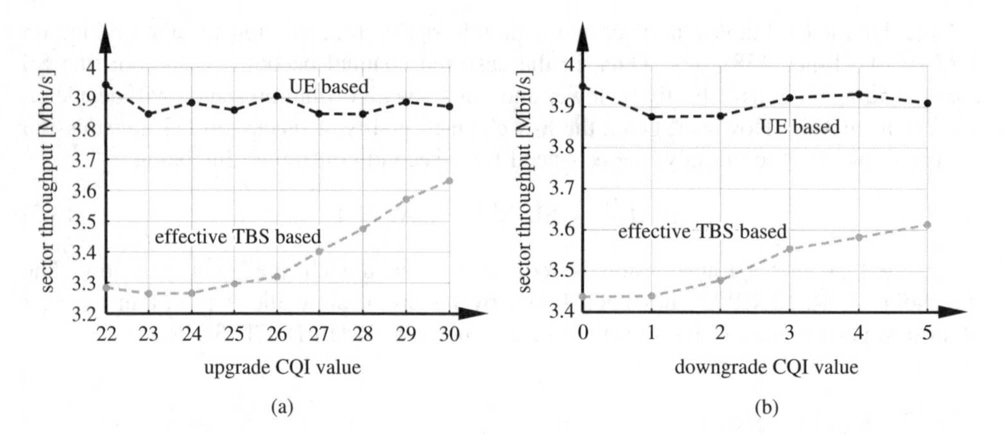

Figure 13.7 Performance comparison of the UE based and the NodeB based stream number decision. Performance comparison for (a) varying upgrade threshold when $c^{(D)} = 2$ and (b) varying downgrade threshold when $c^{(U)} = 28$.

up- and down-grade values, however, the resulting sector throughputs fluctuate slightly. Figure 13.7 shows clearly that, given the assumptions on the SINR-to-CQI mapping and the network behavior as a function of the feedback CQI, the effective TBS-based algorithm is outperformed. Even if the best combination of the threshold values $c^{(D)} = 5$ and $c^{(U)} = 30$ is chosen, the UE-based decision achieves by 200 kbit/s higher average sector throughput.

The main reason is that, in the case of the network-based decision, the NodeB has to decide on the transmission format, or accordingly the transmit CQI value $c^{(T)}$, to be utilized after a change from SS to DS or vice versa. Since no UE CQI report is available for this initial phase of the changed transmission mode, the network has to estimate the optimum transmission format for the UE until new feedback values for the new transmission mode arrive. This is the reason for the observable performance gap. Further details on this issue can be found in [13].

Figure 13.8 finally shows the performance of the effective TBS-based network stream decision when the UE CQI feedback is assumed to be outdated and the CQI mismatch is equal to one. This means that in case a CQI value of 17 would be optimum for the given channel conditions, the UE had reported a CQI of 16 or 18. Such a scenario occurs, for example, when the channel coherence time becomes shorter such that (given the fixed delay constraint of the feedback) the CQI feedback no longer perfectly represents the channel conditions during the transmission.

It can be observed that the effective TBS-based network stream decision outperforms the UE-based stream decision by approximately 500 kbit/s, where the up- and downgrade CQI values were set to 30 and 5 respectively. The main reason for this improved performance is that, in the case of nonperfect UE feedback, no control mechanism is left to ensure an average ACK/NACK ratio of 0.1. The effective TBS-based algorithm, however, implicitly tries to maximize the sector throughput that can be delivered successfully. With

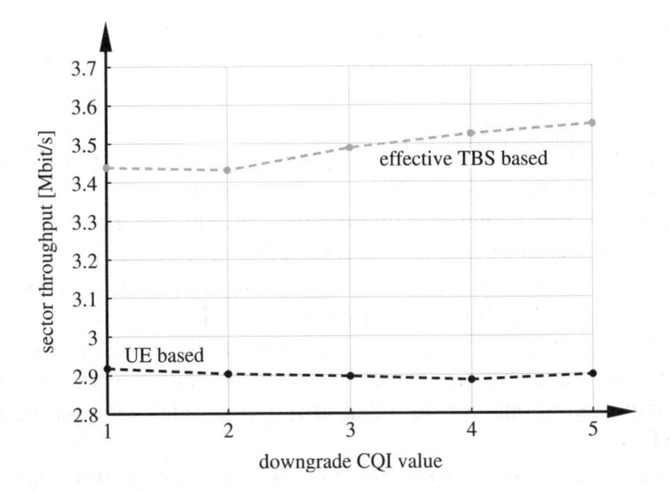

Figure 13.8 Robustness performance comparison of the UE-based and the NodeB-based stream number decision when the UE CQI reporting is nonoptimal by one.

the up- and down-grade thresholds set accordingly, the network is able to stabilize effects that would increase the ACK/NACK ratio.

13.3 Content-Aware Scheduling

As already mentioned at the beginning of this chapter, system-level modeling and simulation can also be utilized to perform cross-layer optimizations. The increase in available Downling (DL) data-rate and the developments in the core network architecture make Internet Protocol (IP) traffic in wireless networks increasingly important. This traffic consists of a variety of different applications with different degrees of interactivity and technical needs. Thus, 3rd Generation Partnership Project (3GPP) standards define four classes of IP traffic [4]. These classes have been introduced to ensure certain QoS goals of the customers. However, the QoS does not necessarily reflect the Quality of Experience (QoE) [12] of the user. In particular for video applications, stringent constraints in average data-rate, delay, jitter, or data-rate variance do not necessarily map one to one to the QoE. Since the QoE finally also determines the users' willingness to pay for certain entertainment services, applying it as a metric for network optimization seems intuitive. Accordingly, this section is devoted to introducing a Content Aware (CA) scheduler trying to maximize the QoE of the user and to assess its performance.

13.3.1 Video Packet Prioritization in HSDPA

Content awareness [22] of the NodeB scheduler requires the signaling of content information to the Medium Access Control for High-Speed Downlink Packet Access (MAC-hs) layer. In particular, the importance of the encoded slice can then also be signaled, exploiting the fields of the Network Abstract Layer (NAL) unit header [19]. At the video streaming server, this priority information can be conveyed to the IP layer. The IP header contains a byte originally designed to specify the Type Of Service (TOS) which is currently used for Differentiate Service (DiffServ) marking [18]. This specifies how a packet has to be handled by each network element.

Universal mobile telecommunications system Terrestrial Radio Access Network (UTRAN) networks provide so-called Packet Data Protocol (PDP) contexts to identify and arrange different packet-switched connections. Within this method, the user terminal at one side and the Gateway-General packet radio service Support Node (GGSN) at the other side agree on several parameters; for example, the IP address, the PDP type, and the QoS profile of the following packets. Each PDP context supports one QoS parameter setting at a time. The QoS information of the corresponding PDP contexts is also available to the MAC-hs layer. To allow for multiple services with different requirements dedicated to one user terminal to run in parallel, multiple PDP contexts are allowed in UMTS.

The basic idea of the cross-layer signalization of video content information is to encapsulate the video packet information in the DiffServ field of the IP header. This information is then passed to the GGSN, which, when establishing the PDP contexts to the desired user, will establish a set of PDP contexts, one for each priority class of the video content. According to the Traffic Flow Template (TFT) filter[4] required at the GGSN for the operation of the multiple PDP contexts, the incoming IP packets will be filled in the corresponding flows. The MAC-hs scheduler then can treat these packets arriving through different flows differently, enabling CA scheduling. Further details on the content information delivery from IP to the MAC-hs layer can be found in [26].

13.3.2 Content-Aware Scheduler

With the IP packet priority information made available at the MAC-hs scheduler in the NodeB, the question remains how to exploit this information. The basic idea is to protect packets of higher importance for the QoE better when decoded and displayed at the user terminal against transmission errors. Simultaneously, the scheduler shall have the flexibility to downgrade the error protection on the physical layer for packets with less priority.

As explained in Chapter 1 and elaborated in Section 13.2, in HSDPA the NodeB has to adapt the transmission settings according to the feedback information of the UEs, $c^{(\mathrm{F})}$. If the same quantized set of transmission formats as the table of feedback CQI values is utilized, then the NodeB has to perform the mapping

$$\mathcal{T}: c^{(\mathrm{F})} \rightarrow c^{(\mathrm{T})}, \tag{13.10}$$

from feedback CQI $c^{(\mathrm{F})}$ to the transmit CQI $c^{(\mathrm{T})}$, representing the applied code rate and modulation alphabet utilized for the transmission. The scheduler presented introduces an adaptive mapping, depending on the video packet priority. Better protection can be achieved by remapping the transmit CQI to lower values and vice versa. However, if the transmit CQI is lowered too much, the average sector throughput would

[4] A TFT filter specifies a set of rules to assign incoming packets to PDP contexts. Details about the design of these filters can be found in [27].

decrease considerably; this, in general, is undesired because spectral resources would be wasted.

When transmitting video data, packets belonging to I-frames contribute significantly to the overall video quality, whereas the influence of packets belonging to P-frames is not that prominent. In particular, this holds true for packets of P-frames that are at the end of a Group Of Pictures (GOP). Also note that, on average, one I-frame corresponds to four P-frames in size. Given this QoE priority information, three scheduling priority classes can be defined:

- priority $\nu = 2$, packets belonging to I-frames;
- priority $\nu = 1$, packets belonging to the first $(l_{\mathrm{GOP}} - 4)$ P-frames of the GOP; and
- priority $\nu = 0$, packets belonging to the last 4 P-frames of the GOP.

Here, L_{GOP} denotes the length of the GOP in frames. Then the presented transmission mapping $\mathcal{T}$ of the MAC-hs scheduler is given by

$$\mathcal{T}_{\nu} : c^{(\mathrm{T})} = c^{(\mathrm{F})} + 1 - \nu \tag{13.11}$$

To ensure fairness among all active users in the cell, the scheduler selects the served user according to an RR strategy. In addition, the maximum packet delay of the network has been limited by allowing a maximum number of one Hybrid Automatic Repeat reQuest (HARQ) retransmission.

13.3.3 Simulation Results

To assess the performance gain of the scheduler presented, system-level simulations of a Single-Input Single-Output (SISO) HSDPA network when transmitting the standard test sequence "Foreman" in Quarter Common Intermediate Format (QCIF) resolution have been conducted. The results obtained are then compared against the typical RR scheduler QoE. The simulation parameters are summarized in Table 13.2.

Figure 13.9 shows the results in terms of the ACK/NACK ratio and the luminance Peak Signal to Noise Ratio (PSNR), Y-PSNR. For the classical RR scheduler no distinction has been made between transport blocks containing I- or P-encoded frames. As expected, the average ACK/NACK ratio lies around 10 %. For the CA scheduling mechanism presented, the error probabilities associated with packets containing I- and P-frames have been presented separately. The CQI mapping presented allows for the error probability of the transport blocks containing I-frames to decrease by a factor of four, resulting in an average ACK/NACK ratio of around 2.7 %. The Y-PSNR is an indicator for the QoE performance. Three different GOP sizes have been investigated: 30, 45, and 60 frames. As a consequence of the smaller ACK/NACK ratio, Figure 13.9 shows that the quality of I-frames has increased by more than 0.5 dB when using the CA scheduler. Such an increase, however, is not only beneficial in terms of contribution to the average frame quality, but rather also has to be considered advantageous for the quality of the following P-frames. On the one hand, a valid source of prediction is offered to the following P-frames and, on the other hand, in the case that the previous GOP was damaged, the temporal error propagation of the error is terminated. The overall sequence quality of the CA scheduling mechanism is 0.6 dB higher than of a classical RR scheduler.

Table 13.2 System-level simulation settings for the performance assessment of the CA scheduler

Parameter	Value
Interference scenario	Homogeneous network
Number of cells	19, layout Type I
NodeB distance (m)	750
Carrier frequency (GHz)	1.9
Total transmit power available at NodeB (W)	20
Total CPICH power (W)	0.8
Power of other channels (W)	1.2
Available HS-PDSCHs	15
Large-scale pathloss model	Urban micro
Antenna utilization	$N_T \times N_R = 1 \times 1$
Channel type	ITU PedA
Active users in target sector	25
User mobility	3 km/h, random direction
UE capability class	10
UE receiver type	MMSE, $L_f = 30$ chips, $\tau = 15$ chips
Feedback delay	4 acTTI
HARQ type	Incremental Redundancy (IR)

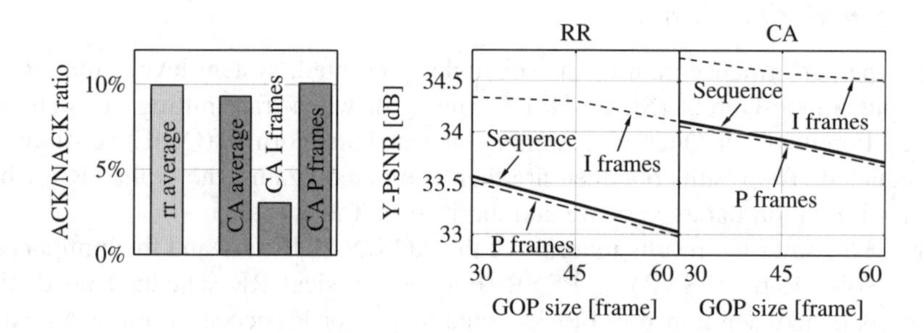

Figure 13.9 System-level simulation results for the CA scheduler.

13.4 CPICH Power Optimization

The DL performance of wireless networks depends not only on its data channels, as for the HS-DSCH in the case of HSDPA, but also on the configuration and performance of the signaling and feedback channels. As a matter of fact, many design goals as the target ACK/NACK ratio or the maximum load of the cell depend on the construction of the signaling and synchronization channels. In particular, the CPICH is important for the network planning and its overall performance; see also Chapter 1. The two main duties of it are to

- provide reference symbols for the channel estimation at the receiver side and
- provide a way for the UE to assess the best service NodeB, thus determining the coverage of the cells.

The cell coverage is an important network planning parameter, because, depending on the coverage area, the load of the individual cells can be managed. In addition, UMTS-based networks have to deal with "cell breathing," [28][5] which complicates matters here.

The task to provide means for channel estimation heavily influences the DL throughput performance as well. Obviously, the channel estimation directly determines the performance of the receiver; for example, the quality of the equalization in the case of an MMSE equalizer. A higher transmission power of the CPICH would lead, on average, to a better Signal to Noise Ratio (SNR) and, thus, could be conjectured to lead to better channel estimation and correspondingly to better equalization. This would deliver better throughput and error performances. On the other hand, however, an increase in the CPICH power increases the interference imposed on the HS-DSCH, which counterbalances the previous effect. This raises the question of how to find the optimum CPICH power value.

The optimization problem of the CPICH power configuration has attracted a lot of attention in the scientific community. Most recent studies, however, focus on the task of network load optimization when specific traffic density maps are available [7–9, 23, 24] and utilizing CPICH SNR targets as a side constraint. Earlier studies focus on UMTS systems with active power control [11, 25] – which is not active in the HSDPA DL – and also do not take the interference effect of the CPICH into account.

13.4.1 System-Level Modeling of the CPICH Influence

UMTS-based networks offer two different CPICHs, namely the primary and the secondary CPICH, where the secondary one is designed to serve dedicated hot-spot areas and thus will be neglected in the following. The primary CPICH has the following features:

- the spreading sequence for the primary CPICH is always the same, with length 256;
- there is only one primary CPICH in the cell that is broadcasted over the whole cell area; and
- 4-Quadrature Amplitude Modulation (QAM) modulation is utilized, which results in a bit-rate of 30 kbit/s.

A detailed description of the primary and secondary CPICH, including the channel coding and scrambling sequence details, can be found in [3]. In the following, the primary CPICH is denoted only by the name CPICH for the sake of notational simplicity.

[5] Cell breathing is the expansion or contraction of the effective coverage of a cell in response to the number of active mobiles in a network by redirecting UEs to surrounding cells.

13.4.1.1 Link-Quality Model Enhancement

To integrate the CPICH influence in the link-quality model, the receive filter employed – in this work the MMSE equalizer from Equation (12.12) – has to be based on the estimated channel. In particular, if the estimated channel is denoted by $\hat{\tilde{\mathbf{H}}}_{\mathrm{w}}$, the MMSE equalizer is given by

$$\hat{\mathbf{F}} = \mathbf{R}_{\mathbf{x}^{\mathrm{S}}_{i-\tau}\tilde{\mathbf{x}}_i} \hat{\tilde{\mathbf{H}}}_{\mathrm{w}}^{\mathrm{H}} (\hat{\tilde{\mathbf{H}}}_{\mathrm{w}} \mathbf{R}_{\tilde{\mathbf{x}}_i} \hat{\tilde{\mathbf{H}}}_{\mathrm{w}}^{\mathrm{H}} + \mathbf{R}_{\tilde{\mathbf{n}}_i})^{-1} \tag{13.12}$$

As investigated in [15],[6] the estimated channel $\hat{\tilde{\mathbf{H}}}_{\mathrm{w}}$ can statistically be represented by the matrix

$$\hat{\tilde{\mathbf{H}}}_{\mathrm{w}} = \tilde{\mathbf{H}}_{\mathrm{w}} + \mathbf{H}_{\Delta}, \tag{13.13}$$

with the error matrix $\mathbf{H}_{\Delta}$ being of the same structure as $\tilde{\mathbf{H}}_{\mathrm{w}}$. This means that $\mathbf{H}_{\Delta}$ has to be generated according to $\mathbf{H}_{\Delta} \triangleq \sigma^2_{\mathrm{MSE}}\mathbf{G}$, with $\mathbf{G}$ being a random Gaussian matrix of the same frequency selectivity structure as $\tilde{\mathbf{H}}_{\mathrm{w}}$ with variance one. The variance σ^2_{MSE} specifies the weighting of the error matrix and is a function of the CPICH transmission power. Accordingly, the equivalent fading parameters can easily be generated utilizing $\hat{\mathbf{F}}$ instead of $\mathbf{F}$ for various CPICH transmission power values.

13.4.1.2 Modeling Validation

To validate the new equivalent fading parameters, a set of link-level simulations has been performed for different antenna configurations, comparing the true SINR (see (3.71) in [29][7]) with the SINR as predicted by the modeling. Figure 13.10 shows the resulting model fittings for varying CPICH $E_{\mathrm{c}}/I_{\mathrm{or}}$ when an interference situation of $I_{\mathrm{oc}}/\hat{I}_{\mathrm{or}} = -6\,\mathrm{dB}$ is assumed at the UE[8] and a standard LS channel estimator is utilized [17]. It can be observed that the simplified equivalent fading parameter modeling based on the statistical error matrix modeling of Equation (13.13) is able to describe the CPICH influence nearly perfect. Details on the dependency on the interference power situation $I_{\mathrm{oc}}/\hat{I}_{\mathrm{or}}$ which effectively shifts the CPICH SINR as a function of the $E_{\mathrm{c}}/I_{\mathrm{or}}$ can be found in [15].

13.4.2 *CPICH Optimization in the Cellular Context*

As observable in Figure 13.10, for a given interference scenario defined by $I_{\mathrm{oc}}/\hat{I}_{\mathrm{or}}$, an optimum CPICH $E_{\mathrm{c}}/I_{\mathrm{or}}$ that maximizes the HS-DSCH SINR can be found. The interference situation for the CPICH, however, depends in the position of the UE in the cell.

Thus, by evaluating the pre-equalization SINR (see Figure 4.6 in [29]) $I_{\mathrm{oc}}/\hat{I}_{\mathrm{or}}$ can be calculated for every position in the cell. The optimization procedure then is as follows:

[6] Download the thesis from http://publik.tuwien.ac.at/files/PubDat_174816.pdf.

[7] Download the thesis from http://publik.tuwien.ac.at/files/PubDat_181165.pdf.

[8] The parameter $E_{\mathrm{c}}/I_{\mathrm{or}}$ specifies the fraction of the total available transmit power utilized for the transmission of the CPICH and $I_{\mathrm{oc}}/\hat{I}_{\mathrm{or}}$ is the ratio between signal power from the serving NodeB and total interference power. Note that, in this section, the average channel power is assumed to be one.

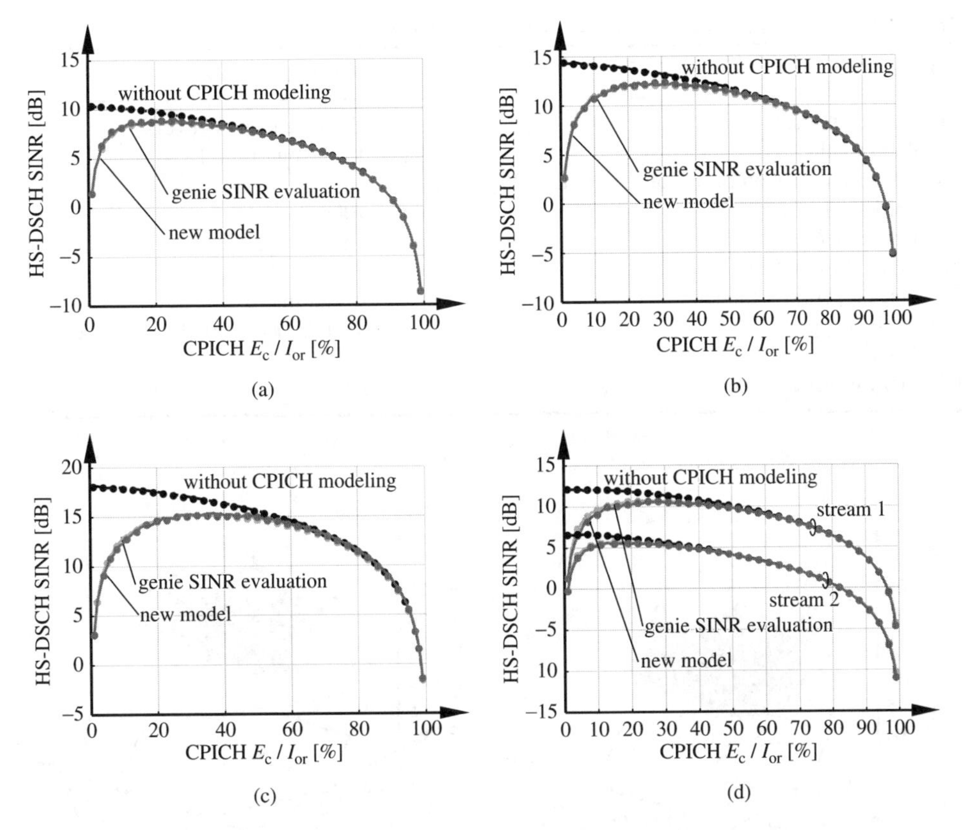

Figure 13.10 Enhanced link-quality model validation for different antenna configurations and stream utilization. (a) SISO $N_T \times N_R = 1 \times 1$ system utilizing one stream in a PedA channel environment. (b) MISO 2×1 system utilizing one stream in a PedA channel environment. (c) MIMO 2×2 system utilizing one stream in a PedA channel environment. (d) MIMO 2×2 system utilizing two streams in a PedA channel environment.

1. calculate the pre-equalization SINR map of the desired network;
2. generate a regular grid of positions within the target sector to be evaluated and evaluate the interference situation $I_{oc}/\hat{I}_{or}$ for all grid points;
3. calculate the HS-DSCH SINR curves as function of the CPICH E_c/I_{or} given all stored $I_{oc}/\hat{I}_{or}$ values, and
4. find the optimum CPICH E_c/I_{or} value of every calculated HS-DSCH SINR curve according to

$$\left(\frac{E_c}{I_{or}}\right)^*_{\text{CPICH}} = \frac{1}{N_S}\sum_{n=1}^{N_S} \arg \max_{\left(\frac{E_c}{I_{or}}\right)_{\text{CPICH}}} \text{SINR}_n^{\text{HS-DSCH}} \tag{13.14}$$

Table 13.3 lists the system-level simulation settings for the optimization procedure. The resulting optimum CPICH E_c/I_{or} values for all positions evaluated in the target sector when utilizing two streams in an $N_T \times N_R = 2 \times 2$ channel are depicted in Figure 13.11.

Table 13.3 System-level simulation settings for the CPICH power optimization

Parameter	Value
Interference scenario	Homogeneous network
Number of cells	19, layout Type I
Available transmission power at the NodeB (W)	20
NodeB distance (m)	1000
Carrier frequency (GHz)	1.9
Available HS-PDSCHs	15
Large-scale pathloss model	Urban micro
UE receiver type	MMSE, $L_f = 30$ chips, $\tau = 15$ chips
Feedback delay	4 TTI
Channel estimation type	LS

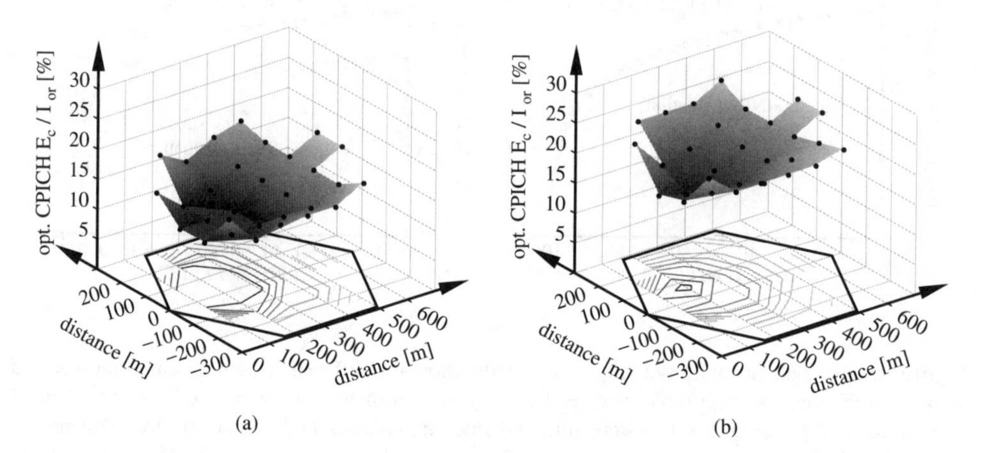

Figure 13.11 DS utilization CPICH power optimization results for different antenna configurations experiencing PedA and PedB channel environments. (a) MIMO $N_T \times N_R = 2 \times 2$ network utilizing two streams in a PedA channel environment. (b) MIMO 2×2 network utilizing two streams in a PedB channel environment.

It can be seen that the optimum CPICH E_c/I_{or} value increases towards the sector edge, which is due to the higher interference power. Furthermore, the PedB channel environment requires significantly higher E_c/I_{or} values to achieve the optimum performance. This is due to the greatly increased delay spread of the PedB channel when compared with the PedA channel, which makes the channel estimation and the equalization much more complicated.

Setting the CPICH E_c/I_{or} to different values depending on the UE position in the sector is, unfortunately, impossible. Thus, for the network operator the CPICH power has to be set to an average value. A simple option is to perform a linear averaging of all evaluated

Table 13.4 Resulting optimum average CPICH E_c/I_{or} values for different network configurations (TxAA and D-TxAA) and channel profiles

Network configuration	$\overline{E_c/I_{or}}$ (%)
MISO 2 × 1, TxAA, PedA	11.00
MISO 2 × 1, TxAA, PedB	19.92
MIMO 2 × 2, TxAA, PedA	13.08
MIMO 2 × 2, TxAA, PedB	21.08
MIMO 2 × 2, D-TxAA, PedA	14.67
MIMO 2 × 2, D-TxAA, PedB	21.75

grid points within the target sector,

$$\overline{E_c/I_{or}} = \frac{1}{|\mathcal{G}|} \sum_{\mathcal{G}} \left(\frac{E_c}{I_{or}} \right)^{*}_{\text{CPICH}}, \tag{13.15}$$

with $\mathcal{G}$ denoting the grid of target sector positions at which the optimum CPICH $\frac{E_c}{I_{or}}$ has been evaluated. In case of a known traffic density map, the CPICH power averaging could also consider this density distribution. The resulting optimum average CPICH E_c/I_{or} values for all network scenarios evaluated are listed in Table 13.4.

Note that the optimization procedure presented neglects CPICH pollution effects [20], which can cause problems for the UMTS soft handover. To reflect these optimization constraints, target CPICH, SINR values at the cell edge would have to be defined. Nevertheless, the resulting optimum CPICH power values can serve as a design guideline for HSDPA networks given different antenna configurations and may be particularly helpful for when MIMO networks are put in trial.

References

[1] 3GPP (2007) Technical Specification TS 25.214 Version 7.4.0 'Physical layer procedures (FDD),' www.3gpp.org.

[2] 3GPP (2007) Technical Specification TS 25.996 Version 7.0.0 'Spatial channel model for multiple-input multiple-output (MIMO) simulations,' www.3gpp.org.

[3] 3GPP (2009) Technical Specification TS 25.211 Version 8.4.0 'Physical channels and mapping of transport channels onto physical channels (FDD),' www.3gpp.org.

[4] 3GPP (2008) Technical Specification TS 23.107 Version 8.0.0 'Quality of service (QoS) concept and architecture,' www.3gpp.org.

[5] Ahmed, M. H., Yanikomeroglu, H., and Mahmoud, S. (2003) 'Fairness enhancement of link adaptation techniques in wireless access networks,' in *Proceedings of IEEE Vehicular Technology Conference Fall (VTC)*, volume 3, pp. 1554–1557.

[6] Chao, H., Liang, Z., Wang, Y., and Gui, L. (2004) 'A dynamic resource allocation method for HSDPA in WCDMA system,' in *Proceedings of IEE International Conference on 3G Mobile Communication Technologies*, pp. 569–573.

[7] Chen, L. and Yuan, D. (2008) 'Automated planning of CPICH power for enhancing HSDPA performance at cell edges with preserved control of R99 soft handover,' in *Proceedings of IEEE International Conference on Communications (ICC)*, pp. 2936–2940.

[8] Chen, L. and Yuan, D. (2008) 'CPICH power planning for optimizing HSDPA and R99 SHO performance: Mathematical modelling and solution approach,' in *Proceedings of IFIP 1st Wireless Days Conference (WD)*, pp. 1–5.

[9] Chen, L. and Yuan, D. (2009) 'Achieving higher HSDPA performance and preserving R99 soft handover control by large scale optimization in CPICH coverage planing,' in *Proceedings of IEEE Wireless Telecommunications Symposium (WTS)*, pp. 1–6.

[10] De Bruin, I., Heijenk, G., El Zarki, M., and Zan, L. (2003) 'Fair channel-dependent scheduling in CDMA systems,' in *Proceedings of IST Mobile & Wireless Communications Summit*, pp. 737–741.

[11] Garcia-Lozano, M., Ruiz, S., and Olmos, J. (2003) 'CPICH power optimisation by means of simulated annealing in an UTRA-FDD environment,' *Electronics Letters*, 39 (23), 1676–1677.

[12] Gómez, G. and Sanchez, R. (2006) *End-to-End Quality of Service over Cellular Networks*, John Wiley & Sons, Ltd.

[13] Lilley, G., Wrulich, M., and Rupp, M. (2009) 'Network based stream-number decision for MIMO HSDPA,' in *Proceedings of ITG International Workshop on Smart Antennas (WSA)*. Available from http://publik.tuwien.ac.at/files/PubDat_174264.pdf

[14] Mäder, A., Staehle, D., and Spahn, M. (2007) 'Impact of HSDPA radio resource allocation schemes on the system performance of UMTS networks,' in *Proceedings of IEEE 66th Vehicular Technology Conference Fall (VTC)*, pp. 315–319.

[15] Mateu-Torelló, A. (2009) 'CPICH power optimization for MIMO HSDPA,' Master's thesis, Vienna University of Technology. Available from http://publik.tuwien.ac.at/files/PubDat_174816.pdf.

[16] Mehlführer, C., Caban, S., Wrulich, M., and Rupp, M. (2008) 'Joint throughput optimized CQI and precoding weight calculation for MIMO HSDPA,' in *Proceedings of 42nd Asilomar Conference on Signals, Systems and Computers*, Pacific Grove, USA. Available from http://publik.tuwien.ac.at/files/PubDat_167015.pdf

[17] Mehlführer, C. and Rupp, M. (2008) 'Novel tap-wise LMMSE channel estimation for MIMO W-CDMA,' in *Proceedings of IEEE 51st Global Telecommunications Conference (GLOBECOM)*. doi: 10.1109/GLOCOM.2008.ECP.829, Available from http://publik.tuwien.ac.at/files/PubDat_169129.pdf

[18] Network Working Group (1998) 'An architecture for differentiated services,' Technical Report RFC 2475, Internet Engineering Task Force.

[19] Network Working Group (2004) 'RTP payload format for H.264 video,' Technical Report RFC 3984, Internet Engineering Task Force.

[20] Niemelä, J. and Lempäinen, J. (2004) 'Mitigation of pilot pollution through base station antenna configuration in WCDMA,' in *Proceedings of IEEE 60th Vehicular Technology Conference (VTC) Fall*, volume 6, pp. 4270–4274.

[21] Pedersen, K. I., Lootsma, T. F., Stottrup, M. *et al.* (2004) 'Network performance of mixed traffic on high speed downlink packet access and dedicated channels in WCDMA,' in *Proceedings of IEEE 60th Vehicular Technology Conference (VTC)*, volume 6, pp. 4496–4500.

[22] Rupp, M. (2009) *Video and Multimedia Transmissions over Cellular Networks: Analysis, Modeling and Optimization in Live 3G Mobile Networks*, John Wiley & Sons, Ltd.

[23] Siomina, I. and Yuan, D. (2004) 'Optimization of pilot power for load balancing in WCDMA networks,' in *Proceedings of IEEE Global Telecommunications Conference (GLOBECOM)*, volume 6, pp. 3872–3876.

[24] Siomina, I. and Yuan, D. (2008) 'Enhancing HSDPA performance via automated and large-scale optimization of radio base station antenna configuration,' in *Proceedings of IEEE Vehicular Technology Conference Spring (VTC)*, pp. 2061–2065.

[25] Sun, Y., Gunnarsson, F., and Hiltunen, H. (2003) 'CPICH power settings in irregular WCDMA macro cellular networks,' in *Proceedings of IEEE 14th International Symposium on Personal, Indoor and Mobile Radio Communications (PIMRC)*, volume 2, pp. 1176–1180.

[26] Superiori, L., Wrulich, M., Svoboda, P., and Rupp, M. (2009) 'Cross-layer optimization of video services over HSDPA networks,' in *Proceedings of ICST International Conference on Mobile Lightweight Wireless Systems*, Athens, Greece. Available from http://publik.tuwien.ac.at/files/PubDat_175881.pdf

[27] Superiori, L., Wrulich, M., Svoboda, P. *et al.* (2009) 'Content-aware scheduling for video streaming over HSDPA networks,' in *Proceedings of IEEE Second International Workshop on Cross-Layer Design*, Mallorca, Spain. Available from http://publik.tuwien.ac.at/files/PubDat_175882.pdf

[28] Thng, K. L., Yeo, B. S., and Chew, Y. H. (2005) 'Performance study on the effects of cell-breathing in WCDMA,' in *Proceedings of IEEE 2nd International Symposium on Wireless Communication Systems*, pp. 44–49.
[29] Wrulich, M. (2009) 'System-level modeling and optimization for MIMO HSDPA networks,' Ph.D. thesis, Vienna University of Technology. Available from http://publik.tuwien.ac.at/files/PubDat_181165.pdf.
[30] Wrulich, M., Weiler, W., and M. Rupp (2008) 'HSDPA performance in a mixed traffic network,' in *Proceedings of IEEE Vehicular Technology Conference Spring (VTC)*, Marina Bay, Singapore, pp. 2056–2060. Available from http://publik.tuwien.ac.at/files/pub-et_13769.pdf.

14

Optimal Multi-User MMSE Equalizer

Interference[1] is said to be the most important performance-limiting factor of modern communication systems. This especially holds true for Wideband Code-Division Multiple Access (WCDMA) networks, where frequency selectivity in the Downlink (DL) channel causes a loss in orthogonality between the spreading codes utilized and imposes restrictive throughput constraints. This issue has already been identified in Chapter 12 for the development of the link-quality model, as well as in the simulation results in Chapter 13. However, besides the unavoidable interference caused by spreading code crosstalk, the question arises as to whether possibilities exist to exploit structural properties in the interference for Multiple-Input Multiple-Output (MIMO) High-Speed Downlink Packet Access (HSDPA). The information theoretic principles for this kind of interference problem are represented by the MIMO broadcast channel [14, 37], which is related to the MIMO multiple-access channel in the Uplink (UL) [36]. Although the capacity regions in the case of fading channels are not known yet, all of the results so far indicate the need for interference mitigation to come close to the upper bound on the sum-capacity [12, 30].

Interference can be combated at the transmitter and/or the receiver side. In particular, in the DL each receiver needs to detect a "single" desired signal, while experiencing two main types of interference. These are caused by the serving Base station (NodeB), called "intra-cell interference," and by a few dominant neighboring NodeBs, called "inter-cell interference." In the UL, on the other hand, the base station receiver has to detect "all" desired users in the cell while having to suppress neighboring interference from many different sources. When interference mitigation is performed at the transmitter, accurate channel state information from all users is needed [43]. The optimum solution known so far for the multi-user case is the utilization of Dirty Paper Coding (DPC) [13], which requires full channel and transmit signal knowledge at a central coordinating entity. A generalization of the classical DPC idea to the case of multiple base stations has been made in [25]. This requires lots of signaling and feedback information exchange,

[1] Reproduced from [40], M. Wrulich, C. Mehlführer, and M. Rupp, "Managing the interference structure of MIMO HSDPA: a multi-user interference aware MMSE receiver with moderate complexity," *IEEE Transactions on Wireless Communications*, vol. 9, no. 4, pp. 1472–1482, Apr. 2010 by permission of © 2010 IEEE.

Evaluation of HSDPA and LTE: From Testbed Measurements to System Level Performance, First Edition.
Sebastian Caban, Christian Mehlführer, Markus Rupp and Martin Wrulich.

which is typically not available in a cellular context. Beamforming may serve as an approximation to the optimum interference mitigation at the transmitter, but faces some theoretical difficulties [31].

Handling the interference on the receiver side is a difficult job as well, especially in the multi-user case, where classical approaches result in complex receiver structures [9, 27, 35]. Moreover, it is also not known theoretically to what extent receiver cooperation is needed to achieve a close-to-capacity performance, apart from the fact that it is not known what the optimum receiver cooperation really looks like, as in [16, 24]. More practical investigations on this subject are also conducted by 3rd Generation Partnership Project (3GPP) [2], but so far none of these recommendations have been implemented. Given the limited battery capabilities of today's handsets, complexity is also an important issue [20]. A good overview of the different practical interference situations together with some well-known solutions for them can be found in [8].

Coming back to MIMO HSDPA systems, the classical Minimum Mean Square Error (MMSE) equalizers recommended by 3GPP do not take any special knowledge of the interference structure into account [21, 32, 44]. The only possibility to include some knowledge about the interference structure in these approaches would be to estimate the noise covariance matrix in a way that reflects the interference structure. However, noise covariance estimation is usually performed very rough, if not at all treated as white, to obtain simpler algorithms due to complexity arguments. In [44], Multi-User (MU) interference terms are considered but the authors investigated a nonstandard conform precoding scheme and the proposed equalizer works on the symbol level as opposed to the chip-level-based receiver presented here. Moreover, no pilot interference structure is taken into account. The Double Transmit Antenna Array (D-TxAA) MIMO HSDPA operation, however, implies a spatial structure of the intra-cell interference due to its precoding. If the channel quality is low, and thus only one stream is supported, the resulting transmission mode is Transmit Antenna Array (TxAA),[2] which is also the supported mode when the UE has only one receive antenna available. In the case that multi-user scheduling for TxAA users takes place, the spatial structure can be exploited to achieve a higher throughput. The same argument also holds for the D-TxAA operation; unfortunately the spatial structure does not allow directly for an exploitation because all Degrees of Freedom (DoFs) have already been utilized, as described in Theorem 12.1.[3]

14.1 System Model

Figure 14.1 shows the TxAA transmission scheme for one receive antenna, when U users are simultaneously served. In TxAA HSDPA there is only one stream active for every user. Accordingly, the spread and scrambled chip stream of user u at time instant i can

[2] TxAA has been introduced already with Universal Mobile Telecommunications System (UMTS) in 1999. In contrast to the Double-Stream (DS) operation of MIMO HSDPA, TxAA allows for the same MU scheduling techniques as in the classical Single-Input Single-Output (SISO) HSDPA case. For the single-antenna operation, MU scheduling – also called "multi-code scheduling" – is well suited to work optimally in terms of the sum-rate throughput and the short-term fairness tradeoff [19]. Also note that the TxAA scheme allows for an arbitrary number of receive antennas at the User Equipment (UE).

[3] Note that this holds because of the quantized precoding codebook of 3GPP [3], but is not necessarily the case if a different codebook is employed.

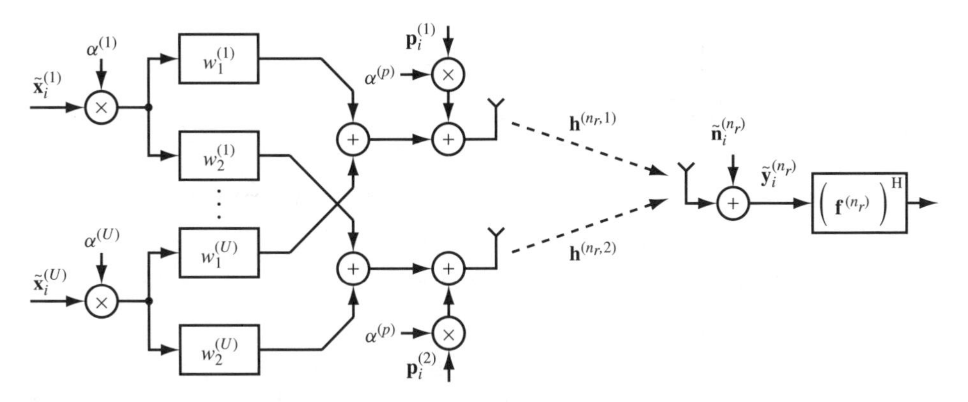

Figure 14.1 Multi-user transmission in TxAA, for a total number U of simultaneously served users. The precoding is conducted individually for every user. At the receiver, only one receive antenna is depicted, although the scheme allows for an arbitrary number of receive antennas.

be defined as

$$\tilde{\mathbf{x}}_i^{(u)} \triangleq \left[x_i^{(u)} \quad \cdots \quad x_{i-L_\mathrm{h}-L_\mathrm{f}+2}^{(u)}\right]^\mathrm{T}, \tag{14.1}$$

which corresponds to the "stacked" transmit chip vector in Chapter 5. For the sake of notational simplicity, the NodeB or sector index b has been omitted in this section. The statements and derivations presented also only take the intra-cell interference components explicitly into account (in contrast to Chapter 12); thus, there is no need to complicate the mathematical notation unnecessarily. The vector $\tilde{\mathbf{x}}_i^{(u)}$ contains the $L_\mathrm{h} + L_\mathrm{f} - 1$ most recent transmitted chips. This notation serves to represent the convolution of the transmit signal and the frequency-selective MIMO channel in vector matrix notation. Furthermore, note that this chip stream contains the sum of all spreading sequences utilized for user u; thus, the potential multi-code utilization of HSDPA is also modeled. Without loss of generality, the energy of the chip stream $\tilde{\mathbf{x}}_i^{(u)}$ of each user u can be normalized to one:

$$(\sigma_x^{(u)})^2 \equiv \sigma_x^2 = 1 \tag{14.2}$$

Thus, by multiplying $\tilde{\mathbf{x}}_i^{(u)}$ with a factor $\alpha^{(u)}$, the NodeB can allocate a certain amount of transmit power to each served user. After the power allocation, the chip streams are weighted by the user-dependent complex precoding coefficients $w_1^{(u)}$ and $w_2^{(u)}$ at the first and second transmit antennas respectively. This description holds for arbitrary precoding weights, although the 3GPP recommendation is strongly quantized. The weighted chip streams of all users are then added to the sequences $\alpha^{(\mathrm{p})}\mathbf{p}_i^{(1)}$ and $\alpha^{(\mathrm{p})}\mathbf{p}_i^{(2)}$, representing the sum of all channels that are transmitted without precoding; that is, the Common Pilot CHannel (CPICH), the High-Speed Shared Control CHannels (HS-SCCHs), and other signaling channels – which will be referred to as "non-data" channels.

The frequency-selective channel between the n_tth transmit and the n_rth receive antenna in Figure 14.1 is represented by the vector $\mathbf{h}^{(n_\mathrm{r},n_\mathrm{t})}$, $n_\mathrm{t} = 1, 2$, composed of the taps of the channel. If the non-data channels $\alpha^{(\mathrm{p})}\mathbf{p}_i^{(1)}$ and $\alpha^{(\mathrm{p})}\mathbf{p}_i^{(2)}$ are neglected for the moment – they

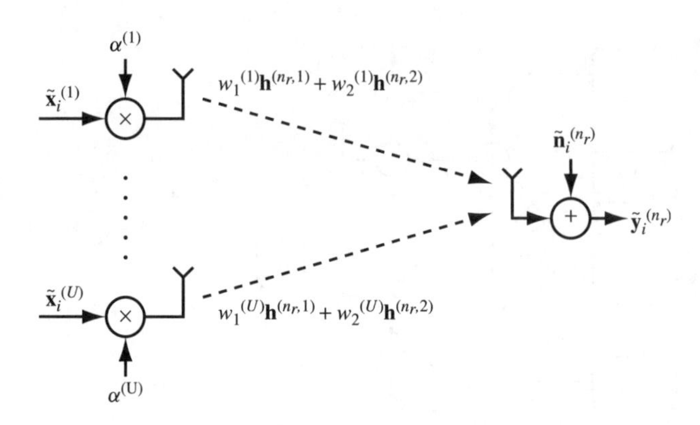

Figure 14.2 Equivalent representation of the multi-user TxAA transmission, showing U virtual antennas and U virtual channels.

will be included for the main derivation again – the multi-user transmission of Figure 14.1 can be represented by U virtual antennas, one for each active user, as illustrated in Figure 14.2. The resulting equivalent (also called "virtual") channels between user $u = 1, \ldots, U$ and receive antenna $n_\mathrm{r} = 1, \ldots, N_\mathrm{R}$ are then given by

$$w_1^{(u)}\mathbf{h}^{(n_\mathrm{r},1)} + w_2^{(u)}\mathbf{h}^{(n_\mathrm{r},2)} \tag{14.3}$$

From this description it can be seen that the intra-cell interference caused by other users and observed by the desired user u can be treated as being transmitted over up to $U-1$ different channels. This somehow represents a Single-Input Multiple-Output (SIMO) multiple access channel, instead of an MIMO broadcast channel. Of course, if two users utilize the same precoding, then their channels cannot be distinguished anymore. The classical MMSE approach, however, would be determined only by the precoding weights $w_1^{(u)}$ and $w_2^{(u)}$ and does not consider the structure of the interference. Thus, the degraded transmission scheme of TxAA imposes an interference situation which cannot be handled well by the classical MMSE equalizer that is only matched to the channel of the desired user. This is in contrast to the SISO HSDPA case, where, owing to the lack of precoding, the interference of simultaneously served users is transmitted over the same channel as that of the desired user. Thus, equalization of the desired user's signal also equalizes the signal of the simultaneously served users.

For the following evaluations, the $L_\mathrm{f} \times (L_\mathrm{h} + L_\mathrm{f} - 1)$-dimensional band matrix models the channel between the n_tth transmit and the n_rth receive antennas,

$$\mathbf{H}^{(n_\mathrm{r},n_\mathrm{t})} = \begin{bmatrix} h_0^{(n_\mathrm{r},n_\mathrm{t})} & \cdots & h_{L_\mathrm{h}-1}^{(n_\mathrm{r},n_\mathrm{t})} & 0 \\ \ddots & & \ddots & \\ 0 & h_0^{(n_\mathrm{r},n_\mathrm{t})} & \cdots & h_{L_\mathrm{h}-1}^{(n_\mathrm{r},n_\mathrm{t})} \end{bmatrix} \tag{14.4}$$

The entire frequency-selective MIMO channel can then be represented by a block matrix $\mathbf{H}$ consisting of $N_\mathrm{R} \times N_\mathrm{T}$ band matrices,

$$\mathbf{H} = \begin{bmatrix} \mathbf{H}^{(1,1)} & \mathbf{H}^{(1,2)} \\ \vdots & \vdots \\ \mathbf{H}^{(N_\mathrm{R},1)} & \mathbf{H}^{(N_\mathrm{R},2)} \end{bmatrix}, \tag{14.5}$$

which explicitly shows the restriction of TxAA to utilize only $N_\mathrm{T} = 2$ transmit antennas [3]. By stacking the received signal vectors of all N_R receive antennas,

$$\tilde{\mathbf{y}}_i = \left[(\tilde{\mathbf{y}}_i^{(1)})^\mathrm{T} \; \cdots \; (\tilde{\mathbf{y}}_i^{(N_\mathrm{R})})^\mathrm{T} \right]^\mathrm{T}, \tag{14.6}$$

and by stacking the transmitted signal vectors of all U users and the auxiliary channel signal vectors $\mathbf{p}_i^{(1)}$ and $\mathbf{p}_i^{(2)}$,

$$\tilde{\mathbf{x}}_i = \left[(\tilde{\mathbf{x}}_i^{(1)})^\mathrm{T} \; \cdots \; (\tilde{\mathbf{x}}_i^{(U)})^\mathrm{T} \; (\mathbf{p}_i^{(1)})^\mathrm{T} \; (\mathbf{p}_i^{(2)})^\mathrm{T} \right]^\mathrm{T}, \tag{14.7}$$

the compact system description follows as

$$\tilde{\mathbf{y}}_i = \mathbf{H}(\mathbf{W}^{(\mathrm{MU})} \otimes \mathbf{I}_{L_\mathrm{h}+L_\mathrm{f}-1})\tilde{\mathbf{x}}_i + \tilde{\mathrm{n}}_i = \mathbf{H}_\mathrm{w}\tilde{\mathbf{x}}_i + \tilde{\mathrm{n}}_i \tag{14.8}$$

In this formulation, $\tilde{\mathrm{n}}_i$ can be used to incorporate both the thermal noise and the inter-cell interference from other base stations, which would allow the resulting receive filter some possibilities to combat the inter-cell interference as well.

The $2 \times (U + 2)$-dimensional matrix $\mathbf{W}^{(\mathrm{MU})}$ contains the precoding coefficients $w_1^{(u)}$ and $w_2^{(u)}$ of all users, as well as the power coefficients $\alpha^{(u)}$:

$$\mathbf{W}^{(\mathrm{MU})} \triangleq \begin{bmatrix} \alpha^{(1)} w_1^{(1)} & \cdots & \alpha^{(U)} w_1^{(U)} & \alpha^{(\mathrm{p})} & 0 \\ \alpha^{(1)} w_2^{(1)} & \cdots & \alpha^{(U)} w_2^{(U)} & 0 & \alpha^{(\mathrm{p})} \end{bmatrix} \tag{14.9}$$

This matrix reflects the premise that non-data channels are not precoded; thus, the two columns on the right side are specified solely by the single parameter $\alpha^{(\mathrm{p})}$, which controls the total power spent on these channels. In general, the coefficients α are prone to a sum-power constraint,

$$\sum_{u=1}^{U} (\alpha^{(u)})^2 + 2(\alpha^{(\mathrm{p})})^2 = P_\mathrm{Tx} \tag{14.10}$$

Furthermore, it can be assumed that the power control is completely included in the power coefficients $\alpha^{(u)}$ and $\alpha^{(\mathrm{p})}$, which imposes an additional power constraint on the precoding coefficients, $\|\mathbf{w}^{(u)}\|^2 = 1$ with $\mathbf{w}^{(u)} \triangleq [w_1^{(u)}, w_2^{(u)}]^\mathrm{T}$.

14.2 Intra-Cell Interference Aware MMSE Equalization

Having the system model being specified, the resulting MMSE equalizer can be derived. Without loss of generality, it can be assumed that the transmit data sequence of user 1 has to be reconstructed. Accordingly, the MMSE equalizer coefficients can be calculated by minimizing the quadratic cost function [29]

$$J(\mathbf{f}) = \mathbb{E}\left\{\left|\mathbf{f}^{\mathrm{H}}\tilde{\mathbf{y}}_i - x_{i-\tau}^{(1)}\right|^2\right\}, \tag{14.11}$$

with τ again specifying the delay of the equalized signal, and fulfilling $\tau \geq L_{\mathrm{h}}$ due to causality arguments. Furthermore, perfect channel knowledge at the receiver side will be assumed.

The cost function minimizes the distance between the equalized chip stream and the transmitted chip stream in the Euclidean distance sense. In Equation (14.11), the vector $\mathbf{f}$ defines N_{R} equalization filters:

$$\mathbf{f} = \left[(\mathbf{f}^{(1)})^{\mathrm{T}} \ \cdots \ (\mathbf{f}^{(N_{\mathrm{R}})})^{\mathrm{T}}\right]^{\mathrm{T}} \tag{14.12}$$

Each filter $\mathbf{f}^{(n_{\mathrm{r}})} = [f_0^{(n_{\mathrm{r}})} \ \cdots \ f_{L_{\mathrm{f}}-1}^{(n_{\mathrm{r}})}]$ has a length of L_{f} chips, similar to the equalizer derived in Chapter 5. Note that, because of the definition of $\mathbf{f}$ and $\tilde{\mathbf{y}}_i$, the inner product $\mathbf{f}^{\mathrm{H}}\tilde{\mathbf{y}}_i$ can be implemented by summation of the outputs of the N_{R} equalization filters $\mathbf{f}^{(n_{\mathrm{r}})}$. This sum then yields the MMSE estimate of the transmitted chip sequence. To obtain the optimum receive filter, the cost function $J(\mathbf{f})$ has to be minimized,

$$\mathbf{f}^{(\mathrm{opt})} \triangleq \arg\min_{\mathbf{f}} J(\mathbf{f}), \tag{14.13}$$

which, owing to the design of the cost function, is a convex nonconstraint optimization problem [10, 23]. Accordingly, it can be solved by finding the point at which the gradient of the cost function is equal to zero:

$$\nabla J(\mathbf{f}^{(\mathrm{opt})}) \stackrel{!}{=} \mathbf{0}_{N_{\mathrm{R}}L_{\mathrm{f}}}, \tag{14.14}$$

where $\mathbf{0}_{N_{\mathrm{R}}L_{\mathrm{f}}}$ denotes a zero vector of length $N_{\mathrm{R}}L_{\mathrm{f}}$. Using the same definition of the complex-valued derivative as in [15],[4]

$$z = x + \mathrm{i}y \rightarrow \frac{\partial}{\partial z} = \frac{1}{2}\left(\frac{\partial}{\partial x} - \mathrm{i}\frac{\partial}{\partial y}\right), \ \frac{\partial}{\partial z^*} = \frac{1}{2}\left(\frac{\partial}{\partial x} + \mathrm{i}\frac{\partial}{\partial y}\right) \tag{14.15}$$

[4] Note that this definition also implies that

$$\frac{\partial z}{\partial z} = 1, \ \frac{\partial z}{\partial z^*} = \frac{\partial z^*}{\partial z} = 0$$

for complex-valued numbers z, as well as

$$\frac{\partial \mathbf{z}}{\partial \mathbf{z}} = \mathbf{I}, \ \frac{\partial \mathbf{z}}{\partial \mathbf{z}^*} = \frac{\partial \mathbf{z}^*}{\partial \mathbf{z}} = \mathbf{0}$$

for complex-valued vectors $\mathbf{z}$. This definition extends the differentials to nonhomomorphic functions in the complex numbers plane, where the Cauchy–Riemann equations no longer hold. A definition such as this is necessary because most functions encountered in physical sciences and engineering are not analytic. In the mathematical literature, these relationships are often called "Wirtinger calculus" [38].

Then it can be shown that the gradient problem can be reformulated to

$$\nabla J(\mathbf{f}) = 2\frac{\partial}{\partial \mathbf{f}^*} J(\mathbf{f}) \tag{14.16}$$

By applying linearity of the expectation and the partial derivative operator, the gradient of the cost function can be shown to be

$$\frac{\partial}{\partial \mathbf{f}^*} J(\mathbf{f}) = \mathbb{E}\left\{\frac{\partial}{\partial \mathbf{f}^*} \mathbf{f}^{\mathrm{H}}\tilde{\mathbf{y}}_i\tilde{\mathbf{y}}_i^{\mathrm{H}}\mathbf{f} - x_{i-\tau}^{(1)}\tilde{\mathbf{y}}_i^{\mathrm{H}}\mathbf{f} - \mathbf{f}^{\mathrm{H}}\tilde{\mathbf{y}}_i(x_{i-\tau}^{(1)})^* + \left|x_{i-\tau}^{(1)}\right|^2\right\}, \tag{14.17}$$

which becomes

$$\frac{\partial}{\partial \mathbf{f}^*} J(\mathbf{f}) = \mathbb{E}\left\{\tilde{\mathbf{y}}_i\tilde{\mathbf{y}}_i^{\mathrm{H}}\mathbf{f} - \tilde{\mathbf{y}}_i(x_{i-\tau}^{(1)})^*\right\} \tag{14.18}$$

Inserting the input–output relation from Equation (14.8) and assuming uncorrelated data and noise samples, the derivative of the cost function can be further simplified to

$$\frac{\partial}{\partial \mathbf{f}^*} J(\mathbf{f}) = [\mathbf{H}_{\mathrm{w}}\mathbb{E}\left\{\tilde{\mathbf{x}}_i\tilde{\mathbf{x}}_i^{\mathrm{H}}\right\}\mathbf{H}_{\mathrm{w}}^{\mathrm{H}} + \mathbb{E}\left\{\tilde{\mathbf{n}}_i\tilde{\mathbf{n}}_i^{\mathrm{H}}\right\}]\mathbf{f} - \mathbf{H}_{\mathrm{w}}\mathbb{E}\left\{\tilde{\mathbf{x}}_i(x_{i-\tau}^{(1)})^*\right\}, \tag{14.19}$$

which, by defining $\mathbf{R}_{xx} \triangleq \mathbb{E}\left\{\tilde{\mathbf{x}}_i\tilde{\mathbf{x}}_i^{\mathrm{H}}\right\}$ and $\mathbf{R}_{\mathrm{nn}} \triangleq \mathbb{E}\left\{\tilde{\mathbf{n}}_i\tilde{\mathbf{n}}_i^{\mathrm{H}}\right\}$, as well as assuming the individual transmit chips to be uncorrelated,

$$\mathbb{E}\left\{\tilde{\mathbf{x}}_i\left(x_{i-\tau}^{(1)}\right)^*\right\} = \mathbb{E}\left\{\left|x_{i-\tau}^{(1)}\right|^2\right\}\mathbf{e}_\tau = \sigma_x^2\mathbf{e}_\tau, \tag{14.20}$$

leads to

$$\frac{\partial J}{\partial \mathbf{f}^*} = (\mathbf{H}_{\mathrm{w}}\mathbf{R}_{xx}\mathbf{H}_{\mathrm{w}}^{\mathrm{H}} + \mathbf{R}_{\mathrm{nn}})\mathbf{f} - \sigma_x^2\mathbf{H}_{\mathrm{w}}\mathbf{e}_\tau \stackrel{!}{=} \mathbf{0} \tag{14.21}$$

The matrices $\mathbf{R}_{xx}$ and $\mathbf{R}_{\mathrm{nn}}$ are the signal and noise covariance matrices respectively, and the vector $\mathbf{e}_\tau$ is a zero vector of length $(U+2)(L_{\mathrm{h}}+L_{\mathrm{f}}-1)$ with a single "1" at position τ. The equalizer coefficients for the data stream of the first user are therefore given by

$$\mathbf{f} = \sigma_x^2(\mathbf{H}_{\mathrm{w}}\mathbf{R}_{xx}\mathbf{H}_{\mathrm{w}}^{\mathrm{H}} + \mathbf{R}_{\mathrm{nn}})^{-1}\mathbf{H}_{\mathrm{w}}\mathbf{e}_\tau \tag{14.22}$$

If the transmitted data signals of the users are uncorrelated with equal powers σ_x^2, the covariance matrix $\mathbf{R}_{xx}$ becomes $\sigma_x^2\mathbf{I}$, and if the noise vector $\tilde{\mathrm{n}}_i$ is assumed white with variance σ_{n}^2, the noise covariance matrix becomes $\sigma_{\mathrm{n}}^2\mathbf{I}$. As already mentioned, the transmit signal covariance is assumed to be equal to one, because the individual transmit powers of the users are anyway determined by the power coefficients $\alpha^{(u)}$. The variance σ_{n}^2, on the other hand, is specified by the thermal noise power and the sum interference power from the neighboring base stations. Note that if the receiver ideally takes the structure of the inter-cell interference into account, then effort would have to be put into obtaining an accurate estimation of the covariance matrix $\mathbf{R}_{\mathrm{nn}}$, which would no longer be a scaled identity matrix. Since this equalizer considers the interference of all users in the cell due to the full knowledge of the matrix $\mathbf{W}^{(\mathrm{MU})}$, it can be called an "intra-cell interference-aware MMSE equalizer." The standard equalizer is a special case of this solution and neglects the interference structure imposed by the other users, which will be called a Single-User

(SU) equalizer in the following. It can be calculated from Equation (14.22) by utilizing the SU precoding weight matrix of rank 1,

$$\mathbf{W}^{(\mathrm{SU})} = \begin{bmatrix} \alpha^{(1)} w_1^{(1)} \\ \alpha^{(1)} w_2^{(1)} \end{bmatrix} \mathbf{e}_1^{\mathrm{T}}, \tag{14.23}$$

instead of the multi-user matrix $\mathbf{W}^{(\mathrm{MU})}$. Here, $\mathbf{e}_1$ is a zero column-vector of length $U+2$ with a "1" at the first position. Note that if only a single user is receiving data in the cell, then both equalizers are very similar. The only difference is that the intra-cell interference-aware equalizer also considers the interference generated by the non-data channels.

14.2.1 Interference Suppression Capability

Having derived the solution of the interference-aware MMSE equalizer, it is of interest to analytically assess its interference-suppression capabilities as well as its performance bounds. Using the system-level link-quality model introduced in Chapter 11, which is capable of describing the post-equalization and despreading Signal to Interference and Noise Ratio (SINR) for arbitrary linear receivers in a potentially multi-stream closed-loop MIMO Code-Division Multiple Access (CDMA) system, the remaining intra-cell interference after equalization – for a specific channel realization and precoding state – generated by the desired user and all other active users is explicitly given by

$$P^{\mathrm{intra}} = \sum_{\substack{m=0 \\ m\neq\tau}}^{L_{\mathrm{h}}+L_{\mathrm{f}}-2} \left|\mathbf{f}^{\mathrm{H}}\mathbf{h}_m^{(1)}\right|^2 + \sum_{u=2}^{U} \sum_{\substack{m=0 \\ m\neq\tau}}^{L_{\mathrm{h}}+L_{\mathrm{f}}-2} \left|\mathbf{f}^{\mathrm{H}}\mathbf{h}_m^{(u)}\right|^2, \tag{14.24}$$

where, in reference to Equation (12.35), $N_{\mathrm{S}} = 1$ due to the restriction to the TxAA transmission mode and the total transmission power per stream $P^{(u,b_0)}/N_{\mathrm{S}}$ is now incorporated in the power coefficients $\alpha^{(u)}$. These, furthermore, are contained in the equivalent MIMO channel matrix columns $\mathbf{h}_m^{(u)}$ of the user-dedicated channel matrix

$$\mathbf{H}^{(u)} = \mathbf{H}\left(\begin{bmatrix} \alpha^{(u)} w_1^{(u)} & \alpha^{(u)} w_2^{(u)} \end{bmatrix} \otimes \mathbf{I}_{L_{\mathrm{h}}+L_{\mathrm{f}}-1}\right), \tag{14.25}$$

where perfect channel and noise power knowledge at the receiver are assumed. Equation (14.24) depends on the current realization of the precoding state, in particular the precoding choices of the interfering users. Thus, to assess the average interference suppression capability of the proposed equalizer, the expected intra-cell interference is utilized, which for a certain number of active users in the cell becomes

$$\mathbb{E}_w\left\{P^{\mathrm{intra}}\right\} = \underbrace{\sum_{\substack{m=0 \\ m\neq\tau}}^{L_{\mathrm{h}}+L_{\mathrm{f}}-2} \left|\mathbf{f}^{\mathrm{H}}\mathbf{h}_m^{(1)}\right|^2}_{f_{\mathrm{self}}} + \underbrace{\sum_{u=2}^{U} \mathbb{E}_w\left\{\sum_{\substack{m=0 \\ m\neq\tau}}^{L_{\mathrm{h}}+L_{\mathrm{f}}-2} \left|\mathbf{f}^{\mathrm{H}}\mathbf{h}_m^{(u)}\right|^2\right\}}_{f_{\mathrm{other}}}, \tag{14.26}$$

with f_{self} and f_{other} denoting the determinative factors for the self- and other-user intra-cell interference remaining after equalization, and $\mathbb{E}_w\{\cdot\}$ being the expectation with respect to the precoding coefficients w_1 and w_2 of the other users.

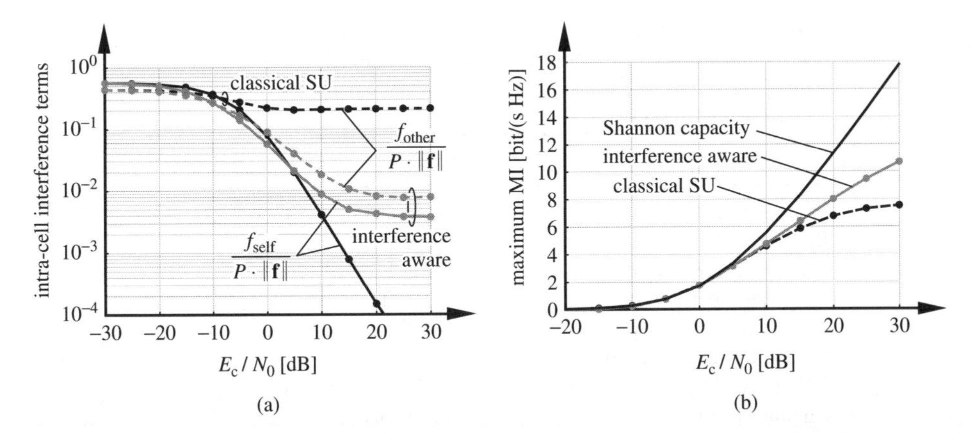

Figure 14.3 Theoretical performance investigation in terms of the interference suppression capabilities and the maximum Mutual Information (MI). (a) Intra-cell interference terms f_{self} and f_{other} for the classical SU equalizer and the proposed interference-aware equalizer, assuming perfect knowledge of the cell's precoding state. (b) Comparison of the maximum MI performance of the classical SU equalizer and the proposed interference-aware equalizer, assuming Gaussian post-equalization interference.

By utilizing the analytical description of the post-equalization intra-cell interference, the interference suppression capabilities and the theoretical performance bounds of the proposed equalizer can be evaluated. Table 14.1 lists the simulation parameters applied to assess the capability of the proposed equalizer to suppress the intra-cell interference caused due to the other active users in the cell. Figure 14.3(a) shows the performance in terms of the interference suppression capabilities, both for the self-interference f_{self}, and the other-user interference f_{other}, assuming perfect knowledge of the cell's precoding state. Note that the two coefficients are normalized by the total received interference power, which, since the channel (by assumption) is normalized to one, corresponds to dividing by the norm of the equalizer $\|\mathbf{f}\|$ and the total transmitted intra-cell power P_{Tx}.

It can be seen that the proposed equalizer is able to outperform the classical SU by significantly reducing the interference term f_{other} of the other users. The self-interference term f_{self}, on the other hand, becomes larger around 5 dB E_c/N_0. As specified by the cost function in Equation (14.11), the interference-aware equalizer minimizes the overall interference. Figure 14.3(a) thus illustrates that, at higher E_c/N_0, the equalizer sacrifices self-interference cancelation performance for the sake of a lower overall intra-cell interference.

Based on these results, it is also possible to evaluate bounds for the spectral efficiency. With the expected intra-cell interference being given by Equation (14.24), and considering the desired signal power being proportional to $|\mathbf{f}^{\text{H}}\mathbf{h}_{\tau}^{(u_0)}|^2$ [41], the equivalent DL data transmission channel including the equalization can be represented as an SISO Additive White Gaussian Noise (AWGN) channel. The corresponding Signal to Noise Ratio (SNR)

Table 14.1 Simulation parameters for the link-quality model examination of the interference suppression capabilities of the multi-user intra-cell interference-aware MMSE equalizer under ITU PedB channel model

Parameter	Value
Fading model	Improved Zheng model [42, 45]
Receive antennas	$N_R = 2$
Precoding codebook	3GPP TxAA [3]
Equalizer span (chips)	$L_f = 40$
Equalizer delay (chips)	$\tau = 20$
Precoding delay (slots)	11
UE speed (km/h)	3
Channel profile	ITU PedB [22]
Active users	$U = 4$

is thus given by

$$\mathrm{SNR} = \frac{\mathrm{SF}|\mathbf{f}^{\mathrm{H}}\mathbf{h}_{\tau}^{(u_0)}|^2}{f_{\mathrm{self}} + f_{\mathrm{other}} + N_0}, \tag{14.27}$$

when considering only intra-cell interference and with SF denoting the spreading factor of the High-Speed Physical Downlink Shared CHannel (HS-PDSCH). This corresponds to a single-cell scenario. Based on this SNR, the maximum MI can be evaluated as $\mathrm{MI} = \log_2(1 + \mathrm{SNR})$, which denotes an upper bound if perfect channel coding were to be utilized. Note, however, that this bound assumes a Gaussian distribution of the post-equalization interference, which in practice is not necessarily the case. Nevertheless, the so-derived maximum MI can serve as a figure of merit to assess the performance gain achievable by the interference-aware equalizer. Figure 14.3(b) shows the maximum MI for the classical SU equalizer and the interference-aware equalizer together with the Shannon channel capacity of an $N_T \times N_R = 2 \times 2$ channel. It can be seen that the proposed equalizer offers significant potential performance gains in the higher E_c/N_0 region.

14.3 The Cell Precoding State

In TxAA HSDPA, the MIMO channel is estimated by utilizing the CPICH, similar to UMTS. To be able to calculate the receive filter for the High-Speed Downlink Shared CHannel (HS-DSCH) data channel, however, the UE needs to know at least (i) the power offset of the individual HS-PDSCHs compared with the power level of the CPICH and (ii) the precoding coefficients that the NodeB applied for the transmission.[5] The power offset is signaled by higher layers [7], and the precoding coefficients of all simultaneous

[5] This is due to the fact that the channel estimation in UMTS and in HSDPA is based on the non-precoded CPICH; see Chapter 13. Accordingly, at least the precoding coefficients for the desired user transmission are needed to evaluate the MMSE equalizer weights for the classical SU equalizer.

transmissions are signaled on the respective HS-SCCHs [1, 6], where every active user has their own channel. This unfortunately makes things difficult for the proposed equalizer, because the HS-SCCHs are scrambled with user-specific scrambling sequences,[6] thus making it impossible to monitor the precoding state of the other users active in the cell.

In order to overcome this problem, three different strategies are possible:

1. change the signaling scheme in the HS-SCCH such that all active users know about the complete precoding state in the cell;
2. include some training data in the HS-PDSCHs of the users to be able to estimate the precoding state; or
3. blindly estimate the precoding state.

(1) is an obvious solution that needs no further explanation. Thus, in the following, a possible solution for (2) will be discussed, but the main focus will be dedicated to the blind estimation (3), because it can be implemented without changes in the current transmission standard.

The principal estimation problem is the following. According to Figure 14.1, every active user can have their own precoding coefficient pair $\{w_1^{(u)}, w_2^{(u)}\}$ and their own power factor $\alpha^{(u)}$, where only the coefficients dedicated to the user themselves are known. In addition, it is also not known how many users U are currently active. Estimation problems of this kind can be investigated within the framework of random set theory [9], leading to optimum Bayesian Maximum Likelihood (ML) estimators. However, these solutions require a joint estimation of the data sequences of all users and the precoding state, which is typically very complex and thus disadvantageous for battery-powered mobile devices. Accordingly, the efforts in this work are restricted to classical approaches with moderate complexity. According to [3], the precoding codebook that is utilized in practice is strongly quantized and, in addition, several users may share the same precoding vector. The codebook of precoding vectors can be defined as

$$\mathcal{W} \triangleq \{\mathbf{v}_1, \ldots, \mathbf{v}_{|\mathcal{W}|}\}, \mathbf{w}^{(u)} \in \mathcal{W}, \tag{14.28}$$

where $|\mathcal{W}|$ specifies the cardinality of the codebook. Furthermore, the set of users being served with the same precoding vector can be denoted as $\mathcal{U}_k = \{u : \mathbf{w}^{(u)} = \mathbf{v}_k\}$, with $k = 1, \ldots, |\mathcal{W}|$ denoting the codebook index of the corresponding precoding vector. Note that $\bigcup_{k=1}^{|\mathcal{W}|} \mathcal{U}_k$ does not necessarily have to be equal to $\mathcal{W}$, and that in case no user utilizes the precoding vector $\mathbf{v}_k$ the corresponding set is empty, $\mathcal{U}_k = \emptyset$.

Definition 14.1 *The precoding state of an HSDPA cell is defined as*

$$\mathcal{P} \triangleq \{\tilde{\alpha}^{(1)}, \ldots, \tilde{\alpha}^{(k)}, \ldots, \tilde{\alpha}^{(|\mathcal{W}|)}, \alpha^{(p)}\}, \tag{14.29}$$

[6] The scrambling sequence in HSDPA is a function of the user identification number, known only to the NodeB and the particular user.

where each $\tilde{\alpha}^{(k)}$ denotes the power coefficient utilized to transmit on a particular precoding vector $\mathbf{w}^{(k)}$,

$$\tilde{\alpha}^{(k)} = \sqrt{\sum_{u \in \mathcal{U}_k} (\alpha^{(u)})^2} \tag{14.30}$$

The coefficients $\alpha^{(u)}$ denote the power coefficient of every user $u = 1, \ldots, U$, as utilized in Equation (14.9).

With this definition, the two options for the precoding state estimation, (2) training sequences based estimation and (3) blind estimation, can be defined and their performance assessed.

14.3.1 Training-Sequence-Based Precoding State Estimation

Before going into the details on the training-sequence-based estimator, note that the input–output relation in Equation (14.8) can be rewritten as

$$\tilde{\mathbf{y}}_i = \mathbf{H} \begin{bmatrix} \mathbf{X}_i \mathbf{D}_1 \\ \mathbf{X}_i \mathbf{D}_2 \end{bmatrix} \boldsymbol{\alpha} + \tilde{\mathbf{n}}_i, \tag{14.31}$$

with the matrices $\mathbf{D}_1$ and $\mathbf{D}_2$ containing the precoding coefficients of the active users,

$$\mathbf{D}_1 = \mathrm{diag}\{w_1^{(1)}, \ldots, w_1^{(U)}, 1, 0\}, \tag{14.32}$$

$$\mathbf{D}_2 = \mathrm{diag}\{w_2^{(1)}, \ldots, w_2^{(U)}, 0, 1\}, \tag{14.33}$$

and the matrix $\mathbf{X}_i$ is the rearranged transmit vector $\tilde{\mathbf{x}}_i$,

$$\mathbf{X}_i = \begin{bmatrix} \tilde{\mathbf{x}}_i^{(1)} & \cdots & \tilde{\mathbf{x}}_i^{(U)} & \mathbf{p}_i^{(1)} & \mathbf{p}_i^{(2)} \end{bmatrix} \tag{14.34}$$

Finally, the vector $\boldsymbol{\alpha}$ lists the utilized power coefficients in the cell,

$$\boldsymbol{\alpha} = \begin{bmatrix} \alpha^{(1)} & \cdots & \alpha^{(U)} & \alpha^{(p)} & \alpha^{(p)} \end{bmatrix} \tag{14.35}$$

Considering the special structure of the problem in Equations (14.8) and (14.31), some redundancy can be observed. In particular, if two users u_1 and u_2 utilize the same precoding vector $\mathbf{w}^{(k)}$, the input–output relation for these two users can also be represented by one “combined” user u', utilizing $\mathbf{w}^{(k)}$ with

$$\alpha^{(u')} = \sqrt{(\alpha^{(u_1)})^2 + (\alpha^{(u_2)})^2} \tag{14.36}$$

as their effective power coefficient. This is in concordance with the precoding state of the cell, as given in Definition 14.1, and assumes that the non-data channels at the two transmit antennas deploy the same power coefficient $\alpha^{(\mathrm{p})}$. Given these arguments, it is sufficient to estimate only the precoding state $\mathcal{P}$ in order to describe the whole precoding situation in the cell. For the interference-aware MMSE equalizer, accordingly, the knowledge of $\mathcal{P}$ is sufficient to be able to suppress the intra-cell interference caused by the multi-user data transmission.

Considering the definition of the precoding state $\mathcal{P}$ in Equation (14.29), the vector $\tilde{\boldsymbol{\alpha}} = [\tilde{\alpha}^{(1)} \; \cdots \; \tilde{\alpha}^{(|\mathcal{W}|)} \; \alpha^{(\mathrm{p})}]^{\mathrm{T}}$ has to be estimated. The estimation error can be defined as

$$C = \left\| \tilde{\boldsymbol{\alpha}} - \hat{\tilde{\boldsymbol{\alpha}}} \right\|_2^2, \tag{14.37}$$

with $\hat{\tilde{\boldsymbol{\alpha}}}$ denoting the estimate of $\tilde{\boldsymbol{\alpha}}$. In the case that training data is available – for example, at the beginning of each transmission frame in HSDPA – a possible estimator is given by the Least Squares (LS) solution. The precoding codebook representing matrices can be set to

$$\tilde{\mathbf{D}}_1 \triangleq \mathrm{diag}\{w_1^{(1)}, \ldots, w_1^{(|\mathcal{W}|)}\}, \tag{14.38}$$

$$\tilde{\mathbf{D}}_1 \triangleq \mathrm{diag}\{w_2^{(1)}, \ldots, w_2^{(|\mathcal{W}|)}\} \tag{14.39}$$

In addition, the NodeB is assumed to provide orthogonal training sequences $\mathbf{t}^{(k)}$ for every precoding vector $\mathbf{w}^{(k)}$, which can be formed into the training matrix

$$\mathbf{S}_{\mathrm{T}} \triangleq \left[\mathbf{t}^{(1)} \; \cdots \; \mathbf{t}^{(|\mathcal{W}|)}\right] \tag{14.40}$$

Then the LS estimator of $\boldsymbol{\theta} = [\tilde{\alpha}^{(1)} \; \cdots \; \tilde{\alpha}^{(|\mathcal{W}|)}]^{\mathrm{T}}$, not including $\alpha^{(\mathrm{p})}$, is given by [23]

$$\hat{\boldsymbol{\theta}}_{\mathrm{LS}} = \Re\left\{ \left(\mathbf{H} \begin{bmatrix} \mathbf{S}_{\mathrm{T}}\tilde{\mathbf{D}}_1 \\ \mathbf{S}_{\mathrm{T}}\tilde{\mathbf{D}}_2 \end{bmatrix} \right)^{\#} \tilde{\mathbf{y}}_i \right\}^{+}, \tag{14.41}$$

where the real-valued operator $\Re\{\cdot\}^{+}$ ensures that the coefficients are real valued and positive, even in the low SNR regime where the noise would potentially cause non-real-valued estimates. The coefficient $\alpha^{(\mathrm{p})}$ can be calculated by utilizing the sum power constraint in Equation (14.10). Accordingly, the augmented LS estimate of $\tilde{\boldsymbol{\alpha}}$ is given by

$$\hat{\tilde{\boldsymbol{\alpha}}}_{\mathrm{LS}} = \left[\hat{\boldsymbol{\theta}}_{\mathrm{LS}}^{\mathrm{T}} \;\; 1 - \hat{\boldsymbol{\theta}}_{\mathrm{LS}}^{\mathrm{T}} \hat{\boldsymbol{\theta}}_{\mathrm{LS}} \right]^{\mathrm{T}} \tag{14.42}$$

14.3.2 Blind Precoding State Estimation

Blind estimation is, in general, a quite challenging task, particularly in the multi-user context. Typical approaches treat the unknown inputs – in this case the unknown transmit data $\tilde{\mathbf{y}}_i$ – as "nuisance parameters" that the estimator has to cope with in order to supply blind estimates of the parameters of interest.

The ML principle provides a systematic way for deducing the Minimum Variance Unbiased (MVU) estimator, maximizing the joint likelihood function $f_{\tilde{\mathbf{y}}}\left(\tilde{\mathbf{y}}; \boldsymbol{\alpha}, \tilde{\mathbf{x}}\right)$ [18]. As discussed in [26, 33], a number of possibilities exist to avoid the joint estimation of all parameters, in this case $\boldsymbol{\alpha}$ and $\tilde{\mathbf{x}}_i$. The "unconditional" or "stochastic" ML criterion models the vector of nuisance parameters as a random vector and maximizes the marginal of the likelihood function conditioned to $\tilde{\mathbf{x}}$,

$$f_{\tilde{\mathbf{y}}}\left(\tilde{\mathbf{y}};\boldsymbol{\alpha}\right) = \mathbb{E}_{\tilde{\mathbf{x}}}\left\{f_{\tilde{\mathbf{y}}|\tilde{\mathbf{x}}}\left(\tilde{\mathbf{y}}|\tilde{\mathbf{x}};\boldsymbol{\alpha}\right)\right\}, \tag{14.43}$$

where $\mathbb{E}_{\tilde{\mathbf{x}}}\{\cdot\}$ stands for the expectation operator with respect to the transmit signal $\tilde{\mathbf{x}}$.

Unfortunately, the unconditional ML estimator is generally unknown, because the expectation with respect to $\tilde{\mathbf{x}}$ typically cannot be solved in closed form. This is also the case for this particular problem. However, in the low SNR regime, the unconditional likelihood function $f_{\tilde{\mathbf{y}}}\left(\tilde{\mathbf{y}};\boldsymbol{\alpha}\right)$ becomes quadratic in the observation with independence of the statistical distribution of the nuisance parameters. This estimator class is also generally difficult to solve and works only reasonably well in the low SNR regime [18].

This fact motivated research in the area of "second-order" estimators; for example, the "conditional" ML criterion that models the nuisance parameters as deterministic unknowns and maximizes the compressed likelihood function $f_{\tilde{\mathbf{y}}}(\tilde{\mathbf{y}};\boldsymbol{\alpha},\hat{\tilde{\mathbf{x}}})$, in which $\hat{\tilde{\mathbf{x}}}$ denotes the ML estimate of $\tilde{\mathbf{x}}$. Unfortunately, for the particular problem of estimating the precoding state of the cell, this estimator class cannot be utilized because it would require the matrix $\mathbf{H}_{\mathrm{w}}$ to be tall [33], which in the signal model used is not the case.

Another approach is the Gaussian ML estimator class which models the nuisance parameters as Gaussian random variables in order to obtain an analytical solution for the expectation in $\mathbb{E}_{\tilde{\mathbf{x}}}\left\{f_{\tilde{\mathbf{y}}|\tilde{\mathbf{x}}}\left(\tilde{\mathbf{y}}|\tilde{\mathbf{x}};\boldsymbol{\alpha}\right)\right\}$ [34]. This assumption seems to fit naturally into the system model because, owing to multi-code operation, many different transmit chips are added up, likely resulting in a near-Gaussian distribution of the receive signal.

The Gaussian ML estimator for the precoding state of an HSDPA cell, is the one minimizing the nonlinear cost function

$$\Lambda_{\mathrm{GML}}\left(\boldsymbol{\alpha}\right) = {}^{\mathrm{T}}\left[\ln \mathbf{R}\left(\boldsymbol{\alpha}\right) + \mathbf{R}^{-1}\left(\boldsymbol{\alpha}\right)\hat{\mathbf{R}}\right] = \ln\det\left[\mathbf{R}\left(\boldsymbol{\alpha}\right)\right] + {}^{\mathrm{T}}\left[\mathbf{R}^{-1}\left(\boldsymbol{\alpha}\right)\hat{\mathbf{R}}\right], \tag{14.44}$$

with $\hat{\mathbf{R}} = \tilde{\mathbf{y}}_i\tilde{\mathbf{y}}_i^{\mathrm{H}}$ denoting the sample covariance matrix and

$$\mathbf{R}\left(\boldsymbol{\alpha}\right) = \mathbf{H}_{\mathrm{w}}\mathbf{H}_{\mathrm{w}}^{\mathrm{H}} + \sigma_{\mathrm{n}}^{2}\mathbf{I} \tag{14.45}$$

being its expected value as a function of $\boldsymbol{\alpha}$. Note that Gaussian white noise with variance σ_{n}^2 is assumed here. The direct application of this estimator would require the number of users U to be known for the estimation, which is not the case in this problem setting. To overcome this problem, note that $\mathbf{W}^{(\mathrm{MU})}$ from Equation (14.9) can be rewritten as

$$\mathbf{W}^{(\mathrm{MU})} = \mathbf{W}\mathrm{diag}\,\boldsymbol{\alpha}, \tag{14.46}$$

with $\mathbf{W}$ containing the same precoding coefficients as $\mathbf{W}^{(\mathrm{MU})}$, but without the $\alpha^{(u)}$ coefficients. Recalling that the goal is to estimate the precoding state $\mathcal{P}$, the matrix $\mathbf{W}$ can be replaced in the context of Equation (14.45) by

$$\tilde{\mathbf{W}} = \left[\mathbf{w}^{(1)} \;\cdots\; \mathbf{w}^{(|\mathcal{W}|)}\right]\mathrm{diag}\left\{\underbrace{\left[\tilde{\alpha}^{(1)} \;\cdots\; \tilde{\alpha}^{(|\mathcal{W}|)}\right]^{\mathrm{T}}}_{\boldsymbol{\theta}}\right\}, \tag{14.47}$$

thus leading to

$$\mathbf{R}(\boldsymbol{\theta}) = \mathbf{H}(\tilde{\mathbf{W}}\tilde{\mathbf{W}}^{\mathrm{H}} \otimes \mathbf{I})\mathbf{H}^{H} + \sigma_{\mathrm{n}}^{2}\mathbf{I}, \tag{14.48}$$

and the new associated cost-function $\tilde{\Lambda}_{\mathrm{GML}}(\boldsymbol{\theta}) = {}^{\mathrm{T}}\,[\ln \mathbf{R}(\boldsymbol{\theta}) + \mathbf{R}^{-1}(\boldsymbol{\theta})\hat{\mathbf{R}}]$. Please note that, as in the training-based estimation, $\alpha^{(\mathrm{p})}$ can be calculated *from* the sum power constraint in Equation (14.10). The minimum of $\tilde{\Lambda}_{\mathrm{GML}}$ can, for example, be found by means of iterative or time-recursive scoring methods [15, 34], based on

$$\hat{\boldsymbol{\theta}}_{l+1} = \hat{\boldsymbol{\theta}}_{l} + \mathbf{J}_{\mathrm{GML}}^{-1}(\hat{\boldsymbol{\theta}})\nabla_{\mathrm{GML}}(\tilde{\mathbf{y}}; \hat{\boldsymbol{\theta}}), \tag{14.49}$$

with $\mathbf{J}_{\mathrm{GML}}^{-1}(\hat{\boldsymbol{\theta}})$ and $\nabla_{\mathrm{GML}}(\tilde{\mathbf{y}}; \hat{\boldsymbol{\theta}})$ being the Fisher information matrix and the gradient respectively. Alternatively, any other known efficient optimization technique, such as for example Sequential Quadratic Programming (SQP) methods [28], can be used. Such convex optimization-based techniques potentially offer large complexity gains if the underlying problem shows some form of sparsity which allows for a suitable factorization [10].

14.3.3 Estimator Performance

Figure 14.4 shows the Mean Square Error (MSE) C of the two estimators for different E_{c}/N_0 values. The simulation parameters are the same as in Table 14.1, except that the performances for one and two receive antennas are presented. For the LS estimation, Hadamard sequences of length 64 have been used for the training. To refine the estimation, both known side constraints, (i) that the coefficients $\tilde{\alpha}^{(k)}$ have to be real valued and strictly positive and (ii) that the sum power constraint in Equation (14.10) cannot be exceeded, have been utilized. It can be seen in Figure 14.4 that the training-based estimator works reasonably well from -20 dB E_{c}/N_0 on, and that the performance saturates at around 10 dB. The blind estimator, on the other hand, is not able to deliver

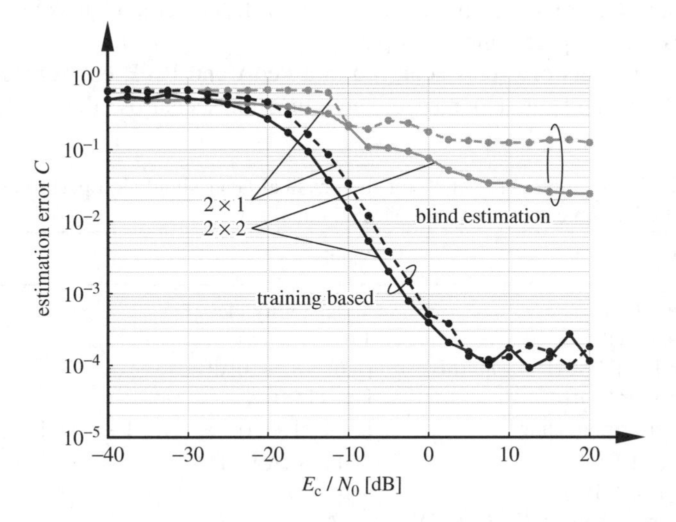

Figure 14.4 Mean quadratic estimation error C – defined in Equation (14.37) – of the LS and the second-order blind precoding state estimators versus E_{c}/N_0.

similar results and shows an operating range starting approximately at $-10\,\mathrm{dB}$ E_c/N_0. For both estimator classes, the availability of a second receive antenna is beneficial for the precoding state estimation. However, for the blind precoding state estimator the gain is even more dramatic. Note, however, that the (poor) performance of the blind estimator is sufficient for the proposed equalizer, as will be shown in the next section.

14.4 Performance Evaluation

In order to assess the performance of the proposed equalizer in comparison with the classical SU equalizer, as well as to evaluate the influence of the precoding state estimation, the simulations are split into two different parts: (i) physical-layer simulations for a fixed transmission setup of TxAA HSDPA and (ii) system-level simulations with adaptive feedback and scheduling. Each simulation approach has a different focus, with the physical-layer simulations covering channel encoding and decoding, WCDMA processing, as well as channel estimation in detail. On the other hand, system-level simulations represent a whole HSDPA network, with adaptive feedback, scheduling, and Radio Resource Control (RRC) algorithms. For the following results it is also assumed that the channel and the noise power are perfectly known at the receiver.

14.4.1 Physical-Layer Simulation Results

Physical-layer simulations have been conducted utilizing a standard-compliant WCDMA simulator [17]. The simulation assumptions in Table 14.2 correspond to a cell in which four users are receiving data simultaneously. User 1 is moving through the cell and obtains precoding coefficients as adaptively requested, according to the definition in Section 1.3.2 [3]. The three interfering users are assumed to be stationary; thus, their precoding coefficients and transmit power do not change. In the physical-layer simulations it is assumed, furthermore, that all users are always scheduled with the same CQI value; thus, no link adaptation besides the precoding takes place.

The data throughputs of user 1 achieved in a PedA and PedB environment are plotted in Figures 14.5(a) and 14.5(b) respectively. In both scenarios, the interference-aware

Table 14.2 Simulation parameters for the physical-layer simulation performance assessment of the interference-aware MMSE equalizer

Parameter	Value
Active users U	4, capability class 6
Desired user CQI	13
Interfering HS-PDSCH $E_\mathrm{c}/I_\mathrm{or}$ (dB)	$[-6, -8, -10]\,\mathrm{dB}$
Interfering user CQIs	[16, 11, 8]
Interfering user precoding	$[1, \frac{1}{\sqrt{2}}(1-\mathrm{i})]$, $[1, \frac{1}{\sqrt{2}}(-1+\mathrm{i})]$, $[1, \frac{1}{\sqrt{2}}(-1-\mathrm{i})]$
Precoding codebook	3GPP TxAA [3]
CPICH $E_\mathrm{c}/I_\mathrm{or}$ (dB)	-10
Other non-data channel $E_\mathrm{c}/I_\mathrm{or}$ (dB)	-12
UE speed (km/h)	3

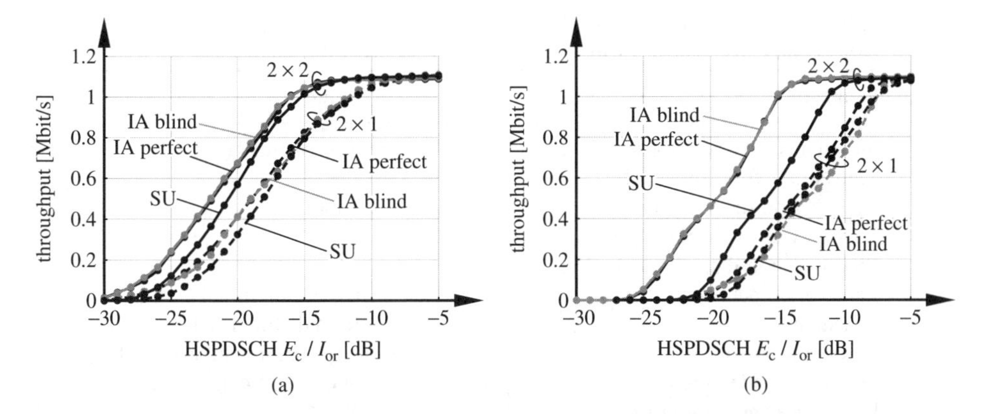

Figure 14.5 Physical-layer simulation results under interference-aware equalizer IA. (a) Physical-layer throughput of the desired user in a spatially uncorrelated ITU PedA channel at CQI 13, corresponding to a maximum throughput of 1.14 Mbit/s. (b) Physical-layer throughput of the desired user in a spatially uncorrelated ITU PedB channel at CQI 13, corresponding to a maximum throughput of 1.14 Mbit/s.

MMSE equalizer with perfect knowledge of the precoding state significantly outperforms the SU MMSE equalizer. If the precoding state of the cell is blindly estimated, then the performance of the MMSE equalizer nearly approaches the performance when $\mathcal{P}$ is perfectly known.

The gain in the PedB channel in Figure 14.5(b) is much larger than the gain in the PedA channel which has a much shorter maximum delay spread. This is caused by the larger loss of orthogonality in the PedB environment and the subsequently larger post-equalization interference. In the 2×1 case, the equalizer applying the blind precoding state estimation loses significantly compared with the equalizer with perfect knowledge of $\mathcal{P}$, which is a result of the considerably worse estimator performance when only one receive antenna is available; see Figure 14.4. The fact that the performance loss is greater in the PedB channel is due to its larger delay spread, which represents a more challenging environment for the equalizer, making it more sensitive to estimation errors in the precoding state. For a larger number of receive antennas, the simulation results show larger performance gains. Especially in Figure 14.5(b), the precoding state estimation becomes significantly better in the 2×2 case, thus closing the gap to the throughput performance of the equalizer utilizing perfect knowledge of $\mathcal{P}$. The interference-aware MMSE equalizer can thus effectively utilize the spatial information to suppress the interfering signals. The largest performance increase of the proposed MMSE equalizer was found for the 2×2 PedB environment with 4 dB.

14.4.2 System-Level Simulation Results

To assess the performance on a network level, a set of system-level simulations with the simulator as described in [39, 41] has also been conducted. The simulation assumptions in Table 14.3 correspond to a 19-site scenario with a homogeneous network load in which the multi-code scheduler serves four active users simultaneously. All 25 simulated users

Table 14.3 Simulation parameters for the system-level simulation performance assessment of the interference-aware MMSE equalizer

Parameter	Value
Simultaneously active users U	4, capability class 10
Transmitter frequency (GHz)	1.9
Base station distance (m)	1000
Total power available at NodeB (W)	20
Power of non-data channels (W)	2
Large-scale pathloss model	Urban micro [11]
Scheduler	Round Robin (RR)
Cell deployment	19 cells, layout Type I [5]
Precoding codebook	3GPP TxAA [3]
Equalizer span (chips)	40
Feedback delay (slots)	11
UE speed (km/h)	3, random direction
Simulation time	25 000 slots, each 2/3 ms

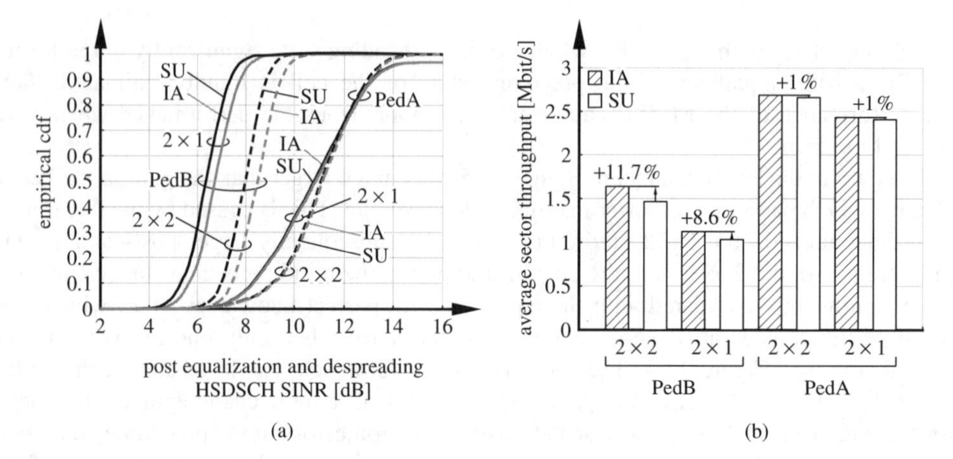

Figure 14.6 System-level simulation results under interference-aware equalizer IA. (a) Empirical cumulative density functions (cdfs) of the post equalization and despreading SINR of the HS-DSCH for ITU PedA and PedB channels, as observed on a network level. (b) Average system-level sector throughput results for ITU PedA and PedB environments.

are moving through the cell with random directions, adaptively reporting their CQI and precoding feedback according to their capability class [4].

The distributions of the SINR for the PedA and PedB channel, averaged over all active users in the cell, are plotted in Figure 14.6(a). It can be seen that the interference-aware MU MMSE equalizer is able to deliver significantly higher SINRs for PedB channels. In the PedA environment, the gain is negligible. Figure 14.6(b) shows the average sector throughput comparison. The interference-aware MU MMSE equalizer outperforms the classical SU MMSE equalizer significantly, with remarkable gains in the PedB environment of up to 11.7 %. Similar to the physical-layer simulation results, for both channels the equalizer is able to utilize the advantage of multiple receive antennas, in this case the 2×2 MIMO channel, to advance the precoding state estimation.

References

[1] 3GPP (2006) Technical Specification TS 25.321 Version 7.0.0 'Medium access control (MAC) protocol specification,' www.3gpp.org.
[2] 3GPP (2007) Technical Specification TS 25.963 Version 7.0.0 'Feasibility study on interference cancellation for UTRA FDD user equipment (UE),' www.3gpp.org.
[3] 3GPP (2007) Technical Specification TS 25.876 Version 7.0.0 'Multiple-input multiple-output UTRA,' www.3gpp.org.
[4] 3GPP (2007) Technical Specification TS 25.214 Version 7.0.0 'Physical layer procedures,' www.3gpp.org.
[5] 3GPP (2007) Technical Specification TS 25.996 Version 7.0.0 'Spatial channel model for multiple-input multiple-output (MIMO) simulations,' www.3gpp.org.
[6] 3GPP (2009) Technical Specification TS 25.212 Version 8.5.0 'Multiplexing and channel coding (FDD),' www.3gpp.org.
[7] 3GPP (2009) Technical Specification TS 25.322 Version 8.4.0 'Radio link control (RLC) protocol specification,' www.3gpp.org.
[8] Andrews, J. (2005) 'Interference cancellation for cellular systems: a contemporary overview,' *IEEE Wireless Communications Magazine*, **12** (2), 19–29.
[9] Biglieri, E. and Lops, M. (2007) 'Multiuser detection in a dynamic environment. Part I: user identification and data detection,' *IEEE Transactions on Information Theory*, **53** (9), 3158–3170.
[10] Boyd, S. and Vandenberghe, L. (2004) *Convex Optimization*, Cambridge University Press.
[11] Cichon, D. J. and Kürner, T. (1998) *COST 231 – Digital Mobile Radio Towards Future Generation Systems*, chapter 4, COST.
[12] Dabora, R. and Goldsmith, A. J. (2008) 'The capacity region of the degraded finite-state broadcast channel,' in *Proceedings of IEEE Information Theory Workshop (ITW)*, pp. 11–15.
[13] Erez, U. and ten Brink, S. (2005) 'A close-to-capacity dirty paper coding scheme,' *IEEE Transactions on Information Theory*, **51** (10), 3417–3432.
[14] Goldsmith, A. J., Jafar, S. A., Jindal, N., and Vishwanath, S. (2003) 'Capacity limits of MIMO channels,' *Journal on Selected Areas in Communications*, **21** (5), pp. 684–702.
[15] Haykin, S. (2002) *Adaptive Filter Theory*, 4th edition, Prentice-Hall.
[16] Holliday, T., Goldsmith, A. J., and Poor, H. V. (2008) 'Joint source and channel coding for MIMO systems: is it better to be robust or quick?' *IEEE Transactions on Information Theory*, **54** (4), 1393–1405.
[17] Kaltenberger, F., Freudenthaler, K., Paul, S. *et al.* (2005) 'Throughput enhancement by cancellation of synchronization and pilot channel for UMTS high speed downlink packet access,' in *Proceedings of the IEEE 6th Workshop on Signal Processing Advances in Wirless Communications (SPAWC)*, pp. 580–584.
[18] Kay, S. M. (1993) *Fundamentals of Statistical Signal Processing. Estimation Theory*, volume 1, Prentice-Hall.
[19] Kim, D. I. and Fraser, S. (2004) 'Two-best user scheduling for high-speed downlink multicode CDMA with code constraint,' in *Proceedings of IEEE Global Telecommunications Conference (GLOBECOM)*, volume 4, pp. 2569–2663.
[20] Mailaender, L. (2005) 'Linear MIMO equalization for CDMA downlink signals with code reuse,' *IEEE Transactions on Wireless Communications*, **4** (5), 2423–2434.
[21] Melvasalo, M., Janis, P., and Koivunen, V. (2006) 'MMSE equalizer and chip level inter-antenna interference canceler for HSDPA MIMO systems,' in *Proceedings of the IEEE 63rd Vehicular Technology Conference Spring (VTC)*, volume 4, pp. 2008–2012.

[22] Members of ITU (1997) 'Recommendation ITU-R M.1225: Guidelines for evaluation of radio transmission technologies for IMT-2000,' Technical report, International Telecommunication Union (ITU).
[23] Moon, T. K. and Stirling, W. C. (2000) *Mathematical Methods and Algorithms for Signal Processing*, Prentice-Hall.
[24] Ng, C. T. K., Jindal, N., Goldsmith, A. J., and Mitra, U. (2007) 'Capacity gain from two-transmitter and two-receiver cooperation,' *IEEE Transactions on Information Theory*, **53** (10), 3822–3827.
[25] Ng, C. T. K., Jindal, N., Goldsmith, A. J., and Mitra, U. (2008) 'Power and bandwidth allocation in cooperative dirty paper coding,' in *Proceedings of IEEE International Conference on Communications (ICC)*, pp. 1018–1023.
[26] Ottersten, B., Viberg, M., and Kailath, T. (1992) 'Analysis of subspace fitting and ML techniques for parameter estimation from sensor array data,' *IEEE Signal Processing Letters*, **40** (3), pp. 590–600.
[27] Petre, F., Engels, M., Bourdoux, A. *et al.* (1999) 'Extended MMSE receiver for multiuser interference rejection in multipath DS-CDMA channels,' in *Proceedings of the IEEE VTS 50th Vehicular Technology Conference Fall (VTC)*, volume 3, pp. 1840–1844.
[28] Powell, M. J. D. (1978) 'A fast algorithm for nonlinearly constrained optimization calculations,' in *Numerical Analysis* (ed. G. A. Watson), volume 630 of *Lecture Notes in Mathematics*, Springer, Berlin, pp. 144–157.
[29] Proakis, J. (2000) *Digital Communications*, 4th edition, McGraw-Hill Science Engineering.
[30] Sato, H. (1978) 'An outer bound to the capacity region of broadcast channels,' *IEEE Transactions on Information Theory*, **24** (3), 374–377.
[31] Sharif, M. and Hassibi, B. (2007) 'A comparison of time-sharing, DPC, and beamforming for MIMO broadcast channels with many users,' *IEEE Communications Letters*, **55** (1), 11–15.
[32] Shenoy, S., Ghauri, M., and Slock, D. (2008) 'Receiver designs for MIMO HSDPA,' in *Proceedings of the IEEE International Conference on Communications (ICC)*, pp. 941–945.
[33] Villares, J. and Vázquez, G. (2005) 'Second-order parameter estimation,' *IEEE Signal Processing Letters*, **53** (7), 2408–2420.
[34] Villares, J. and Vázquez, G. (2007) 'The Gaussian assumption in second-order estimation problems in digital communications,' *IEEE Signal Processing Letters*, **55** (10), 4994–5002.
[35] Virtej, E., Lampinen, M., and Kaasila, V. (2008) 'Performance of an intra- and inter-cell interference mitigation algorithm in HSDPA system,' in *Proceedings of the IEEE 67th Vehicular Technology Conference Spring (VTC)*, pp. 2041–2045.
[36] Viswanath, P. and Tse, D. N. C. (2003) 'Sum capacity of the vector Gaussian broadcast channel and uplink–downlink duality,' *IEEE Transactions on Information Theory*, **49** (8), 1912–1921.
[37] Weingarten, H., Steinberg, Y., and Shamai, S. (2006) 'The capacity region of the Gaussian multiple-input multiple-output broadcast channel,' *IEEE Transactions on Information Theory*, **52** (9), 3936–3964.
[38] Wirtinger, W. (1927) 'Zur Formalen Theorie der Funktionen von mehr Komplexen Veränderlichen,' *Mathematische Annalen*, **97**, 357–375.
[39] Wrulich, M., Eder, S., Viering, I., and Rupp, M. (2008) 'Efficient link-to-system level model for MIMO HSDPA,' in *Proceedings of the IEEE 4th Broadband Wireless Access Workshop*, New Orleans, LA. Available from http://publik.tuwien.ac.at/files/PubDat_170334.pdf.
[40] Wrulich, M., Mehlführer, C., and Rupp, M. (2010) 'Managing the interference structure of MIMO HSDPA: a multi-user interference aware MMSE receiver with moderate complexity,' *IEEE Transactions on Wireless Communications*, **9** (4), 1472–1482, doi: 10.1109/TWC.2010.04.090612. Available from http://publik.tuwien.ac.at/files/PubDat_180743.pdf.
[41] Wrulich, M. and Rupp, M. (2009) 'Computationally efficient MIMO HSDPA system-level modeling,' *EURASIP Journal on Wireless Communications and Networking*, **2009**, article ID 382501, doi: 10.1155/2009/382501.
[42] Zemen, T. and Mecklenbräuker, C. (2005) 'Time-variant channel estimation using discrete prolate spheroidal sequences,' *IEEE Transactions on Signal Processing*, **53** (9), 3597–3607.

[43] Zhang, H. and Dai, H. (2004) 'Cochannel interference mitigation and cooperative processing in downlink multicell multiuser MIMO networks,' *EURASIP Journal on Wireless Communications and Networking*, **2004** (2), 222–235.
[44] Zhang, H., Ivrlac, M., Nossek, J. A., and Yuan, D. (2008) 'Equalization of multiuser MIMO high speed downlink packet access,' in *Proceedings of the IEEE 19th International Symposium on Personal, Indoor and Mobile Radio Communications (PIMRC)*, pp. 1–5.
[45] Zheng, Y. and Xiao, C. (2003) 'Simulation models with correct statistical properties for Rayleigh fading channels,' *IEEE Transactions on Communications*, **51** (6), 920–928.

15

LTE Advanced Versus LTE

Contributed by Stefan Schwarz
Vienna University of Technology (TU Wien), Austria

The development of Long-Term Evolution (LTE) was the first step towards a fourth-generation (4G) radio access technology compatible with International Telecommunication Union (ITU) International Mobile Telecommunications (IMT)-Advanced [10]. LTE-Advanced (LTE-A) was submitted in October 2009 as a candidate radio interface technology for 4G IMT-Advanced and in October 2010 the ITU decided that LTE-A is one of the first two technologies to meet the set requirements, besides IEEE's 802.16m [9]. Although the Downlink (DL) of LTE already surpasses the requirements for a 4G technology, this is not true for the Uplink (UL) direction. Therefore, work on the next evolutionary step for LTE is ongoing in the 3rd Generation Partnership Project (3GPP) and the first backwards compatible version of LTE-A was already released at the end of December 2010 (Rel'10) [3].

This chapter is intended to give an introduction to the technological changes and enhancements introduced in LTE-A to further improve the performance of LTE. In Section 15.1 we discuss the performance requirements for a 4G radio access technology as set by the ITU. Furthermore, we compare these requirements with the performance targets defined by the 3GPP for LTE as well as LTE-A. The key enhancements for the LTE-A radio interface are briefly described in Section 15.2 and, moreover, an outlook on technologies considered for future releases is presented. Section 15.3 offers a more in-depth look on the Multiple-Input Multiple-Output (MIMO) improvements of LTE-A, allowing up to eight antenna transmissions. The proposed eight-antenna codebook design is discussed and its performance for the Single User MIMO (SU-MIMO) and Multi User MIMO (MU-MIMO) spatial multiplexing transmission modes is analyzed. In Section 5.4 the performance of LTE-A is investigated by means of link-level simulations. The single-user throughput performance of the eight-transmit-antenna Closed-Loop Spatial Multiplexing (CLSM) transmission mode is compared for different numbers of receive antennas in Section 15.4.1. Furthermore, the throughput bounds of Section 10.5 are utilized to

Evaluation of HSDPA and LTE: From Testbed Measurements to System Level Performance, First Edition.
Sebastian Caban, Christian Mehlführer, Markus Rupp and Martin Wrulich.

analyze the throughput losses caused by practical system design constraints. In Section 15.4.2 the single-user throughput of several LTE- and LTE-A-defined antenna configurations is cross-compared and finally in Section 15.4.3 we set the SU-MIMO and MU-MIMO spatial multiplexing modes in contrast with each other.

15.1 IMT-Advanced and 3GPP Performance Targets

With LTE-A the 3GPP intends to provide a 4G radio access technology, according to the ITU requirements [10]. We provide a comparison in Table 15.1 to show whether this goal is achieved with the performance targets, set by the 3GPP for LTE-A [1]. Additionally, we also contrast with performance results obtained in LTE Rel'8 according to the RAN document [20]. The table shows that the DL of LTE already achieves the IMT-Advanced goals in terms of peak spectral efficiency, but not in terms of average cell spectral efficiency and cell edge spectral efficiency. The LTE-A DL, on the other hand, surpasses or meets the 4G requirements in all items considered. In UL, LTE does not achieve any of the ITU requirements, whereas LTE-A again exceeds them. For the same antenna configuration, LTE-A is expected to improve the performance of LTE by a factor of 1.4–1.6. It should also be noted that not all User Equipment (UE) categories will be capable of achieving peak spectral efficiencies in the order of 30 bit/(s Hz). Only high-class devices with large antenna numbers will obtain such a performance. Other more

Table 15.1 Comparison of LTE performance [20], LTE-A targets [1] and ITU IMT-Advanced requirements [10]

	Item	$N_T \times N_R$	LTE	LTE-A	IMT-Advanced
	Bandwidth		up to 20 MHz	up to 100 MHz	at least 40 MHz
DOWNLINK	peak spectr. eff. [bit/(s Hz)]		16.3 (4 × 4)	30 (8 × 8)	15 (4 × 4)
	average cell spectr. eff. [bit/(s Hz)]	2 × 2	1.69	2.4	
		4 × 2	1.87	2.6	2.6
		4 × 4	2.67	3.7	
	cell edge spectr. eff. [bit/(s Hz)]	2 × 2	0.05	0.07	
		4 × 2	0.06	0.09	0.075
		4 × 4	0.08	0.12	
UPLINK	peak spectr. eff. [bit/(s Hz)]		4.32 (1 × 1)	15 (4 × 4)	6.75 (2 × 4)
	average cell spectr. eff.	1 × 2	0.735	1.2	
		2 × 4		2	1.8
	cell edge spectr. eff.	1 × 2	0.024	0.04	
		2 × 4		0.07	0.05
	C-plane latency		50 ms	50 ms	< 100 ms
	U-plane latency		4.9 ms	4.9 ms	< 10 ms

general requirements were additionally set by the ITU in [10]; for example, worldwide roaming capability, compatibility with fixed networks, and interworking with other radio systems.

In addition to the 4G performance minimum requirements, the 3GPP imposes several targets in [1] that are taken into account in the development of LTE-A:

- *Backwards compatibility*. LTE-A shall be fully backwards compatible with LTE, meaning that LTE-conforming UEs can operate in LTE-A networks.
- *Peak data rate*. The DL targets a peak data rate of 1 Gbit/s and the UL aims at 500 Mbit/s.
- *Mobility*. Mobility across the cellular network shall be supported for mobile speeds up to 350 km/h (or even up to 500 km/h in specific frequency bands), aiming at high-speed train communication.
- *Spectrum flexibility*. In addition to the bands currently defined for LTE, several new frequency bands in the range from 450 MHz to 5 GHz are identified in [1]. To achieve the targeted 100 MHz system bandwidth, in many cases it will be necessary to combine noncontiguous frequency bands, which is enabled by means of carrier aggregation (see Section 15.2.1).

15.2 Radio Interface Enhancements

In order to meet and surpass the requirements set by the ITU for a 4G wireless network and to enable future competitiveness, the radio access technology of LTE is continuously evolving. With Rel'10 the 3GPP incorporated the necessary enhancements into the LTE specifications to achieve these goals. This section provides a short overview about the most important features of LTE-A, which are:

- bandwidth extension through carrier aggregation;
- enhanced MIMO support;
- improvements in the UL.

Furthermore, we have a look at technologies that are considered important for future releases and evolutionary steps in the development of LTE. The section is based on the corresponding LTE-A specifications summarized in [3]. More details can be found in [2] and in several company white papers; for example, [6, 13, 15].

15.2.1 Bandwidth Extension

One possibility to achieve higher cell throughputs and peak data rates is extending the bandwidth of a wireless communication system. LTE restricts the maximum channel bandwidth to 20 MHz. In order to enable backwards compatibility between LTE and LTE-A and to support the reuse of spectrum already owned by network operators for LTE, larger channel bandwidths are supported in LTE-A by means of carrier aggregation [2]. Carrier aggregation allows the creation of wider channel bandwidths by jointly utilizing several LTE bands for transmission. Depending on the number of UE transceivers, up to five component carriers (LTE carriers) can be aggregated, enabling a maximum of

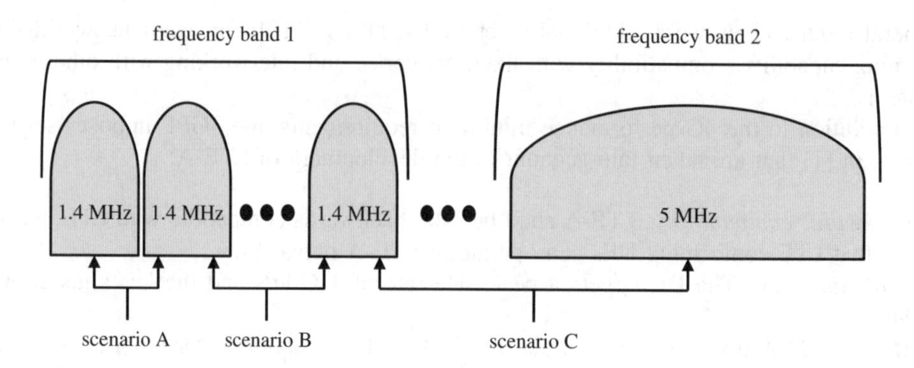

Figure 15.1 Visualization of carrier aggregation scenarios defined for LTE-A.

100 MHz bandwidth for transmission. This technique enables Rel'8 UEs to be operated in the same spectrum as LTE-A UEs, but only on single-component carriers. Different aggregation scenarios are distinguished, posing increasingly demanding requirements on the transceiver of a UE [6, 13, 15]:

- Scenario A intra-band aggregation with contiguous carriers – contiguous component carriers within a single frequency band are combined.
- Scenario B intra-band aggregation with non-contiguous carriers – non-contiguous component carriers within a single frequency band are combined.
- Scenario C inter-band aggregation – component carriers from different frequency bands are combined.

Examples of these three scenarios are shown in Figure 15.1. To account for the commonly asymmetric traffic load in UL and DL, different amounts of carriers can be aggregated for each direction. Each scheduled component carrier is served via its own Hybrid Automatic Repeat reQuest (HARQ) process and Physical (PHY) unit, with a common Medium Access Control (MAC) layer interface for all carriers. The PHY unit is LTE compatible and comprises the signal processing described in Section 2.3, thus enabling backwards compatibility. This structure has the disadvantage that each codeword in LTE-A is distributed over the same frequency span as in LTE, thus attaining the same frequency diversity gain, though the total transmission bandwidth increases.

15.2.2 *Enhanced MIMO*

LTE confines the maximum number of spatial transmission layers per UE (SU-MIMO) in the DL to four, whereas in the UL only single-stream transmission is supported. Spatial multiplexing in the UL is only supported by means of MU-MIMO (see Section 15.3), which improves the UL capacity but not the single-user peak data rate. This is the main reason for not achieving the IMT-Advanced peak spectral efficiency requirements in the LTE UL (see Section 15.1).

To improve the performance, LTE-A supports up to eight-layer spatial multiplexing in the DL and up to four layers in the UL. This puts strong demands on the complexity of the

UEs, requiring eight antennas plus receivers and four transmitters to enable the full MIMO gains. Therefore, Rel'10 emphasizes lower order spatial multiplexing in combination with beam steering, rather than a pure eight-layer spatial multiplexing mode [6]. The possible gains of higher order MIMO are the following:

- *Improved peak spectral efficiency*. By transmitting up to eight spatial streams in parallel to a single UE, the peak spectral efficiency of LTE can almost be doubled (there are some inevitable losses due to an increasing number of pilot symbols).
- *Gains in the cell edge spectral efficiency*. A large number of antennas at the Evolved base station (eNodeB) allows the forming of a very sharp beam, enabling a Signal to Noise Ratio (SNR) gain that improves the performance of cell edge users.
- *Better cell spectral efficiency*. Both of the gains described lead to an improvement in the average cell spectral efficiency. Additionally, the large number of eNodeB antennas enables one to serve multiple users in parallel employing MU-MIMO transmission. This adds one additional degree of freedom for UE scheduling and provides a spatial multi-user diversity gain, besides frequency and temporal diversity.

LTE Rel'8 and Rel'9 rely on densely packed cell-specific reference signals for channel estimation and calculation of Channel State Information (CSI). With an increasing number of transmit antennas, the overhead for these pilots also grows, as shown in Figure 2.6. For higher order spatial multiplexing this approach loses its appeal, because the number of pilots does not grow linearly with the instantaneous transmission rank L, but rather with the maximum possible transmission rank, which is equal to the number of transmit antennas N_{T}. For data detection, the UE does not require knowledge of the full channel matrix $\mathbf{H}_r$, but only of the compound channel matrix $\mathbf{H}_r\mathbf{W}_r$, composed of the channel matrix and the precoder (see Equation (10.3) in Section 10.2.2). Estimation of the compound channel requires pilot symbols on each layer only. LTE-A exploits this potential overhead reduction by defining UE-specific Demodulation Reference Signals (DMRS) [4], which realize exactly these precoded pilots. Additionally, the DMRS can be utilized to employ non-codebook-based precoding, as explained in Section 15.3.2.

CSI estimation still requires reference signals on each antenna port in order to decide for the preferred number of spatial streams, signaled by means of the Rank Indicator (RI) (see Section 10.4). For that purpose, cell-specific CSI Reference Signals (CSI-RSs) are defined that are placed sparsely in time, frequency, and space to keep the overhead small [4].

15.2.3 Uplink Improvements

The UL of LTE Rel'8 is based on localized Single-carrier FDMA (SC-FDMA), also known as Discrete Fourier Transform (DFT) spread Orthogonal Frequency-Division Multiple Access (OFDMA) [14]. It has the advantage of a reduced Peak-to-Average Power Ratio (PAPR) compared with OFDMA, enabling more efficient power amplifier implementations, but on the downside it requires carrier allocation across a contiguous block of spectrum, preventing the scheduling flexibility of OFDMA. To enable frequency-selective scheduling, LTE-A therefore introduces clustered SC-FDMA, which enables UE scheduling on multiple noncontiguous blocks/clusters of subcarriers. This improves the system performance by providing a frequency diversity gain, but also increases the PAPR.

15.2.4 Beyond Release 10

Several advanced techniques are under consideration for possible future implementation in the LTE-A specifications. In this section we discuss two of the most important enhancements as presented in [6, 13, 15, 17].

15.2.4.1 Coordinated Multipoint Transmission and Reception

Achieving the improvements in spectral efficiency from Table 15.1, especially for cell edge users, requires one to consider advanced techniques for intercell-interference coordination [17]. Interference within a cell is avoided by assigning orthogonal resources in either space, time and/or frequency employing OFDMA/SC-FDMA combined with SU-MIMO/MU-MIMO. Still, the interference between neighboring cells strongly degrades the performance of cell edge users, which experience comparable signal strength from the serving and the interfering eNodeBs.

LTE Rel'8 already introduced simple methods for interference coordination between neighboring eNodeBs utilizing the X2 interface, called autonomous intercell radio resource management. It enables simple multicell radio resource management comprising several autonomous eNodeBs; for instance, semi-statically configured partial frequency reuse. The main issue in this respect is the delay of the X2 interface, which is on the order of 10 ms, preventing it from being employed for more sophisticated techniques [17].

LTE-A studies Coordinated Multi-Point (CoMP) transmission and reception as a means to increase the cell throughput, to improve the coverage of high data rates, and to enhance the cell edge performance. The goal of this technique is to transform the interference from other cells into a useful signal for the UE. This requires fast intercell radio resource management, which cannot be handled with the current X2 interface delay. One possibility for solving this problem is to employ so-called centralized intercell radio resource management. In this case, a group of Remote Radio Equipments (RREs) is connected to a central eNodeB via optical fiber links, enabling fast transmission of the required baseband data between the RREs. The eNodeB centrally manages the radio resources of all RREs (see [17] for details). Different transmission/reception schemes are under investigation for CoMP, which are visualized in Figure 15.2:

Coherent transmission/reception: In this case all antennas of multiple base stations are utilized as one single multi-antenna element (virtual MIMO) to transmit data to a single UE. This requires joint processing of the signals transmitted from all antennas at a central node and distribution of the baseband signal to the corresponding eNodeBs (eNodeB 1 and 2 in Figure 15.2). In the UL, the signals received at multiple base stations are jointly processed to maximize the received Signal to Interference and Noise Ratio (SINR); for example, by means of a linear filter (MMSE, ZF). Only one single UE can be served at a given resource by all involved eNodeBs (UE1 in Figure 15.2).

Fast selection/dynamic cell selection: This technique allows the switching of a UE between several eNodeBs (eNodeB 2 and 3 in Figure 15.2), without requiring to perform a handover. Depending on the SINR, the signal is transmitted/received at the best eNodeB and forwarded to the serving eNodeB. The signaling information is still only received from the serving base station. Again, a single UE occupies resources at several eNodeBs.

Interference coordinated beamforming/scheduling: Interference coordinated beamforming is shown for UE5 and UE6 in Figure 15.2. The transmit beams at each eNodeB are formed such that the SINR of not only the own user is maximized, but also the SINR of the interfering user from the other base station. Interference-coordinated scheduling, on the other hand, jointly allocates resources to users from multiple cells in order to improve the cell edge performance, by minimizing the interference. These two techniques have the advantage that each UE occupies resources of a single eNodeB only.

15.2.4.2 Relaying

Relaying techniques are employed to enhance the coverage and capacity of cellular networks. Relay nodes improve mainly urban or indoor throughput and extend/provide coverage in rural areas or otherwise dead zones. Typical scenarios are shown in Figure 15.3. A Relay Node (RN) is connected wirelessly to a so-called donor eNodeB. This link can be in-band, in which case the eNodeB to RN connection is handled in the same frequency band as the eNodeB to UE connections, occupying the same resources, or out-band, for example

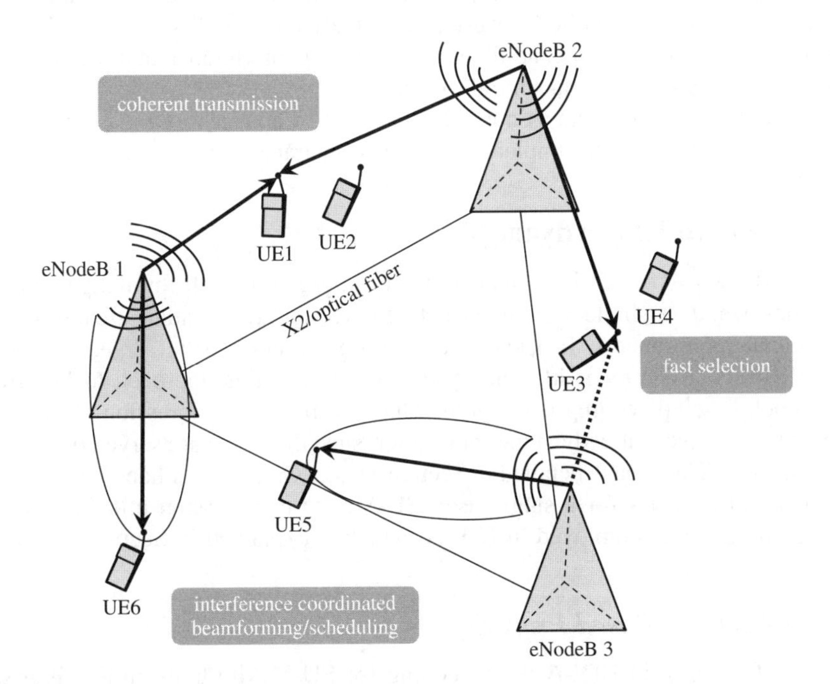

Figure 15.2 Visualization of different coordinated multipoint transmission/reception scenarios.

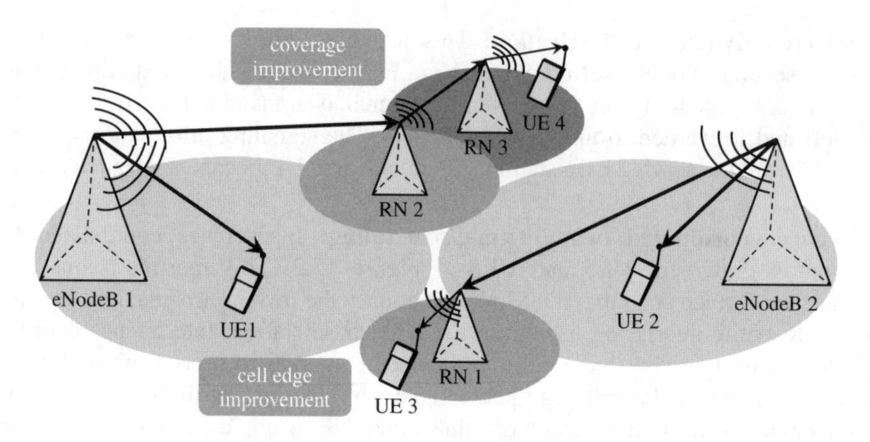

Figure 15.3 Visualization of different relaying scenarios.

over a dedicated microwave connection. In Figure 15.3, RN 1 is utilized to improve the cell edge performance, whereas RN 2 and RN 3 improve the coverage of the network.

A relay node can implement different functionalities.

- *Layer 1 relays*. Such relays simply receive, amplify, and retransmit the signal in the DL and UL. They devices are relatively simple, operating purely at the Radio Frequency (RF) level.
- *Layer 2 relays*. These devices are capable of decoding and re-encoding the data before retransmission. Traffic can be forwarded selectively only to UEs that would otherwise experience too poor channel conditions, thus reducing interference that would be caused by layer 1 relays which retransmit all traffic.
- *Layer 3 relays*. They operate as normal eNodeBs and appear to UEs as ordinary cells. This simplifies the terminal implementation and guarantees backwards compatibility.

15.3 MIMO in LTE Advanced

In LTE-A, MU-MIMO is an important means to improve the cell throughput, especially in situations where a single user does not require multiple spatial streams to satisfy its throughput requirements. In such cases, multiple users can be served on the same Resource Blocks (RBs) over different spatial layers. For this approach to be effective, non-codebook-based precoding is required, which enables the suppression of interference between multiple layers already at the transmitter, such that users are served on orthogonal spatial resources. This idea is detailed in Section 15.3.2. On the other hand, to achieve high peak spectral efficiencies for a single user, SU-MIMO transmission utilizing codebook-based precoding is still employed in LTE-A, which is explained in Section 15.3.1.

15.3.1 Codebook-Based Precoding

Similar to LTE Rel'8, in LTE-A the precoding for SU-MIMO transmission is based on a prespecified codebook [4]. If LTE-compliant antenna configurations are utilized, then

Table 15.2 Summary of the number of precoders defined in the LTE-A codebook for eight transmit antennas

Rank	Number of wideband precoders	Number of subband precoders	Total number of precoder combinations
1	16	16	256
2	16	16	256
3	4	16	64
4	4	8	32
5	4	1	4
6	4	1	4
7	4	1	4
8	1	1	1

the LTE codebook is reused. Otherwise, if eight transmit antennas are employed, a new precoder codebook is defined in [4], consisting in total of 621 elements. In contrast to the LTE codebook, this eight-antenna codebook is not indexed with a single Precoding Matrix Indicator (PMI), but with two indices $[i_1, i_2]$, which we call the wideband and subband PMIs for reasons that will become clear soon. The number of precoders per transmission rank possibility is summarized in Table 15.2.

The reason for indexing the codebook with a two-label index is explained in [5]. This document states that the wideband precoder is kept constant for the total system bandwidth, while the subband precoder can vary between different subbands. Therefore, just a single wideband PMI must be signaled by the UE to the eNodeB and additionally distinct subband PMIs for each subband. It is shown in [22] that this eight-antenna LTE-A codebook can be generated by the multiplication of two matrices $\mathbf{W} = \mathbf{W}^{(1)} \cdot \mathbf{W}_s^{(2)}$, each stemming from an individual codebook. In this notation, the index i_1 of the wideband precoder corresponds to $\mathbf{W}^{(1)}$ and the index i_2 refers to the subband precoder $\mathbf{W}_s^{(2)}$. In Equation (10.8) of Section 10.4.2, we employ the same partitioning during feedback calculation for LTE, in order to distinguish the cases of wideband- and subband-specific PMI feedback. In this case, one of the two precoders $\mathbf{W}^{(1)}$ or $\mathbf{W}_s^{(2)}$ is chosen from the trivial codebook $\{1\}$ (depending on whether wideband or subband feedback is considered) and the other one is optimized over the standard defined codebook. In LTE-A, both the subband and wideband precoder codebooks are defined by different choices of indices i_1 and i_2. Therefore, optimization has to be performed over both precoders during feedback calculation to obtain the optimal wideband and subband PMIs. This can be handled with the same algorithms for feedback computation as employed for LTE (see Section 10.4.2), simply by identifying the wideband precoder with index i_1 and the subband precoder with i_2.

To investigate the influence of the wideband and subband precoder choice on the beam pattern of a ULA, we consider an eight-element $\lambda/2$-spaced ULA and compute the antenna gain pattern for different rank 1 wideband/subband precoder combinations. Figure 15.4(a) shows the beam patterns obtained with four different wideband precoders for a fixed subband precoder. It is observed that the wideband precoder determines the steering angle of the ULA. In Figure 15.4(b), the wideband precoder is fixed to obtain a center beam and the subband precoder is varied. The subband precoder keeps the coarse steering

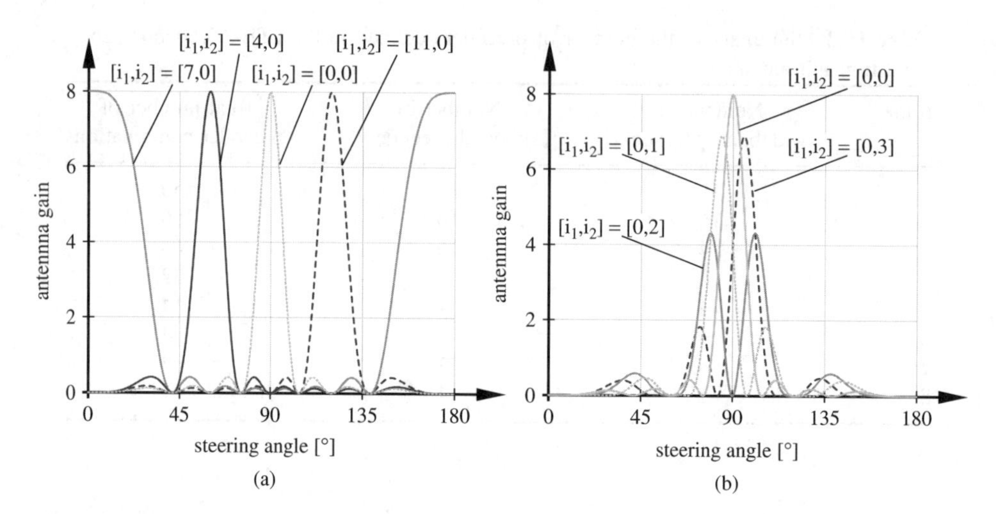

Figure 15.4 Antenna gain obtained for a $\lambda/2$-spaced Uniform Linear Array (ULA) with different wideband and subband precoder combinations. (a) Varying wideband and fixed subband precoder. (b) Fixed wideband and varying subband precoder.

angle of the antenna beam constant, but changes the shape of the beam pattern, as shown in the figure.

15.3.2 Non-Codebook-Based Precoding

Non-codebook-based precoding is employed in combination with MU-MIMO techniques, with the goal to achieve a spatial multiplexing gain in case not all Degrees of Freedom (DoFs) are being employed by a single UE; that is, for example, single receive antenna users or UEs needing only few layers to satisfy their rate requirements. In general, achieving the capacity of such a MIMO broadcast channel requires nonlinear dirty paper coding [7]. This method necessitates full channel knowledge at the eNodeB and involves transmitter/receiver signal processing with very high computational complexity. The most promising approach for practical non-codebook-based precoding, in terms of the tradeoff between complexity and performance, is simple linear precoding based on a Zero Forcing (ZF) or regularized ZF criterion [16]. In these cases, a linear precoder is computed at the transmitter in order to null the interference between multiple spatial layers (ZF approach). Owing to its practical relevance, this approach is considered in detail in the following.

To compute the ZF precoder, the transmitter/eNodeB requires channel knowledge for all UEs. In current proposals, this knowledge is obtained by means of Channel Vector Quantization (CVQ) [8, 23, 24]. CVQ in LTE-A utilizes the same standard defined codebook as is employed for SU-MIMO precoding, but for the purpose of obtaining quantized channel knowledge at the transmitter. Thereby, the UE chooses one of the matrices from the codebook to acquire the "best" (in a sense that is defined later) quantized representation of the current channel realization. The index of this matrix is then signaled to the eNodeB as Channel Direction Indicator (CDI) and utilized to compute the linear ZF precoder. Because the precoder is based on quantized channel knowledge, there is some residual interference between the spatial layers at the receivers, which leads to an interference-limited multiple access system, as shown in Section 15.4.3. LTE-A limits the number of layers assigned to a single UE in MU-MIMO to one [5].

An additional difficulty in MU-MIMO is the calculation of the Channel Quality Indicators (CQIs) at the receiver. In SU-MIMO the CQI signals the preferred Modulation and Coding Scheme (MCS), in order to achieve a given target Block Error Ratio (BLER) (see Section 10.4). This is no longer possible in MU-MIMO, because a UE does not know in advance how many other users will be served in parallel. This means that the UE also does not know the applied precoder, which would be required to obtain an accurate estimate of the CQI. The current state- of-the-art technique for CQI calculation is to employ a lower bound on the SINR experienced by the receiver, which assumes transmission with the largest possible transmission rank (equal to the number of transmit antennas N_{T}). This lower bound is then fed back to the eNodeB, updated during scheduling to account for the actual transmission rank, and then used to choose the appropriate MCS.

In the following we provide a mathematical formulation of this procedure, according to [21]. We start by deriving a lower bound on the SINR experienced by user k. The precoder for user k is denoted $\mathbf{w}_k \in \mathbb{C}^{N_{\mathrm{T}}\times 1}$, which is just a single column because we assume rank 1 transmission to each user. With knowledge of the precoder the SINR of user k equals

$$\mathrm{SINR}_k = \frac{P_{\mathrm{Tx},k}|\mathbf{h}_k\mathbf{w}_k^*|^2}{\sigma_{\mathrm{n}}^2 + \sum_{i\in\mathcal{S}\backslash\{k\}} p_i|\mathbf{h}_k\mathbf{w}_i^*|^2} \tag{15.1}$$

Here, $P_{\mathrm{Tx},k}$ denotes the transmit power of user k, $\mathbf{h}_k \in \mathbb{C}^{1\times N_{\mathrm{T}}}$ refers to the effective channel experienced by user k (including the receive filter if the UE is equipped with multiple antennas), $\mathcal{S}$ is the set of served users, and σ_{n}^2 denotes the receiver noise variance (also including the noise enhancement effect of the receive filter; see Section 10.3.1). Assuming equal power allocation for all users, the total transmit power P_{Tx} is divided to the users according to

$$P_{\mathrm{Tx},k} = \frac{P_{\mathrm{Tx}}}{|\mathcal{S}|\cdot||\mathbf{w}_k||^2} \tag{15.2}$$

We denote the normalized channel vector as $\tilde{\mathbf{h}}_k = \mathbf{h}_k/||\mathbf{h}_k||$. The user provides a quantized version of this vector, $\hat{\mathbf{h}}_k \in \mathcal{W}$, via the feedback channel to the eNodeB. Here, $\mathcal{W}$ denotes the codebook utilized for CVQ. With this vector $\hat{\mathbf{h}}_k$ the normalized channel can be decomposed as

$$\tilde{\mathbf{h}}_k = \tilde{\mathbf{h}}_k[\hat{\mathbf{h}}_k^{\mathrm{H}}\hat{\mathbf{h}}_k] + \mathbf{e}_k = \cos\theta_k\hat{\mathbf{h}}_k + \mathbf{e}_k \tag{15.3}$$

Here, $\mathbf{e}_k$ denotes the error caused by the quantization, which is perpendicular to $\hat{\mathbf{h}}_k$. The cosine of the angle between the normalized channel vector and its quantized version is defined as $\cos\theta_k = \tilde{\mathbf{h}}_k\hat{\mathbf{h}}_k^H$. Furthermore, the ZF precoder orthogonalizes different layers of the quantized channels: $\hat{\mathbf{h}}_k\mathbf{w}_i^* = 0, \forall i \in \mathcal{S}\backslash\{k\}$. With these definitions and equalities, the SINR can be rewritten in the following way:

$$\mathrm{SINR}_k = \frac{P_{\mathrm{Tx},k}||\mathbf{h}_k||^2||\mathbf{w}_k||^2(|\cos\theta_k\hat{\mathbf{h}}_k\tilde{\mathbf{w}}_k^* + \mathbf{e}_k\tilde{\mathbf{w}}_k^*|)^2}{\sigma_{\mathrm{n}}^2 + \frac{P_{\mathrm{Tx}}}{|\mathcal{S}|}||\mathbf{h}_k||^2\sin^2\theta_k\sum_{i\in\mathcal{S}\backslash\{k\}}|\tilde{\mathbf{e}}_k\tilde{\mathbf{w}}_i^*|^2} \tag{15.4}$$

In this equation, $\tilde{\mathbf{w}}_k$ and $\tilde{\mathbf{e}}_k$ denote the normalized versions of the precoder and quantization error, respectively. The norm of the quantization error can be calculated according to $||\tilde{\mathbf{e}}_k||^2 = 1 - \cos^2\theta_k = \sin^2\theta_k$. Assuming ZF precoding, where the precoder is computed from the quantized channel knowledge $\hat{\mathbf{h}}_k$, it follows that

$$|\hat{\mathbf{h}}_k \tilde{\mathbf{w}}_k^*| = \frac{1}{||\mathbf{w}_k||} \tag{15.5}$$

Until now, the exact SINR expression of user k was just manipulated. Without knowledge of the precoder $\mathbf{w}_k$, the UE cannot compute this value and, therefore, is not able to deliver a CQI. To overcome this problem, a lower bound on the SINR is utilized instead. For deriving the lower bound on the SINR, several assumptions are necessary. First, the rightmost expression in the numerator of Equation (15.4), $\mathbf{e}_k \tilde{\mathbf{w}}_k^*$, is set equal to zero.[1] Furthermore, the terms $|\tilde{\mathbf{e}}_k \tilde{\mathbf{w}}_i^*|^2$ are Beta-distributed with mean value $1/(N_\mathrm{T} - 1)$ [12]. Utilizing these results, by taking the expectation of the multi-user interference term in the denominator (justified by the fact that the actual interference realization is not known at the UEs) and applying Jensen's inequality, the following lower bound on the SINR is obtained:

$$\mathrm{SINR}_k \geq \frac{P_{\mathrm{Tx},k} ||\mathbf{h}_k||^2 \cos^2\theta_k}{\sigma_\mathrm{n}^2 + \frac{P_\mathrm{Tx}}{|\mathcal{S}|}\frac{|\mathcal{S}|-1}{N_\mathrm{T}-1} ||\mathbf{h}_k||^2 \sin^2\theta_k} \tag{15.6}$$

Still, a UE cannot compute this value because $|\mathcal{S}|$ and $P_{\mathrm{Tx},k}$ are not known. Assuming the worst case for a UE, of a fully loaded MU-MIMO system, $|\mathcal{S}| = N_\mathrm{T}$, and additionally $P_{\mathrm{Tx},k} = P_\mathrm{Tx}/N_\mathrm{T}$, the final lower bound on the SINR of user k equals

$$\mathrm{SINR}_k \geq \frac{\frac{P_\mathrm{Tx}}{N_\mathrm{T}} ||\mathbf{h}_k||^2 \cos^2\theta_k}{\sigma_\mathrm{n}^2 + \frac{P_\mathrm{Tx}}{N_\mathrm{T}} ||\mathbf{h}_k||^2 \sin^2\theta_k} \tag{15.7}$$

This bound depends on the quantized channel vector $\hat{\mathbf{h}}_k$ via the angle θ_k. The user chooses the preferred CVQ representation of the normalized channel $\tilde{\mathbf{h}}_k$ by maximizing this expression over all codebook entries:

$$\hat{\mathbf{h}}_k = \underset{\hat{\mathbf{h}} \in \mathcal{W}}{\operatorname{argmax}} \frac{\frac{P_\mathrm{Tx}}{N_\mathrm{T}} ||\mathbf{h}_k||^2 (\tilde{\mathbf{h}}_k \hat{\mathbf{h}}^H)^2}{\sigma_\mathrm{n}^2 + \frac{P_\mathrm{Tx}}{N_\mathrm{T}} ||\mathbf{h}_k||^2 (1 - (\tilde{\mathbf{h}}_k \hat{\mathbf{h}}^H)^2)} \tag{15.8}$$

The index of the best quantization vector can be signaled to the eNodeB by employing the otherwise unused PMI. The CQI is obtained by plugging the quantized channel vector back into the SINR lower bound:

$$\mathrm{CQI}_k = \frac{\frac{P_\mathrm{Tx}}{N_\mathrm{T}} ||\mathbf{h}_k||^2 (\tilde{\mathbf{h}}_k \hat{\mathbf{h}}_k^H)^2}{\sigma_\mathrm{n}^2 + \frac{P_\mathrm{Tx}}{N_\mathrm{T}} ||\mathbf{h}_k||^2 [1 - (\tilde{\mathbf{h}}_k \hat{\mathbf{h}}_k^H)^2]} \tag{15.9}$$

The feedback values, which are signaled to the eNodeB, are therefore given by a quantized version of the normalized channel vector and a lower bound on the SINR. The eNodeB

[1] In [21] it has been shown that this value tends to zero if the number of users is sufficiently large, such that nearly orthogonal users are chosen for transmission.

utilizes these values to decide for the set $\mathcal{S}$ of users that are served in parallel on multiple spatial layers and then determines the ZF precoder for these users by computing the right pseudo inverse of the combined channel matrix $\hat{\mathbf{H}}(\mathcal{S})$:

$$\hat{\mathbf{H}}(\mathcal{S}) = [\hat{\mathbf{h}}_{s_1}^T, \ldots, \hat{\mathbf{h}}_{s_{|\mathcal{S}|}}^T]^T; \qquad \mathcal{S} = \{s_1, \ldots, s_{|\mathcal{S}|}\}, \tag{15.10}$$

$$\mathbf{W}(\mathcal{S}) = \hat{\mathbf{H}}(\mathcal{S})^H (\hat{\mathbf{H}}(\mathcal{S})\hat{\mathbf{H}}(\mathcal{S})^H)^{-1} \tag{15.11}$$

In the derivation of the SINR lower bound Equation (15.7), it was assumed that the maximum number of $|\mathcal{S}| = N_\mathrm{T}$ users is served and the transmit power of user k equals $P_{\mathrm{Tx},k} = P_\mathrm{Tx}/N_\mathrm{T}$. As soon as the scheduling decision is obtained, the eNodeB knows how many users are actually served and can compute their transmit powers according to Equation (15.2). Utilizing these powers $P_{\mathrm{Tx},k}$, a tighter lower bound for the SINR can be computed, by updating the feedback values according to

$$\mathrm{SINR}_k \geq \mathrm{CQI}_k \frac{P_{\mathrm{Tx},k}}{\frac{P_\mathrm{Tx}}{N_\mathrm{T}}} = \mathrm{CQI}_k \frac{N_\mathrm{T}}{|\mathcal{S}| \cdot ||\mathbf{w}_k||^2} \tag{15.12}$$

This value is then employed to choose the MCSs of the served users such that the BLER target is satisfied; for example, by utilizing a mapping from SINR to BLER such as the one shown in Figure 10.2.

The UEs require knowledge of the precoders, utilized during transmission, to coherently detect the data symbols. For that purpose, the precoders must be signaled to the UEs. To avoid additional signaling overhead, this is handled by means of UE-specific reference signals. These reference signals are precoded with the same precoder as used for the data of the corresponding UE. At the UE the concatenation of the wireless channel and the precoder is jointly estimated by the reference signals.

15.4 Physical-Layer Throughput Simulation Results

In this section the performance of the LTE-A DL is evaluated by means of single-user link-level simulations, utilizing the "Vienna LTE Link Level Simulator" (see Chapter 11). Specifically, in Section 15.4.1 the focus is on systems with eight transmit antennas, as this feature is novel in LTE-A. For that case the throughput gain obtained by increasing the number of receive antennas is determined. Furthermore, an investigation similar to that in Section 10.6 is conducted to analyze the reasons for the throughput loss of the simulated system compared to channel capacity, utilizing the throughput bounds derived in Section 10.5. Afterwards, in Section 15.4.2, different LTE and LTE-A antenna configurations are contrasted. Finally, in Section 15.4.3, the performance of SU-MIMO and MU-MIMO, employing the feedback methods described in Section 15.3, is presented, assuming a single eNodeB that serves 10 users.

15.4.1 Eight-Antenna Transmission

The support of eight transmit antennas enables LTE-A to obtain higher peak spectral efficiencies (compared with LTE) and to improve the cell edge user performance, by providing a large beamforming SNR gain. In the following, simulated throughput results for

Table 15.3 Simulation parameters for the simulations of Section 15.4.1

Parameter	Value
Channel model	ITU-T VehA [11]
System bandwidth (MHz)	1.4
Subcarrier number	72
Receiver	Zero Forcing (ZF)
Antennas $N_\mathrm{R} \times N_\mathrm{T}$	$\{1, \ldots, 8\} \times 8$
Transmission mode	CLSM
CQI granularity	Wideband
Subband PMI granularity	Resource block specific
Simulation time	1000 subframes
Feedback delay	0 TTI

that case are presented, applying the simulation parameters from Table 15.3. Additionally, perfect channel knowledge at the receiver and perfect timing and frequency synchronization of the transmitter and receiver is assumed. The UE provides feedback indicators for adaptation of the transmission parameters (see Section 10.4), which we assume to happen before transmission (zero TTI feedback delay), in order to avoid influences of outdated feedback. The temporal granularity of the feedback values equals one TTI (1 ms), meaning that new feedback values are computed every TTI. The CQI, RI, and wideband PMI frequency granularity are set equal to the total system bandwidth, whereas the subband PMI granularity is given by one RB. Furthermore, cell-specific reference signals are employed, by extending the pattern used for four transmit antennas.[2]

15.4.1.1 Increasing the Number of Receive Antennas

In Figure 15.5 the single-user throughput of an LTE-A system employing eight transmit antennas and different numbers of receive antennas is shown. The system with one receive antenna achieves at most a throughput of 4.4 Mbit/s, whereas this value increases by a factor of eight to 35 Mbit/s for the system with eight receive antennas. Note that an unrealistically high SNR of 45 dB is required to fully exploit the possible throughput of eight transmission layers. On the other hand, significant throughput gains are also obtained at moderate SNRs by increasing the number of receive antennas. At an operating point of SNR = 12 dB (detailed in the inset of Figure 15.5) the throughput increases by a factor of 3.25 by utilizing eight receive antennas instead of one. The throughput amount gained depends on the SNR operating point considered and increases with SNR.

15.4.1.2 Throughput Bounds

Next, we compare the throughput of an $N_\mathrm{R} \times N_\mathrm{T} = 8 \times 8$ LTE-A system with the practical throughput bounds derived in Section 10.5. The simulation results of this investigation are shown in Figure 15.6. The leftmost curve corresponds to channel capacity, while the

[2] As we assume perfect channel knowledge, this only influences the amount of system overhead.

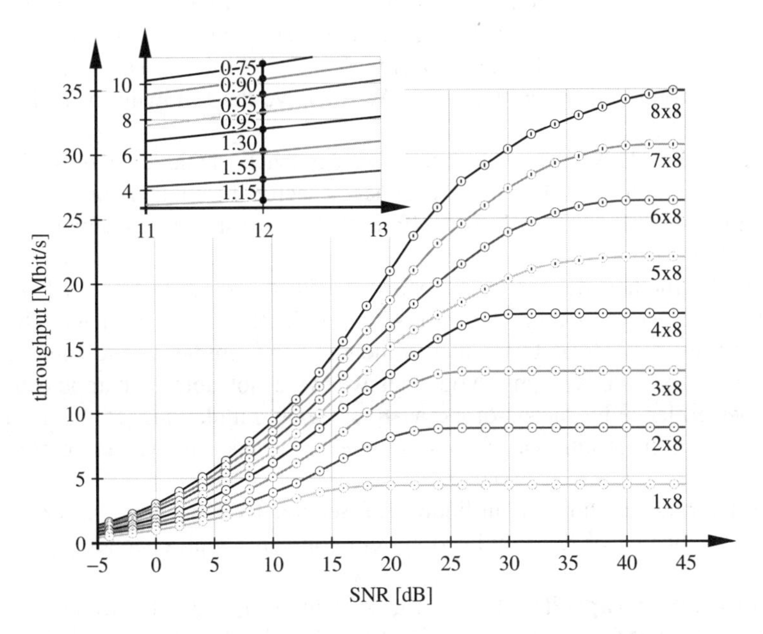

Figure 15.5 Comparison of simulated throughputs for a single-user LTE-A system employing CLSM transmission with different antenna configurations $N_R \times N_T$, $N_T = 8$.

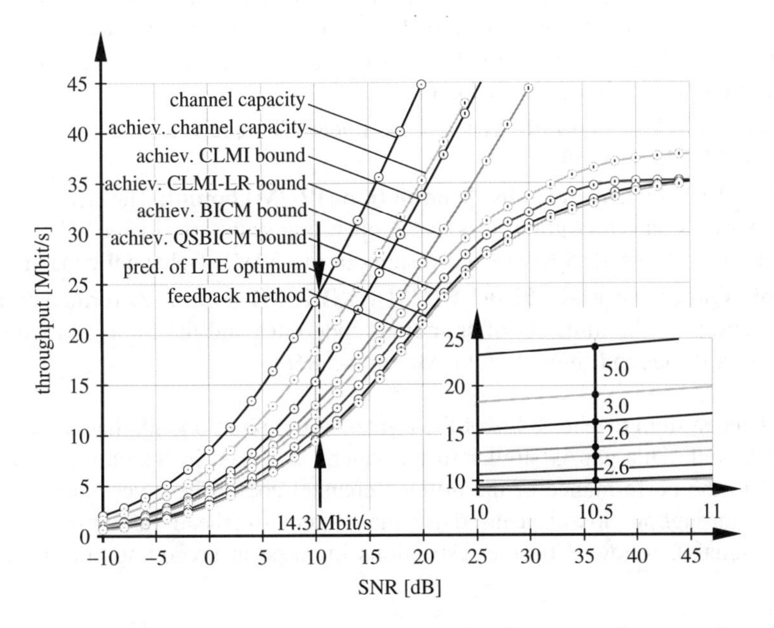

Figure 15.6 Comparison of throughput bounds and simulated system performance for a single-user $N_R \times N_T = 8 \times 8$ CLSM LTE-A system.

rightmost curve shows the simulated performance obtained with the SU-MIMO feedback methods of Section 15.3. We analyze the system performance at an operating point of SNR = 10.4 dB, where the channel capacity equals 24 Mbit/s.[3] The inset in the figure shows a detail around this operating point. We observe the following behavior:

- The simulated LTE-A system, utilizing UE feedback, achieves a throughput of 9.7 Mbit/s, corresponding to 40.5 % of channel capacity. In comparison, a 4 × 4 system obtains a throughput of 5.9 Mbit/s or 49 % of the corresponding channel capacity (see Section 10.6).
- Based on the method described in Section 10.5.6 the predicted optimal performance of the LTE-A system, obtains 9.9 Mbit/s or 41.3 % of channel capacity. The feedback method achieves 98 % of this value, which is very similar to the loss observed in Section 10.6 for LTE systems. Note that we were not able to compute the optimal LTE-A performance by means of exhaustive search simulations, as we did in Section 10.6, due to the huge number of possible precoder, rank, and MCS combinations.

Employing the practical throughput bounds of Section 10.5, we now analyze the reasons for the throughput loss of the simulated system compared with channel capacity:

Achievable channel capacity: Taking the system overhead into account reduces the throughput to 19 Mbit/s, corresponding to 80 % of channel capacity. Note that this assumes cell-specific reference signals. The overhead can be reduced with UE-specific reference signals (see Section 15.3).

Achievable Closed-Loop Mutual Information (CLMI) bound: The simulated system restricts the precoders to the standard defined codebook and employs uniform power allocation. This is considered with the CLMI bound and reduces the possible throughput to 16 Mbit/s or 66.7 % of channel capacity.

Achievable CLMI-LR bound: The simple linear ZF receiver reduces the throughput further to 13.4 Mbit/s (56 %).

Achievable Bit-Interleaved Coded Modulation (BICM) bound: The BICM architecture only causes a small throughput loss at the operating point considered. The value of the BICM bound equals 12.5 Mbit/s, corresponding to 52 % of channel capacity.

Achievable Quantized and Shifted BICM (QSBICM) bound: A further performance loss is caused by the finite set of supported code rates and the finite code blocklength, achieving a throughput bound of 11 Mbit/s (45.8 %).

The simulated system reaches 88 % of the tightest throughput bound, that is the achievable QSBICM bound. This gap is similar to the values observed in Section 10.6 and can be accounted to the performance of the utilized channel code. The investigation shows that the largest throughput loss is caused by the system overhead (assuming cell specific reference signals), followed by the restrictions imposed at precoding and the utilization

[3] Note that at this operating point the throughput of the 4 × 4 LTE system, considered in Section 10.6, is doubled.

of the simple ZF receiver. At the considered operating point the system mainly transmits over three, four or five spatial layers. Therefore, the overhead can be reduced with UE specific reference signals, that require pilot symbols on each layer only. Furthermore, a large performance gain is possible by employing a Maximum Likelihood (ML) receiver instead of the ZF equalizer.

15.4.2 Comparison between LTE and LTE Advanced

In this section, we compare different LTE-compliant antenna configurations (up to four transmit antennas) with LTE-A, supported by eight-transmit-antenna configurations. We reuse the simulation parameters of Table 15.3, but with different antenna configurations. Transmission systems with one or two receive antennas are considered. The corresponding simulation results are shown in Figure 15.7. In the figure, the solid lines represent the results of the single-receive-antenna systems, while the dashed lines correspond to the two-receive-antenna systems. Considering first the former case, we observe the following:

- Increasing the number of transmit antennas from one to eight provides an SNR gain of approximately 5 dB, as shown in the lower-right inset.
- At high SNR the throughput decreases with the number of transmit antennas, because the required amount of pilot symbols increases. This again assumes cell-specific reference signals. With UE-specific reference signals, all antenna configurations considered achieve the same throughput at high SNR, equal to the Single-Input Single-Output (SISO) throughput. Correspondingly, the non-SISO throughput curves must be scaled up to obtain the performance with UE-specific reference signals.

The behavior of the system utilizing two receive antennas is very similar, with the difference that the slope of the 2×2 throughput curve is considerably smaller than that of the other configurations around 20 dB SNR. This is because all antennas are utilized to support the two spatial layers, leaving no DoFs to obtain an additional beamforming or diversity gain.

15.4.3 Comparison of SU-MIMO and MU-MIMO

In this section the throughput performances of the SU-MIMO and MU-MIMO transmission modes, employing the feedback methods described in Section 15.3.1, are compared. For simplicity, a simple frequency-flat Rayleigh fading channel with 1.4 MHz bandwidth is assumed, over which 10 users with equal average receive SNR are served. The eNodeB is equipped with $N_\mathrm{T} = 4$ transmit antennas, while the UEs just have $N_\mathrm{R} = 1$ receive antenna each.

In SU-MIMO mode, multiple access is handled by assigning orthogonal resources in time and/or frequency to the users. We employ a "Best CQI" scheduling strategy [18], which assigns an RB to the user with the highest CQI. Because all users experience the same average SNR, they are also served equally likely. Furthermore, as the users are only equipped with one receive antenna, spatial multiplexing is not possible in that case.

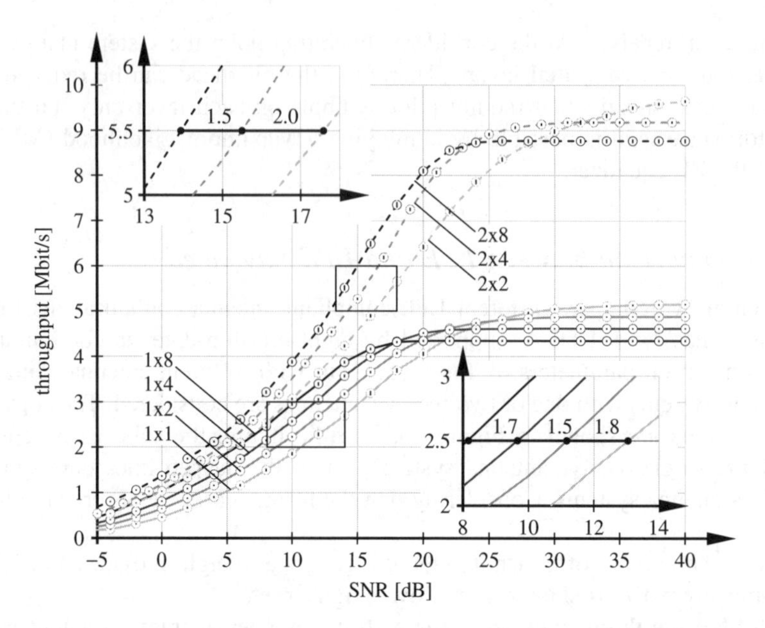

Figure 15.7 Comparison of simulated single-user throughput obtained with different LTE- and LTE-A-defined antenna configurations $N_R \times N_T$, $N_R = \{1, 2\}$, $N_T = \{1, 2, 4, 8\}$.

On the other hand, the MU-MIMO scheme is capable of multiplexing different users not only in time and frequency, but also in space, by assigning orthogonal layers to the users. Because the eNodeB only has quantized channel knowledge, some residual interference between the users is left, leading to an interference-limited multiple access system. At most four UEs can be served in parallel, over the four transmit antennas of the eNodeB. To decide for the set $\mathcal{S}$ of users that is spatially multiplexed, a greedy scheduling algorithm, described in [8], is employed that incrementally adds users to the set $\mathcal{S}$ as long as they lead to an increase in the expected throughput.

For both transmission modes, the LTE precoding codebook is employed. The SU-MIMO mode utilizes this codebook for precoding, while the MU-MIMO mode just uses it for CVQ, to obtain quantized channel knowledge at the eNodeB. Alternatively, we employ randomly generated codebooks, with unit norm codebook entries that are isotropically distributed in $\mathbb{C}^{1 \times N_T}$. With this construction, larger codebooks than the 4 bit LTE codebook can be generated. The simulation results are statistically averaged over codebook realizations, in addition to the averaging over noise and channel realizations.

In Figure 15.8, the simulation results obtained with these parameters and settings are shown. The solid lines correspond to MU-MIMO transmission, whereas the dashed lines show the sum throughput achieved with SU-MIMO. The leftmost curve (denoted as "MU-MIMO Perfect" in the figure) shows the achievable performance of MU-MIMO with perfect channel knowledge at the transmitter, using the greedy suboptimal scheduling algorithm described above for resource allocation. It shows that MU-MIMO has the potential to considerably outperform SU-MIMO, by enabling a spatial multiplexing MIMO

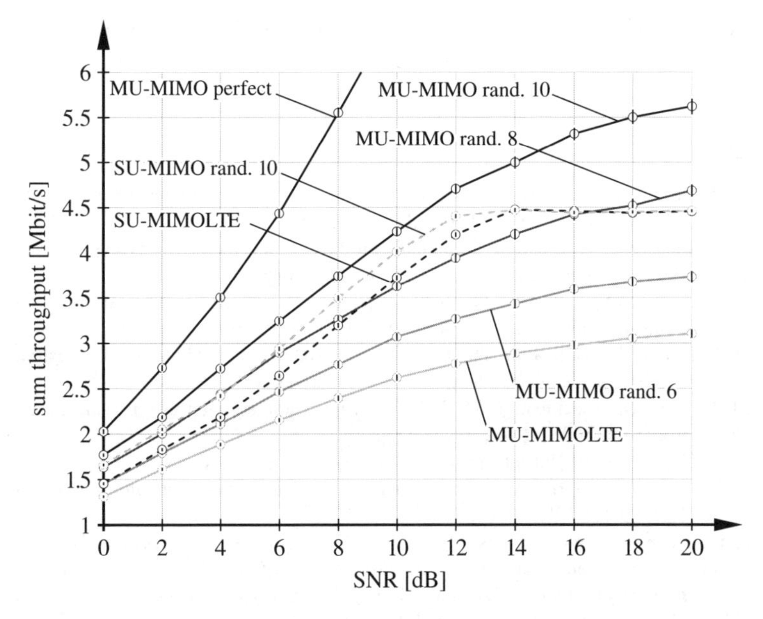

Figure 15.8 Comparison of simulated cell throughput for a 10-user $N_R \times N_T = 4 \times 1$ LTE-A system employing SU-MIMO and MU-MIMO transmission.

gain. Investigating the performance obtained with CVQ reverses this picture. The right-most curve ("MU-MIMO LTE") shows the throughput of MU-MIMO utilizing the LTE codebook for CVQ. This system performs much worse than the corresponding SU-MIMO system ("SU-MIMO LTE"). The reason for this behavior is that the channel knowledge at the eNodeB is very inaccurate and, therefore, strong interference between the spatial layers degrades the performance. Additionally, the SINR estimate of the MU-MIMO feedback scheme is by far not as reliable as the one obtained with the SU-MIMO feedback scheme.

To improve the performance of MU-MIMO, more accurate channel knowledge at the transmitter is required. This can be obtained by increasing the codebook size, which we achieve with the random codebooks described. A codebook size of 256 vectors, requiring 8 bits for indexing (denoted as 'MU-MIMO Rand.8' - indicating that a random codebook of size 2^8 is employed), is necessary to outperform the SU-MIMO mode, at least in a limited SNR regime ($< 6\,\text{dB}$ or $> 16\,\text{dB}$). With a codebook of size 1024 ("MU-MIMO Rand.10"), MU-MIMO performs better than SU-MIMO over the full SNR range, but the feedback overhead of the system is more than doubled (10 bits instead of 4 bits). With an optimized codebook (for example, of Grassmannian type [19]), a slightly better performance can be expected. The figure also shows that the SU-MIMO system gains much less from increasing the codebook size than does the MU-MIMO system. This investigation reveals that simple CVQ might not be the right choice for obtaining quantized channel knowledge at the transmitter, at least not with the amount of feedback overhead available in current wireless communication systems. Further investigations, especially in more realistic scenarios, are still required to confirm this observation.

References

[1] 3GPP (2009) Technical Specification TS 36.913 'Requirements for furhter advancements for evolved universal terrestrial radio access (E-UTRA) (LTE Advanced),' www.3gpp.org.
[2] 3GPP (2010) Technical Specification TS 36.814 'Further advancements for evolved universal terrestrial radio access (E-UTRA); physical layer aspects,' www.3gpp.org.
[3] 3GPP (2010) Technical Specification TS 36.201 Version 10.0.0 'Evolved universal terrestrial radio access (E-UTRA); LTE physical layer-general description,' www.3gpp.org.
[4] 3GPP (2010) Technical Specification TS 36.211 Version 10.0.0 'Evolved universal terrestrial radio access (E-UTRA); physical channels and modulation,' www.3gpp.org.
[5] 3GPP (2010) Technical Specification TS 36.213 'Evolved universal terrestrial radio access (E-UTRA); physical layer procedures (release 10),' www.3gpp.org.
[6] Agilent Technologies (2010) 'Introducing LTE-Advanced,' Application note.
[7] Costa, M. H. M. (1983) 'Writing on dirty paper,' *IEEE Transactions on Information Theory*, **29** (3), 439–441, doi: 10.1109/TIT.1983.1056659.
[8] Dietl, G., Labrèche, O., and Utschick, W. (2009) 'Channel vector quantization for multiuser MIMO systems aiming at maximum sum rate,' in *Proceedings of the 28th IEEE conference on Global Telecommunications (GLOBECOM'09)*, IEEE Press, Piscataway, NJ, USA, pp. 5113–5117. Available from http://portal.acm.org/citation.cfm?id=1811982.1812229.
[9] IEEE and P802.16m (2009) 'The draft IEEE 802.16m System Description Document,' IEEE 802.16m-08/003r8.
[10] International Telecommunication Union ITU 'Requirements related to technical performance for IMT-advanced radio interface(s), ITU-R M.2134,' Online, ITU-R M.2134, http://www.itu.int/pub/R-REP-M.2134-2008/en.
[11] ITU (1997) 'Recommendation ITU-R M.1225: Guidelines for evaluation of radio transmission technologies for IMT-2000,' Technical report, ITU.
[12] Jindal, N. (2006) 'MIMO broadcast channels with finite-rate feedback,' *IEEE Transactions on Information Theory*, **52** (11), 5045–5060.
[13] Kottkamp, M. (2010) 'LTE Advanced Technology Introduction,' White paper, Rohde & Schwarz.
[14] Myung, H. G., Lim, J., and Goodman, D. J. (2006) 'Single carrier FDMA for uplink wireless transmission,' *IEEE Communications Magazine*, **1**, 30–38.
[15] Parkvall, S., Furuskär, A., and Dahlman, E. (2010) 'Next generation LTE, LTE Advanced,' Ericsson review 2.
[16] Peel, C., Hochwald, B., and Swindlehurst, A. (2005) 'A vector-perturbation technique for near-capacity multiantenna multiuser communication – part I: channel inversion and regularization,' *IEEE Transactions on Communications*, **53** (3), 537–544.
[17] Sawahashi, M., Kishiyama, Y., Morimoto, A. *et al.* (2010) 'Coordinated multipoint transmission/reception techniques for LTE-advanced,' *IEEE Wireless Communications*, **17** (3), 26–34.
[18] Schwarz, S., Mehlführer, C., and Rupp, M. (2010) 'Low complexity approximate maximum throughput scheduling for LTE,' in *44th Annual Asilomar Conference on Signals, Systems, and Computers*, Pacific Grove, CA. Available from http://publik.tuwien.ac.at/files/PubDat_187402.pdf.
[19] Shu, F., Gang, W., Yue, X., and Shao-qian, L. (2009) 'Multi-user MIMO linear precoding with Grassmannian codebook,' in *WRI International Conference on Communications and Mobile Computing, (CMC'09)*, volume 1, pp. 250–255, doi: 10.1109/CMC.2009.188.
[20] Technical Specification Group RAN WG1 (2007), 'LS on LTE performance verification work,' Technical report R1-072580. Available from http://www.3gpp.org/ftp/tsg_ran/WG1_RL1/TSGR1_49/Docs/R1-072580.zip.
[21] Technical Specification Group RAN (2006) 'Comparison between MU-MIMO codebook-based channel reporting techniques for LTE downlink,' Technical Report R1 062483, 3GPP,Philips.
[22] Technical Specification Group RAN (2010) 'Way forward on 8Tx codebook for Rel.10 DL MIMO,' Technical Report R1 105011, 3GPP.

[23] Trivellato, M., Boccardi, F., and Huang, H. (2008) 'On transceiver design and channel quantization for downlink multiuser MIMO system with limited feedback,' *IEEE Journal on Selected Areas in Communications*, **26** (8), 1494–1504.

[24] Trivellato, M., Boccardi, F., and Tosate, F. (2007) 'User selection schemes for MIMO broadcast channels with limited feedback,' in *IEEE 65th Vehicular Technology Conference (VTC2007-Spring)*, Dublin.

Index

Evaluation of HSDPA and LTE: From Testbed Measurements to System Level Performance, First Edition.
Sebastian Caban, Christian Mehlführer, Markus Rupp and Martin Wrulich.